Texts in Applied Mathematics

Volume 31

The mathematization of all sciences, the fading of traditional scientific boundaries, the impact of computer technology, the growing importance of computer modelling and the necessity of scientific planning all create the need both in education and research for books that are introductory to and abreast of these developments. The aim of this series is to provide such textbooks in applied mathematics for the student scientist. Books should be well illustrated and have clear exposition and sound pedagogy. Large number of examples and exercises at varying levels are recommended. TAM publishes textbooks suitable for advanced undergraduate and beginning graduate courses, and complements the Applied Mathematical Sciences (AMS) series, which focuses on advanced textbooks and research-level monographs.

More information about this series at http://www.springer.com/series/1214

Pierre Brémaud

Markov Chains

Gibbs Fields, Monte Carlo
Simulation and Queues

Second Edition

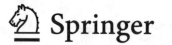 Springer

Pierre Brémaud
Paris, France

ISSN 0939-2475   ISSN 2196-9949   (electronic)
Texts in Applied Mathematics
ISBN 978-3-030-45984-0   ISBN 978-3-030-45982-6   (eBook)
https://doi.org/10.1007/978-3-030-45982-6

Mathematics Subject Classification (2010): 60-XX, 68-XX, 82-XX, 90-XX, 92-XX, 94-XX

This Springer imprint is published by the registered company Springer Nature Switzerland AG
The registered company address is: Gewerbestrasse 11, 6330 Cham, Switzerland

Pour Marion

Preface

From Pushkin to Monte Carlo

When Markov introduced his famous model in 1906, he was not preoccupied by applications. He just wanted to show that independence is not necessary for the law of large numbers (the weak law of large numbers at that time, since Borel proved the strong law for heads and tails only in 1909). An example that he considered was the alternation of consonants and vowels in Pushkin's *Eugene Onegin*, which he described as a two-state chain. (This, however, does not say much about the plot!) Almost at the same time, and faithful to the French tradition of gambling probabilists, Poincaré studied Markov chains on finite groups, with applications to card shuffling. The Austrian physicists Paul and Tatiana Ehrenfest proposed in 1907 a Markov chain model which clarified the controversial issue of thermodynamical irreversibility. Sir Francis Galton, a cousin of Darwin, who was interested in the probability of survival of English peerage, is the inventor of the branching process, another famous Markov model with many applications besides the original one. He posed the problem in the Educational Times in 1873 and in the same year and same journal, Reverend Watson proposed the method of solution which became a textbook classic.

The dates mentioned above show that Markov models were already around even before Markov started the systematic study of this class of random sequences. However, the work of Markov challenged the best probabilists, such as Kolmogorov, Doeblin and Fréchet, just to mention the leading pioneering figures. The outcome was a clean and sound theory ready for applications, and today Markov chains are omnipresent in the applied sciences. For instance, **biology** is an important consumer of Markov models, many of them concerning *genetics* and *population theory*. In the **social sciences**, social mobility can be described in Markovian terms. Quantitative **psychology** uses Markovian models of learning. **Physics** is a major patron of Markov chain theory, and Markov models (for instance, the *Ehrenfest diffusion model*, the *annealing model*, and the Ising–Peierls model of *phase tran-*

sition) have been very useful in understanding qualitatively complex phenomena. Markov chains have found many applications in **electrical engineering** and the **information sciences**, for instance in the performance analysis of multiple access *communications protocols* and of *communications networks*, and in *image processing*. Recently, Markov chain theory has received an additional impetus from the advent of Monte Carlo Markov chain simulation. Markov chains have found a privileged domain of application in **operations research**, in *reliability theory* and *queueing theory* for instance.

A useful, simple, and beautiful theory

The list of applications of Markov chains is virtually infinite, and one is entitled to say that it is the single most successful class of stochastic processes, its success being due to the relative simplicity of its theory and to the fact that simple Markov models can exhibit extremely varied and complex behavior. The modeling power of Markov chains may well be compared to that of ordinary differential equations.

The theory of Markov chains with a countable state space is an ideal introduction to stochastic processes; it is not protected by a wall of technicalities, and therefore the student has quick access to the main results. Indeed, the mathematical equipment required for a rewarding study of this class of models consists only of the notion of conditional independence and the strong law of large numbers.

Another pleasant feature of Markov chain theory is that this classical topic can be presented in terms of the elegant concepts of the modern theory of stochastic processes, such as *reversibility, martingales* and *coupling*.

The second edition

This new edition remains centered on the study of homogeneous Markov chains (HMCs) with a countable state space, in discrete time and in continuous time, and responds to the need for a unified treatment of related topics such as finite Gibbs fields, non-homogeneous Markov chains, discrete-time regenerative processes, Monte Carlo simulation, simulated annealing and queueing theory. It is a thoroughly revised and augmented edition of the book with the same title published in 1999. The main additions are the *exact sampling* algorithm of Propp and Wilson, the *electrical network analogy* of symmetric random walks on graphs, *mixing times* and additional details on the *branching process*. The structure of the book has been modified in order to smoothly incorporate this new material. Among the features that should improve the reader-friendliness, the three main ones are: a shared numbering system for the definitions, theorems and examples;

the attribution of titles to the examples and exercises; the blue highlighting of important terms.

Course suggestions

The probability review of the first chapter and the appendix make this book self-contained and useable in a *self-teaching mode*. For the purpose of devising a *basic course* on discrete-time Markov chains, Chapters 1 to 7 are sufficient. An *advanced course* will include Chapters 8 to 12. Chapters 13 and 14 on the continuous-time Markov chains and Markovian queueing theory are natural complements of the basic course.

Acknowledgements

This book has benefited from the interaction with the students of the École Polytechnique, of the École Normale Supérieure of Paris and of the EPFL (Switzerland), and from the suggestions, comments and criticisms of alert readers who have contributed to many improvements of the manuscript. I wish to thank them all as well as Sébastien Allam, Anne Bouillard, Hervé Carfantan, Philippe Ciuciu, Jim Fill, Jean-François Giovannelli, Jérôme Idier, Takis Konstantopoulos, Torgny Lindvall, Laurent Massoulié, Sibi Raj Pillai, Andrea Ridolfi and Pam Williams, who helped me in many different ways.

Pierre Brémaud
Paris, January 1, 2020

Contents

Chapter 1

Probability Review

To keep the book self-contained, we begin with a review of the elementary notions and results of probability theory which will be directly useful in the sequel when we study Markov chains. The reader familiar with the basic theory of probability can skip this chapter or rapidly browse through it.

1.1 Basic Concepts

1.1.1 Events, Random Variables and Probability

Probability theory provides a mathematical framework for the study of random phenomena. It requires a precise description of the *outcome* of an observation when such a phenomenon is observed. The collection of all possible outcomes ω is called the *sample space* Ω.

EXAMPLE 1.1.1: A DIE, TAKE 1. The experiment consists in tossing a die once. The possible outcomes are $\omega = 1, 2, \ldots, 6$ and the sample space is the set $\Omega = \{1, 2, 3, 4, 5, 6\}$ (see Figure 1.1.1).

EXAMPLE 1.1.2: DARTS, TAKE 1. The experiment consists in throwing a dart at a very large wall. The sample space can be chosen to be the plane $\mathbb{R}^2$, an idealization of the wall, an outcome being the position $\omega = (x, y)$ hit by the dart (see Figure 1.1.1).

© Springer Nature Switzerland AG 2020
P. Brémaud, *Markov Chains*, Texts in Applied Mathematics 31,
https://doi.org/10.1007/978-3-030-45982-6_1

EXAMPLE 1.1.3: COIN TOSSES, TAKE 1. The experiment is an infinite sequence
of coin tosses. One can take for the sample space Ω the collection of all sequences
$\omega = \{x_n\}_{n \geq 1}$, where $x_n = 1$ or 0, depending on whether the n-th toss results in
heads or tails (see Figure 1.1.1).

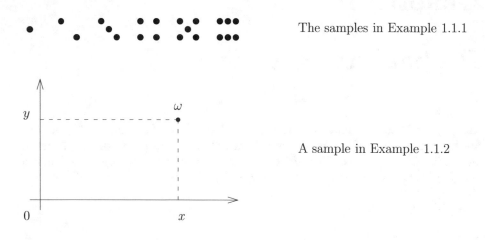

The samples in Example 1.1.1

A sample in Example 1.1.2

HTTTHHTTTHTHTHT...   A sample in Example 1.1.3

Figure 1.1.1. Samples

The Logics of Events

Probabilists have their own interpretation of sets. It is reflected in their terminology, which we shall now review.

Any subset A of the sample space Ω can be regarded as a representation of
some *event*. In Example 1.1.1, the set $A = \{1, 3, 5\}$ represents the event "result is
odd." In Example 1.1.2, the subset $A = \{(x, y); x^2 + y^2 \geq 1\}$ could be the event
"you missed the dartboard" if the dartboard is centered at 0 and of radius 1. In
Example 1.1.3, the subset $A = \{\omega; x_k = 1 \text{ for } k = 1 \text{ to } 1{,}000\}$ is a very lucky event
if you bet on heads.

One says that the outcome ω *realizes* the event A if $\omega \in A$. Obviously, if ω
does not realize A, it realizes $\bar{A}$, the complement of A in Ω.

The event $A \cap B$ is realized by outcome ω if and only if ω realizes both A and B. Similarly, $A \cup B$ is realized by ω if and only if at least one event among A and B is realized.

Two events A and B are called *incompatible* when $A \cap B = \varnothing$. In other words, the event $A \cap B$ is impossible, and so no outcome ω can realize both A and B. For this reason one refers to the empty set $\varnothing$ as the *impossible event*. Naturally, Ω is called the *certain event*.

The notation $A + B$ (the *sum* of A and B) implies by convention that A and B are disjoint and represents the union $A \cup B$. Similarly, the notation $\sum_{k=1}^{\infty} A_k$ is used for $\cup_{k=1}^{\infty} A_k$ only when the A_k's are pairwise disjoint. The equality

$$\sum_{k=0}^{\infty} A_k = \Omega$$

means that the events $A_1, A_2, \ldots$ are *mutually incompatible and exhaustive*. They are exhaustive in the sense that any outcome ω realizes at least one among them. They are mutually exclusive in the sense that any two events among them are incompatible. Therefore, any ω realizes one and only one of the events $A_1, \ldots, A_n$.

If $B \subset A$, event B is said to *imply* event A, because A is realized by ω whenever B is realized by ω. The notation $A - B$ is used only if $B \subset A$, and it stands for $A \cap \bar{B}$. In particular, if $B \subset A$, then $A = B + (A - B)$ (see Figure 1.1.2).

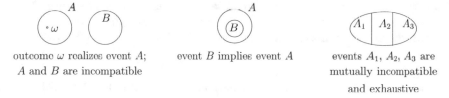

outcome ω realizes event A;　　　　event B implies event A　　　　events A_1, A_2, A_3 are

A and B are incompatible　　　　　　　　　　　　　　　　　　　mutually incompatible

and exhaustive

Figure 1.1.2. The logic of events

Probability theory assigns to an event a number, the *probability* of the said event. For technical reasons, the collection $\mathcal{F}$ of events that are assigned a probability is not always identical to the collection of all subsets of Ω. The requirements on $\mathcal{F}$ are the following:

1. The impossible event $\varnothing$ and the certain event Ω are in $\mathcal{F}$,

2. if A is in $\mathcal{F}$, then so is $\bar{A}$, and

3. if $A_1, A_2, \ldots$ are in $\mathcal{F}$, then so is $\cup_{k=1}^{\infty} A_k$.

One calls the collection of subsets $\mathcal{F}$ a σ-*field* on Ω, here the σ-field of *events*.

If the sample space Ω is finite, one usually considers any subset of Ω to be an event. The same is generally true for a countable sample space.

Random Variables

Definition 1.1.4 *A* random variable *is a function* $X : \Omega \to \overline{\mathbb{R}}$ *such that for all* $a \in \mathbb{R}$, *the event* $\{X \leq a\} := \{\omega; X(\omega) \leq a\}$ *can be assigned a probability, that is,*

$$\{X \leq a\} \in \mathcal{F}.$$

A function $X : \Omega \to E$ *where* E *is a denumerable set is called a* discrete random variable *if for all* $x \in E$

$$\{X = x\} \in \mathcal{F}.$$

If $X(\omega)$ *is real for all* ω, *then* X *is called a* real random variable.

Remark 1.1.5 Since a denumerable set E can be identified with the set of integers $\mathbb{N}$, we may consider a discrete random variable to be a special case of a real random variable.

Remark 1.1.6 Sometimes a random variable is called a *random number*. This is an innocuous habit as long as one is aware that it is not the function X that is random, but the outcome ω, which in turn makes the number $X(\omega)$ random.

EXAMPLE 1.1.7: A DIE, TAKE 2. Continuation of Example 1.1.1. Take for X the identity $X(\omega) = \omega$. In that sense X is a random number obtained by tossing a die.

EXAMPLE 1.1.8: DARTS, TAKE 2. Continuation of Example 1.1.2. Here $\omega := (x, y)$, where $x, y \in \mathbb{R}$. Define the *coordinate random variables* of Ω, X and Y, by

$$X(\omega) := x, Y(\omega) := y.$$

EXAMPLE 1.1.9: COIN TOSSES, TAKE 2. Continuation of Example 1.1.3. Recall that $\omega := \{x_n\}_{n \geq 1}$, a sequence of binary digits (0's and 1's). Define X_n to be the random number obtained at the n-th toss:

$$X_n(\omega) := x_n \,.$$

Probability

The *probability* $P(A)$ of an event $A \in \mathcal{F}$ measures the likeliness of its occurrence. As a function defined on $\mathcal{F}$, the probability P is required to satisfy a few properties, the *axioms of probability*.

Definition 1.1.10 *A probability on* $(\Omega, \mathcal{F})$ *is a mapping* $P : \mathcal{F} \to \mathbb{R}$ *such that*

1. $0 \leq P(A) \leq 1$,

2. $P(\Omega) = 1$, *and*

3. $P(\sum_{k=1}^{\infty} A_k) = \sum_{k=1}^{\infty} P(A_k)$.

Property 3 is called σ-*additivity*. The triple $(\Omega, \mathcal{F}, P)$ is called a *probability space*, or *probability model*.

Remark 1.1.11 The axioms of probability are motivated by the following heuristic interpretation of $P(A)$ as the empirical frequency of occurrence of event A. If n "independent" experiments are performed, among which n_A result in the realization of A, then the empirical frequency

$$F(A) = \frac{n_A}{n}$$

should be close to $P(A)$ if n is "sufficiently large." Clearly, the function F satisfies the axioms.

Remark 1.1.12 The axiomatic presentation of probability theory is nevertheless logically independent of the frequency interpretation. As a matter of fact, its success is due to its apparent ignorance of the frequency interpretation, which blurs the picture because the empirical frequency F depends on too many things: on the number of experiments and on the experiments themselves. The axiomatic theory of probability connects to the frequency interpretation *a posteriori*: the latter appears as a *theorem*, the famous *strong law of large numbers* (SLNN) given in Section 1.4. To obtain it, all that is needed besides the axioms of probability and clever computations is a good definition of what is meant by independent experiments. This definition will be given in the next section.

EXAMPLE 1.1.13: A DIE, TAKE 3. Continuation of Example 1.1.7. For $A \subset \Omega = \{1, 2, 3, 4, 5, 6\}$, the formula

$$P(A) := \frac{|A|}{6},$$

where $|A|$ is the *cardinality* of A, that is, the number of elements in A, defines a probability P.

EXAMPLE 1.1.14: DARTS, TAKE 3. Continuation of Example 1.1.8. Take, for instance,

$$P(A) := \frac{1}{2\pi\sigma^2} \int \int_A e^{-\frac{(x^2+y^2)}{2\sigma^2}} \, dx \, dy.$$

It can be checked that $P(\Omega) = 1$. Indeed,

$$\frac{1}{2\pi\sigma^2} \int_{-\infty}^{+\infty} \int_{-\infty}^{+\infty} e^{-\frac{x^2+y^2}{2\sigma^2}} \, dx \, dy = \frac{1}{\pi} \int_{-\infty}^{+\infty} \int_{-\infty}^{+\infty} e^{-(x^2+y^2)} \, dx \, dy$$

$$= \frac{1}{2\pi} \int_0^{2\pi} \int_0^{+\infty} e^{-\rho^2} 2\rho \, d\rho \, d\theta$$

$$= \frac{1}{2\pi} \int_0^{2\pi} \int_0^{+\infty} e^{-u} \, du \, d\theta = 1.$$

EXAMPLE 1.1.15: COIN TOSSES, TAKE 3. Continuation of Example 1.1.9. Choose probability P such that for any event of the form $A = \{x_1 = a_1, \ldots, x_k = a_k\}$ ($a_1, \ldots, a_k$ in $\{0, 1\}$),

$$P(A) := \frac{1}{2^k}.$$

The choices for probability P in these examples is arbitrary and many other choices are possible. That of Example 1.1.13 suggests an unbiased die (the outcomes have the same probability). Probability P of Example 1.1.15 implies an unbiased coin and independent tosses (see Example 1.1.22).

From the axioms of probability, the following properties are easy to check:

$$P(\bar{A}) = 1 - P(A), \tag{1.1}$$

$$P(\varnothing) = 0, \tag{1.2}$$

$$A \subseteq B \Rightarrow P(A) \leq P(B), \tag{1.3}$$

$$P(\cup_{k=1}^{\infty} A_k) \leq \sum_{k=1}^{\infty} P(A_k). \tag{1.4}$$

Proof. For (1.1), use additivity: $1 = P(\Omega) = P(A + \bar{A}) = P(A) + P(\bar{A})$. For (1.2), use (1.1) with $A = \Omega : P(\varnothing) = 1 - P(\Omega) = 1 - 1 = 0$. For (1.3), write $P(B) = P(A) + P(B - A) \geq P(A)$. Finally, for (1.4), observe that

$$\cup_{k=1}^{\infty} A_k = \sum_{k=1}^{\infty} A'_k,$$

where

$$A'_k = A_k \cap \left\{ \overline{\cup_{i=1}^{k-1} A_i} \right\}.$$

Therefore,

$$P\left(\cup_{k=1}^{\infty} A_k\right) = P\left(\sum_{k=1}^{\infty} A'_k\right) = \sum_{k=1}^{\infty} P(A'_k).$$

But $A'_k \subseteq A_k$, and therefore $P(A'_k) \leq P(A_k)$. $\square$

Property (1.3) is the *monotonicity* property, and (1.4) is the *sub-σ-additivity* property.

Theorem 1.1.16 *Let $\{A_n\}_{n \geq 1}$ be a non-decreasing sequence of events, that is,*

$$A_{n+1} \supseteq A_n \quad (n \geq 1).$$

Then

$$P(\cup_{n=1}^{\infty} A_n) = \lim_{n \uparrow \infty} \uparrow P(A_n) \tag{1.5}$$

(where $\lim \uparrow a_n$ is a notation emphasizing the fact that the sequence $\{a_n\}$ is non-decreasing). Similarly, if $\{B_n\}_{n \geq 1}$ is a non-increasing sequence of events, that is, for all $n \geq 1$,

$$B_{n+1} \subset B_n,$$

then

$$P(\cap_{n=1}^{\infty} B_n) = \lim_{n \uparrow \infty} \downarrow P(B_n). \tag{1.6}$$

Proof. For (1.5), write

$$A_n = A_1 + (A_2 - A_1) + \cdots + (A_n - A_{n-1})$$

and

$$\cup_{k=1}^{\infty} A_k = A_1 + (A_2 - A_1) + (A_3 - A_2) + \cdots.$$

Therefore,

$$P(\cup_{k=1}^{\infty} A_k) = P(A_1) + \sum_{j=1}^{\infty} P(A_j - A_{j-1})$$

$$= \lim_{n\uparrow\infty} \left\{ P(A_1) + \sum_{j=1}^{n} P(A_j - A_{j-1}) \right\} = \lim_{n\uparrow\infty} P(A_n).$$

For (1.6), write

$$P(\cap_{n=1}^{\infty} B_n) = 1 - P\left(\overline{\cap_{n=1}^{\infty} B_n}\right) = 1 - P(\cup_{n=1}^{\infty} \overline{B_n})$$

and apply (1.5) with $A_n = \overline{B_n}$.   □

1.1.2   Independence and Conditional Probability

In the frequency interpretation, the definition of independence (1.7) reads, in rough and imprecise terms and using the notation introduced in Subsection 1.1.1, $n_{A\cap B}/n \approx (n_A/n) \cdot (n_B/n)$, or

$$\frac{n_{A\cap B}}{n_B} \approx \frac{n_A}{n}$$

(here $\approx$ is a "fuzzy" version of the equality sign). Therefore, statistics relative to A do not vary when performed on a neutral sample of population or on a selected sample of population characterized by the property B. For example, the proportion of people with a family name beginning with H is the same among a large population with the usual mix of men and women as it would be among a large all-male population. This is very much the intuitive notion of independence. Dependence between A and B occurs when $P(A \cap B) \neq P(A)P(B)$. In this case the relative frequency $n_{A\cap B}/n_B \approx P(A \cap B)/P(B)$ is different from the frequency n_A/n.

These remarks lead to the forthcoming formal definitions of independence and conditional probability. We start with the notion of independent events.

Definition 1.1.17 *Two events A and B are called* independent *if*

$$P(A \cap B) = P(A)P(B). \qquad (1.7)$$

Two random variables X and Y are called independent if for all $a, b \in \mathbb{R}$,

$$P(X \leq a, Y \leq b) = P(X \leq a)P(Y \leq b). \qquad (1.8)$$

Remark 1.1.18 In the above definition, we used the following notation: $P(X \leq a) = P(\{X \leq a\})$, where $\{X \leq a\} = \{\omega \ ; \ X(\omega) \leq a\}$. Also, the left-hand side of (1.8) is $P(\{X \leq a\} \cap \{Y \leq b\})$. This is a general feature of the notational system: commas often replace intersection signs. For instance, $P(A, B)$ is the probability that both events A and B occur.

EXAMPLE 1.1.19: DARTS, TAKE 4. Continuation of Example 1.1.14. The coordinate random variables X and Y are independent. Indeed,

$$\{(x, y) \in \mathbb{R}^2; x \leq a, y \leq b\} = \{X \leq a\} \cap \{Y \leq b\},$$

and therefore

$$
\begin{aligned}
P(\{X \leq a\} \cap \{Y \leq b\}) &= \frac{1}{2\pi\sigma^2} \int_{-\infty}^{a} \int_{-\infty}^{b} e^{-\frac{x^2 + y^2}{2\sigma^2}} \, dx dy \\
&= \left(\frac{1}{\sigma\sqrt{2\pi}} \int_{-\infty}^{a} e^{-\frac{x^2}{2\sigma^2}} \, dx \right) \left(\frac{1}{\sigma\sqrt{2\pi}} \int_{-\infty}^{b} e^{-\frac{y^2}{2\sigma^2}} \, dy \right) \\
&= P(X \leq a)P(Y \leq b),
\end{aligned}
$$

where the last equality comes from

$$P(X \leq a) = \lim_{n \uparrow \infty} P(X \leq a, Y \leq n)$$

(sequential continuity) and the identity

$$\frac{1}{\sigma\sqrt{2\pi}} \int_{-\infty}^{+\infty} e^{-\frac{y^2}{2\sigma^2}} \, dy = 1.$$

Remark 1.1.20 Incompatibility does not in general imply independence. In fact, two incompatible events A and B are independent if and only if at least one of them has zero probability. Indeed, $P(A \cap B) = P(\emptyset) = 0$ implies that (1.7) holds if and only if $P(A)P(B) = 0$.

The notion of independence carries over to families of events.

Definition 1.1.21 *A family $\{A_i\}_{i \in I}$ of events indexed by an arbitrary set I is called independent if for any finite collection of distinct events $A_{i_1}, \ldots, A_{i_k}$ from this family,*

$$P\left(\cap_{j=1}^{k} A_{i_j}\right) = \prod_{j=1}^{k} P(A_{i_j}). \tag{1.9}$$

A family $\{X_i\}_{i \in I}$ of real random variables indexed by an arbitrary set I is called independent if for any finite collection of distinct random variables $X_{i_1}, \ldots, X_{i_k}$ from this family,

$$P\left(\cap_{j=1}^{k}\{X_{i_j} \leq a_j\}\right) = \prod_{j=1}^{k} P(X_{i_j} \leq a_j) \tag{1.10}$$

for all real numbers $a_1, \ldots, a_k$. The family of real random variables $\{X_i\}_{i \in I}$ is said to be independent from the family of real random variables $\{Y_j\}_{j \in J}$ if

$$P\left((\cap_{\ell=1}^{r}\{X_{i_\ell} \leq a_\ell\}) \cap (\cap_{m=1}^{s}\{Y_{j_m} \leq b_m\})\right)$$
$$= P\left(\cap_{\ell=1}^{r}\{X_{i_\ell} \leq a_\ell\}\right) P\left(\cap_{m=1}^{s}\{Y_{j_m} \leq b_m\}\right)$$

for all indices $i_1, \ldots, i_r \in I$ and $j_1, \ldots, j_s \in J$, where I and J are arbitrary index sets, and all real numbers $a_1, \ldots, a_r$ and $b_1, \ldots, b_s$.

EXAMPLE 1.1.22: COIN TOSSES, TAKE 4.   Continuation of Example 1.1.15. Event $\{X_k = a_k\}$ is the direct sum of events $\{X_1 = a_1, \ldots, X_{k-1} = a_{k-1}, X_k = a_k\}$ for all possible values of $(a_1, \ldots, a_{k-1})$. Since there are 2^{k-1} such values and each one has probability 2^{-k}, we have $P(X_k = a_k) = 2^{k-1}2^{-k}$, that is,

$$P(X_k = 1) = P(X_k = 0) = \frac{1}{2}.$$

Therefore,

$$P(X_1 = a_1, \ldots, X_k = a_k) = P(X_1 = a_1)P(X_k = a_k)$$

for all $a_1, \ldots, a_k \in \{0, 1\}$, from which it follows by definition that $X_1, \ldots, X_k$ are independent random variables, and more generally that $\{X_n\}_{n \geq 1}$ is a family of independent random variables. Therefore, we have a model for *independent* tosses of an *unbiased* coin.

Bayes' Rules

Definition 1.1.23 *The* conditional probability *of A given B is the number*

$$P(A \mid B) := \frac{P(A \cap B)}{P(B)}, \qquad (1.11)$$

defined when $P(B) > 0$.

The quantity $P(A \mid B)$ represents one's expectation that A is realized when the only available information is that B is realized. Indeed, this expectation would then be based upon the relative frequency $n_{A \cap B}/n_B$ alone.

A symmetric form of (1.11) is

$$P(A \cap B) = P(A \mid B)P(B) = P(B \mid A)P(A). \qquad (1.12)$$

Of course, if A and B are independent, then

$$P(A \mid B) = P(A), P(B \mid A) = P(B). \qquad (1.13)$$

Probability theory is primarily concerned with the computation of probabilities of complex events. The following formulas are useful additions to the tool kit already containing rules (1.1) to (1.6).

Theorem 1.1.24

Bayes' rule of retrodiction. *With $P(A) > 0$, we have*

$$P(B \mid A) = \frac{P(A \mid B)P(B)}{P(A)}, \qquad (1.14)$$

a rephrasing of (1.12).

Bayes' rule of total causes. *For $B_1, B_2, \ldots$ such that*

$$\sum_{i=1}^{\infty} B_i = \Omega$$

and for all A,

$$P(A) = \sum_{i=1}^{\infty} P(A \mid B_i)P(B_i). \qquad (1.15)$$

Bayes' sequential formula. *For any sequence of events $A_1, \ldots, A_n$,*

$$P\left(\cap_{i=1}^{k} A_i\right) = P(A_1)P(A_2 \mid A_1)P(A_3 \mid A_1 \cap A_2) \cdots P\left(A_k \mid \cap_{i=1}^{k-1} A_i\right). \qquad (1.16)$$

Proof. For (1.15), just observe that $A = A \cap (\sum_{i=1}^{\infty} B_i)$, so that by σ-additivity:

$$P(A) = P\left(A \cap \left(\sum_{i=1}^{\infty} B_i\right)\right) = P\left(\sum_{i=1}^{\infty}(A \cap B_i)\right)$$

$$= \sum_{i=1}^{\infty} P(A \cap B_i) = \sum_{i=1}^{\infty} P(A \mid B_i)P(B_i).$$

For (1.16), proceed by induction. Suppose that (1.16) is true for k. Write

$$P\left(\cap_{i=1}^{k+1} A_i\right) = P\left(\left(\cap_{i=1}^k A_i\right) \cap A_{k+1}\right) = P\left(A_{k+1} \mid \cap_{i=1}^k A_i\right) P\left(\cap_{i=1}^k A_i\right),$$

and replace $P\left(\cap_{i=1}^k A_i\right)$ by the assumed equality (1.16). $\qquad \square$

EXAMPLE 1.1.25: MEDICAL TESTS. Doctors apply a test which, if the patient is affected by the disease being tracked, gives a positive result in 99% of the cases. However, it happens in 2% of the cases that a healthy patient has a positive test. Statistical data show that one individual out of 1000 has the disease. What is the probability for a patient with a positive test to be affected by the disease?

Solution. Let M be the event "patient is ill," and $+$ and $-$ the events "test is positive" and "test is negative," respectively. We have the data

$$P(M) = 0.001, \ P(+ \mid M) = 0.99, \ P(+ \mid \bar{M}) = 0.02,$$

and we must compute $P(M \mid +)$. By Bayes' retrodiction formula,

$$P(M \mid +) = \frac{P(+ \mid M)P(M)}{P(+)}.$$

By Bayes' formula of total causes,

$$P(+) = P(+ \mid M)P(M) + P(+ \mid \bar{M})P(\bar{M}).$$

Therefore,

$$P(M \mid +) = \frac{(0.99)(0.001)}{(0.99)(0.001) + (0.02)(0.999)} \approx \frac{1}{20}.$$

Remark 1.1.26 The result of the previous example may look surprising since the probability of illness is very low given the fact that the test is positive. However, the situation is not as bad as it seems, because in view of the ambiguous results, all positive patients will be tested once again, this time with a better test (but more expensive, one expects). The gain will be that only a small portion of the population will be given the expensive second test. Indeed, $P(+) = (0.99)(0.001) + (0.002)(0.999) \approx 0.003$. One possible cause of a positive test for a healthy patient is the practice of group testing, when the biological samples of several patients are mixed.

We now proceed to introduce the central concept of Markov chain theory.

Markov Property

Definition 1.1.27 *A and B are said to be* conditionally independent given C *if*

$$P(A \cap B \mid C) = P(A \mid C)P(B \mid C). \tag{1.17}$$

Let X, Y, Z be random variables taking their values in the denumerable sets E, F, G, respectively. One says that X and Y are conditionally independent given Z *if for all x, y, z in E, F, G, respectively, events $\{X = x\}$ and $\{Y = y\}$ are conditionally independent given $\{Z = z\}$.*

Remark 1.1.28 Note that (exercise) for fixed C, $P_C(A) := P(A \mid C)$ defines a probability P_C. Equality (1.17) expresses the independence of A and B with respect to this probability.

Theorem 1.1.29 *Let A_1, A_2, A_3 be three events of positive probability. Events A_1 and A_3 are conditionally independent given A_2 if and only if the "Markov property" holds, that is,*

$$P(A_3 \mid A_1 \cap A_2) = P(A_3 \mid A_2).$$

Proof. Assume conditional independence. Then

$$
\begin{aligned}
P(A_3 \mid A_1 \cap A_2) &= P(A_1 \cap A_2 \cap A_3)/P(A_1 \cap A_2) \\
&= P(A_1 \cap A_3 \mid A_2)P(A_2)/P(A_1 \cap A_2) \\
&= P(A_1 \mid A_2)P(A_3 \mid A_2)P(A_2)/P(A_1 \cap A_2) \\
&= P(A_1 \mid A_2)P(A_3 \mid A_2)/P(A_1 \mid A_2) = P(A_3 \mid A_2).
\end{aligned}
$$

Similar computations yield the converse implication. □

The next result is a characterization of conditional independence for random variables. It is a simple result that will be frequently used in the sequel. Recall once again the notation with commas instead of intersection symbols, for instance, $P(A, B \mid C, D) = P(A \cap B \mid C \cap D)$.

Theorem 1.1.30 *Let X, Y, and Z be three discrete random variables with values in E, F, and G, respectively. If for some function $g : E \times F \to [0,1]$, $P(X = x \mid Y = y, Z = z) = g(x, y)$ for all x, y, z, then $P(X = x \mid Y = y) = g(x, y)$ for all x, y, and X and Y are conditionally independent given Y.*

Proof. We have

$$
\begin{aligned}
P(X = x, Y = y) &= \sum_z P(X = x, Y = y, Z = z) \\
&= \sum_z P(X = x \mid Y = y, Z = z) P(Y = y, Z = z) \\
&= g(x, y) \sum_z P(Y = y, Z = z) \\
&= g(x, y) P(Y = y).
\end{aligned}
$$

Therefore,

$$
P(X = x \mid Y = y) = g(x, y) = P(X = x \mid Y = y, Z = z).
$$

The conclusion then follows from Theorem 1.1.29.   □

EXAMPLE 1.1.31: CHEAP WATCHES.   Two factories A and B manufacture watches. Factory A produces on average one defective item out of 100, and B produces one bad watch out of 200. A retailer receives a container of watches from one of the two above factories, but he does not know which. He checks the first watch. It works! What is the probability that the second watch he will check is good?

Solution. Let X_n be the state of the n-th watch extracted from the container, with $X_n = 1$ if it works and $X_n = 0$ if it does not. Let Y be the factory of origin. We express our a priori ignorance of where the case comes from by

$$
P(Y = A) = P(Y = B) = \frac{1}{2}.
$$

Also, we assume that given $Y = A$ (resp., $Y = B$), the states of the successive watches are independent. For instance,

$$
P(X_1 = 1, X_2 = 0 \mid Y = A) = P(X_1 = 1 \mid Y = A) P(X_2 = 0 \mid Y = A).
$$

We have the data

$$
P(X_n = 0 \mid Y = A) = 0.01, P(X_n = 0 \mid Y = B) = 0.005.
$$

We are required to compute $P(X_2 = 1 \mid X_1 = 1)$, that is,

$$P(X_1 = 1, X_2 = 1)/P(X_1 = 1).$$

By Bayes' formula of total causes, the numerator of the last fraction equals

$$P(X_1 = 1, X_2 = 1 \mid Y = A)P(Y = A) + P(X_1 = 1, X_2 = 1 \mid Y = B)P(Y = B),$$

that is, $\frac{1}{2}(99/100)^2 + \frac{1}{2}(199/200)^2$, and the denominator is

$$P(X_1 = 1 \mid Y = A)P(Y = A) + P(X_1 = 1 \mid Y = B)P(Y = B),$$

that is, $\frac{1}{2}(99/100) + \frac{1}{2}(199/200)$. Therefore,

$$P(X_2 = 1 \mid X_1 = 1) = \frac{\left(\frac{99}{100}\right)^2 + \left(\frac{199}{200}\right)^2}{\frac{99}{100} + \frac{199}{200}}.$$

Note that the states of the two watches are not independent. Indeed, if they were, then

$$P(X_2 = 1 \mid X_1 = 1) = P(X_2 = 1) = \frac{1}{2}\left(\frac{99}{100}\right) + \frac{1}{2}\left(\frac{199}{200}\right),$$

a result different from what we obtained.

Remark 1.1.32 The above example shows that even though for some given event C, two events A and B can be conditionally independent given C and conditionally independent given $\bar{C}$, yet they need *not* be unconditionally independent.

1.1.3 Expectation

Cumulative Distribution Function

Definition 1.1.33 *From the probabilistic point of view, a random variable X is described by its* cumulative distribution function (CDF)

$$F(x) = P(X \le x).$$

Theorem 1.1.34 *The cumulative distribution function has the following properties:*

(i) $F : \mathbb{R} \to [0, 1]$.

(ii) F *is non-decreasing.*

(iii) F is right-continuous.

(iv) $F(+\infty) := \lim_{a\uparrow\infty} F(a) = P(X < \infty)$.

(v) $F(-\infty) := \lim_{a\downarrow-\infty} F(a) = P(X = -\infty)$.

Proof.

(i) Obvious.

(ii) If $a \leq b$, then $\{X \leq a\} \subset \{X \leq b\}$, and therefore, by (1.3), $P(X \leq a) \leq P(X \leq b)$.

(iii) Apply (1.6) with $B_n = \{X \leq a + \frac{1}{n}\}$. Since $\cap_{n\geq1}\{X \leq a + \frac{1}{n}\} = \{X \leq a\}$, we have $\lim_{n\uparrow\infty} P\left(X \leq a + \frac{1}{n}\right) = P(X \leq a)$.

(iv) Apply (1.5) with $A_n = \{X \leq n\}$ and observe that $\cup_{n=1}^{\infty}\{X \leq n\} = \{X < \infty\}$.

(v) Apply (1.6) with $B_n = \{X \leq -n\}$ and observe that $\cap_{n=1}^{\infty}\{X \leq -n\} = \{X = -\infty\}$. $\qquad\square$

Since F is non-decreasing, the following limit exists for all $x \in \mathbb{R}$:

$$F(x-) := \lim_{h\downarrow0} F(x - h).$$

The sequence $B_n = \{a - \frac{1}{n} < X \leq a\}$ is decreasing, and $\cap_{n=1}^{\infty} B_n = \{X = a\}$. Therefore, by the monotone sequential continuity property (1.5), $P(X = a) = \lim_{n\uparrow\infty} P\left(a - \frac{1}{n} < X \leq a\right) = \lim_{n\uparrow\infty} \left(F(a) - F\left(a - \frac{1}{n}\right)\right)$, that is to say,

$$P(X = a) = F(a) - F(a-). \tag{1.18}$$

In particular, F is continuous if and only if $P(X = a) = 0$ for every fixed point $a \in \mathbb{R}$. From (1.18) and the definition of F, we have

$$\begin{aligned}
P(a < X \leq b) &= F(b) - F(a), \\
P(a \leq X < b) &= F(b-) - F(a-), \\
P(a \leq X \leq b) &= F(b) - F(a-), \\
P(a < X < b) &= F(b-) - F(a).
\end{aligned}$$

Also, recall that

$$\begin{aligned}
P(X = +\infty) &= 1 - F(+\infty), \\
P(X = -\infty) &= F(-\infty).
\end{aligned}$$

A special case of a continuous CDF is

$$F(x) = \int_{-\infty}^{x} f(y)\mathrm{d}y \tag{1.19}$$

for some function $f \geq 0$ called the *probability density function* (PDF) of X. Note that a continuous CDF F need not admit a representation such as (1.19). If (1.19) holds, the CDF and the random variable are both called *absolutely continuous*.

Expectation, Mean, and Variance

The set of discontinuity points of a CDF are denumerable, as is the case for any bounded non-decreasing function. Calling $\{d_n\}_{n \geq 1}$ the sequence of discontinuity points, define

$$F_d(t) := \sum_{d_n \leq t}\{F(d_n) - F(d_n{}^-)\}$$

and

$$F_c(t) := F(t) - F_d(t)\,,$$

the discontinuous and continuous components of F, respectively. The symbol

$$\int_{-\infty}^{+\infty} g(x)\mathrm{d}F(x) \tag{1.20}$$

represents the *Stieltjes–Lebesgue integral* with respect to F of the function g (Section A.3). In the special case where the continuous component of the CDF is in fact absolutely continuous, that is,

$$F_c(x) = \int_{-\infty}^{x} f_c(y)\mathrm{d}y,$$

it suffices for our purpose to interpret (1.20) as

$$\int_{-\infty}^{+\infty} g(x)\mathrm{d}F(x) = \sum_{n=1}^{\infty} g(d_n)(F(d_n) - F(d_n{}_-)) + \int_{-\infty}^{+\infty} g(x)f_c(x)\mathrm{d}x.$$

The most frequent cases arising are the purely discontinuous case where $F(t) = F_d(t)$, for which

$$\int_{-\infty}^{+\infty} g(x)\mathrm{d}F(x) = \sum_{n=1}^{\infty} g(d_n)\{F(d_n) - F(d_{n-})\}, \tag{1.21}$$

and the absolutely continuous case, for which

$$\int_{-\infty}^{+\infty} g(x)\mathrm{d}F(x) = \int_{-\infty}^{+\infty} g(x)f(x)\mathrm{d}x. \tag{1.22}$$

If X takes the value $+\infty$ with positive probability, $+\infty$ should be considered conventionally as a discontinuity point, and a sum such as (1.21) will include the term $g(\infty)(1 - F(\infty))$. Of course, $g(\infty)$ must be well-defined.

EXAMPLE 1.1.35: SUM OF INDEPENDENT UNIFORMS, TAKE 1. For a random point inside the unit square $[0, 1]^2 = [0, 1] \times [0, 1]$, the following model is chosen: $\Omega = [0, 1]^2, \mathrm{P}(A) = $ area of A. Let X and Y be the coordinate random variables, and let $Z = X + Y$ (see Figure 1.1.3).

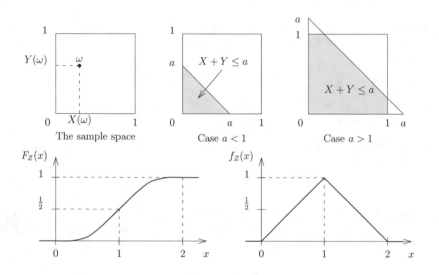

Figure 1.1.3. Sum of 2 independent uniform random variables

We compute the CDF of Z. For instance, when $a > 1$,

$$F_Z(a) := \mathrm{P}(Z \leq a) = \mathrm{P}(X + Y \leq a) = \mathrm{P}(\{\omega \ ; \ X(\omega) + Y(\omega) \leq a\})$$

is just the area of the domain $[0, 1]^2 \cap \{x + y \leq a\}$ (see Figure 1.1.3), and

$$F_Z(a) = \left\{ \begin{array}{ll} \frac{1}{2}a^2 & \text{if } 0 \leq a \leq 1, \\ 1 - \frac{1}{2}(2 - a)^2 & \text{if } 1 \leq a \leq 2. \end{array} \right.$$

The random variable Z admits a probability density given by

$$f_Z(x) = \frac{\mathrm{d}F_Z(x)}{\mathrm{d}x}$$

that is

$$f_Z(a) = \begin{cases} x & \text{if } 0 \le x \le 1, \\ 2 - x & \text{if } 1 \le x \le 2. \end{cases}$$

When a random variable takes its values in a denumerable set, its CDF reduces to the discontinuous part F_d, and the discontinuity points d_n are the values taken by the random variable. In particular,

$$P(X = d_n) = F(d_n) - F(d_{n-}).$$

For such random variables, the *probability distribution* $\{p(d_n)\}_{n \ge 1}$, where

$$p(d_n) := P(X = d_n),$$

is usually preferred to the cumulative distribution function to describe its probabilistic behavior. In this notation,

$$\int_{-\infty}^{+\infty} g(x) \mathrm{d}F(x) = \sum_{n=1}^{\infty} g(d_n) p(d_n).$$

EXAMPLE 1.1.36: COIN TOSSES, TAKE 5. Continuation of Example 1.1.22. The number of occurrences of heads in n is $S_n = X_1 + \cdots + X_n$ tosses. This random variable takes the integer values 0 to n. The event $\{S_n = k\}$ is "k among $X_1, \ldots, X_n$ are equal to 1". There are $\binom{n}{k}$ distinct ways of assigning k values of 1 and $n - k$ values of 0 to $X_1, \ldots, X_n$, and all have the same probability 2^{-n}. Therefore,

$$P(S_n = k) = \binom{n}{k} \frac{1}{2^n}.$$

Definition 1.1.37 *Let X be a random variable with the CDF $F(x) = P(X \le x)$ and let $g : \bar{\mathbb{R}} \to \mathbb{R}$ be such that*

$$\int_{-\infty}^{+\infty} |g(x)| \mathrm{d}F(x) < \infty. \tag{1.23}$$

Then one defines the expectation *of $g(X)$ by the formula*

$$E[g(X)] := \int_{-\infty}^{+\infty} g(x) \mathrm{d}F(x). \tag{1.24}$$

EXAMPLE 1.1.38: SUM OF INDEPENDENT UNIFORMS, TAKE 2. Continuation of Example 1.1.35.

$$E[Z^2] = \int_{-\infty}^{+\infty} x^2 f(x) \mathrm{d}x = \int_0^1 x^3 \mathrm{d}x + \int_1^2 x^2 (2 - x) \mathrm{d}x \,.$$

Therefore, $E[Z^2] = 7/6$.

EXAMPLE 1.1.39: COIN TOSSES, TAKE 6. Continuation of Example 1.1.36. We compute the expectation of the sum $S_n = X_1 + \cdots + X_n$.

$$E[S_n] = \sum_{k=0}^{n} k P(S_n = k)$$

$$= \frac{1}{2^n} \sum_{k=1}^{n} k \frac{n!}{k!(n-k)!}$$

$$= \frac{n}{2^n} \sum_{k=1}^{n} \frac{(n-1)!}{(k-1)!((n-1)-(k-1))!}$$

$$= \frac{n}{2^n} \sum_{k=0}^{n-1} \frac{(n-1)!}{j!(n-1-j)!} = \frac{n}{2^n} 2^{n-1} \,.$$

Therefore, $E[S_n] = n/2$.

We now list the elementary property of expectation. Expectation inherits the linearity property of the Stieltjes–Lebesgue integral:

$$E[\lambda_1 g_1(X) + \lambda_2 g_2(X)] = \lambda_1 E[g_1(X)] + \lambda_2 E[g_2(X)], \tag{1.25}$$

where $\lambda_1, \lambda_2 \in \mathbb{R}$, and g_1 and g_2 satisfy the integrability condition (1.23). Also, expectation is monotone, in the sense that $g_1(x) \le g_2(x)$ for all x implies

$$E[g_1(X)] \le E[g_2(X)] \,. \tag{1.26}$$

Consider now the *indicator function* of an event A:

$$1_A(\omega) := \begin{cases} 1 & \text{if } \omega \in A, \\ 0 & \text{if } \omega \notin A. \end{cases}$$

The random variable $X = 1_A$ takes the value 1 with probability $P(X = 1) = P(A)$ and the value 0 with probability $P(X = 0) = P(\bar{A}) = 1 - P(A)$. Therefore, $E[X] = 0 \times P(X = 0) + 1 \times P(X = 1) = P(X = 1) = P(A)$, that is to say,

$$E[1_A] = P(A). \tag{1.27}$$

In particular, $E[1] = 1$.

Sometimes, one needs to define $E[g(X)]$ for a complex function $g : \mathbb{R} \to \mathbb{C}$, that is, $g(x) = g_R(x) + ig_I(x)$ where g_R and g_I take real values. The definition of expectation is now

$$E[g(X)] = E[g_R(X)] + iE[g_I(X)],$$

provided that the expectations on the right-hand side are finite.

The triangle inequality

$$|E[g(X)]| \leq E[|g(X)|] \tag{1.28}$$

is useful, and its proof follows from the analogous inequality for the Stieltjes–Lebesgue integral:

$$\left| \int_{-\infty}^{+\infty} g(x) \mathrm{d}F(x) \right| \leq \int_{-\infty}^{+\infty} |g(x)| \mathrm{d}F(x).$$

Definition 1.1.40 *The* mean *m and* variance *σ^2 of a real random variable X are defined by*

$$m = E[X] = \int_{-\infty}^{+\infty} x \mathrm{d}F(x),$$

$$\sigma^2 = E[(X - m)^2] = \int_{-\infty}^{+\infty} (x - m)^2 \mathrm{d}F(x).$$

Of course, the integrals involved must be well defined. The variance is also denoted by $\mathrm{var}(X)$. From the linearity of expectation, it follows that $E[(X-m)^2] = E[X^2] - 2mE[X] + m^2$, that is,

$$\mathrm{var}(X) = E[X^2] - m^2. \tag{1.29}$$

EXAMPLE 1.1.41: MAXIMUM. Let $X_1, X_2, \ldots, X_n$ be independent random variables uniformly distributed on $[0, 1]$, that is to say, with the probability density $f(x) = 1_{[0,1]}(x)$. We compute the expectation of $Z = \max(X_1, \ldots, X_n)$. For $z \in [0, 1]$,

$$P(\max(X_1, \ldots, X_n) \leq z) = P(X_1 \leq z, \ldots, X_n \leq z) = \prod_{k=1}^{n} P(X_1 \leq z) = z^n.$$

Therefore, Z has the probability density $\frac{d}{dz}z^n = nz^{n-1}$ for $z \in [0,1]$, and 0 otherwise, and

$$E[Z] = \int_0^1 z(nz^{n-1})dz = \frac{n}{n+1}.$$

The following is the *telescope formula*, to be used repeatedly in the sequel.

Theorem 1.1.42 *For a random variable X taking its values in $\mathbb{N}$,*

$$E[X] = \sum_{n=1}^{\infty} P(X \geq n). \tag{1.30}$$

Proof.

$$\begin{aligned}
E[X] &= P(X = 1) + 2P(X = 2) + 3P(X = 3) + \ldots \\
&= P(X = 1) \; + P(X = 2) + P(X = 3) + \ldots \\
&\qquad\qquad\quad + P(X = 2) + P(X = 3) + \ldots \\
&\qquad\qquad\qquad\qquad\quad + P(X = 3) + \ldots
\end{aligned}$$

$\square$

Theorem 1.1.43 *Let $\{X_n\}_{n\geq 1}$ be a sequence of integrable random variables such that $E[X_n] = E[X_1]$ for all $n \geq 1$. Let T be an integer-valued random variable such that for all $n \geq 1$, the event $\{T \geq n\}$ is independent of X_n. Then*

$$E[\sum_{n=1}^{T} X_n] = E[X_1]E[T].$$

Proof. Let $S = \sum_{n=1}^{T} X_n$. Then by dominated convergence (Theorem A.3.2)

$$E[S] = E\left[\sum_{n=1}^{\infty} X_n 1_{\{n \leq T\}}\right] = \sum_{n=1}^{\infty} E[X_n 1_{\{n \leq T\}}].$$

But $E[X_n 1_{\{n \leq T\}}] = E[X_n]E[1_{\{n \leq T\}}] = E[X_1]P(n \leq T\})$. The result then follows from the telescope formula. $\square$

Famous Random Variables

Consider a sequence $\{X_n\}_{n \geq 1}$ of random variables taking their values in $\{0, 1\}$, with the common probability distribution

$$P(X_n = 1) = p,$$

where $p \in (0, 1)$. Suppose in addition that the X_n's are independent. Since $P(X_j = a_j) = p$ or $1 - p$ depending upon whether $a_i = 1$ or 0, and since there are exactly $\sum_{j=1}^{k} a_j$ numbers among $a_1, \ldots, a_k$ that are equal to 1,

$$P(X_1 = a_1, \ldots, X_k = a_k) = p^{\sum_{j=1}^{k} a_j} (1-p)^{k - \sum_{j=1}^{k} a_j}. \tag{1.31}$$

This is a model for a game of "heads and tails" with a biased coin if $p \neq \frac{1}{2}$. It provides a framework that shelters three famous discrete random variables: the binomial, the geometric and the Poisson random variables.

Binomial. Define

$$S_n = X_1 + \cdots + X_n.$$

This random variable takes the values $0, 1, \ldots, n$. To obtain $S_n = i$ where $i \in [0, n]$, one must have $X_1 = a_1, \ldots, X_n = a_n$ with $\sum_{j=1}^{n} a_j = i$. There are $\binom{n}{i}$ distinct ways of having this, and each occurs with probability $p^i (1-p)^{n-i}$. Therefore, for $i \in [0, n]$,

$$P(S_n = i) = \binom{n}{i} p^i (1-p)^{n-i}. \tag{1.32}$$

Definition 1.1.44 *A random variable X with distribution (1.32) is called a binomial random variable of size n and parameter $p \in (0, 1)$.*

A direct computation gives for the mean and the variance of the binomial random variable

$$E[X] = np, \ \text{var}(X) = np(1-p). \tag{1.33}$$

(Do Exercise 1.5.19 for an alternative to the direct computation via generating functions.)

Geometric. Define the random variable T to be the first time of occurrence of 1 in the sequence $X_1, X_2, \ldots$, that is,

$$T = \inf\{n \geq 1; X_n = 1\},$$

with the convention that if $X_n = 0$ for all $n \geq 1$, then $T = \infty$ (actually, we shall show that this event has probability 0). The event $\{T = k\}$ is exactly $\{X_1 = 0, \ldots, X_{k-1} = 0, X_k = 1\}$, and therefore,

$$P(T = k) = P(X_1 = 0) \cdots P(X_{k-1} = 0) P(X_k = 1),$$

that is, for $k \geq 1$,

$$P(T = k) = (1 - p)^{k-1}p. \tag{1.34}$$

In particular, $P(T < \infty) = \sum_{k=1}^{\infty} P(T = k) = \sum_{k=1}^{\infty} p(1-p)^{k-1} = 1$ (recall that $p > 0$).

Definition 1.1.45 *A random variable with the distribution (1.34) is called a ge-ometric random variable with parameter p. A direct computation (or Exercise 1.5.19) gives*

$$E[T] = \frac{1}{p}. \tag{1.35}$$

Poisson. Suppose you play heads and tails for a large number of turns N with a coin such that

$$P(X_n = 1) = \frac{\alpha}{N}.$$

The number S_N of tails you observe is therefore distributed according to

$$P(S_N = k) = \binom{N}{k} \left(\frac{\alpha}{N}\right)^k \left(1 - \frac{\alpha}{N}\right)^{N-k}.$$

From (1.33), the average number of tails is $N.\frac{\alpha}{N} = \alpha$, a constant. It is interesting to find the limit of the distribution of S_N as $N \to \infty$. Denoting by $p_N(k)$ the probability of $S_N = k$, we have that

$$p_N(0) = \left(1 - \frac{\alpha}{N}\right)^N$$

goes to $e^{-\alpha}$ as N goes to ∞. Also,

$$\frac{p_N(k + 1)}{p_N(k)} = \frac{\frac{N-k}{k+1} \frac{\alpha}{N}}{1 - \frac{\alpha}{N}}$$

goes to $\frac{\alpha}{k+1}$. Therefore, for all $k \geq 0$,

$$\lim_{n\uparrow\infty} p_N(k) = e^{-\alpha}\frac{\alpha^k}{k!},$$

with the convention $0! = 1$. By definition, a *Poisson random variable with param-eter* $\theta > 0$ is a random variable such that for all $k \geq 0$,

$$P(X = k) = e^{-\theta}\frac{\theta^k}{k!}. \tag{1.36}$$

Therefore, the probability distribution of S_N tends as $N \to \infty$ to the probability distribution of a Poisson r.v. of parameter α. This phenomenon is called the

Poisson law of rare events, because as $N \to \infty$, obtaining tails is an increasingly rare event, of probability $\frac{\alpha}{N}$.

A direct computation (or Exercise 1.5.19) gives for the Poisson random variable

$$E[X] = \theta, \operatorname{var}(X) = \theta.$$

EXAMPLE 1.1.46: SUM OF INDEPENDENT POISSON VARIABLES. Let X_1 and X_2 be two *independent* Poisson random variables with means $\theta_1 > 0$ and $\theta_2 > 0$, respectively. Show that $X = X_1 + X_2$ is a Poisson random variable with mean $\theta = \theta_1 + \theta_2$.

Solution. For $k \geq 0$,

$$
\begin{aligned}
P(X = k) &= P(X_1 + X_2 = k) = P\left(\sum_{i=0}^{k}\{X_1 = i, X_2 = k - i\}\right) \\
&= \sum_{i=0}^{k} P(X_1 = i, X_2 = k - i) = \sum_{i=0}^{k} P(X_1 = i)P(X_2 = k - i) \\
&= \sum_{i=0}^{k} e^{-\theta_1}\frac{\theta_1^i}{i!} e^{-\theta_2}\frac{\theta_2^{k-i}}{(k-i)!} = \frac{e^{-(\theta_1+\theta_2)}}{k!}(\theta_1 + \theta_2)^k.
\end{aligned}
$$

EXAMPLE 1.1.47: POISSON SUMS OF IID VARIABLES. Let $\{X_n\}_{n\geq 1}$ be independent random variables taking the values 0 and 1 with probability $q = 1 - p$ and p, respectively, where $p \in (0, 1)$. Let T be a Poisson random variable with mean $\theta > 0$, independent of $\{X_n\}_{n\geq 1}$. Define

$$S = X_1 + \cdots + X_T$$

(that is, $S(\omega) = X_1(\omega) + \cdots + X_n(\omega)$ if $T(\omega) = n$). We show that S is a Poisson random variable with mean $p\theta$:

$$
\begin{aligned}
P(S = k) &= P(X_1 + \cdots + X_T = k) = P\left(\sum_{n=k}^{\infty}\{X_1 + \cdots + X_n = k, T = n\}\right) \\
&= \sum_{n=k}^{\infty} P(X_1 + \cdots + X_n = k, T = n) \\
&= \sum_{n=k}^{\infty} P(X_1 + \cdots + X_n = k)P(T = n) \\
&= \sum_{n=k}^{\infty} \frac{n!}{k!(n-k)!} p^k q^{n-k} e^{-\theta} \frac{\theta^n}{n!} \\
&= e^{-\theta}\frac{(p\theta)^k}{k!} \sum_{n=k}^{\infty} \frac{(q\theta)^{n-k}}{(n-k)!} \\
&= e^{-\theta}\frac{(p\theta)^k}{k!} \sum_{i=0}^{\infty} \frac{(q\theta)^i}{i!} = e^{-\theta}\frac{(p\theta)^k}{k!} e^{q\theta} = e^{p\theta}\frac{(p\theta)^k}{k!}.
\end{aligned}
$$

Thus, if one "thins out" with thinning probability $1 - p$ a population sample of Poissonian size, the remaining sample also has a Poissonian size, with a mean p times that of the original sample.

————

EXAMPLE 1.1.48: BERNOULLI–POISSON EGGS. The bluepinko bird lays T eggs, blue or pink, with probability p of laying a blue egg. The previous exercise showed that if the number of eggs is Poisson with mean θ, then the number of blue eggs is Poisson with mean θp and the number of pink eggs is Poisson with mean θq. We now show that the number of blue eggs and the number of pink eggs are independent random variables.

If S is the number of blue eggs, $T - S$ is the number of pink eggs. One must show that for any integers $k \geq 0, \ell \geq 0$,

$$
\begin{aligned}
P(S = k, T - S = \ell) &= P(S = k)P(T - S = \ell) \\
&= e^{-\theta p}\frac{(\theta p)^k}{k!} e^{-\theta q}\frac{(\theta q)^\ell}{\ell!}.
\end{aligned}
$$

But

$$
\begin{aligned}
P(S = k, T - S = \ell) &= P(S = k, T = k + \ell) \\
&= P(X_1 + \cdots + X_{k+\ell} = k, T = k + \ell) \\
&= P(X_1 + \cdots + X_{k+\ell} = k)P(T = k + \ell) \\
&= \frac{(k + \ell)!}{k!\ell!}p^k q^\ell e^{-\theta}\frac{\theta^{k+\ell}}{(k + \ell)!} \\
&= e^{-p\theta}\frac{(p\theta)^k}{k!}e^{-q\theta}\frac{(q\theta)^\ell}{\ell!}.
\end{aligned}
$$

Our small list of random variables continues with absolutely continuous ones. Recall that an *absolutely continuous* random variable is by definition a real random variable with a probability density, that is,

$$
P(X \le x) = \int_{-\infty}^{x} f(x)\mathrm{d}x,
$$

where $f(x) \ge 0$, and since X is real,

$$
\int_{-\infty}^{+\infty} f(x)\mathrm{d}x = 1.
$$

Uniform.

Definition 1.1.49 *A random variable* X *with probability density function*

$$
f(x) = \frac{1}{b - a}1\{x \in [a, b]\}
$$

is called a uniform random variable *on* $[a, b]$.

Its mean and variance are

$$
E[X] = \frac{a + b}{2}, \quad \mathrm{var}(X) = \frac{(b - a)^2}{12},
$$

as simple computations reveal.

Gaussian.

Definition 1.1.50 *The random variable* X *with* PDF

$$
f(x) = \frac{1}{\sigma\sqrt{2\pi}}e^{-\frac{1}{2}\frac{(x-m)^2}{\sigma^2}},
$$

where $m \in \mathbb{R}$ *and* $\sigma > 0$, *is called a* Gaussian random variable.

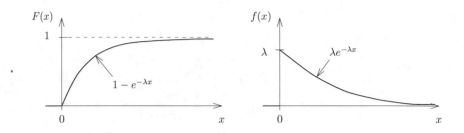

Figure 1.1.4. Exponential random variable

One can check that $E[X] = m$ and $\text{var}(X) = \sigma^2$.

Exponential.

Definition 1.1.51 *The random variable X with* CDF

$$F(x) = (1 - e^{-\lambda x})1_{\{x \geq 0\}}$$

and PDF

$$f(x) = \lambda e^{-\lambda x}1_{\{x \geq 0\}}$$

(see Figure 1.1.4) is called an exponential random variable with parameter λ.

Its mean is

$$E[X] = 1/\lambda.$$

Absolutely Continuous Random Vectors

A *random vector* of dimension n is a collection of n real random variables

$$X = (X_1, \ldots, X_n).$$

Remark 1.1.52 Each of the random variables $X_1, \ldots, X_n$ can be characterized from a probabilistic point of view by its cumulative distribution function. However, the CDFs of each coordinate of a random vector do not completely describe the probabilistic behavior of the whole vector. For instance, if U_1 and U_2 are two independent random variables with the same CDF $G(x)$, the vectors $X = (X_1, X_2)$ defined respectively by $X_1 = U_1, X_2 = U_2$ and $X_1 = U_1, X_2 = U_1$ have each of their coordinates with the same CDF, and they are quite different.

The CDF of the vector $X = (X_1, \ldots, X_n)$ is the function $F : \mathbb{R}^n \to [0, 1]$ defined by

$$F(x_1, \ldots, x_n) = P(X_1 \leq x_1, \ldots, X_n \leq x_n).$$

Definition 1.1.53 *An absolutely continuous random vector is one that admits a* probability density function, *that is,*

$$F(x_1, \ldots, x_n) := \int_{-\infty}^{x_1} \cdots \int_{-\infty}^{x_n} f(y_1, \ldots, y_n) \, dy_1 \cdots dy_n$$

for some non-negative function $f : \mathbb{R}^n \to \mathbb{R}$ *such that*

$$\int_{-\infty}^{+\infty} \cdots \int_{-\infty}^{+\infty} f(y_1, \ldots, y_n) \, dy_1 \cdots dy_n = 1 \,.$$

For a function $g : \mathbb{R}^n \to \mathbb{R}$, the *expectation* of $g(X)$ is by definition

$$\mathrm{E}[g(X)] = \int_{-\infty}^{+\infty} \cdots \int_{-\infty}^{+\infty} g(x_1, \ldots, x_n) f(x_1, \ldots, x_n) dx_1 \cdots dx_n \,,$$

where it is required that $\mathrm{E}[|g(X)|] < \infty$ in order to make the integral on the right-hand side meaningful.

Expectation so defined enjoys, *mutatis mutandis*, the properties mentioned for the scalar case: linearity (see (1.25)), monotonicity (see (1.26)), and the triangle inequality (see (1.28)).

Consider a 2-dimensional vector $X = (X_1, X_2)$ with PDF $f_{X_1, X_2}(x_1, x_2)$. The PDF of X_1 is obtained by integrating out x_2:

$$f_{X_1}(x_1) = \int_{-\infty}^{+\infty} f_{X_1, X_2}(x_1, x_2) \, dx_2 \,. \tag{1.37}$$

In fact,

$$\mathrm{P}(X_1 \leq a) = \mathrm{P}((X_1, X_2) \in (-\infty, a] \times \mathbb{R})$$

$$= \int_{-\infty}^{a} \int_{-\infty}^{+\infty} f_{X, X_2}(x_1, x_2) dx_1 dx_2$$

$$= \int_{-\infty}^{a} \left(\int_{-\infty}^{+\infty} f_{X_1, X_2}(x_1, x_2) dx_2 \right) dx_1 \,.$$

Formula (1.37) extends in an obvious manner to vectors.

If $X_1, \ldots, X_n$ are absolutely continuous random variables of respective probability density functions $f_1, \ldots, f_n$, and if, moreover, $X_1, \ldots, X_n$ are independent, then

$$\mathrm{P}(X_1 \leq x_1, \ldots, X_n \leq x_n) = \mathrm{P}(X_1 \leq x_1) \cdots \mathrm{P}(X_n \leq x_n)$$

$$= \left(\int_{-\infty}^{x_1} f_1(y_1) \, dy_1 \right) \cdots \left(\int_{-\infty}^{x_n} f_n(y) \, dy_n \right)$$

$$= \int_{-\infty}^{x_1} \cdots \int_{-\infty}^{x_n} f_1(y_1) \cdots f_n(y_n) \, dy_1 \cdots dy_n \,.$$

The PDF of the vector $(X_1, \ldots, X_n)$ is therefore the product of the PDFs of its coordinates

$$f(x_1, \ldots, x_n) = f_1(x_1) \cdots f_n(x_n). \tag{1.38}$$

Conversely,

Theorem 1.1.54 *If the vector X has a PDF that factors as in (1.38), where $f_1, \ldots, f_n$ are PDFs, then $X_1, \ldots, X_n$ are independent r.v.s with respective PDFs $f_1, \ldots, f_n$.*

Proof. Indeed,

$$P(X_1 \in A_1, \ldots, X_n \in A_n) = \int_{A_1} \cdots \int_{A_n} f_1(y_1) \cdots f_n(y_n) dy_1 \cdots dy_n;$$

that is, by Fubini's theorem,

$$P(X_1 \in A_1, \ldots, X_n \in A_n) = \left(\int_{A_1} f_1(y_1) dy_1 \right) \times \cdots \times \left(\int_{A_n} f_2(y_n) dy_n \right).$$

Letting $A_2 = \cdots = A_n = \mathbb{R}$ in the last identity yields

$$P(X_1 \in A_1) = \int_{A_1} f_1(y_1) dy_1,$$

which proves that X_1 has the PDF f_1. Similarly, $P(X_i \in A_i) = \int_{A_i} f_i(y_i) dy_i$, and therefore

$$P(X_1 \in A_1, \ldots, X_n \in A_n) = P(X_1 \in A_1) \cdots P(X_n \in A_n),$$

which proves independence, since the A_i's are arbitrary. $\qquad\square$

EXAMPLE 1.1.55: NON-COINCIDENCE. We prove that $P(X = Y) = 0$ for any two *independent and absolutely continuous* random variables X and Y. In fact

$$P(X = Y) = E[1_{\{X=Y\}}] = \int_{-\infty}^{+\infty} \int_{-\infty}^{+\infty} g(x, y) \, dx \, dy,$$

where $g(x, y) := 1_{\{x=y\}} f_X(x) f_Y(y)$ is null outside the diagonal. Since the diagonal has null Lebesgue measure, the integral is null.

EXAMPLE 1.1.56: FREEZING. Let $X_1, \ldots, X_n$ be independent random variables with respective PDFs $f_1, \ldots, f_n$. We show that

$$E[g(X_1, \ldots, X_n)] = \int_{-\infty}^{+\infty} E[g(y, X_2, \ldots, X_n)] f_1(y) dy$$

and that

$$P(X_1 \leq X_2, \ldots, X_1 \leq X_n, X_1 \leq x) = \int_{-\infty}^{x} P(X_2 \geq y) \cdots P(X_n \geq y) f_1(y) \, dy.$$

We do the case $n = 2$ for simplicity. We have

$$\begin{aligned} E[g(X_1, X_2)] &= \int_{-\infty}^{+\infty} \int_{-\infty}^{+\infty} g(x_1, x_2) f_1(x_1) f_2(x_2) \, dx_1 \, dx_2 \\ &= \int_{-\infty}^{+\infty} f_1(x_1) \left\{ \int_{-\infty}^{+\infty} g(x_1, x_2) f_2(x_2) \, dx_2 \right\} dx_1 \\ &= \int_{-\infty}^{+\infty} f_1(x) E[g(x, X_2)] \, dx. \end{aligned}$$

The second equality is obtained from the first one by letting (case $n = 3$ this time) $g(X_1, X_2, X_3) := 1_{\{X_1 \leq X_2\}} 1_{\{X_1 \leq X_3\}} 1_{\{X_1 \leq x\}}$ and observing that

$$\begin{aligned} E[g(y, X_2, X_3)] &= E[1_{\{y \leq x\}} 1_{\{X_2 \geq y\}} 1_{\{X_3 \geq y\}}] \\ &= 1_{\{y \leq x\}} P(X_2 \geq y, X_3 \geq y) = 1_{\{y \leq x\}} P(X_2 \geq y) P(X_3 \geq y). \end{aligned}$$

EXAMPLE 1.1.57: SUM OF INDEPENDENT ABSOLUTELY CONTINUOUS VARIABLES. We show that the probability density function of the random variable $Z = X + Y$, where X and Y are independent random variables with respective probability densities f_X and f_Y, is given by the *convolution formula*

$$f_Z(z) = \int_{-\infty}^{+\infty} f_Y(z - y) f_X(y) \, dy.$$

The PDF of vector (X, Y) is $f_X(x) f_Y(y)$, and therefore, for all $a \in \mathbb{R}$,

$$\begin{aligned} P(Z \leq a) &= P(X + Y \leq a) = E[1_{\{X+Y \leq a\}}] \\ &= \int_{-\infty}^{+\infty} \int_{-\infty}^{+\infty} 1_{\{x+y \leq a\}} f_X(x) f_Y(y) \, dx \, dy. \end{aligned}$$

The latter integral can be written, by Fubini's theorem,

$$\int_{-\infty}^{+\infty} \left\{ \int_{-\infty}^{+\infty} 1_{\{y \leq a - x\}} f_Y(y) \, dy \right\} f_X(x) dx = \int_{-\infty}^{+\infty} \left\{ \int_{-\infty}^{a-x} f_Y(y) \, dy \right\} f_X(x) \, dx,$$

that is, after a change of variable,

$$P(Z \leq a) = \int_{-\infty}^{a} \left\{ \int_{-\infty}^{+\infty} f_Y(z - x) f_X(x) dx \right\} dz.$$

Discrete Random Vectors

In what follows, only the case of a discrete random vector $X = (X_1, \ldots, X_n)$ where all the random variables X_i take their values in the *same* denumerable space E will be considered. This restriction is not essential, but it simplifies the notation.

The statistical behavior of X is described by its distribution $(p(u); u \in E^n)$ where

$$p(u) := P(X_1 = u_1, \ldots, X_n = u_n),$$

and the expectation of $g(X)$ is defined by

$$E[g(X)] := \sum_{u \in E^n} g(u), p(u)$$

as long as g is non-negative or the sum on the right-hand side is absolutely convergent.

Here again, as for the scalar case, the expectation so defined has the linearity and monotonicity properties (see (1.25) and (1.26)), and the triangle inequality is true (see (1.28)).

EXAMPLE 1.1.58: THE MULTINOMIAL DISTRIBUTION. n balls are placed independently of one another in K boxes $B_1, \ldots, B_K$, with the probability p_i for a given ball to be assigned to box B_i. Of course,

$$\sum_{i=1}^{K} p_i = 1.$$

After placing all the balls in the boxes, there are X_i balls in box B_i, where

$$\sum_{i=1}^{K} X_i = n.$$

The probability distribution of the random vector $X = (X_1, \ldots, X_n)$ is the *multinomial distribution* of size (n, K) and parameters $p_1, \ldots, p_K$

$$P(X_1 = m_1, \ldots, X_K = m_K) = \frac{n!}{\prod_{i=1}^{K} (m_i)!} \prod_{i=1}^{K} p_i^{m_i}, \tag{1.39}$$

where $m_1 + \cdots + m_K = n$.

To prove this, observe that there are $n!/\prod_{i=1}^{K}(m_i)!$ distinct ways of placing n balls in K boxes in such a manner that m_1 balls are in box B_1, m_2 are in B_2, etc., and that each of these distinct ways occurs with the same probability $\prod_{i=1}^{K} p_i^{m_i}$.

EXAMPLE 1.1.59: THE FIRST COORDINATE OF A MULTINOMIAL VECTOR. Continuation of Example 1.1.58. We show that X_1 is a binomial random variable of size n and parameter p_1. For this, put together the boxes $B_2, \ldots, B_K$ so that they form a single box. The whole process is now one of placing n balls, independently of one another. A given ball is put in the first box B_1 with probability p_1 and in box $B_2 + \cdots + B_K$ with probability $p_2 + \cdots + p_K = 1 - p_1$. But when $K = 2$, and therefore $m_2 = n - m_1, p_2 = 1 - p_1$, (1.39) reduces to

$$P(X_1 - m_1, X_2 - m_2) = P(X_1 = m_1) = \frac{n!}{m_1!(n - m_1)!} p_1^{m_1}(1 - p_1)^{n-m_1}.$$

The Product Formula for Expectation

Let Y and Z be independent random vectors with respective probability densities f_Y and f_Z. Arguments similar to those leading to formula (1.38) give for the vector $X = (Y, Z)$ the probability density

$$f_{Y,Z}(y, z) = f_Y(y)f_Z(z).$$

In particular, if $g_1 : \mathbb{R}^p \to \mathbb{R}$ and $g_2 : \mathbb{R}^q \to \mathbb{R}$ are such that

$$E[|g_1(Y)|] < \infty, E[|g_2(Z)|] < \infty, \tag{1.40}$$

then

$$E[g_1(Y)g_2(Z)] = E[g_1(Y)]E[g_2(Z)], \tag{1.41}$$

as follows from Fubini's theorem.

The analogous result holds when X and Z are discrete random vectors. It turns out that (1.41) is true for *any* random vectors X and Z, as long as the integrability conditions (1.40) are satisfied. Formula (1.41) is also true when the integrability conditions are replaced by the hypothesis that g_1 and g_2 are non-negative.

1.1.4   Conditional Expectation

This section introduces the notion of conditional expectation for *discrete* random variables and gives the results that are needed in the study of Markov chains with a countable state space.

Let X and Y be two discrete random variables taking their values in the denumerable sets F and G, respectively, and let $g : F \times G \to \mathbb{R}$ be a non-negative function. Let $\psi : G \to \mathbb{R}$ be the function defined by

$$\psi(y) = \sum_{x \in F} g(x,y) \mathrm{P}(X = x \mid Y = y).$$

Definition 1.1.60 *For each $y \in G$, $\psi(y)$ is called the conditional expectation of $g(X,Y)$ given $Y = y$, and is denoted by $\mathrm{E}^{Y=y}[g(X,Y)]$ or $\mathrm{E}[g(X,Y) \mid Y = y]$:*

$$\mathrm{E}^{Y=y}[g(X,Y)] = \psi(y).$$

The random variable $\psi(Y)$ is called the conditional expectation of $g(X,Y)$ given Y, and is denoted by $\mathrm{E}^{Y}[g(X,Y)]$ (or $\mathrm{E}[g(X,Y) \mid Y]$):

$$\mathrm{E}^{Y}[g(X,Y)] = \psi(Y). \tag{1.42}$$

Note that if $g \geq 0$ and $\mathrm{E}[g(X,Y)] < \infty$, then $\psi(y) < \infty$ for all $y \in G$ such that $\mathrm{P}(Y = y) > 0$, that is, $\psi(Y) < \infty$ almost surely. This follows from

$$\sum_{y \in G} \psi(y) \mathrm{P}(Y = y) = \mathrm{E}[g(X,Y)] < \infty.$$

Therefore, if $g : F \times G \to \mathbb{R}$ is a function of arbitrary sign such that $\mathrm{E}[|g(X,Y)|] < \infty$, one can define $\mathrm{E}^{Y=y}[g^{\pm}(X,Y)]$, since both terms on the right-hand side are finite, in view of the previous remark and of the fact that $\mathrm{E}[g^{\pm}(X,Y)] < \infty$. One may then define

$$\mathrm{E}^{Y=y}[g(X,Y)] := \sum_{x \in F} g(x,y) \mathrm{P}(X = x \mid Y = y) = \psi(y).$$

The conditional expectation of $g(X,Y)$ given Y is then defined by (1.42) as for the non-negative case.

Finally, for any event A, we have that $\mathrm{P}(A \mid Y)$ or $\mathrm{P}^{Y}(A)$, the probability of A given Y, is equal by definition to $\mathrm{E}^{Y}[1_A]$.

EXAMPLE 1.1.61: A POISSON EXAMPLE. Let X_1 and X_2 be two independent Poisson random variables with respective means $\theta_1 > 0$ and $\theta_2 > 0$. We seek to compute $\mathrm{E}^{X_1+X_2}[X_1]$, or $\mathrm{E}^{Y}[X]$ where $X = X_1$, $Y = X_1 + X_2$. Following the instructions of Definition 1.1.60, we must first compute

$$
\begin{aligned}
\mathrm{P}(X = x \mid Y = y) &= \frac{\mathrm{P}(X = x, Y = y)}{\mathrm{P}(Y = y)} = \frac{\mathrm{P}(X_1 = x, X_1 + X_2 = y)}{\mathrm{P}(X_1 + X_2 = y)} \\
&= \frac{\mathrm{P}(X_1 = x, X_2 = y - x)}{\mathrm{P}(X_1 + X_2 = y)} = \frac{\mathrm{P}(X_1 = x)\mathrm{P}(X_2 = y - x)}{\mathrm{P}(X_1 + X_2 = y)} \\
&= \binom{y}{x} \left(\frac{\theta_1}{\theta_1 + \theta_2} \right)^x \left(\frac{\theta_2}{\theta_1 + \theta_2} \right)^{y-x} 1_{\{y \geq x\}}.
\end{aligned}
$$

Therefore, setting $\alpha = \frac{\theta_1}{\theta_1+\theta_2}$,

$$\psi(y) = E^{Y=y}[X] = \sum_{x=0}^{y} x \binom{y}{x} \alpha^x (1-\alpha)^{y-x} = \alpha y\,.$$

Finally, $E^Y[X] = \psi(Y) = \alpha Y$, that is,

$$E^{X_1+X_2}[X_1] = \frac{\theta_1}{\theta_1+\theta_2}(X_1+X_2)\,.$$

_____

The first property of conditional expectation, linearity, is obvious from the definition:

Theorem 1.1.62 *For all* $\lambda_1, \lambda_2 \in \mathbb{R}$,

$$E^Y[\lambda_1 g_1(X,Y) + \lambda_2 g_2(X,Y)] = \lambda_1 E^Y[g_1(X,Y)] + \lambda_2 E^Y[g_2(X,Y)]$$

whenever the conditional expectations thereof are well defined and do not produce $\infty - \infty$ *forms.*

Monotonicity is equally obvious:

Theorem 1.1.63 *If* $g_1(x,y) \le g_2(x,y)$, *then*

$$E^Y[g_1(X,Y)] \le E^Y[g_2(X,Y)]\,.$$

Next, we have

Theorem 1.1.64
$$E[E^Y[g(X,Y)]] = E[g(X,Y)]\,.$$

Proof. The left-hand side is

$$\sum_{y \in G} \psi(y)P(X=y) = \sum_{y \in G}\sum_{x \in F} g(x,y)P(X=x \mid Y=y)P(Y=y)$$

$$= \sum_{x}\sum_{y} g(x,y)P(X=x, Y=y)\,.$$

$\square$

The next result follows from the definitions.

Theorem 1.1.65 *Assuming that the left-hand sides of the equalities below are well defined,*

$$E^Y[w(Y)] = w(Y) \,,$$

and more generally,

$$E^Y[w(Y)h(X,Y)] = w(Y)E^Y[h(X,Y)] \,.$$

Theorem 1.1.66 *If X and Y are independent and if $v : F \to \mathbb{R}$ is such that $E[|v(X)|] < \infty$, then*

$$E^Y[v(X)] = E[v(X)] \,.$$

Proof. Indeed,

$$E^{Y=y}[v(X)] = \sum_{x \in F} v(x)P(X = x \mid Y = y) = \sum_{x \in F} v(x)P(X = x).$$

$\square$

The Rule of Successive Conditioning

In the definition of conditional expectation, we assumed G to be denumerable but otherwise arbitrary. Take G to be the product $G_1 \times G_2$, that is, $Y = (Y_1, Y_2)$, where Y_1 and Y_2 take their values in the denumerable sets G_1 and G_2, respectively. In this situation, we use the more developed notation

$$E^Y[g(X,Y)] = E^{Y_1,Y_2}[g(X,Y_1,Y_2)] \,.$$

Theorem 1.1.67 *Under either one of the following conditions:*

(i) $g(X,Y) \geq 0$, a.s.,

(ii) $E[|g(X,Y)|] < \infty$,

we have

$$E^{Y_2}[E^{Y_1,Y_2}[g(X,Y_1,Y_2)]] = E^{Y_2}[g(X,Y_1,Y_2)] \,.$$

Proof. Set

$$\psi(Y_1, Y_2) = E^{Y_1,Y_2}[g(X,Y_1,Y_2)] \,.$$

We must show that

$$E^{Y_2}[\psi(Y_1,Y_2)] = E^{Y_2}[g(X,Y_1,Y_2)] \,.$$

But

$$\psi(y_1, y_2) = \sum_x g(x, y_1, y_2) \mathrm{P}(X = x \mid Y_1 = y_1, Y_2 = y_2)$$

and

$$\mathrm{E}^{Y_2 = y_2}[\psi(Y_1, Y_2)] = \sum_{y_1} \psi(y_1, y_2) \mathrm{P}(Y_1 = y_1 \mid Y_2 = y_2),$$

that is,

$$\mathrm{E}^{Y_2 = y_2}[\psi(Y_1, Y_2)]$$
$$= \sum_{y_1} \sum_x g(x, y_1, y_2) \mathrm{P}(X = x \mid Y_1 = y_1, Y_2 = y_2) \mathrm{P}(Y_1 = y_1 \mid Y_2 = y_2)$$
$$= \sum_{y_1} \sum_x g(x, y_1, y_2) \mathrm{P}(X = x, Y_1 = y_1 \mid Y_2 = y_2) = \mathrm{E}^{Y_2 = y_2}[g(X, Y_1, Y_2)].$$

□

EXAMPLE 1.1.68: ONE AMONG TWO. Let X_1 and X_2 be two independent identically distributed random variables with values in the denumerable set E. Assume that $\mathrm{E}[\|X_1\|] < \infty$. We show that

$$\mathrm{E}^{X_1 + X_2}[X_1] = \frac{X_1 + X_2}{2}.$$

In fact, $\mathrm{E}^{X_1 + X_2}[X_1] = \psi(X_1 + X_2)$, and by symmetry, $\mathrm{E}^{X_1 + X_2}[X_2] = \psi(X_1 + X_2)$. Therefore,

$$2\psi(X_1 + X_2) = \mathrm{E}^{X_1 + X_2}[X_1] + \mathrm{E}^{X_1 + X_2}[X_2] = \mathrm{E}^{X_1 + X_2}[X_1 + X_2] = X_1 + X_2.$$

1.2 Transforms of Probability Distributions

1.2.1 Generating Functions

Let $\bar{D}(0; 1)$ be the complex closed unit disk centered at 0.

Definition 1.2.1 *The generating function* (GF) *of the integer-valued random variable X (or, of its distribution) is the function $g_X : \bar{D}(0; 1) \to \mathbb{C}$ defined by*

$$g_X(z) := \mathrm{E}[z^X] = \sum_{k=0}^{\infty} \mathrm{P}(X = k) z^k. \tag{1.43}$$

EXAMPLE 1.2.2: THE GF OF THE BINOMIAL RANDOM VARIABLE. For the binomial random variable of size n and parameter p,

$$\sum_{k=0}^{n} P(X = k)z^k = \sum_{k=0}^{n} \binom{n}{k}(zp)^k(1-p)^{n-k},$$

and therefore

$$g_X(z) = (1 - p + pz)^n.$$

EXAMPLE 1.2.3: THE GF OF THE POISSON RANDOM VARIABLE. For the Poisson random variable of mean θ,

$$\sum_{k=0}^{\infty} P(X = k)z^k = e^{-\theta}\sum_{k=0}^{\infty} \frac{(\theta z)^k}{k!},$$

and therefore

$$g_X(z) = e^{\theta(z-1)}.$$

EXAMPLE 1.2.4: RANDOM SUMS, TAKE 1. Let $\{Y_n\}_{n\geq 1}$ be an IID sequence of integer-valued random variables with the common generating function g_Y. Let T be another integer-valued random variable of generating function g_T, independent of the sequence $\{Y_n\}_{n\geq 1}$. We compute the generating function of the random sum

$$X := \sum_{n=1}^{T} Y_n,$$

where by convention $\sum_{n=1}^{0} = 0$:

$$z^X = z^{\sum_{n=1}^{T} Y_n} = \sum_{k=0}^{\infty}\left\{\left(z^{\sum_{n=1}^{T} Y_n}\right)1_{\{T=k\}}\right\} = \sum_{k=0}^{\infty}\left(z^{\sum_{n=1}^{k} Y_n}\right)1_{\{T=k\}}.$$

Therefore,

$$E[z^X] = \sum_{k=0}^{\infty} E\left[1_{\{T=k\}}\left(z^{\sum_{n=1}^{k} Y_n}\right)\right] = \sum_{k=0}^{\infty} E[1_{\{T=k\}}]E[z^{\sum_{n=1}^{k} Y_n}],$$

where we have used independence of T and $\{Y_n\}_{n \geq 1}$. Now, $\mathrm{E}[1_{\{T=k\}}] = \mathrm{P}(T = k)$, and $\mathrm{E}[z^{\sum_{n=1}^{k} Y_n}] = g_Y(x)^k$, and therefore $\mathrm{E}[z^X] = \sum_{k=0}^{\infty} \mathrm{P}(T = k) g_Y(z)^k$. That is, finally,

$$g_X(z) = g_T(g_Y(z)) \ .$$

The power series associated with the sequence $\{\mathrm{P}(X = n)\}_{n \geq 0}$ has radius of convergence $R \geq 1$, since $\sum_{n=0}^{\infty} \mathrm{P}(X = n) = 1 < \infty$. Therefore, the domain of definition of g_X contains the open unit disk. Inside this open disk, differentiation term by term is possible, for instance,

$$g_X'(z) = \sum_{n=1}^{\infty} n\mathrm{P}(X = n)z^{n-1}, \tag{1.44}$$

$$g_X''(z) = \sum_{n=2}^{\infty} n(n - 1)\mathrm{P}(X = n)z^{n-2}. \tag{1.45}$$

The right-hand side of (1.44) is well defined at $z = 1$, being equal to $\sum_{n=1}^{\infty} n\mathrm{P}(X = n)$, a non-negative quantity, possibly infinite. The left-hand side of (1.44) is, however, not always defined by formula (1.43) for $z = 1$. So we *define* $g'(1) = \sum_{n=1}^{\infty} n\mathrm{P}(X = n)$, that is,

$$g'(1) = \mathrm{E}[X]. \tag{1.46}$$

By *Abel's theorem* (Theorem A.1.6), the limit as the *real* variable x increases to 1 of $\sum_{n=1}^{\infty} n\mathrm{P}(X = n)x^{n-1}$ is $\sum_{n=1}^{\infty} n\mathrm{P}(X = n)$, and therefore g_X, as a function on the real interval $[0, 1)$, can be extended to $[0, 1]$ by (1.46), and this extension preserves continuity. This is used as follows: Suppose $\mathrm{E}[X]$ is not known but that you have an expression of $g_X'(x)$ in $[0, 1)$ for which you can compute $\lim g_X'(x)$ as x increases through real values to 1. Then you know that this limit equals $\mathrm{E}[X]$.

Similarly, starting from (1.45), and defining

$$g_X''(1) = \mathrm{E}[X(X - 1)], \tag{1.47}$$

we make $g_X''(x)$ a continuous function on $[0, 1]$.

EXAMPLE 1.2.5: RANDOM SUMS, TAKE 2. Continuation of Example 1.2.4. The generating function of $X = \sum_{n=1}^{T} Y_n$ was found to be $g_X(z) = g_T(g_Y(z))$. Therefore $g_X'(x) = g_T'(g_Y(x))g_Y'(x)$ for all $x \in [0, 1)$. Letting $x \to 1$, $g_X'(1) = g_T'(g_Y(1))g_Y'(1) = g_T'(1)g_Y'(1)$, and therefore $\mathrm{E}[X] = \mathrm{E}[T]\mathrm{E}[Y]$.

Theorem 1.2.6 *The generating function characterizes the distribution of a random variable.*

This means the following. Suppose that, without knowing the distribution of X, you have been able to compute its generating function $g(z)$, and that, moreover, you are able to give its power series expansion in a neighborhood of the origin:

$$g(z) = \sum_{n=0}^{\infty} a_n z^n.$$

Since $g(z)$ is the generating function of Z,

$$g(z) = \sum_{n=0}^{\infty} \mathrm{P}(X = n) z^n$$

and since the power series expansion around the origin is unique, the distribution of X is identified as

$$\mathrm{P}(X = n) = a_n$$

for all $n \geq 0$. Similarly, if two $\mathbb{N}$-valued random variables have the same g.f., they have the same distribution.

EXAMPLE 1.2.7: THE LOTTERY. Let X_1, X_2, X_3, X_4, X_5, and X_6 be independent random variables uniformly distributed over $\{0, 1, \ldots, 9\}$. We compute the generating function of $Y = 27 + X_1 + X_2 + X_3 - X_4 - X_5 - X_6$ and we use this to compute the probability that in a 6-digit lottery the sum of the first three digits equals the sum of the last three digits.

$$
\begin{aligned}
\mathrm{E}[z^{X_i}] &= \frac{1}{10}(1 + z + \cdots + z^9) = \frac{1}{10}\frac{1 - z^{10}}{1 - z}, \\
\mathrm{E}[z^{-X_i}] &= \frac{1}{10}\left(1 + \frac{1}{z} + \cdots + \frac{1}{z^9}\right) = \frac{1}{10}\frac{1 - z^{-10}}{1 - z^{-1}} = \frac{1}{10}\frac{1}{z^9}\frac{1 - z^{10}}{1 - z}, \\
\mathrm{E}[z^Y] &= \mathrm{E}\left[z^{27 + \sum_{i=1}^{3} X_i - \sum_{i=4}^{6} X_i}\right] = \mathrm{E}\left[z^{27} \prod_{i=1}^{3} z^{X_i} \prod_{i=4}^{6} z^{X_i}\right] \\
&= z^{27} \prod_{i=1}^{3} \mathrm{E}[z^{X_i}] \prod_{i=4}^{6} \mathrm{E}[z^{-X_i}].
\end{aligned}
$$

Therefore,

$$g_Y(z) = \frac{1}{10^6}\frac{(1 - z^{10})^6}{(1 - z)^6}.$$

But $P(X_1 + X_2 + X_3 = X_4 + X_5 + X_6) = P(Y = 27)$ is the factor of z^{27} in the power series expansion of $g_Y(z)$. Since

$$(1 - z^{10})^6 = 1 - \binom{6}{1} z^{10} + \binom{6}{2} z^{20} + \cdots$$

and

$$(1 - z)^{-6} = 1 + \binom{6}{5} z + \binom{7}{5} z^2 + \binom{8}{5} z^3 + \cdots,$$

we find that

$$P(Y = 27) = \frac{1}{10^6} \left(\binom{32}{5} - \binom{6}{1}\binom{22}{5} + \binom{6}{2}\binom{12}{5} \right).$$

Theorem 1.2.8 (α) *Let $g : [0, 1] \to \mathbb{R}$ be defined by $g(x) - E[x^X]$, where X is a non-negative integer-valued random variable. Then g is non-decreasing and convex. Moreover, if $P(X = 0) < 1$, then g is strictly increasing, and if $P(X \le 1) < 1$, it is strictly convex.*

(β) *Suppose $P(X \le 1) < 1$. If $E[X] \le 1$, the equation $x = g(x)$ has a unique solution $x \in [0, 1]$, namely $x = 1$. If $E[X] > 1$, it has two solutions in $[0, 1]$, $x = 1$ and $x = x_0 \in (0, 1)$.*

Proof. Just observe that for $x \in [0, 1]$,

$$g'(x) = \sum_{n=1}^{\infty} nP(X = n)x^{n-1} \ge 0,$$

and therefore g is non-decreasing, and

$$g''(x) = \sum_{n=2}^{\infty} n(n-1)P(X - n)x^{n-2} \ge 0,$$

and therefore g is convex up. For $g'(x)$ to be null for some $x \in (0, 1)$, it is necessary to have $P(X = n) = 0$ for all $n \ge 1$, and therefore $P(X = 0) = 1$. For $g''(x)$ to be null for some $x \in (0, 1)$, one must have $P(X = n) = 0$ for all $n \ge 2$, and therefore $P(X = 0) + P(X = 1) = 1$.

The graph of $g : [0, 1] \to \mathbb{R}$ has, in the strictly increasing strictly convex case $P(X = 0) + P(X = 1) < 1$, the general shape shown in Figure 1.2.1, where we distinguish two cases: $E[X] = g'(1) \le 1$, and $E[X] = g'(1) > 1$. The rest of the proof is then easy. $\qquad \square$

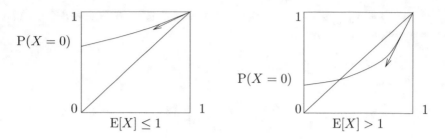

Figure 1.2.1. Two aspects of the generating function

1.2.2 Characteristic Functions

Definition 1.2.9 *The characteristic function (CF) $\psi_X : \mathbb{R}^n \to \mathbb{C}$ of a real random vector $X = (X_1, \ldots, X_n)$ is defined by*

$$\psi_X(u) = \mathrm{E}[e^{iu^T X}]. \tag{1.48}$$

EXAMPLE 1.2.10: THE CF OF THE GAUSSIAN AND EXPONENTIAL VARIABLES. One can check that the following formulas give the characteristic functions of the two main continuous random variables:

(i) Gaussian

$$f(x) = \frac{1}{\sqrt{2\pi}} e^{-\frac{1}{2}\frac{(x-m)^2}{\sigma^2}} \longleftrightarrow \psi(u) = e^{imu - \frac{1}{2}\sigma^2 u^2}.$$

(ii) Exponential

$$f(x) = \lambda e^{-\lambda x} 1_{x>0} \longleftrightarrow \psi(u) = \frac{\lambda}{\lambda - iu}.$$

Theorem 1.2.11 *The characteristic function determines the distribution of a random vector (just as generating functions determine the distribution of integer-valued random variables).*

This will be admitted without proof. However, for continuous random vectors it is an easy consequence of the classical Fourier theory. Indeed, suppose that f_X and f_Y are the probability distribution functions of two random vectors X and Y on $\mathbb{R}^n$ and suppose that their characteristic functions are the same. Then $f_X - f_Y$ admits the Fourier transform $\psi_X - \psi_Y \equiv 0$, and therefore, by the Fourier inversion theorem, $\int_C f_X(x)\mathrm{d}x = \int_C f(y)\mathrm{d}y$ for all intervals $C \subset \mathbb{R}^n$.

The Fourier inversion formula gives, when $\psi_X(u)$ is integrable,

$$f_X(x) = \frac{1}{(2\pi)^n} \int_{\mathbb{R}^n} e^{-iu^T x} \psi_X(u) du$$

for the PDF of X.

EXAMPLE 1.2.12: THE CF OF THE CAUCHY VARIABLE. We show that the characteristic function of the Cauchy random variable, with the PDF

$$f(x) = \frac{1}{\pi} \frac{1}{1+x^2},$$

is $\psi_X(u) = e^{-|u|}$. We first observe that the Fourier transform of $u \to e^{-|u|}$ is $x \to \int_{-\infty}^{+\infty} e^{-|u|} e^{-iux} du = \frac{2}{1+x^2} = g(x)$. Since $g(x)$ is integrable, the Fourier inversion formula applies, and we get

$$e^{-|u|} = \frac{1}{2\pi} \int_{-\infty}^{+\infty} g(x) e^{iux} dx = \int_{-\infty}^{+\infty} \frac{1}{\pi} \frac{1}{1+x^2} e^{iux} dx.$$

Theorem 1.2.13 *Suppose that Y and Z are two random vectors of respective dimensions p and q, and that for all $v \in \mathbb{R}^p$, $w \in \mathbb{R}^q$, it holds that*

$$E[e^{i(v^T Y + w^T Z)}] = \psi_1(v)\psi_2(w), \tag{1.49}$$

where $\psi_1(v)$ and $\psi_2(w)$ are the characteristic functions of some random vectors $\widetilde{Y}$ and $\widetilde{Z}$ of appropriate dimensions. Then Y and Z are independent, Y has the same distribution as $\widetilde{Y}$, and Z has the same distribution as $\widetilde{Z}$.

Proof. Define $X = (Y, Z)$ and $u = (v, w)$, so that (1.49) reads

$$E[e^{iu^T X}] = \psi(u) = \psi_1(v)\psi_2(w).$$

If one can find a vector $\hat{X} = (\hat{Y}, \hat{Z})$ such that $\psi_{\hat{X}}(u) = \psi(u)$, then, since the c.f. characterizes the CDF, $X = (Y, Z)$ has the same CDF as $\hat{X} = (\hat{Y}, \hat{Z})$. Take $\hat{Y}$ distributed as $\widetilde{Y}$, $\hat{Z}$ distributed as $\widetilde{Z}$, and $\hat{Y}$ and $\hat{Z}$ independent. Then, using the product formula,

$$E[e^{iu^T \hat{X}}] = E[e^{iv^T \hat{Y}} e^{iw^T \hat{Z}}] = E[e^{iv^T \hat{Y}}] E[e^{iw^T \hat{Z}}] = \psi_1(v)\psi_2(w).$$

Therefore, (Y, Z) has the same distribution as $(\hat{Y}, \hat{Z})$ and in particular, Y and Z are independent. $\qquad\square$

Theorem 1.2.14 *Let A be an event and X a random variable such that for all $u \in \mathbb{R}$,*

$$\mathrm{E}[1_A e^{iuX}] = \mathrm{P}(A)\mathrm{E}[e^{iuX}].$$

Then A and X are independent; that is, A and $\{X \leq a\}$ are independent for all $a \in \mathbb{R}$.

Proof. The last statement means that X and $Y = 1_A$ are independent. We must therefore check that for all $u, v \in \mathbb{R}$,

$$\mathrm{E}[e^{iuX} e^{ivY}] = \mathrm{E}[e^{iuX}]\mathrm{E}[e^{ivY}].$$

But

$$e^{ivY} = 1 - 1_A + 1_A e^{iv} = 1 + 1_A(e^{iv} - 1),$$

and therefore

$$\mathrm{E}[e^{iuX} e^{ivY}] = \mathrm{E}[e^{iuX}] + (e^{iv} - 1)\mathrm{E}[1_A e^{iuX}]$$

and

$$\mathrm{E}[e^{iuX}]\mathrm{E}[e^{ivY}] = \mathrm{E}[e^{iuX}] + (e^{iv} - 1)\mathrm{P}(A)\mathrm{E}[e^{iuX}].$$

The hypothesis concludes the proof.   □

Let A be some event of positive probability, and let P_A denote the probability P conditioned by A, that is,

$$\mathrm{P}_A(\cdot) := \mathrm{P}(\cdot \mid A).$$

Definition 1.2.15 *The random variables X and Y are said to be conditionally independent given A if they are independent with respect to probability P_A.*

By Theorem 1.2.13, a necessary and sufficient condition for this is that for all $u, v \in \mathbb{R}$,

$$\mathrm{E}_A[e^{iuX} e^{ivY}] = \mathrm{E}_A[e^{iuX}]\mathrm{E}_A[e^{ivY}].$$

Observe that for an integrable or non-negative random variable Z,

$$\mathrm{P}(A)\mathrm{E}_A[Z] = \mathrm{E}[Z 1_A].$$

The following result is then immediate. It is recorded for future reference.

Theorem 1.2.16 *Let A be an event of positive probability. The random variables X and Y are conditionally independent given A if and only if for all $u, v \in \mathbb{R}$,*

$$\mathrm{P}(A)\mathrm{E}[e^{iuX} e^{ivY} 1_A] = \mathrm{E}[e^{iuX} 1_A]\mathrm{E}[e^{ivY} 1_A].$$

1.3 Transformations of Random Vectors

1.3.1 Smooth Change of Variables

Let $X = (X_1, \ldots, X_n)$ be a random vector with the PDF f_X, and define the random vector

$$Y := g(X),$$

where $g : \mathbb{R}^n \to \mathbb{R}^n$. More explicitly,

$$\begin{cases} Y_1 = g_1(X_1, \ldots, X_n) \\ \vdots \\ Y_n = g_n(X_1, \ldots, X_n). \end{cases}$$

Under smoothness assumptions on g, the random vector Y is absolutely continuous, and its PDF can be explicitly computed from g and the PDF and f_X. The precise result is recalled below.

Let g be a function from an open set $U \subset \mathbb{R}^n$ into $\mathbb{R}^n$,

$$g : U \to \mathbb{R}^n,$$

and the coordinate functions g_i are continuously differentiable on U. Denote the Jacobian matrix of the function g by

$$J_g(x_1, \ldots, x_n) := \left\{ \frac{\partial g_i}{\partial x_j}(x_1, \ldots, x_n) \right\}_{1 \leq i,j \leq n}$$

and assume that on U,

$$|\det J_g(x_1, \ldots, x_n)| > 0.$$

Then $V = g(U)$ is an open subset of $\mathbb{R}^n$ and there exists an inverse $g^{-1} : V \to \mathbb{R}$ of g with the same properties as g. In particular, on V,

$$|\det J_{g^{-1}}(y_1, \ldots, y_n)| > 0.$$

Also,

$$J_{g^{-1}}(y) = J_g(g^{-1}(y))^{-1}.$$

Theorem 1.3.1 *Under the conditions just stated for X, g, and U, and if moreover*

$$P(X \in U) = 1,$$

then Y admits the density

$$f_Y(y) = f_X(g^{-1}(y))|\det J_g(g^{-1})(y)|^{-1} 1_V(y). \tag{1.50}$$

Proof. The proof consists in checking that for any bounded function $h : \mathbb{R} \to \mathbb{R}$,

$$E[h(Y)] = \int_{\mathbb{R}^n} h(y)\psi(y)\,dy, \tag{1.51}$$

where ψ is the function on the right-hand side of (1.50). In fact, with $h(y) := 1_{\{y \leq a\}} = 1_{\{y_1 \leq a_1\}} \cdots 1_{\{y_n \leq a_n\}}$, (1.51) reads

$$P(Y_1 \leq a_1, \ldots, Y_n \leq a_n) = \int_{-\infty}^{a_1} \cdots \int_{-\infty}^{a_n} \psi(y_1, \ldots, y_n)\,dy_1 \cdots dy_n.$$

To prove that (1.51) holds with the appropriate ψ, one just uses the basic rule of change of variables of calculus:[1]

$$\int_U u(x)dx = \int_{g(U)} u(g^{-1}(y))|\det J_{g^{-1}}(y)|\,dy.$$

Indeed,

$$\begin{aligned}
E[h(Y)] &= E[h(g(X))] \\
&= \int_U h(g(x))f_X(x)\,dx \\
&= \int_V h(y)f_X(g^{-1}(y))|\det J_{g^{-1}}(y)|\,dy.
\end{aligned}$$

$\square$

EXAMPLE 1.3.2: INVERTIBLE AFFINE TRANSFORMATIONS. Here $U = \mathbb{R}^n$ and g is an affine function $g(x) = Ax + b$, where A is an $n \times n$ invertible matrix and $b \in \mathbb{R}^n$. Then $|\det J_{g^{-1}}(y)| = \frac{1}{|\det A|}$. Therefore, the random vector

$$Y = AX + B$$

admits the density

$$f_Y(y) = f_X(A^{-1}(y - b))\frac{1}{|\det A|}.$$

---

[1]See Rudin, *Principles of Mathematical Analysis*, McGraw-Hill, 3rd edition, 1976, 10.9.

1.3.2   Order Statistics

Let $X_1, \ldots, X_n$ be independent random variables with the same PDF $f(x)$. It was shown in Example 1.1.55 that the probability that two or more among $X_1, \ldots, X_n$ take the same value is null. Thus one can define unambiguously the random variables $Z_1, \ldots, Z_n$ obtained by reordering $X_1, \ldots, X_n$ in increasing order:

$$\begin{cases} Z_i \in \{X_1, \ldots, X_n\} \\ Z_1 < Z_2 < \cdots < Z_n. \end{cases}$$

In particular, $Z_1 = \min(X_1, \ldots, X_n)$ and $Z_n = \max(X_1, \ldots, X_n)$.

Theorem 1.3.3 *The probability density of $Z = (Z_1, \ldots, Z_n)$ is*

$$f_Z(z_1, \ldots, z_n) = n! \left\{ \prod_{j=1}^{n} f(z_j) \right\} 1_C(z_1, \ldots, z_n), \qquad (1.52)$$

where

$$C := \{ (z_1, \ldots, z_n) \in \mathbb{R}^n \; ; \; z_1 < z_2 < \cdots < z_n \}.$$

Proof. Let σ be the permutation of $\{1, \ldots, n\}$ that orders $X_1, \ldots, X_n$ in ascending order, that is,

$$X_{\sigma(i)} = Z_i$$

(note that σ is a *random* permutation). For any set $A \subset \mathbb{R}^n$, noting that $Z \in C$ and therefore $\{Z \in A\} \equiv \{Z \in A \cap C\}$,

$$P(Z \in A) = P(Z \in A \cap C) = P(X_\sigma \in A \cap C) = \sum_{\sigma_o} P(X_{\sigma_o} \in A \cap C, \sigma = \sigma_o),$$

where the sum is over all permutations of $\{1, \ldots, n\}$. Observing that $X_{\sigma_o} \in A \cap C$ implies $\sigma = \sigma_o$,

$$P(X_{\sigma_o} \in A \cap C, \sigma = \sigma_o) = P(X_{\sigma_o} \in A \cap C)$$

and therefore since the probability distribution of X_{σ_o} does not depend upon a fixed permutation σ_o (here we need independence and equidistribution of the X_i's),

$$P(X_{\sigma_o} \in A \cap C) = P(X \in A \cap C).$$

Therefore,

$$P(Z \in A) = \sum_{\sigma_o} P(X \in A \cap C) = n! P(X \in A \cap C)$$

$$= n! \int_{A \cap C} f_X(x) dx = \int_A n! f_X(x) 1_C(x) dx .$$

$\square$

EXAMPLE 1.3.4: ORDERING IID UNIFORM RANDOM VARIABLES. Applying formula (1.52) to the situation where the X_i's are uniformly distributed over $[a, b]$ gives

$$f_Z(z_1, \ldots, z_n) = \frac{n!}{(b-a)^n} 1_{[a,b]^n}(z_1, \ldots, z_n) 1_C(z_1, \ldots, z_n) .$$

In particular, since $\int_{\mathbb{R}^n} f_Z(z) dz = 1$,

$$\int_a^b \cdots \int_a^b 1_C(z_1, \ldots, z_n) dz_1 \cdots dz_n = \frac{(b-a)^n}{n!} .$$

1.4   Almost-Sure Convergence

1.4.1   Two Basic Tools

Definition 1.4.1 *A sequence $\{Z_n\}_{n \geq 1}$ of real random variables is said to converge P-almost surely (P-a.s.) to the real random variable Z if*

$$P(\lim_{n \uparrow \infty} Z_n = Z) = 1 .$$

(Paraphrasing: For all ω outside a set N of null probability, $\lim_{n \uparrow \infty} Z_n(\omega) = Z(\omega)$).

In order to prove almost sure convergence of a random sequence, the Borel–Cantelli lemma and Markov's inequality (or some form of the latter) are the basic tools.

The Borel–Cantelli Lemma

Consider a sequence of events $\{A_n\}_{n \geq 1}$ where the index n can be, if one wishes, interpreted as time. One is interested in the probability that A_n occurs infinitely often, that is, the probability of the event

$$\{\omega; \omega \in A_n \text{ for an infinity of indices } n\} , \tag{1.53}$$

denoted by $\{A_n \text{ i.o.}\}$, where *i.o.* means *infinitely often.*

Lemma 1.4.2 *Let $\{A_n\}_{n\geq 1}$ be a sequence of events such that*

$$\sum_{n=1}^{\infty} P(A_n) < \infty. \tag{1.54}$$

Then

$$P(A_n \text{ i.o.}) = 0.$$

Proof. We first give a manageable expression for the event in (1.53). We have

$$\{A_n \text{ i.o.}\} = \bigcap_{n=1}^{\infty} \bigcup_{k\geq n} A_k.$$

Indeed, if ω belongs to the set on the right-hand side, then for *all* $n \geq 1$, ω belongs to at least one among $A_n, A_{n+1}, \ldots$, which implies that ω is in A_n for an infinite number of indices n. Conversely, if ω is in A_n for an infinite number of indices n, it is for *all* $n \geq 1$ in at least one of the sets $A_n, A_{n+1}, \ldots$

Let now B_n be the set $\cup_{k\geq n} A_k$, which decreases as n increases, so that by the sequential continuity property of probability,

$$P(A_n \text{ i.o.}) = \lim_{n\uparrow\infty} P\left(\bigcup_{k\geq n} A_k\right).$$

But by sub-σ-additivity,

$$P\left(\bigcup_{k\geq n} A_k\right) \leq \sum_{k\geq n} P(A_k),$$

and by hypothesis (1.54), the right-hand side of this inequality goes to 0 as n goes to ∞. $\qquad\square$

EXAMPLE 1.4.3: SEQUENCES OF BINARY DIGITS. Let $\{X_n\}_{n\geq 1}$ be an independent family of $\{0, 1\}$-valued random variables such that

$$\sum_{n=1}^{\infty} P(X_n = 1) < \infty. \tag{1.55}$$

Then, in view of the Borel–Cantelli lemma, $P(X_n = 1 \text{ i.o.}) = 0$. But the event $\{X_n = 1 \text{ i.o.}\}$ is exactly $\{\omega; \lim_{n\uparrow\infty} X_n(\omega) = 0\}$, since the process takes only the values 0 or 1. Therefore, (1.55) implies that $\lim_{n\uparrow\infty} X_n = 0$ almost surely.

Markov's Inequality

Theorem 1.4.4 *Let X be a random variable with values in $\mathbb{R}$, $f : \mathbb{R} \to \mathbb{R}_+$, and $a > 0$. We then have*

$$P(f(X) \geq a) \leq \frac{E[f(X)]}{a}.$$

Proof. From the inequality

$$f(X) \geq a 1_{\{f(X) \geq a\}}$$

it follows by taking expectations that

$$E[f(X)] \geq a E[1_{\{f(X) \geq a\}}] = a P(f(X) \geq a).$$

$\square$

Specializing Markov's inequality to $f(x) = (x - m)^2, a = \epsilon^2 > 0$, we obtain *Chebyshev's inequality*

$$P(|X - m| \geq \epsilon) \leq \frac{\sigma^2}{\epsilon^2}.$$

EXAMPLE 1.4.5: CHERNOFF'S BOUND. Let X be a real-valued random variable such that for all $t \in \mathbb{R}$, $E[e^{tX}] < \infty$. Set $\psi(t) = \log E[e^{tX}]$. Observing that for $t \geq 0$, $a \in \mathbb{R}$, $X > a$ is equivalent to $e^{tX} > e^{ta}$, and using Markov's inequality, we find that

$$P(X > a) = P(e^{tX} > e^{ta}) \leq \frac{E[e^{tX}]}{e^{ta}} = \exp\{-(at - \psi(t))\}.$$

This being true for all $t \geq 0$, it follows that

$$P(X > a) \leq e^{-h(a)}, \tag{1.56}$$

where

$$h(a) = \sup_{t \geq 0}\{at - \psi(t)\}.$$

Theorem 1.4.6 *Let $X_1, \ldots, X_n$ be IID discrete real-valued random variables and let $a \in \mathbb{R}$. Then*

$$P\left(\sum_{i=1}^{n} X_i \geq na\right) \leq e^{-nh^+(a)},$$

where

$$h^+(a) := \sup_{t > 0}\{at - \log E\left[e^{tX_1}\right]\}. \tag{1.57}$$

Proof.   First observe that since the X_i's are independent and identically distributed,

$$\mathrm{E}\left[\exp\left\{t\sum_{i=1}^{n}X_i\right\}\right] = \mathrm{E}\left[\exp\{tX_1\}\right]^n .$$

For all $t > 0$, Markov's inequality gives

$$\mathrm{P}\left(\sum_{i=1}^{n}X_i \geq na\right) = \mathrm{P}\left(e^{t\sum_{i=1}^{n}X_i} \geq e^{nta}\right)$$

$$\leq \mathrm{E}\left[e^{t\sum_{i=1}^{n}X_i}\right] \times e^{-nta}$$

$$= \mathrm{E}\left[e^{tX_1}\right]^n \times e^{-nta}$$

$$= \exp\{-n\left(at - \log\mathrm{E}\left[e^{tX_1}\right]\right)\},$$

from which the result follows by optimizing this bound with respect to $t > 0$. $\square$

Remark 1.4.7 Of course this bound is useful only if $h^+(a)$ is positive. Suppose for instance that the X_i's are bounded. Let x_i $(i \geq 1)$ be an enumeration of the values taken by X_1, and define $p_i = \mathrm{P}(X = x_i)$, so that

$$at \quad \log\mathrm{E}\left[c^{tX_1}\right] = at \quad \log\left(\sum_{i\geq 1}p_i c^{tx_i}\right) .$$

The derivative with respect to t of this quantity is

$$a - \frac{\sum_{i\geq 1}p_i x_i e^{tx_i}}{\sum_{i\geq 1}p_i e^{tx_i}}$$

and therefore the function $t \to at - \log\mathrm{E}\left[e^{tX_1}\right]$ is finite and differentiable on $\mathbb{R}$, with derivative at 0 equal to $a - \mathrm{E}\left[X_1\right]$, which implies that when $a > \mathrm{E}\left[X_1\right]$, $h^+(a)$ is positive.[2]

Similarly to (1.57), we obtain that

$$\mathrm{P}\left(\sum_{i=1}^{n}X_i \leq na\right) \leq e^{-nh^-(a)}, \tag{1.58}$$

where $h^-(a) := \sup_{t<0}\{at - \log\mathrm{E}\left[e^{tX_1}\right]\}$, and moreover, $h^-(a)$ is positive if $a < \mathrm{E}[X_1]$.

---

[2]In fact, the boundedness assumption can be relaxed and replaced by $\mathrm{E}\left[e^{tX_1}\right] < \infty$ for all $t \geq 0$, and even by any assumption guaranteeing that $t \to \sum_{i\geq 1}p_i e^{tx_i}$ is differentiable in a neighborhood of zero with a derivative equal to $\sum_{i\geq 1}p_i x_i e^{tx_i}$. See Exercise 1.5.27.

1.4.2   The Strong Law of Large Numbers

We now give a sufficient condition for almost-sure convergence, using the Borel–Cantelli lemma. Let $\{\epsilon_n\}_{n\geq 1}$ be a sequence of positive numbers converging to 0. If for a given ω, $|Z_n(\omega) - Z(\omega)| < \epsilon_n$ for all but a finite number of indices n, then $\lim_{n\uparrow\infty} Z_n(\omega) = Z(\omega)$. Therefore, by the Borel–Cantelli lemma, we have the following:

Theorem 1.4.8 *Let $\{Z_n\}_{n\geq 1}$ and Z be random variables. If*

$$\sum_{n\geq 1} P(|Z_n - Z| \geq \epsilon_n) < \infty$$

for some sequence of positive numbers $\{\epsilon_n\}_{n\geq 1}$ converging to 0, then the sequence $\{Z_n\}_{n\geq 1}$ converges almost surely to Z.

The next result is similar, but this time we have a necessary and sufficient condition.

Theorem 1.4.9 *The sequence $\{Z_n\}_{n\geq 1}$ of real random variables converges almost surely to the real random variable Z if and only if for all $\epsilon > 0$,*

$$P(|Z_n - Z| \geq \epsilon \ i.o.) = 0 \,. \tag{1.59}$$

Proof. For the necessity, observe that

$$\{|Z_n - Z| \geq \epsilon \ \text{i.o.}\} \subset \overline{\{\omega; \lim_{n\uparrow\infty} Z_n(\omega) = Z(\omega)\}},$$

and therefore

$$P(|Z_n - Z| \geq \epsilon \ \text{i.o.}) \leq 1 - P(\lim_{n\uparrow\infty} Z_n = Z) = 0 \,.$$

For the sufficiency, let N_k be the last index n such that $|Z_n - Z| \geq \frac{1}{k}$ (set $N_k = \infty$ if $|Z_n - Z| \geq \frac{1}{k}$ for all $n \geq 1$). By (1.59) with $\epsilon = \frac{1}{k}$, we have $P(N_k < \infty) = 0$. By sub-σ-additivity, $P(\cup_{k\geq 1}\{N_k < \infty\}) = 0$. Equivalently,

$$P(N_k = \infty, \ \text{for all } k \geq 1) = 1,$$

which implies $P(\lim_{n\uparrow\infty} Z_n = Z) = 1$. $\square$

A strong law of large numbers is a statement about the almost sure convergence of the empirical average

$$\frac{S_n}{n} = \frac{X_1 + \cdots + X_n}{n},$$

where $\{X_n\}_{n\geq 1}$ is an independent and identically distributed sequence of random variables.

Theorem 1.4.10 *Let* $\{X_n\}_{n\geq 1}$ *be an* IID *sequence of random variables such that*

$$E[|X_1|] < \infty. \tag{1.60}$$

Then, P-a.s.

$$\lim_{n\uparrow\infty} \frac{S_n}{n} = E[X_1].$$

In other words, there exists a set N of probability 0 such that if ω is not in N, then

$$\lim_{n\uparrow\infty} \frac{S_n(\omega)}{n} = E[X_1].$$

The empirical average is asymptotically equal to the probabilistic average. Such is the "physical" content of Kolmogorov's strong law of large numbers (1933). Émile Borel proved the strong law of large numbers (SLNN) in 1909 in the special case where $X_n = 0$ or 1, with $P(X_n = 1) = p$, thus showing in particular that the average fraction of heads in a fair game of coins should tend to $\frac{1}{2}$ as the number of tosses increases indefinitely.

There are numerous versions of the SLNN extending Kolmogorov's result to situations where independence of the sequence $\{X_n\}_{n\geq 1}$ is not required. As a matter of fact, the ergodic theorem for irreducible positive recurrent Markov chains (see Chapter 3) is one of these extensions.

We shall give the proof of Kolmogorov's SLNN in the last subsection. Before that we shall prove Borel's SLNN. It is based on the Borel–Cantelli lemma and on Markov's inequality.

EXAMPLE 1.4.11: BOREL'S STRONG LAW OF LARGE NUMBERS. This SLNN concerns the case where the X_n's are bounded, say, by 1. First, we shall bound the probability that $\left|\frac{S_n}{n} - m\right|$ exceeds some $\epsilon > 0$ where $m = E[X_1]$. For this we apply Markov's inequality

$$P\left(\left|\frac{S_n}{n} - m\right| \geq \epsilon\right) = P\left(\left(\frac{S_n}{n} - m\right)^4 \geq \epsilon^4\right) \leq \frac{E\left[\left(\frac{S_n}{n} - m\right)^4\right]}{\epsilon^4}.$$

Now,

$$\left(\frac{S_n}{n} - m\right)^4 = \frac{\left(\sum_{i=1}^n (X_i - m)\right)^4}{n^4},$$

and therefore

$$P\left(\left|\frac{S_n}{n} - m\right| \geq \epsilon\right) \leq \frac{E\left[\left(\sum_{i=1}^n (X_i - m)\right)^4\right]}{n^4 \epsilon^4}.$$

If we can prove that

$$E\left[\left(\sum_{i=1}^{n}(X_i - m)\right)^4\right] \leq Kn^2 \tag{1.61}$$

for some finite K, then

$$P\left(\left|\frac{S_n}{n} - m\right| \geq \epsilon\right) \leq \frac{K}{n^2\epsilon^4},$$

and in particular, with $\epsilon = n^{-\frac{1}{8}}$,

$$P\left(\left|\frac{S_n}{n} - m\right| \geq n^{-\frac{1}{8}}\right) \leq \frac{K}{n^{\frac{3}{2}}},$$

from which it follows that

$$\sum_{n=1}^{\infty} P\left(\left|\frac{S_n}{n} - m\right| \geq n^{-\frac{1}{8}}\right) < \infty.$$

Therefore, by Theorem 8.1, $\left|\frac{S_n}{n} - m\right|$ converges to 0 P-a.s.

It remains to prove (1.61). To simplify the notation, call Y_i the random variable $X_i - m$, and remember that $E[Y_i] = 0$. Also, in view of the independence hypothesis, $E[Y_1 Y_2 Y_3 Y_4] = E[Y_1]E[Y_2]E[Y_3]E[Y_4] = 0, E[Y_1 Y_2^3] = E[Y_1]E[Y_2^3] = 0$, and the like. Finally, in the development

$$E\left[\left(\sum_{i=1}^{n} Y_i\right)^4\right] = \sum_{i,j,k,\ell=1}^{n} E[Y_i Y_j Y_k Y_\ell],$$

only the terms of the form $E[Y_i^4]$ and $E[Y_i^2 Y_j^2](i \neq j)$ remain. There are n terms of the first type and $3n(n-1)$ terms of the second type. Therefore,

$$E\left[\left(\sum_{i=1}^{n} Y_i\right)^4\right] = nE[Y_1^4] + 3n(n-1)E[Y_1^2 Y_2^2],$$

which is smaller than Kn^2 for some finite K.

———

Proof of Kolmogorov's Strong Law of Large Numbers

We assume without loss of generality that $E[X_1] = 0$.

Lemma 1.4.12 *Let $X_1, \ldots, X_n$ be independent random variables such that for all i $(1 \le i \le n)$,*

$$E[|X_i|^2] < \infty, \quad E[X_i] = 0.$$

Then for all $\lambda > 0$,

$$P(\max_{1 \le k \le n} |S_k| \ge \lambda) \le \frac{E[S_n^2]}{\lambda^2},$$

where $S_k = X_1 + \ldots + X_k$.

Proof. Let T be the first (random) index k $(1 \le k \le n)$ such that $|S_k| \ge \lambda$, with $T = \infty$ if $\max_{1 \le k \le n} |S_k| < \lambda$. For $k \le n$,

$$
\begin{aligned}
E[S_n^2 1_{\{T=k\}}] &= E\left[1_{\{T=k\}} \left\{ (S_n - S_k)^2 + 2S_k(S_n - S_k) + S_k^2 \right\} \right] \\
&= E\left[1_{\{T=k\}} \left\{ (S_n - S_k)^2 + S_k^2 \right\} \right] \ge E[1_{\{T-k\}} S_k^2],
\end{aligned}
$$

where we used the fact that $1_{\{T=k\}} S_k$ is a function of $X_1, \ldots, X_k$ and therefore independent of $S_n - S_k$, so that $E[1_{\{T=k\}} S_k (S_n - S_k)] = E[1_{\{T=k\}} S_k] E[S_n - S_k] = 0$. Therefore,

$$
\begin{aligned}
E[|S_n|^2] &\ge \sum_{k=1}^{n} E[1_{\{T=k\}} S_k^2] \ge \sum_{k=1}^{n} E[1_{\{T=k\}} \lambda^2] - \lambda^2 \sum_{k=1}^{n} P(T-k) \\
&= \lambda^2 P(T \le n) = \lambda^2 P(\max_{1 \le k \le n} |S_k| \ge \lambda).
\end{aligned}
$$

$\square$

The following corollary contains a proof of the SLNN when $E[|X_n|^2] < \infty$.

Corollary 1.4.13 *Let $\{X_n\}_{n \ge 1}$ be a sequence of independent random variables such that for all $n \ge 1$,*

$$E[|X_n|^2] < \infty, \quad E[X_n] = 0.$$

If

$$\sum_{n \ge 1} \frac{E[X_n^2]}{n^2} \le \infty, \tag{1.62}$$

then

$$\lim_{n \uparrow \infty} \frac{1}{n} \sum_{k=1}^{n} X_k = 0, \; P\text{-a.s.}$$

Proof. If $2^{k-1} \leq n \leq 2^k$, then $\frac{|S_n|}{n} \geq \epsilon$ implies $\frac{|S_n|}{2^{k-1}} \geq \epsilon$. Therefore, for all $\epsilon > 0$, and all $k \geq 1$,

$$
P\left(\frac{|S_n|}{n} \geq \epsilon \text{ for some } n \in [2^{k-1}, 2^k]\right)
$$
$$
\leq P\left(|S_n| \geq \epsilon 2^{k-1} \text{ for some } n \in [2^{k-1}, 2^k]\right)
$$
$$
\leq P\left(|S_n| \geq \epsilon 2^{k-1} \text{ for some } n \in [1, 2^k]\right)
$$
$$
= P\left(\max_{1 \leq n \leq 2^k} |S_n| \geq \epsilon 2^{k-1}\right) \leq \frac{4}{\epsilon^2} \frac{1}{(2^k)^2} \sum_{n=1}^{2^k} E[X_n^2],
$$

where the last inequality follows from Kolmogorov's inequality. But defining $m = m(n)$ by $2^{m-1} \leq n < 2^m$, we have

$$
\sum_{k=1}^{\infty} \frac{1}{(2^k)^2} \sum_{n=1}^{2^k} E[X_n^2] = \sum_{n=1}^{\infty} E[X_n^2] \sum_{j=m}^{\infty} \frac{1}{(2^j)^2},
$$

which is bounded by

$$
\sum_{n=1}^{\infty} E[X_n^2] \frac{K}{(2^m)^2} \leq \sum_{n=1}^{\infty} E[X_n^2] \frac{K}{n^2}
$$

for some finite K. Therefore, by (1.62),

$$
\sum_{k=1}^{\infty} P\left(\frac{|S_n|}{n} \geq \epsilon \text{ for some } n; \ 2^{k-1} < n \leq 2^k\right) < \infty,
$$

and by the Borel–Cantelli lemma,

$$
P\left(\frac{|S_n|}{n} \geq \epsilon \text{ i.o.}\right) = 0.
$$

The result then follows from Theorem 1.4.9.   □

Having proved the corollary, it now remains to get rid of the assumption $E[|X_n|^2] < \infty$, and the natural technique for this is truncation. Define

$$
\widetilde{X}_n = \begin{cases} X_n & \text{if } |X_n| \leq n, \\ 0 & \text{otherwise.} \end{cases}
$$

Since $E[|X_1|] < \infty$, we have by dominated convergence (Theorem A.3.2) that $\lim_{n \uparrow \infty} E[X_1 1_{\{|X_1| \leq n\}}] = E[X_1] = 0$. Since X_n has the same distribution as X_1,

$$
\lim_{n \uparrow \infty} E[\widetilde{X}_n] = \lim_{n \uparrow \infty} E[X_n 1_{\{|X_n| \leq n\}}] = \lim_{n \uparrow \infty} E[X_1 1_{\{|X_1| \leq n\}}] = E[X_1].
$$

In particular (Cesàro's Lemma, Theorem A.1.7),

$$\lim_{n\uparrow\infty} \frac{1}{n} \sum_{k=1}^{n} \mathrm{E}[\widetilde{X}_k] = 0 \,.$$

Also,

$$\mathrm{E}[|X_1|] \;=\; \sum_{n=0}^{\infty} \mathrm{E}[|X_1|1_{(n,n+1]}(|X_1|)] \geq \sum_{n=0}^{\infty} \mathrm{E}[n1_{(n,n+1]}(|X_1|)]$$

$$= \sum_{n=0}^{\infty} n(\mathrm{P}(|X_1| > n) - \mathrm{P}(|X_1| > n+1)) = \sum_{n=1}^{\infty} \mathrm{P}(|X_1| > n) \,.$$

In particular, using the telescope formula,

$$\sum_{n=1}^{\infty} \mathrm{P}(|X_n| > n) = \sum_{n=1}^{\infty} \mathrm{P}(|X_1| > n) \leq \mathrm{E}[|X_1|] < \infty,$$

and therefore, by the Borel–Cantelli lemma,

$$\mathrm{P}(\widetilde{X}_n \neq X_n \text{ i.o.}) = \mathrm{P}(X_n > n \text{ i.o.}) = 0 \,.$$

Therefore, to prove the SLNN it suffices to show that

$$\lim_{n\uparrow\infty} \frac{1}{n} \sum_{k=1}^{n} (\widetilde{X}_k - \mathrm{E}[\widetilde{X}_k]) = 0 \,.$$

In view of the preceding corollary, it suffices to prove that

$$\sum_{n=1}^{\infty} \frac{\mathrm{E}[(\widetilde{X}_n - \mathrm{E}[\widetilde{X}_n])^2]}{n^2} < \infty \,.$$

But

$$\mathrm{E}[(\widetilde{X}_n - \mathrm{E}[\widetilde{X}_n])^2] \leq \mathrm{E}[\widetilde{X}_n^2] = \mathrm{E}[X_1^2 1_{\{|X_1| \leq n\}}] \,.$$

It is therefore enough to show that

$$\sum_{n=1}^{\infty} \frac{\mathrm{E}[X_1^2 1_{\{|X_1| \leq n\}}]}{n^2} < \infty \,.$$

The left-hand side of the above inequality is equal to

$$\sum_{n=1}^{\infty} \frac{1}{n^2} \sum_{k=1}^{n} \mathrm{E}[X_1^2 1_{\{k-1 < |X_1| \leq k\}}] = \sum_{k=1}^{\infty} \sum_{n=k}^{\infty} \frac{1}{n^2} \mathrm{E}[X_1^2 1_{\{k-1 < |X_1| \leq k\}}] \,.$$

Using the fact that

$$\sum_{n=k}^{\infty} \frac{1}{n^2} \le \frac{1}{k^2} + \int_{k}^{\infty} \frac{1}{x^2}\mathrm{d}x = \frac{1}{k^2} + \frac{1}{k} \le \frac{2}{k}$$

(draw the graph of $x \to x^{-2}$), this is less than or equal to

$$\sum_{k=1}^{\infty} \frac{2}{k} E[X_1^2 1_{\{k-1<|X_1|\le k\}}] \ = \ 2\sum_{k=1}^{\infty} E[\frac{X_1^2}{k} 1_{\{k-1<|X_1|\le k\}}]$$

$$\le \ 2\sum_{k=1}^{\infty} E[|X_1| 1_{\{k-1<|X_1|\le k\}}]$$

$$= \ 2E[|X_1|] < \infty .$$

1.5   Exercises

Exercise 1.5.1. SOME IDENTITIES
Prove and generalize the following identities:

$$P(A \cup B) = P(A) + P(B) - P(A \cap B) .$$

$$P(A \cup B \cup C) = P(A) + P(B) + P(C)$$
$$+ P(A \cap B) - P(B \cap C) - P(C \cap A)$$
$$+ P(A \cap B \cap C).$$

Exercise 1.5.2. $P(A \cup B) = 1 - P(\bar{A} \cap \bar{B})$
Prove the identity in the title of the exercise.

Exercise 1.5.3. THREE COIN TOSSES
Give a probability model for three successive tosses of an unbiased die. What is the probability that one of these tosses results in a number that is the sum of the two other numbers?

Exercise 1.5.4. THE END OF "HEADS"
In the coin model (Examples 1.3, 1.6, and 1.9) show that $\cup_{n=1}^{\infty} \cap_{k=n}^{\infty} \{X_k = 1\}$ is the event "after some finite random time, *all* tosses result in heads" (recall that "heads $= 1$").

Exercise 1.5.5. RANDOM POINT IN THE SQUARE

Consider the following probabilistic model: $\Omega = [0, 1]^2$, $P(A) = $ area of A. Thus $\omega = (x, y)$, where $x, y \in [0, 1]$. Define $X(\omega) = x$, $Y(\omega) = y$. Show that X and Y are independent random variables.

Exercise 1.5.6. RANDOM POINT IN THE DISK

Consider the following probability model: $\Omega = \{(x, y) \in \mathbb{R}^2, x^2 + y^2 \leq 1\}$, $P(A) = \frac{1}{\pi} \times ($ area of A$)$. Thus $\omega = (x, y) \in \Omega$ is a point uniformly distributed inside the unit disk. Defining $X(\omega) = x$ and $Y(\omega) = x$, show that X and Y are *not* independent random variables.

Exercise 1.5.7. PAIRWISE INDEPENDENT BUT NOT INDEPENDENT

Give a simple example of a probability space $(\Omega, \mathcal{F}, P)$ with three events A_1, A_2, A_3 that are pairwise independent, but *not* globally independent (that is, the family $\{A_1, A_2, A_3\}$ is not independent).

Exercise 1.5.8. ABOUT INDEPENDENCE

Show that if $\{A_i\}_{i \in I}$ is an independent collection of events, then $\{\widetilde{A}_i\}_{i \in I}$ is also, where, independently for each $i \in I$, $\widetilde{A}_i = A_i$ or $\bar{A}_i$ (for instance, with $I = \mathbb{N}$, $\widetilde{A}_0 = A_0, \widetilde{A}_1 = \bar{A}_1, \widetilde{A}_3 = A_3, \ldots$).

Exercise 1.5.9. GUESS THE COLOR

There are 3 cards. The first one has both faces red, the second one has both faces white, and the third one is white on one face, red on the other. A card is drawn at random, and the color of a randomly selected face of this card is announced. What is the winning strategy if you must bet on the color of the hidden face?

Exercise 1.5.10. THE SCHOOL FOR INTELLECTUAL APARTHEID

In the School for Intellectual Apartheid, students have been separated into three groups for pedagogical purposes. In group A, one finds students who individually have a probability of passing equal to 0.95. In group B this probability is 0.75, and in group C only 0.65. What is the probability that a student passing the course comes from group A? B? C?

Exercise 1.5.11. THE ENTOMOLOGIST
A given insect of a specific breed has the probability θ of being a male. An
entomologist seeks to collect exactly $M > 1$ males, and therefore stops hunting as
soon as she captures M males. She has to capture an insect in order to determine
its gender. What is the distribution of X, the number of insects she must catch
to collect *exactly* M males?

Exercise 1.5.12. THE GAUSSIAN VARIABLE SQUARED
Find the CDF and the PDF of $Y = X^2$, where X is a Gaussian random variable
with mean 0 and variance $\sigma^2 > 0$.

Exercise 1.5.13. LACK OF MEMORY OF THE EXPONENTIAL DISTRIBUTION
Let T be a geometric random variable. Show that for any integers $k, k_0 \geq 1$, we
have $P(T = k + k_0 \mid T > k_0) = P(T = k)$. Let X be an exponential random
variable with mean $1/\lambda$. Show that for all $t, t_0 \in \mathbb{R}_+$, we have

$$P(X \geq t_0 + t \mid X \geq t_0) = P(X \geq t).$$

Exercise 1.5.14. $Z = X + Y$
Let X be a real valued random variable with a PDF $f(x)$ and let Y be an integer-
valued random variable with the distribution $P(X = k) = p_k$, $k \geq 0$. Find the
PDF of $Z = X + Y$ when X and Y are independent.

Exercise 1.5.15. THE INFIMUM OF INDEPENDENT EXPONENTIAL VARIABLES
Let X_i $(0 \leq i \leq n)$ be independent exponential random variables with the respective parameters λ_i $(0 \leq i \leq n)$. Define $Z = \inf(X_1, \ldots, X_n)$ and let J be
the (random) index such that $X_J = Z$ (J is for almost all $\omega \in \Omega$ unambiguously
defined in view of Example 1.1.55). Show that Z and J are independent, and give
their respective distributions.

Exercise 1.5.16.
Let $(Y_1, \ldots, Y_k)$ be a vector of independent exponential random variables with
parameters $\lambda_1, \ldots, \lambda_k$, respectively, that is independent of and has the same distribution as the random vector $(X_1, \ldots, X_k)$. Let $Z = \inf(X_1, \ldots, X_k)$ and let J be
the random index such that $X_J = Z$. Define $R_1, \ldots, R_k$ by $R_J = Y_J, R_i = X_i - X_J$
$(i \neq J)$. Show that $R_1, \ldots, R_k$ are independent exponential random variables with
parameters $\lambda_1, \ldots, \lambda_k$, respectively, and are independent of (Z, J).

Exercise 1.5.17. SUM OF MULTINOMIALS
Let $X = (X_1, \ldots, X_k)$ and $Y = (Y_1, \ldots, Y_k)$ be two independent multinomial random vectors of sizes (n, K) and (m, K), respectively, and with the same parameters $p_1, \ldots, p_K$. What is the distribution of $Z = X + Y$?

Exercise 1.5.18. THINNING IN A SEQUENCE OF EXPONENTIALS
Let $\{X_n\}_{n \geq 1}$ be an IID sequence of exponential random variables with mean $1/\theta$, where $\theta \in (0, \infty)$. Let $\{Y_n\}_{n \geq 1}$ be an IID sequence with $P(Y_1 = 1) = 1 - P(Y_1 = 0) = p \in (0, 1)$, independent of $\{X_n\}_{n \geq 1}$. Let $U_1, U_2, \ldots$ be the successive indices n for which $Y_n = 1$, and define $S_1 = X_1 + \cdots + X_{U_1}$, $S_2 = X_{U_1+1} + \cdots + X_{U_2}$, $S_3 = X_{U_2+1} + \cdots + X_{U_3}$, etc. Show that $\{S_n\}_{n \geq 1}$ is an IID sequence of exponential random variables with mean $1/(p\theta)$.

Exercise 1.5.19. MEAN AND VARIANCE VIA GENERATING FUNCTIONS
Compute the mean and variance of the binomial random variable from its generating function. Do the same for the geometric random variable and the Poisson random variable.

Exercise 1.5.20. THE INFIMUM OF INDEPENDENT UNIFORM VARIABLES
Find the probability distribution of the random variable Z_i, the ith smallest among $X_1, \ldots, X_n$, when the X_i's are independent $[0, 1]$-uniform random variables.

Exercise 1.5.21. THE QUOTIENT OF INDEPENDENT EXPONENTIALS
Let X and Y be two independent random variables with a common exponential distribution of mean θ^{-1}. Give the PDF of the vector $(X/Y, Y)$ and of the variable X/Y.

Exercise 1.5.22. RANDOMLY CUTTING A RANDOM SEGMENT
Consider a random segment of length U uniformly distributed on $[0, 1]$, and make of it two random pieces of length $X = UV$, $Y = U(1 - V)$, respectively, where V is uniformly distributed on $[0, 1]$ and independent of U. Compute the PDF of the length UV of the first piece.

Exercise 1.5.23. $\mathrm{E}^{\max(X_1, X_2)}[X_1]$
Let X_1 and X_2 be two independent random variables taking their values in $\{1, 2, \ldots, N\}$, and uniformly distributed, that is, $P(X_1 = k) = P(X_2 = k) = \frac{1}{N}$ $(1 \leq k \leq N)$. Compute $\mathrm{E}^{\max(X_1, X_2)}[X_1]$.

Exercise 1.5.24. BIJECTIONS AND DISCRETE RANDOM VARIABLES
Let X and Y be two discrete random variables with values in $\mathbb{N}$, and let $h : \mathbb{N} \to \mathbb{N}$ be one-to-one and onto. Show that for all $v : \mathbb{N} \to \mathbb{R}$ such that $\mathrm{E}[|v(X)|] < \infty$, $\mathrm{E}^Y[v(X)] = \mathrm{E}^Z[v(X)]$, where $Z = h(Y)$.

Exercise 1.5.25. CONDITIONING WITH MULTINOMIALS
Let $(X_1, \ldots, X_k)$ be a multinomial random vector of size (n, K) and parameters $p_1, \ldots, p_K$. Compute $\mathrm{E}^{X_1}[X_2 + \cdots + X_{k-1}]$ and $\mathrm{E}^{X_1}[X_2]$.

Exercise 1.5.26. $\lim_{t\to\infty} \frac{N(t)}{t}$
Let $\{S_n\}_{n\geq 1}$ be an IID sequence of real random variables such that $\mathrm{P}(S_1 \in (0, \infty)) = 1$ and $\mathrm{E}[S_1] < \infty$, and let for each $t \geq 0$, $N(t) = \sum_{n\geq 1} 1_{(0,t]}(T_n)$, where $T_n = S_1 + \cdots + S_n$. Prove that $\lim_{t\to\infty} \frac{N(t)}{t} = \frac{1}{\mathrm{E}[S_1]}$.

Exercise 1.5.27. DERIVATIVE OF THE LAPLACE TRANSFORM
Let X be a discrete random variable with values $x_i \in \mathbb{R}_+$ $(i \geq 1)$ and of distribution $p_i = \mathrm{P}(X = x_i)$ $(i \geq 1)$. Suppose that for some $t_0 > 0$, $\sum_{i\geq 1} p_i e^{t_0 x_i} < \infty$. Prove that the function $g : t \to \sum_{i\geq 1} p_i e^{tx_i}$ is differentiable in a neighborhood of 0, with derivative $\sum_{i\geq 1} p_i x_i e^{tx_i}$.

Chapter 2

Discrete-Time Markov Chains

Sequences of independent and identically distributed random variables *are* stochastic processes, but they are not always interesting as stochastic models because they behave more or less in the same way. In order to introduce more variability, one can allow for some dependence on the past in the manner of deterministic recurrence equations. Discrete-time homogeneous Markov chains possess the required feature since they can always be represented—at least distributionwise—by a stochastic recurrence equation $X_{n+1} = f(X_n, Z_{n+1})$, where $\{Z_n\}_{n\geq 1}$ is an IID sequence, independent of the initial state X_0.

The probabilistic dependence on the past is only through the previous state, but this limited amount of memory suffices to produce a great diversity of behaviors. For this reason Markov chains have found applications in many domains, including biology, physics, sociology, operations research, and engineering, where they provide qualitative and quantitative answers as well as precious insights for systems design.

2.1 The Transition Matrix

2.1.1 The Markov Property

A sequence $\{X_n\}_{n\geq 0}$ of random variables with values in a set E is called a *discrete-time stochastic process* with *state space* E. In this book, the state space is countable, and its elements are denoted by $i, j, k,\ldots$ If $X_n = i$, the process is said to be in state i at time n, or to visit state i at time n.

Definition 2.1.1 *Let $\{X_n\}_{n\geq 0}$ be a discrete-time stochastic process with countable state space E. If for all integers $n \geq 0$ and all states $i_0, i_1,\ldots, i_{n-1}, i, j,$*

$$P(X_{n+1} = j \mid X_n = i, X_{n-1} = i_{n-1},\ldots, X_0 = i_0) = P(X_{n+1} = j \mid X_n = i) \quad (2.1)$$

© Springer Nature Switzerland AG 2020
P. Brémaud, *Markov Chains*, Texts in Applied Mathematics 31,
https://doi.org/10.1007/978-3-030-45982-6_2

whenever both sides are well-defined, this stochastic process is called a Markov chain. *It is called a* homogeneous *Markov chain* (HMC) *if in addition, the right-hand side of (2.1) is independent of n.*

Property (2.1) is the *Markov property*. The matrix $P = \{p_{ij}\}_{i,j\in E}$, where

$$p_{ij} = P(X_{n+1} = j \mid X_n = i),$$

is the *transition matrix* of the HMC. Since its entries are probabilities, and since a transition from any state i must be to *some* state, it follows that

$$p_{ij} \geq 0, \sum_{k\in E} p_{ik} = 1$$

for all states i, j. A matrix P indexed by E and satisfying the above properties is called a *stochastic matrix*.

Remark 2.1.2 The state space may be infinite, and therefore such a matrix is in general not of the kind studied in linear algebra. However, the basic operations of addition and multiplication will be defined by the same formal rules. For instance, with $A = \{a_{ij}\}_{i,j\in E}$ and $B = \{b_{ij}\}_{i,j\in E}$, the product $C = AB$ is the matrix $\{c_{ij}\}_{i,j\in E}$, where $c_{ij} = \sum_{k\in E} a_{ik}b_{kj}$. The notation $x = \{x_i\}_{i\in E}$ formally represents a *column* vector, and x^T is a row vector, the transpose of x. For instance, $y = \{y_i\}_{i\in E}$ given by $y^T = x^T A$ is defined by $y_i = \sum_{k\in E} x_k a_{ki}$. Similarly, $z = \{z_i\}_{i\in E}$ given by $z = Ax$ is defined by $z_i = \sum_{k\in E} a_{ik}z_k$.

Proving the Markov property is not, in general, a difficult task, and Theorems 2.2.1 and 2.2.3 below will suffice in most situations. However, there are cases outside their scope, and the following one is quite important, both in theory and in applications.

EXAMPLE 2.1.3: MACHINE REPLACEMENT. Let $\{U_n\}_{n\geq 1}$ be a sequence of IID random variables taking their values in $\{1, 2, \ldots, +\infty\}$. The random variable U_n can be interpreted as the lifetime of some machine, the n-th one, which is replaced by the $(n+1)$-st one upon failure. Thus at time 0, machine 1 is put in service until it breaks down at time U_1, whereupon it is immediately replaced by machine 2, which breaks down at time $U_1 + U_2$, and so on. The elapsed time in service of the current machine at time n is denoted by X_n. Thus, the process $\{X_n\}_{n\geq 0}$ takes its values in $E = \mathbb{N}$ and increases linearly from 0 at time $R_k = \sum_{i=1}^{k} U_i$ to $U_{k+1} - 1$ at time $R_{k+1} - 1$. The sequence $\{R_k\}_{k\geq 0}$ defined in this way, with $R_0 = 0$, is called a *renewal sequence*, and X_n is called the *backward recurrence time* at time n (see

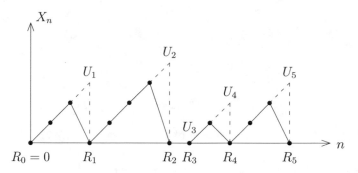

Figure 2.1.1. The backward recurrence HMC

Figure 2.1.1). There is a rich and useful theory associated with renewal sequences, the so-called *renewal theory*. It will be developed in Chapter 5.

Exercise 2.6.1 asks you to prove formally that the sequence $\{X_n\}_{n \geq 0}$ is an HMC with state space $E = \mathbb{N}$ and that the non-null entries of its transition matrix are of the form $p_{i,i+1}$ and $p_{i,0} = 1 - p_{i,i+1}$, where

$$p_{i,i+1} = \frac{P(U_1 > i + 1)}{P(U_1 > i)}. \tag{2.2}$$

The Transition Graph

A transition matrix P is sometimes represented by its *transition graph* G, a graph having for nodes (or vertices) the states of E. This graph has an oriented edge from i to j if and only if $p_{ij} > 0$, in which case this edge is adorned with the label p_{ij}.

The transition graph of the Markov chain of Example 2.1.3 is shown in Figure 2.1.2, where

$$p_i = \frac{P(U_1 = i + 1)}{P(U_1 > i)}.$$

2.1.2 The Distribution of an HMC

The random variable X_0 is called the *initial state*, and its probability distribution

$$\nu(i) := P(X_0 = i) \quad (i \in E)$$

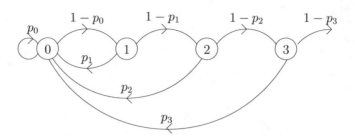

Figure 2.1.2. Transition graph of the backward recurrence chain

is the *initial distribution*. From Bayes' sequential rule,

$$P(X_0 = i_0, X_1 = i_1, \ldots, X_k = i_k)$$
$$= P(X_0 = i_0)P(X_1 = i_1 \mid X_0 = i_0) \cdots P(X_k = i_k \mid X_{k-1} = i_{k-1}, \ldots, X_0 = i_0),$$

and therefore, in view of the homogeneous Markov property and the definition of the transition matrix,

$$P(X_0 = i_0, X_1 = i_1, \ldots, X_k = i_k) = \nu(i_0)p_{i_0 i_1} \cdots p_{i_{k-1} i_k} .$$

This data for all $k \geq 0$, all states $i_0, i_1, \ldots, i_k$, constitute the *probability law*, or *distribution* of the HMC. Therefore we have the following result.

Theorem 2.1.4 *The distribution of a discrete-time HMC is determined by its initial distribution and its transition matrix.*

The distribution at time n of the chain is the vector ν_n, where

$$\nu_n(i) = P(X_n = i) .$$

From Bayes' rule of exclusive and exhaustive causes, $\nu_{n+1}(j) = \sum_{i \in E} \nu_n(i)p_{ij}$, that is, in matrix form, $\nu_{n+1}^T = \nu_n^T P$. Iteration of this equality yields

$$\nu_n^T = \nu_0^T \mathbf{P}^n . \tag{2.3}$$

The matrix $\mathbf{P}^n$ is called the *n-step transition matrix* because its general term is

$$p_{ij}(m) = P(X_{n+m} = j \mid X_n = i) .$$

Indeed, Bayes' sequential rule and the Markov property give for the right-hand side of the last equality

$$\sum_{i_1, \ldots, i_{m-1} \in E} p_{i i_1} p_{i_1 i_2} \cdots p_{i_{m-1} j} ,$$

and this is the general term of the m-th power of P.

The Markov property (2.1) extends to

$$P(X_{n+1} = j_1, \ldots, X_{n+k} = j_k \mid X_n = i, X_{n-1} = i_{n-1}, \ldots, X_0 = i_0)$$

$$= P(X_{n+1} = j_1, \ldots, X_{n+k} = j_k \mid X_n = i)$$

for all $i_0, \ldots, i_{n-1}, i, j_1, \ldots, j_k$ such that both sides of the equality are defined (Exercise 2.6.3). Writing

$$A = \{X_{n+1} = j_1, \ldots, X_{n+k} = j_k\}, B = \{X_0 = i_0, \ldots, X_{n-1} = i_{n-1}\},$$

the last equality reads $P(A \mid X_n = i, B) = P(A \mid X_n = i)$, which is in turn equivalent to

$$P(A \cap B \mid X_n = i) = P(A \mid X_n = i)P(B \mid X_n = i).$$

Remark 2.1.5 In words: The future at time n and the past at time n are conditionally independent given the present state $X_n = i$. This shows in particular that the Markov property is independent of the direction of time.

Notation: We shall abbreviate $P(A \mid X_0 = i)$ as $P_i(A)$. If μ is a probability distribution on E, then $P_\mu(A) = \sum_{i \in E} \mu(i)P_i(A)$ is the probability of A given that the initial state is distributed according to μ.

2.2   Markov Recurrences

2.2.1   A Canonical Representation

Many HMCs receive a natural description in terms of a recurrence equation driven by "white noise".

Theorem 2.2.1 *Let $\{Z_n\}_{n \geq 1}$ be an IID sequence of random variables with values in an arbitrary space F. Let E be a countable space, and $f : E \times F \to E$ be some function. Let X_0 be a random variable with values in E, independent of $\{Z_n\}_{n \geq 1}$. The recurrence equation*

$$X_{n+1} = f(X_n, Z_{n+1}) \tag{2.4}$$

then defines an HMC.

Remark 2.2.2 The phrase *white noise* comes from signal theory and refers to the fact that the driving sequence $\{Z_n\}_{n \geq 1}$ is IID.

Proof. Iteration of recurrence (2.4) shows that for all $n \geq 1$, there is a function g_n such that $X_n = g_n(X_0, Z_1, \ldots, Z_n)$, and therefore

$$P(X_{n+1} = j \mid X_n = i, X_{n-1} = i_{n-1}, \ldots, X_0 = i_0)$$
$$= P(f(i, Z_{n+1}) = j \mid X_n = i, X_{n-1} = i_{n-1}, \ldots, X_0 = i_0) = P(f(i, Z_{n+1}) = j),$$

since the event $\{X_0 = i_0, \ldots, X_{n-1} = i_{n-1}, X_n = i\}$ is expressible in terms of $X_0, Z_1, \ldots, Z_n$ and is therefore independent of Z_{n+1}. Similarly, $P(X_{n+1} = j \mid X_n = i) = P(f(i, Z_{n+1}) = j)$. We therefore have a Markov chain, and it is homogeneous, since the right-hand side of the last equality does not depend on n. Explicitly

$$p_{ij} = P(f(i, Z_1) = j). \tag{2.5}$$

$\square$

Not all homogeneous Markov chains are naturally described by the model of Theorem 2.2.1. A slight modification of it considerably enlarges its scope.

Theorem 2.2.3 *Let things be as in Theorem 2.2.1 except for the statistics of $X_0, Z_1, Z_2, \ldots$. Suppose instead that for all $n \geq 0$, Z_{n+1} is conditionally independent of $Z_n, \ldots, Z_1, X_{n-1}, \ldots, X_0$ given X_n, that is, for all $k, k_1, \ldots, k_n \in F$, $i_0, i_1, \ldots, i_{n-1}, i \in E$,*

$$P(Z_{n+1} = k \mid X_n = i, X_{n-1} = i_{n-1}, \ldots, X_0 = i_0, Z_n = k_n, \ldots, Z_1 = k_1)$$
$$= P(Z_{n+1} = k \mid X_n = i),$$

where the latter quantity is independent of n. Then $\{X_n\}_{n \geq 0}$ is an HMC, *with transition matrix* P *given by*

$$p_{ij} = P(f(i, Z_1) = j \mid X_0 = i).$$

Proof. The proof is analogous to that of Theorem 2.2.1 (Exercise 2.6.12). $\square$

Remark 2.2.4 As we mentioned a few lines above not all homogeneous Markov chains receive a "natural" description of the type featured in Theorems 2.2.1 as Example 2.1.3 (machine replacement) shows. However, for any transition matrix P on E, there exists a homogeneous Markov chain $\{X_n\}_{n \geq 0}$ with this transition matrix and with a representation such as in Theorem 2.2.1, namely,

$$X_{n+1} = j \text{ if } Z_{n+1} \in \left[\sum_{k=0}^{j-1} p_{X_n k}, \sum_{k=0}^{j} p_{X_n k} \right],$$

where $\{Z_n\}_{n \geq 1}$ is IID, uniform on $[0, 1]$. We can apply Theorem 2.2.1, and check that this HMC has the announced transition matrix. This artificial representation is useful for simulating small Markov chains and can also be helpful for the theory.

The examples below will often be used to illustrate the theory.

EXAMPLE 2.2.5: 1-D RANDOM WALK. Let X_0 be a random variable with values in $\mathbb{Z}$. Let $\{Z_n\}_{n\geq 1}$ be a sequence of IID random variables, independent of X_0, taking the values $+1$ or -1, and with the probability distribution

$$P(Z_n = +1) = p\,,$$

where $p \in (0,1)$. The process $\{X_n\}_{n\geq 1}$ defined by

$$X_{n+1} = X_n + Z_{n+1}$$

is, in view of Theorem 2.2.1, an HMC, called the *random walk* on $\mathbb{Z}$.

EXAMPLE 2.2.6: THE REPAIR SHOP, TAKE 1. During day n, Z_{n+1} machines break down, and they enter the repair shop on day $n+1$. Every day one machine among those waiting for service is repaired. Therefore, denoting by X_n the number of machines in the shop on day n,

$$X_{n+1} = (X_n - 1)^+ + Z_{n+1}\,, \tag{2.6}$$

where $a^+ = \max(a, 0)$. In particular, if $\{Z_n\}_{n\geq 1}$ is an IID sequence independent of the initial state X_0, then $\{X_n\}_{n\geq 0}$ is a homogeneous Markov chain. In terms of the probability distribution

$$P(Z_1 = k) = a_k \quad (k \geq 0)\,, \tag{2.7}$$

its transition matrix is

$$P = \begin{pmatrix} a_0 & a_1 & a_2 & a_3 & \cdots \\ a_0 & a_1 & a_2 & a_3 & \cdots \\ 0 & a_0 & a_1 & a_2 & \cdots \\ 0 & 0 & a_0 & a_1 & \cdots \\ \vdots & \vdots & \vdots & \vdots & \end{pmatrix}.$$

Indeed, from (2.5) and (2.7),

$$p_{ij} = P((i-1)^+ + Z_1 = j) = P(Z_1 = j - (i-1)^+) = a_{j-(i-1)^+}\,.$$

EXAMPLE 2.2.7: THE INVENTORY. A given commodity is stocked in order to satisfy a continuing demand. The aggregated demand between time n and time $n+1$ is Z_{n+1} units, and it is assumed that $\{Z_n\}_{n\geq1}$ is IID, and independent of the initial value X_0 of the stock. Replenishment of the stock takes place at times $n+0$ (that is to say, immediately after time n) for all $n \geq 1$.

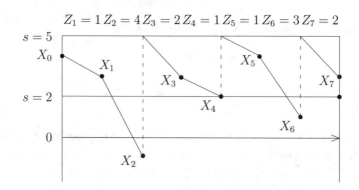

Figure 2.2.1. A sample path of the inventory Markov chain

A popular management strategy is the so-called (s, S)-strategy, where s and S are integers such that $0 < s < S$. Under this inventory policy, if the level of the stock at time n is found not larger than s, then it is brought to level S at time $n + 0$. Otherwise, nothing is done. The initial stock X_0 is assumed not greater than S, and therefore $\{X_n\}_{n\geq1}$ takes its value in $E = \{S, S - 1, S - 2, \ldots\}$. (See Figure 2.2.1.) Negative values of the stock are allowed, with the interpretation that an unfilled demand is immediately satisfied upon restocking. With the above rules of operation, the evolution of the stock is governed by the dynamic equation

$$X_{n+1} = \begin{cases} X_n - Z_{n+1} & \text{if } s < X_n \leq S, \\ S - Z_{n+1} & \text{if } X_n \leq s. \end{cases}$$

In view of this and of Theorem 2.2.1, $\{X_n\}_{n\geq1}$ is a homogeneous Markov chain.

EXAMPLE 2.2.8: STOCHASTIC AUTOMATA. A finite automaton $(E, \mathcal{A}, f)$ can read sequences of letters from a finite alphabet $\mathcal{A}$ written on some infinite tape. It can be in any state of a finite set E, and its evolution is governed by a function $f : E \times \mathcal{A} \to E$, as follows. When the automaton is in state $i \in E$ and reads letter $a \in \mathcal{A}$, it switches from state i to state $j = f(i, a)$ and then reads on the tape the next letter to the right.

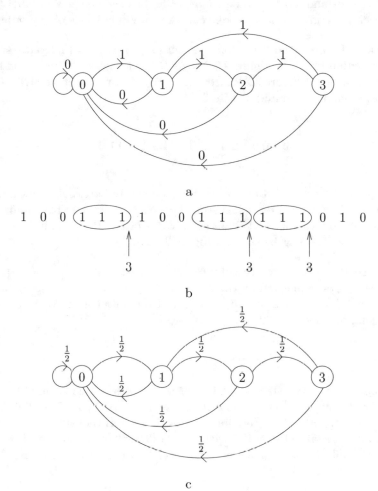

a

$$1 \quad 0 \quad 0 \quad \boxed{1 \quad 1 \quad 1} \quad 1 \quad 0 \quad 0 \quad \boxed{1 \quad 1 \quad 1}\boxed{1 \quad 1 \quad 1} \quad 0 \quad 1 \quad 0$$

b

c

Figure 2.2.2. The automaton: the recognition process and the Markov chain

An automaton can be represented by its transition graph G having for nodes the states of E. There is an oriented edge from the node (state) i to the node j if and only if there exists an $a \in \mathcal{A}$ such that $j = f(i, a)$, and this edge then receives label a. If $j = f(i, a_1) = f(i, a_2)$ for $a_1 \neq a_2$, then there are two edges from i to j with labels a_1 and a_2, or, more economically, one such edge with label (a_1, a_2). More generally, a given oriented edge can have multiple labels of any order.

Consider, for instance, the automaton with alphabet $\mathcal{A} = \{0, 1\}$ corresponding to the transition graph of Figure 2.2.2a. As the automaton, initialized in state 0, reads the sequence of Figure 2.2.2b from left to right, it passes successively through the states (including the initial state 0)

$$0\ 1\ 0\ 0\ 1\ 2\ 3\ 1\ 0\ 0\ 1\ 2\ 3\ 1\ 2\ 3\ 0\ 1\ 0.$$

Rewriting the sequence of states below the sequence of letters, it appears that the automaton is in state 3 after it has seen three consecutive 1's. This automaton is therefore able to recognize and count such blocks of 1's. However, it does not take into account overlapping blocks (see Figure 2.2.2b).

If the sequence of letters read by the automaton is $\{Z_n\}_{n \geq 1}$, the sequence of states $\{X_n\}_{n \geq 0}$ is then given by the recurrence equation $X_{n+1} = f(X_n, Z_{n+1})$ and therefore, if $\{Z_n\}_{n \geq 1}$ is IID and independent of the initial state X_0, then $\{X_n\}_{n \geq 1}$ is, according to Theorem 2.2.1, an HMC.

EXAMPLE 2.2.9: THE URN OF EHRENFEST, TAKE 1. This simplified model of diffusion through a porous membrane was proposed in 1907 by the Austrian physicists Tatiana and Paul Ehrenfest to describe in terms of statistical mechanics the exchange of heat between two systems at different temperatures. Their model also considerably helped our understanding of thermodynamic irreversibility (we shall discuss this later).

There are N particles that can be either in compartment A or in compartment B. Suppose that at time $n \geq 0$, $X_n = i$ particles are in A. One then chooses a particle at random, and this particle is moved at time $n + 1$ from where it is to the other compartment. Thus, the next state X_{n+1} is either $i - 1$ (the displaced particle was found in compartment A) with probability $\frac{i}{N}$, or $i + 1$ (it was found in B) with probability $\frac{N-i}{N}$.

This model pertains to Theorem 2.2.3. For all $n \geq 0$, $X_{n+1} = X_n + Z_{n+1}$, where $Z_n \in \{-1, +1\}$ and $P(Z_{n+1} = -1 \mid X_n = i) = \frac{i}{N}$. The non-null entries of the transition matrix are therefore

$$p_{i,i+1} = \frac{N-i}{N}, \ p_{i,i-1} = \frac{i}{N}.$$

2.2.2 First-Step Analysis

Absorption Probability

Many functionals of homogeneous Markov chains, in particular probabilities of absorption by a closed set (A is called *closed* if $\sum_{j \in A} p_{ij} = 1$ for all $i \in A$) and average times before absorption, can be evaluated by a technique called *first-step analysis*. This technique, which is the motor of most computations in Markov chain theory, is best illustrated by the following example.

EXAMPLE 2.2.10: THE GAMBLER'S RUIN, TAKE 1. Two players A and B play "heads or tails", where heads occur with probability $p \in (0, 1)$, and the successive outcomes form an IID sequence. Calling X_n the fortune in dollars of player A at time n, then $X_{n+1} = X_n + Z_{n+1}$, where $Z_{n+1} = +1$ (resp., -1) with probability p (resp., $q = 1 - p$), and $\{Z_n\}_{n \geq 1}$ is IID. In other words, A bets \$1 on heads at each toss, and B bets \$1 on tails. The respective initial fortunes of A and B are a and b. The game ends when a player is ruined, and therefore the process $\{X_n\}_{n \geq 1}$ is a random walk as described in Example 2.2.5, except that it is restricted to $E = \{0, \ldots, a, a + 1, \ldots, a + b = c\}$. The duration of the game is T, the first time n at which $X_n - 0$ or c, and the probability of winning for A is $u(a) - P(X_T = c \mid X_0 = a)$.

Instead of computing $u(a)$ alone, first-step analysis computes

$$u(i) = P(X_T = c \mid X_0 = i)$$

for all states i ($1 \leq i \leq c$) and for this, it first generates a recurrence equation for the $u(i)$'s by breaking down the event "A wins" according to what can happen after the first step (the first toss) and using the rule of exclusive and exhaustive causes. If $X_0 = i \in [1, c-1]$, then $X_1 = i+1$ (resp., $X_1 = i-1$) with probability p (resp., q), and the probability of ruin of B starting with A's initial fortune $i+1$ (resp., $i-1$) is $u(i+1)$ (resp., $u(i-1)$). Therefore, for i ($1 \leq i \leq c$),

$$u(i) = pu(i+1) + qu(i-1), \tag{2.8}$$

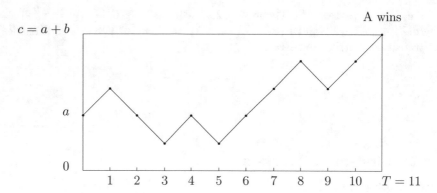

Figure 2.2.3. The basic random walk and the gambler's ruin

with the boundary conditions

$$u(0) = 0, u(c) = 1.$$

The characteristic equation associated with this linear recurrence equation is $pr^2 - r + q = 0$. It has two distinct roots $r_1 = 1$ and $r_2 = \frac{q}{p}$ if $p \neq q$, and a double root $r_1 = 1$ if $p = q = \frac{1}{2}$. Therefore the general solution is $u(i) = \lambda r_1^i + \mu r_2^i = \lambda + \mu \left(\frac{q}{p}\right)^i$ when $p \neq q$, and $u(i) = \lambda r_1^i + \mu i r_1^i = \lambda + \mu i$ when $p = q = \frac{1}{2}$. Taking into account the boundary conditions, one can determine the values of λ and μ. The result is, for $p \neq q$,

$$u(i) = \frac{1 - \left(\frac{q}{p}\right)^i}{1 - \left(\frac{q}{p}\right)^c},$$

and for $p = q = \frac{1}{2}$,

$$u(i) = \frac{i}{c}. \tag{2.9}$$

In the case $p = q = \frac{1}{2}$, the probability $v(i)$ that B wins when the initial fortune of B is $c - i$ is obtained by replacing i by $c - i$ in expression (2.9): $v(i) = \frac{c-i}{c} = 1 - \frac{i}{c}$. One checks that $u(i) + v(i) = 1$, which means in particular that the probability that the game lasts forever is null. The reader is invited to check that the same is true in the case $p \neq q$.

Remark 2.2.11 A justification of Bayes' rule of total causes in the obtention of the recurrence equation (2.8) is needed since we are not exactly in the standard framework of application (why?). The same kind of proof can be performed for every instance of first-step analysis. Let $Y = \{Y_n\}_{n\geq 0}$ denote the Markov chain obtained by delaying $X = \{X_n\}_{n\geq 0}$ by one time unit: $Y_n = X_{n+1}$. When $1 \leq X_0 \leq c - 1$, the events "X is absorbed by 0" and "Y is absorbed by 0" are identical and therefore

$$P(X \text{ is absorbed by } 0, X_1 = i \pm 1, X_0 = i)$$
$$= P(Y \text{ is absorbed by } 0, X_1 = i \pm 1, X_0 = i).$$

Since $\{Y_n\}_{n\geq 0}$ and X_0 are independent given X_1, the right-hand side of the above equality is equal to

$$P(X_0 = i, X_1 = i \pm 1)P(Y \text{ is absorbed by } 0 \mid Y_0 = i \pm 1).$$

The two chains have the same transition matrix, and therefore when they have the same initial state, they have the same distributions. Hence

$$P(Y \text{ is absorbed by } 0 \mid Y_0 = i \pm 1) = P(X \text{ is absorbed by } 0 \mid X_0 = i \pm 1).$$

Combining the above equalities gives the announced result.

EXAMPLE 2.2.12: CAT EATS MOUSE EATS CHEESE. A merry mouse moves in a maze. If it is at time n in a room with k adjacent rooms, it will be at time $n + 1$ in one of the k adjacent rooms, choosing one at random, each with probability $\frac{1}{k}$. A fat lazy cat remains all the time in a given room, and a piece of cheese waits for the mouse in another room. (See Figure 2.2.4.) If the mouse enters the room inhabited by the cat, the cat will eat it. What is the probability that the mouse ever gets to eat the cheese when starting from room 1, the cat and the cheese being in rooms 3 and 5, respectively?

To apply first-step analysis, call $u(i)$ the probability that the mouse initially in room i reaches the cheese without being murdered by the cat. The boundary conditions

$$u(3) = 0 \text{ and } u(5) = 1$$

are clear. If the mouse is in room 1, its first move will take it to room 2 (resp., 4) with probability $\frac{1}{2}$, and its chance of tasting the cheese will then be $u(2)$ (resp., $u(4)$). Therefore,

$$u(1) = \frac{1}{2}u(2) + \frac{1}{2}u(4).$$

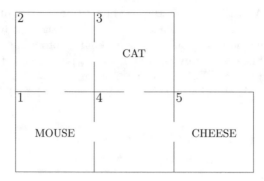

Figure 2.2.4. Maze, mouse, and murder

Similarly,

$$u(2) \;=\; \frac{1}{2}u(1) + \frac{1}{2}u(3) = \frac{1}{2}u(1),$$

$$u(4) \;=\; \frac{1}{3}u(1) + \frac{1}{3}u(3) + \frac{1}{3}u(5) = \frac{1}{3}u(1) + \frac{1}{3}.$$

The solution is $u(1) = \frac{2}{7}$.

EXAMPLE 2.2.13: TENNIS. Ignoring tie-breaks, a game can be modeled as a
Markov chain with the transition graph of Figure 2.2.5a, where p is the probability
that the server A wins the point, and $q = 1 - p$. We compute the probability that
B wins the game. Clearly, from Figure 2.2.5a, a game can be decomposed into two
stages: it first reaches the five upper states of the graph, and it then evolves in
the upper states until absorption in "game A" or "game B." With an appropriate
change of the labels of the upper states, one obtains the transition graph of the
Markov chain corresponding to the second stage (Figure 2.2.5b).

This is a familiar picture already encountered in the gambler's ruin problem.
However, a direct first-step analysis gives for b_i, the probability that B wins given
that the game starts from upper state i,

$$b_1 = 1 \;,\; b_5 = 0$$

and

$$b_2 \;=\; q + pb_3,$$
$$b_3 \;=\; qb_2 + pb_4,$$
$$b_4 \;=\; qb_3.$$

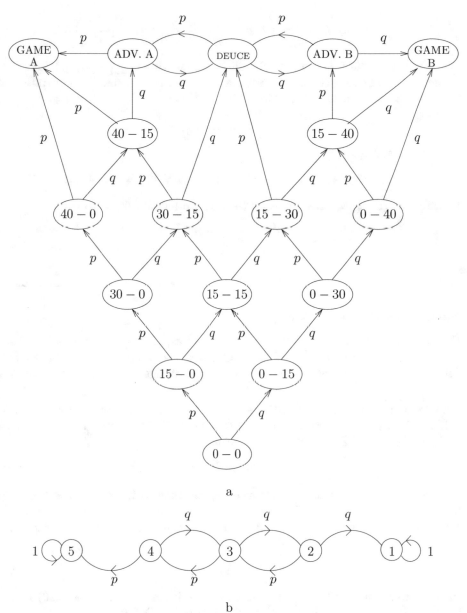

a

b

Figure 2.2.5. Flushing Meadows

For $p \neq q$,

$$(b_1, b_2, b_3, b_4, b_5) = \left(1, q\frac{1-pq}{1-2pq}, \frac{q^2}{1-2pq}, \frac{q^3}{1-2pq}, 0\right).$$

Starting from state $0-0$, the probability that B wins is $\sum_{i=1}^{5} p(i)b_i$, where $p(i)$ is the probability that the first upper state reached is i. A simple enumeration of the paths from $0-0$ to upper state 1 gives $p(1) = q^4 + q^3pq + q^2pq^2 + qpq^3 + pq^4$, that is,

$$p(1) = q^4(1+4p).$$

Similar calculations give

$$p(2) = 4q^3p^2, \; p(3) = 6p^2q^2, \; p(4) = 4p^3q^2, \; p(5) = p^4(1+4q).$$

Putting everything together, we find that the probability for B to win is

$$\frac{q^4(1+2p)(1+4p^2)}{1-2pq}.$$

Mean Time to Absorption

Examples 2.2.10, 2.2.12 and 2.2.13 deal with basically the same problem, that of computing the probability of reaching a state before another state is visited. First-step analysis can also be used to compute average times before absorption.

EXAMPLE 2.2.14: THE GAMBLER'S RUIN, TAKE 2. Continuation of Example 2.2.10. The average duration $m(i) = \mathrm{E}[T \mid X_0 = i]$ of the game when the initial fortune of player A is i satisfies the recurrence equation

$$m(i) = 1 + pm(i+1) + qm(i-1) \quad (1 \leq i \leq c-1). \tag{2.10}$$

Indeed, the coin will be tossed at least once, and then with probability p (resp., q) the fortune of player A will be $i+1$ (resp., $i-1$), and therefore $m(i+1)$ (resp., $m(i-1)$) more tosses will be needed on average before one of the players goes broke. The boundary conditions are

$$m(0) = 0 \text{ and } m(c) = 0. \tag{2.11}$$

In order to solve (2.10) with the boundary conditions (2.11), write (2.10) in the form $-1 = p(m(i+1) - m(i)) - q(m(i) - m(i-1))$. Defining

$$y_i = m(i) - m(i-1),$$

we have

$$-1 = py_{i+1} - qy_i \quad (1 \le i \le c - 1) \tag{2.12}$$

and

$$m(i) = y_1 + y_2 + \cdots + y_i. \tag{2.13}$$

We now solve (2.12) with $p = q = \frac{1}{2}$. From (2.12),

$$
\begin{aligned}
-1 &= \frac{1}{2}y_2 - \frac{1}{2}y_1, \\
-1 &= \frac{1}{2}y_3 - \frac{1}{2}y_2, \\
&\;\;\vdots \\
-1 &= \frac{1}{2}y_i - \frac{1}{2}y_{i-1},
\end{aligned}
$$

and therefore, summing up,

$$-(i - 1) = \frac{1}{2}y_i - \frac{1}{2}y_1 ,$$

that is,

$$y_i = y_1 - 2(i - 1) \quad (1 \le i \le c) .$$

Reporting this expression in (2.13), and observing that $y_1 = m(1)$, we obtain

$$m(i) = im(1) - 2[1 + 2 + \cdots + (i - 1)] = im(1) - i(i - 1).$$

The boundary condition $m(c) = 0$ gives $cm(1) = c(c - 1)$ and therefore, finally,

$$m(i) = i(c - i) .$$

Remark 2.2.15 First-step analysis leads to necessary conditions in the form of a system of linear equations. In the above examples, it turns out that the system in question has a unique solution, a situation that prevails when the state space is finite but that is not the general case with an infinite state space. The issue of uniqueness, and of which solution to choose in case of nonuniqueness, is addressed in Chapter 6, where absorption is studied in more detail.

2.3 Topology of the Transition Matrix

2.3.1 Communication

All the properties defined in the present section are *topological* in the sense that they concern only the *naked* transition graph (without the labels).

Definition 2.3.1 *State j is said to be* accessible *from state i if there exists an $M \geq 0$ such that $p_{ij}(M) > 0$. In particular, a state i is always accessible from itself since $p_{ii}(0) = 1$. States i and j are said to* communicate *if i is accessible from j and j is accessible from i, and this is denoted by $i \leftrightarrow j$.*

For $M \geq 1$, $p_{ij}(M) = \sum_{i_1,\ldots,i_{M-1}} p_{ii_1} \cdots p_{i_{M-1}j}$, and therefore $p_{ij}(M) > 0$ if and only if there exists at least one path $i, i_1, \ldots, i_{M-1}, j$ from i to j such that

$$p_{ii_1} p_{i_1 i_2} \cdots p_{i_{M-1}j} > 0$$

or, equivalently, if there is an oriented path from i to j in the transition graph G. Clearly,

$$i \leftrightarrow i \qquad \text{(reflexivity)},$$
$$i \leftrightarrow j \Rightarrow j \leftrightarrow i \qquad \text{(symmetry)},$$
$$i \leftrightarrow j, j \leftrightarrow k \Rightarrow i \leftrightarrow k \qquad \text{(transitivity)}.$$

Therefore, the communication relation ($\leftrightarrow$) is an equivalence relation, and it generates a partition of the state space E into disjoint equivalence classes called *communication classes*.

Definition 2.3.2 *A state i such that $p_{ii} = 1$ is called* closed. *More generally, a set C of states such that for all $i \in C$, $\sum_{j \in C} p_{ij} = 1$, is called* closed.

EXAMPLE 2.3.3: COMMUNICATION CLASSES. The transition graph of Figure 2.3.1 has 3 communication classes: $\{1, 2, 3, 4\}$, $\{5, 7, 8\}$, and $\{6\}$. State 6 is closed. The communication class $\{5, 7, 8\}$ is not closed, but the set $\{5, 6, 7, 8\}$ is.

Remark 2.3.4 Observe in this example that there may exist oriented edges linking two different communication classes E_k and E_ℓ. However, all the oriented edges between these two communication classes have the same orientation (all from E_k to E_ℓ or all from E_ℓ to E_k). Why?

Definition 2.3.5 *If there exists only one communication class, then the chain, its transition matrix and its transition graph are said to be* irreducible.

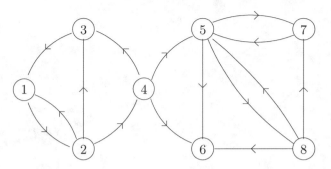

Figure 2.3.1. A transition graph with 3 communication classes

2.3.2 Period

Consider the random walk on $\mathbb{Z}$ (Example 2.2.5). Since $p \in (0,1)$, it is irreducible. Observe that $E = C_0 + C_1$, where C_0 and C_1, the set of even and odd relative integers respectively, have the following property. If you start from $i \in C_0$ (resp., C_1), then in one step you can go only to a state $j \in C_1$ (resp., C_0). The chain $\{X_n\}$ passes alternately from one cyclic class to the other. In this sense, the chain has a periodic behavior, corresponding to the period 2. More generally, we have the following.

Theorem 2.3.6 *For any* irreducible *Markov chain, one can find a* unique partition *of E into d classes C_0, C_1, ..., C_{d-1} such that for all k and all $i \in C_k$,*

$$\sum_{j \in C_{k+1}} p_{ij} = 1 \,,$$

where by convention $C_d = C_0$, and where d is maximal (that is, there is no other such partition C_0', $C_1', \ldots, C_{d'-1}'$ with $d' > d$).

Proof. A direct consequence of Theorem 2.3.10 below. □

The chain therefore moves from one class to the other at each transition, and this cyclically, as shown in Figure 2.3.2.

The number $d \geq 1$ is called the *period* of the chain (resp., of the transition matrix, of the transition graph). The classes $C_0, C_1, \ldots, C_{d-1}$ are called the *cyclic classes*.

EXAMPLE 2.3.7: CYCLES. The chain with the transition graph depicted in Figure 2.3.3 is irreducible and has period $d = 3$, with cyclic classes $C_0 = \{1,2\}, C_1 = \{4,7\}, C_3 = \{3,5,6\}$.

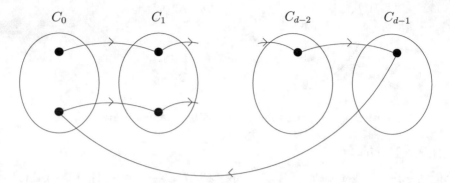

Figure 2.3.2. Cycles

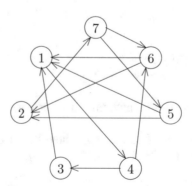

Figure 2.3.3. An irreducible transition graph with period 3

Consider now an irreducible chain of period d with cyclic classes $C_0, C_1, \ldots, C_d$. Renumbering the states of E if necessary, the transition matrix has the block structure below (where $d = 4$, to be explicit),

$$
\mathbf{P} = \begin{array}{c} \\ C_0 \\ C_1 \\ C_2 \\ C_3 \end{array} \begin{array}{cccc} C_0 & C_1 & C_2 & C_3 \\ \left(\begin{array}{cccc} 0 & A_0 & 0 & \\ 0 & 0 & A_1 & 0 \\ 0 & 0 & 0 & A_2 \\ A_3 & 0 & 0 & 0 \end{array} \right) \end{array},
$$

and therefore $\mathbf{P}^2$, $\mathbf{P}^3$, and $\mathbf{P}^4$ also have a block structure corresponding to C_0, C_1, C_2, C_3:

$$
\mathbf{P}^2 = \begin{pmatrix} 0 & 0 & B_0 & 0 \\ 0 & 0 & 0 & B_1 \\ B_2 & 0 & 0 & 0 \\ 0 & B_3 & 0 & 0 \end{pmatrix},
$$

$$
\mathbf{P}^3 = \begin{pmatrix} 0 & 0 & 0 & D_0 \\ D_1 & 0 & 0 & 0 \\ 0 & D_2 & 0 & 0 \\ 0 & 0 & D_3 & 0 \end{pmatrix},
$$

$$
\mathbf{P}^4 = \begin{pmatrix} E_0 & 0 & 0 & 0 \\ 0 & F_1 & 0 & 0 \\ 0 & 0 & E_2 & 0 \\ 0 & 0 & 0 & E_3 \end{pmatrix}.
$$

We observe two phenomena: block-shifting and the block-diagonal form of $\mathbf{P}^4$. This is, of course, general: $\mathbf{P}^d$ has a block-diagonal form corresponding to the cyclic classes $C_0, C_1, \ldots, C_{d-1}$:

$$
\mathbf{P}^d = \begin{array}{c} \\ C_0 \\ C_1 \\ \vdots \\ C_{d-1} \end{array} \begin{array}{cccc} C_0 & C_1 & \cdots & C_{d-1} \\ \left(\begin{array}{cccc} E_0 & & & 0 \\ & E_1 & & \\ & & \ddots & \\ 0 & & & E_{d-1} \end{array} \right) \end{array}. \tag{2.14}
$$

The d-step transition matrix $\mathbf{P}^d$ is also a stochastic matrix, and obviously the $\mathbf{P}$-cyclic classes $C_0, C_1, \ldots, C_{d-1}$ are all in different $\mathbf{P}^d$-communication classes, as the diagonal block structure shows.

The question is this: Is there, in C_0 for instance, more than one $\mathbf{P}^d$-communication class? The answer is no, and therefore matrix E_0 in (2.14) is an irreducible stochastic matrix. To see this, take two different states $i, j \in C_0$. Since they $\mathbf{P}$-communicate ($\mathbf{P}$ was assumed irreducible), there exist $m > 0$ and $n > 0$ such that $p_{ij}(m) > 0$ and $p_{ji}(n) > 0$. But since $\mathbf{P}$ has period d, necessarily $m = Md, n = Nd$ for some $M > 0$ and $N > 0$. Therefore, $p_{ij}(Md) > 0$ and $p_{ji}(Nd) > 0$. But $p_{ij}(Md)$ is the (i, j) term of $(\mathbf{P}^d)^M$, and similarly for $p_{ji}(Nd)$. We therefore have proven that i and j $\mathbf{P}^d$-communicate.

For an arbitrary transition matrix, not necessarily irreducible, the formal definition of *period* is the following.

Definition 2.3.8 *The period d_i of state $i \in E$ is, by definition,*

$$d_i = \gcd\{n \geq 1 \; ; \; p_{ii}(n) > 0\}, \tag{2.15}$$

with the convention $d_i = +\infty$ if there is no $n \geq 1$ with $p_{ii}(n) > 0$. If $d_i = 1$, the state i is called aperiodic.

Theorem 2.3.9 *If states i and j communicate they have the same period.*

Proof. As i and j communicate, there exist integers N and M such that $p_{ij}(M) > 0$ and $p_{ji}(N) > 0$. For any $k \geq 1$,

$$p_{ii}(M + nk + N) \geq p_{ij}(M)(p_{jj}(k))^n p_{ji}(N)$$

(indeed, the path $X_0 = i, X_M = j, X_{M+k} = j, \ldots, X_{M+nk} = j, X_{M+nk+N} = i$ is just one way of going from i to i in $M + nk + N$ steps).

Therefore, for any $k \geq 1$ such that $p_{jj}(k) > 0$, we have $p_{ii}(M + nk + N) > 0$ for all $n \geq 1$. Therefore, d_i divides $M + nk + N$ for all $n \geq 1$, and in particular, d_i divides k. We have therefore shown that d_i divides all k such that $p_{jj}(k) > 0$, and in particular, d_i divides d_j. By symmetry, d_j divides d_i, and therefore, finally, $d_i = d_j$. $\square$

We can therefore speak of the *period of a communication class* or of an irreducible chain.

The important result concerning periodicity is the following.

Theorem 2.3.10 *Let $\mathbf{P}$ be an irreducible stochastic matrix with period d. Then for all states i, j there exist $m \geq 0$ and $n_0 \geq 0$ (m and n_0 possibly depending on i, j) such that*

$$p_{ij}(m + nd) > 0 \quad (n \geq n_0). \tag{2.16}$$

Proof. First observe that it suffices to prove this for $i = j$. Indeed, there exists an m such that $p_{ij}(m) > 0$, because j is accessible from i, the chain being irreducible, and therefore, if for some $n_0 \geq 0$ we have $p_{jj}(nd) > 0$ for all $n \geq n_0$, then $p_{ij}(m + nd) > p_{ij}(m)p_{jj}(nd) > 0$ for all $n \geq n_0$. The gcd of the set $A = \{k \geq 1; p_{jj}(k) > 0\}$ is d, and A is closed under addition. The set A therefore contains all but a finite number of the positive multiples of d (see Subsection A.1.1). In other words, there exists an n_0 such that $n > n_0$ implies $p_{jj}(nd) > 0$. $\qquad\square$

2.4 Steady State

2.4.1 Stationarity

We now introduce the central notion of the stability theory of discrete-time HMCs.

Definition 2.4.1 *A probability distribution π satisfying*

$$\pi^T = \pi^T \mathbf{P} \tag{2.17}$$

is called a stationary distribution *of the transition matrix* $\mathbf{P}$*, or of the corresponding* HMC.

The *global balance equation* (2.17) says that for all states i,

$$\pi(i) = \sum_{j \in E} \pi(j)p_{ji} \, .$$

Iteration of (2.17) gives $\pi^T = \pi^T \mathbf{P}^n$ for all $n \geq 0$, and therefore, in view of (2.3), if the initial distribution $\nu = \pi$, then $\nu_n = \pi$ for all $n \geq 0$. Thus, if a chain is started with a stationary distribution, it keeps the same distribution forever. But there is more, because then,

$$P(X_n = i_0, X_{n+1} = i_1, \ldots, X_{n+k} = i_k) = P(X_n = i_0)p_{i_0 i_1} \ldots p_{i_{k-1} i_k}$$
$$= \pi(i_0)p_{i_0 i_1} \ldots p_{i_{k-1} i_k}$$

does not depend on n. In this sense the chain is *stationary*. One also says that the chain is in a *stationary regime*, or in *equilibrium*, or in *steady state*. In summary:

Theorem 2.4.2 *A chain starting with a stationary distribution is stationary.*

Remark 2.4.3 The balance equation $\pi^T \mathbf{P} = \pi^T$, together with the requirement that π be a probability vector, that is, $\pi^T \mathbf{1} = 1$ (where $\mathbf{1}$ is a column vector with all its entries equal to 1), constitute when E is finite, $|E| + 1$ equations for $|E|$ unknown variables. One of the $|E|$ equations in $\pi^T \mathbf{P} = \pi^T$ is superfluous given the constraint $\pi^T \mathbf{1} = 1$. Indeed, summing up all equalities of $\pi^T \mathbf{P} = \pi^T$ yields the equality $\pi^T \mathbf{P} \mathbf{1} = \pi^T \mathbf{1}$, that is, $\pi^T \mathbf{1} = 1$.

EXAMPLE 2.4.4: THE TWO-STATE MARKOV CHAIN. Take $E = \{1, 2\}$ and define the transition matrix

$$
\mathbf{P} = \begin{matrix} & 1 & 2 \\ \begin{matrix} 1 \\ 2 \end{matrix} & \begin{pmatrix} 1 - \alpha & \alpha \\ \beta & 1 - \beta \end{pmatrix} \end{matrix},
$$

where $\alpha, \beta \in (0, 1)$. The global balance equations are

$$
\begin{aligned}
\pi(1) &= \pi(1)(1 - \alpha) + \pi(2)\beta, \\
\pi(2) &= \pi(1)\alpha + \pi(2)(1 - \beta).
\end{aligned}
$$

This is a dependent system which reduces to the single equation $\pi(1)\alpha = \pi(2)\beta$, to which must be added $\pi(1) + \pi(2) = 1$ expressing that π is a probability vector. We obtain

$$
\pi(1) = \frac{\beta}{\alpha + \beta}, \quad \pi(2) = \frac{\alpha}{\alpha + \beta}.
$$

EXAMPLE 2.4.5: THE URN OF EHRENFEST, TAKE 2. Continuation of Example 2.2.9. The global balance equations are, for $i \in [1, N - 1]$,

$$
\pi(i) = \pi(i - 1)\left(1 - \frac{i - 1}{N}\right) + \pi(i + 1)\frac{i + 1}{N}
$$

and, for the boundary states,

$$
\pi(0) = \pi(1)\frac{1}{N}, \quad \pi(N) = \pi(N - 1)\frac{1}{N}.
$$

Leaving $\pi(0)$ undetermined, one can solve the balance equations for $i = 0, 1, \ldots, N$ successively, to obtain

$$
\pi(i) = \pi(0)\binom{N}{i}.
$$

The value of $\pi(0)$ is then determined by writing that π is a probability vector:

$$1 = \sum_{i=0}^{N} \pi(i) = \pi(0) \sum_{i=0}^{N} \binom{N}{i} = \pi(0)2^N .$$

This gives for π the binomial distribution of size N and parameter $\frac{1}{2}$:

$$\pi(i) = \frac{1}{2^N} \binom{N}{i} .$$

This is the distribution one would obtain by placing independently each particle in the compartments, with probability $\frac{1}{2}$ for each compartment.

EXAMPLE 2.4.6: THE SYMMETRIC RANDOM WALK. A symmetric random walk on $\mathbb{Z}$ cannot have a stationary distribution. Indeed, the solution of the balance equation

$$\pi(i) = \frac{1}{2}\pi(i-1) + \frac{1}{2}\pi(i+1)$$

for $i \geq 0$, with initial data $\pi(0)$ and $\pi(1)$, is

$$\pi(i) = \pi(0) + (\pi(1) - \pi(0))\, i .$$

Since $\pi(i) \in [0,1]$, necessarily $\pi(1) - \pi(0) = 0$. Therefore, $\pi(i)$ is a constant, necessarily 0 because the total mass of π is finite. Thus for all $i \geq 0$, and therefore, by translation invariance, for all i, $\pi(i) = 0$, a contradiction if we want π to be a probability distribution.

EXAMPLE 2.4.7: STATIONARY DISTRIBUTIONS MAY BE MANY. Take the identity as transition matrix. Then any probability distribution on the state space is a stationary distribution.

EXAMPLE 2.4.8: THE LAZY MARKOV CHAIN. Let $\mathbf{P}$ be the transition matrix of an HMC with state space E. The matrix

$$\mathbf{Q} := \frac{I + \mathbf{P}}{2}$$

is clearly a transition matrix, that of an HMC called the lazy version of the original one. In the lazy version, a move is decided after tossing a fair coin. If heads, the lazy traveler stays still, otherwise, he moves according to $\mathbf{P}$. Clearly, a stationary distribution of $\mathbf{P}$ is also a stationary distribution of $\mathbf{Q}$.

Both chains are simultaneously irreducible or not irreducible. However, in the irreducible case, the lazy chain is always aperiodic (since $q_{ii} > 0$) whereas the original chain may be periodic.

___

Recurrence equations can be used to obtain the stationary distribution when the latter exists and is unique. Generating functions sometimes usefully exploit the dynamics.

___

EXAMPLE 2.4.9: THE REPAIR SHOP, TAKE 2. Continuation of Example 2.2.6. For any complex number z with modulus not larger than 1, it follows from the recurrence equation (2.6) that

$$
\begin{aligned}
z^{X_{n+1}+1} &= \left(z^{(X_n-1)^++1} \right) z^{Z_{n+1}} \\
&= \left(z^{X_n} 1_{\{X_n>0\}} + z 1_{\{X_n=0\}} \right) z^{Z_{n+1}} \\
&= \left(z^{X_n} - 1_{\{X_n=0\}} + z 1_{\{X_n=0\}} \right) z^{Z_{n+1}},
\end{aligned}
$$

and therefore

$$
z z^{X_{n+1}} - z^{X_n} z^{Z_{n+1}} = (z-1) 1_{\{X_n=0\}} z^{Z_{n+1}}.
$$

From the independence of X_n and Z_{n+1}, $\mathrm{E}[z^{X_n} z^{Z_{n+1}}] = \mathrm{E}[z^{X_n}] g_Z(z)$, where $g_Z(z)$ is the generating function of Z_{n+1}, and $\mathrm{E}[1_{\{X_n=0\}} z^{Z_{n+1}}] = \pi(0) g_Z(z)$, where $\pi(0) = \mathrm{P}(X_n = 0)$. Therefore,

$$
z \mathrm{E}[z^{X_{n+1}}] - g_Z(z) \mathrm{E}[z^{X_n}] = (z-1) \pi(0) g_Z(z).
$$

But in steady state, $\mathrm{E}[z^{X_{n+1}}] = \mathrm{E}[z^{X_n}] = g_X(z)$, and therefore

$$
g_X(z)(z - g_Z(z)) = \pi(0)(z-1) g_Z(z). \tag{$\star$}
$$

This gives the generating function $g_X(z) = \sum_{i=0}^{\infty} \pi(i) z^i$, as long as $\pi(0)$ is available. To obtain $\pi(0)$, differentiate $(\star)$:

$$
g_X'(z)(z - g_Z(z)) + g_X(z)(1 - g_Z'(z)) = \pi(0)(g_Z(z) + (z-1)g_Z'(z)),
$$

and let $z = 1$, to obtain, taking into account the equalities $g_X(1) = g_Z(1) = 1$ and $g_Z'(1) = \mathrm{E}[Z]$,

$$
\pi(0) = 1 - \mathrm{E}[Z]. \tag{2.18}
$$

Since $\pi(0)$ must be non-negative, this immediately gives the necessary condition $\mathrm{E}[Z] \le 1$. Actually, one must have, if the trivial case $Z_{n+1} \equiv 1$ is excluded,

$$
\mathrm{E}[Z] < 1. \tag{2.19}
$$

Indeed, if $E[Z] = 1$, implying $\pi(0) = 0$, it follows from $(\star)$ that

$$g_X(x)(x - g_Z(x)) = 0$$

for all $x \in [0, 1]$. But excluding the case $Z_{n+1} \equiv 1$ (that is, $g_Z(x) \equiv x$), the equation $x - g_Z(x) = 0$ has only $x = 1$ for a solution when $g'_Z(1) = E[Z] \leq 1$ (Theorem (1.2.8)). Therefore, $g_X(x) \equiv 0$ for $x \in [0, 1)$, and consequently $g_X(z) \equiv 0$ on $\{|z| < 1\}$. This leads to a contradiction, since the generating function of an integer-valued random variable cannot be identically null.

We shall prove later that $E[Z] < 1$ is also a sufficient condition for the existence of a steady state. For the time being, we learn from $(\star)$ and (2.18) that, if the stationary distribution exists, then its generating function is given by the formula

$$\sum_{i=0}^{\infty} \pi(i) z^i = (1 - E[Z]) \frac{(z - 1) g_Z(z)}{z - g_Z(z)}. \tag{2.20}$$

EXAMPLE 2.4.10: BIRTH AND DEATH HMC WITH TWO REFLECTING BARRIERS. The Ehrenfest model is a special case of a birth-and-death Markov chain with reflecting barriers at 0 and N. The generalization is an HMC with state space $E = \{0, 1, \ldots, N\}$ and transition matrix

$$\mathbf{P} = \begin{pmatrix} 0 & 1 & & & & & & \\ q_1 & r_1 & p_1 & & & & & \\ & q_2 & r_2 & p_2 & & & & \\ & & \ddots & & & & & \\ & & & q_i & r_i & p_i & & \\ & & & & \ddots & \ddots & \ddots & \\ & & & & & q_{N-1} & r_{N-1} & p_{N-1} \\ & & & & & & 1 & 0 \end{pmatrix},$$

where $p_i > 0$, $q_i > 0$, and $p_i + q_i + r_i = 1$ for all states i ($1 \leq i \leq N - 1$). The global balance equations for the states $i \in [1, N - 1]$ are

$$\pi(i) = p_{i-1}\pi(i - 1) + r_i\pi(i) + q_{i+1}\pi(i + 1) \quad (1 \leq i \leq N - 1),$$

and for the boundary states,

$$\pi(0) = \pi(1)q_1, \ \pi(N) = \pi(N - 1)p_{N-1}.$$

Of course, π must be a probability, whence

$$\sum_{n=0}^{N} \pi(i) = 1 \,.$$

Writing $r_i = 1 - p_i + q_i$ and regrouping terms gives

$$\pi(i+1)q_{i+1} - \pi(i)p_i = \pi(i)q_i - \pi(i-1)p_{i-1} \quad (1 \le i \le N - 2)$$

and

$$\begin{aligned}
\pi(1)q_1 - \pi(0) &= 0, \\
\pi(2)q_2 - \pi(1)p_1 &= \pi(1)q_1 - \pi(0).
\end{aligned}$$

Therefore, $\pi(1)q_1 = \pi(0)$ and

$$\pi(i)q_i = \pi(i-1)p_{i-1} \quad (1 \le i \le N - 2)\,.$$

This gives

$$\pi(1) = \pi(0)\frac{1}{q_1},$$

and

$$\pi(i) = \pi(0)\frac{p_1 p_2 \cdots p_{i-1}}{q_1 q_2 \cdots q_i} \quad (2 \le i \le N)\,. \tag{2.21}$$

The unknown $\pi(0)$ is obtained by $\sum_{i=0}^{N} \pi(i) = 1$, that is,

$$\pi(0)\left\{ 1 + \frac{1}{q_1} + \frac{p_1}{q_1 q_2} + \cdots + \frac{p_1 p_2 \cdots p_{N-1}}{q_1 q_2 \cdots q_{N-1} q_N} \right\} = 1\,. \tag{2.22}$$

EXAMPLE 2.4.11: BIRTH AND DEATH HMC WITH ONE REFLECTING BARRIER. The model is the same as in the previous example, except that the state space is $E = \mathbb{N}$, and therefore the upper barrier is at infinity. The same computations as above lead to the expression (2.21) for the general solution of $\pi^T \mathbf{P} = \pi^T$, which depends on the initial condition $\pi(0)$. For this solution to be a probability, we must have $\pi(0) > 0$. Also, writing $\sum_{i=1}^{\infty} \pi(i) = 1$,

$$\pi(0)\left\{ 1 + \frac{1}{q_1} + \sum_{j=1}^{\infty} \frac{p_1 p_2 \cdots p_j}{q_1 q_2 \cdots q_{j+1}} \right\} = 1\,. \tag{2.23}$$

Thus a stationary distribution exists if and only if

$$\sum_{j=1}^{\infty} \frac{p_1 p_2 \cdots p_j}{q_1 q_2 \cdots q_{j+1}} < \infty. \tag{2.24}$$

In this case $\pi(i)$ is given by the expressions in (2.21), where $\pi(0)$ is determined by (2.23).

2.4.2 Time Reversal

Reversed Chain

The notions of time-reversal and time-reversibility are very productive, in particular in the theory of Markov chains, and especially in Monte Carlo simulation (Chapter 11).

Let $\{X_n\}_{n \geq 0}$ be an HMC with transition matrix $\mathbf{P}$ and admitting a stationary distribution π such that

$$\pi(i) > 0 \tag{2.25}$$

for all states i. Define the matrix $\mathbf{Q}$, indexed by E, by

$$\pi(i) q_{ij} - \pi(j) p_{ji}. \tag{2.26}$$

This matrix is stochastic, since

$$\sum_{j \in E} q_{ij} - \sum_{j \in E} \frac{\pi(j)}{\pi(i)} p_{ji} = \frac{1}{\pi(i)} \sum_{j \in E} \pi(j) p_{ji} = \frac{\pi(i)}{\pi(i)} = 1,$$

where the third equality uses the balance equations. Its interpretation is the following: Suppose that the initial distribution of $\{X_n\}$ is π, in which case for all $n \geq 0$, all $i \in E$,

$$P(X_n = i) = \pi(i). \tag{2.27}$$

Then, from Bayes' retrodiction formula,

$$P(X_n = j \mid X_{n+1} = i) = \frac{P(X_{n+1} = i \mid X_n = j) P(X_n = j)}{P(X_{n+1} = i)},$$

that is, in view of (2.26) and (2.27),

$$P(X_n = j \mid X_{n+1} = i) = q_{ij}. \tag{2.28}$$

We see that $\mathbf{Q}$ is the transition matrix of the initial chain when time is reversed.

The following is a very simple observation that will be promoted to the rank of a theorem in view of its usefulness and also for the sake of easy reference.

Theorem 2.4.12 *Let* **P** *be a stochastic matrix indexed by a countable set E, and let π be a probability distribution on E. Let* **Q** *be a stochastic matrix indexed by E such that for all $i, j \in E$,*

$$\pi(i)q_{ij} = \pi(j)p_{ji}. \tag{2.29}$$

Then π is a stationary distribution of **P**.

Proof. For fixed $i \in E$, sum equalities (2.29) with respect to $j \in E$ to obtain

$$\sum_{j \in E} \pi(i)q_{ij} = \sum_{j \in E} \pi(j)p_{ji}.$$

But the left-hand side is equal to $\pi(i) \sum_{j \in E} q_{ij} = \pi(i)$, and therefore, for all $i \in E$,

$$\pi(i) = \sum_{j \in E} \pi(j)p_{ji}.$$

$\square$

EXAMPLE 2.4.13: EXTENSION OF A STATIONARY CHAIN TO NEGATIVE TIMES. Time reversal can also be used to extend to negative times a chain $\{X_n\}_{n \geq 0}$ in steady state corresponding to a stationary distribution π such that $\pi(i) > 0$ for all $i \in E$ (Exercise 2.6.28).

Time Reversibility

Definition 2.4.14 *One calls* reversible *a stationary Markov chain with initial distribution π (a stationary distribution) if*

$$\pi(i)p_{ij} = \pi(j)p_{ji} \quad (i, j \in E). \tag{2.30}$$

In this case, $q_{ij} = p_{ij}$, and therefore the chain and the time-reversed chain are statistically the same, since the distribution of a homogeneous Markov chain is entirely determined by its initial distribution and its transition matrix. Equations (2.30) are called the *detailed balance equations*. The following is an immediate corollary of Theorem 2.4.12.

Corollary 2.4.15 *Let* **P** *be a transition matrix on the countable state space E, and let π be some probability distribution on E. If for all $i, j \in E$, the detailed balance equations (2.30) are satisfied, then π is a stationary distribution of* **P**.

EXAMPLE 2.4.16: THE URN OF EHRENFEST, TAKE 3. Continuation of Examples 2.2.9 and 2.4.5. Recall that we obtained the expression

$$\pi(i) = \frac{1}{2^N}\binom{N}{i}$$

for the stationary distribution. Checking the detailed balance equations

$$\pi(i)p_{i,i+1} = \pi(i+1)p_{i+1,i}$$

is immediate.

EXAMPLE 2.4.17: REVERSIBILITY OF THE BIRTH AND DEATH HMC. One verifies that for both Examples 2.4.10 and 2.4.11, when the stationary distribution π exists, the detailed balance equations $\pi(i)p_i = \pi(i+1)q_{i+1}$ hold for all $i \in E$.

Random Walk on a Graph

Definition 2.4.18 *A (finite non-oriented) graph $(V, \mathcal{E})$ consists of a finite collection V of vertices v and of a collection $\mathcal{E}$ of unordered pairs of distinct vertices, $\langle u, v \rangle$, called the edges. If $\langle u, v \rangle \in \mathcal{E}$, then u and v are called neighbors, and this is also denoted by $u \sim v$. The degree of vertex $v \in V$ is the number of edges stemming from it.*

Consider a finite non-oriented graph and call E the set of vertices of this graph. Denote by d_i the degree of vertex i. Transform this graph into an oriented graph by splitting each edge into two oriented edges of opposite directions, and make it a transition graph by associating to the oriented edge from i to j the transition probability $\frac{1}{d_i}$ (see Figure 2.4.1).

This is an abstract form of the motion of a mouse in a maze (see Example 2.2.12).

It will be assumed, as is the case in Figure 2.4.1, that $d_i > 0$ for all states i. A stationary distribution (in fact, *the* stationary distribution, as we shall see later, in Chapter 8) is given by

$$\pi(i) = \frac{d_i}{\sum_{j \in E} d_j}.$$

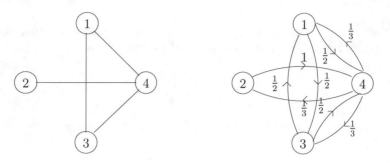

Figure 2.4.1. A random walk on a graph

For this, we can use Corollary 2.4.15, making the insider's guess that the chain is reversible. We just have to check that

$$\pi(i)\frac{1}{d_i} = \pi(j)\frac{1}{d_j}.$$

Hence we have obtained the stationary distribution and proved the reversibility of the chain.

EXAMPLE 2.4.19: THE LAZY WALK ON THE HYPERCUBE, TAKE 1. The N-hypercube is a graph whose set of vertices is $V = \{0, 1\}^N$ and its set of edges $\mathcal{E}$ consists of the pairs of vertices $\langle x, y \rangle$ that are adjacent in the sense that there exists an index i ($1 \le i \le N$) such that $y = x^{(i)} := (x_1, \ldots, x_{i-1}, 1 - x_i, x_{i+1}, \ldots, x_N)$. The (pure) random walk on the hypercube is the HMC describing the motion of a particle along the edges at random. That is to say, if the position at a given time is x, the next position is $x^{(i)}$ where i is chosen uniformly at random among $\{1, 2, \ldots, N\}$ independently of all that happened before.

To discard periodicity, we consider the *lazy random walk*, for which the decision to move depends on the result of a fair coin toss. More precisely, $p_{x,x} = \frac{1}{2}$ and if y is adjacent to x, $p_{xy} = \frac{1}{2N}$. This modification does not change the stationary distribution, which is the uniform distribution.

We may always describe, distributionwise, the HMC $\{X_n\}_{n \ge 0}$ in the manner of Theorem 2.2.3, that is $X_{n+1} = f(X_n, Z_{n+1})$ where $\{Z_n\}_{n \ge 1}$ is an IID sequence of random variables uniformly distributed on $\{1, \ldots, N\}$ independent of the initial state X_0: take $Z_n = (U_n, B_n)$ where the sequence $\{(U_n, B_n)\}_{n \ge 1}$ is IID and uniformly distributed on $\{1, 2, \ldots, N\} \times \{0, 1\}$. The position at time $n + 1$ is that of X_n except that the bit in position U_{n+1} is replaced by B_{n+1}.

EXAMPLE 2.4.20: RANDOM WALK ON A GROUP. Let G be a finite associative group with respect to the operation $*$ and let the inverse of $a \in G$ be denoted by a^{-1} and the identity by id. Let μ be a probability distribution on G. Let X_0 be an arbitrary random element of G, and let $\{Z_n\}_{n \geq 1}$ be a sequence of IID random elements of G, independent of X_0, with common distribution μ. The recurrence equation

$$X_{n+1} = Z_{n+1} * X_n \tag{2.31}$$

defines according to Theorem 2.2.1 an HMC whose transition probabilities are

$$P_{g,h*g} = \mu(h)$$

for all $g, h \in G$.

For $H \subset G$, denote by $\langle H \rangle$ the smallest subgroup of G containing H. Recall that $\langle H \rangle$ consists of all elements of the type $b_r * b_{r-1} * \cdots * b_1$ where the b_i's are elements of H or inverses of elements of H. Let $S = \{g \in G;\ \mu(g) > 0\}$.

Theorem 2.4.21 *(a) The random walk is irreducible if and only if S generates G, that is, $\langle S \rangle = G$.*

(b) The uniform distribution U on G is a stationary distribution of the chain.

Proof. (a) Assume irreducibility. Let $a \in G$. There exists an $r > 0$ such that $p_{e,a}(r) > 0$, that is, there exists a sequence $s_1, \ldots, s_r$ of S such that $a = s_r * \cdots * s_1$. Therefore $a \in \langle S \rangle$. Conversely, suppose that S generates G. Let $a, b \in G$. The element $b * a^{-1}$ is therefore of the type $u_r * u_{r-1} * \cdots * u_1$ where the u_i's are elements of S or inverses of elements of S. Now, every element of G is of finite order, that is, can be written as a power of some element of G. Therefore $b * a^{-1}$ can be written as $b * a^{-1} = s_r * \cdots * s_1$ where the s_i's are in S. In particular, $p_{a,b}(r) > 0$.

(b) In fact

$$\sum_{g \in G} U(g) p_{g,f} = \frac{1}{|G|} \sum_{h \in G} p_{h^{-1}*f,f} = \frac{1}{|G|} \sum_{h \in G} \mu(h) = \frac{1}{|G|}.$$

$\square$

The probability distribution μ on G is called symmetric iff $\mu(g) = \mu(g^{-1})$ for all $g \in G$. If this is the case, then the chain is reversible. We just have to check the detailed balance equations

$$U(g) p_{g,h} = U(h) p_{h,g}$$

that is

$$\frac{1}{|G|}\mu(h * g^{-1}) = \frac{1}{|G|}\mu(g * h^{-1}),$$

which is true because of the assumed symmetry of μ.

2.5   Regeneration

2.5.1   The Strong Markov Property

We first introduce a fundamental notion of the theory of stochastic processes.

Definition 2.5.1 *A stopping time with respect to a stochastic process $\{X_n\}_{n\geq 0}$ is, by definition, a random variable τ taking its values in $\mathbb{N} \cup \{+\infty\}$ and such that for all integers $m \geq 0$, the event $\{\tau = m\}$ can be expressed in terms of $X_0, X_1, \ldots, X_m$.*

The latter property is symbolized by the notation

$$\{\tau = m\} \in X_0^m. \tag{2.32}$$

When the state space is countable, (2.32) means that

$$1_{\{\tau=m\}} = \psi_m(X_0, \ldots, X_m),$$

for some function ψ_m with values in $\{0, 1\}$.

EXAMPLE 2.5.2: RETURN TIMES. In the theory of Markov chains, a typical and most important stopping time is the *return time* to state $i \in E$,

$$T_i = \inf\{n \geq 1; \ X_n = i\}, \tag{2.33}$$

where $T_i = \infty$ if $X_n \neq i$ for all $n \geq 1$. It is indeed a stopping time. Do a direct proof, or wait for Example 2.5.7.

Note that $T_i \geq 1$, and in particular, $X_0 = i$ does *not* imply $T_i = 0$. This is why T_i is called the *return* time to i, and not the *hitting time* of i. The latter is $S_i = T_i$ if $X_0 \neq i$, and $S_i = 0$ if $X_0 = i$. It is also a stopping time.

EXAMPLE 2.5.3: DETERMINISTIC TIMES. A constant time is a stopping time.

EXAMPLE 2.5.4: DELAYED STOPPING TIMES. If τ is a stopping time and n_0 a non-negative deterministic time, then $\tau + n_0$ is a stopping time. Indeed, $\{\tau + n_0 = m\} \equiv \{\tau = m - n_0\}$ is expressible in terms of $X_0, X_1, \ldots, X_{m-n_0}$.

Remark 2.5.5 For a given stopping time τ, one can decide whether $\tau = m$ just by observing $X_0, X_1, \ldots, X_m$. This is why stopping times are said to be *non-anticipative*.

EXAMPLE 2.5.6: COUNTEREXAMPLE. The random time

$$\tau = \inf\{n \geq 0;\ X_{n+1} = i\}\,,$$

where $\tau = \infty$ if $X_{n+1} \neq i$ for all $n \geq 0$, is anticipative because $\{\tau = m\} = \{X_1 \neq i, \ldots, X_m \neq i, X_{m+1} = i\}$ for all $m \geq 0$. Knowledge of this random time provides information about the value of the process just after it. It is *not* a stopping time.

EXAMPLE 2.5.7: SUCCESSIVE RETURNS. Let $\tau_1 = T_i, \tau_2, \ldots$ be the successive return times to state i. If there are only r returns to state i, let $\tau_{r+1} = \tau_{r+2} = \cdots = \infty$. These random times are stopping times with respect to $\{X_n\}_{n \geq 0}$, since for any $m \geq 1$,

$$\{\tau_k = m\} = \left\{ \sum_{n=1}^{m} 1_{\{X_n = i\}} = k, X_m = i \right\}$$

is indeed expressible in terms of $X_0, \ldots, X_m$.

Let τ be a random time taking its values in $\mathbb{N} \cup \{+\infty\}$, and let $\{X_n\}_{n \geq 0}$ be a stochastic process with values in the countable set E. In order to define X_τ when $\tau = \infty$, one must decide how to define X_∞. This is done by taking some arbitrary element Δ not in E, and setting

$$X_\infty = \Delta.$$

By definition, the *process* $\{X_n\}$ *after* τ is the stochastic sequence

$$\{X_{n+\tau}\}_{n > 0}\,.$$

The *process* $\{X_n\}$ *before* τ, or the *process* $\{X_n\}$ *stopped at time* τ, is the stochastic sequence

$$\{X_{n \wedge \tau}\}_{n \geq 0}\,,$$

which freezes at time τ at the value X_τ.

The main result of the present section roughly says that the Markov property, that is, the independence of past and future given the present state, extends to the situation where the present time is a stopping time. More precisely:

Theorem 2.5.8 *Let $\{X_n\}_{n\geq 0}$ be an* HMC *with countable state space E and transition matrix* **P**. *Let τ be a stopping time with respect to this chain. Then for any state $i \in E$, given that $X_\tau = i$ (in particular, $\tau < \infty$, since $i \neq \Delta$),*

   (α) the process after τ and the process before τ are independent, and

   (β) the process after τ is an HMC *with transition matrix* **P**.

Proof.   (α) One must show that for all times $k \geq 1, n \geq 0$, and all states $i_0, \dots, i_n, i, j_1, \dots, j_k$,

$$P(X_{\tau+1} = j_1, \dots, X_{\tau+k} = j_k \mid X_\tau = i, X_{\tau \wedge 0} = i_0, \dots, X_{\tau \wedge n} = i_n)$$

$$= P(X_{\tau+1} = j_1, \dots, X_{\tau+k} = j_k \mid X_\tau = i). \qquad (2.34)$$

We shall prove a simplified version of the above equality, namely

$$P(X_{\tau+k} = j \mid X_\tau = i, X_{\tau \wedge n} = i_n) = P(X_{\tau+k} = j \mid X_\tau = i). \qquad (2.35)$$

The general case is obtained by the same arguments. The left-hand side of the above equality is equal to

$$\frac{P(X_{\tau+k} = j, X_\tau = i, X_{\tau \wedge n} = i_n)}{P(X_\tau = i, X_{\tau \wedge n} = i_n)} .$$

The numerator of the above expression can be developed as

$$\sum_{r \geq 0} P(\tau = r, X_{r+k} = j, X_r = i, X_{r \wedge n} = i_n). \qquad (2.36)$$

But

$$P(\tau = r, X_{r+k} = j, X_r = i, X_{r \wedge n} = i_n)$$
$$= P(X_{r+k} = j \mid X_r = i, \ X_{r \wedge n} = i_n, \tau = r) \, P(\tau = r, X_{r \wedge n} = i_n, X_r = i),$$

and since $r \wedge n \leq r$ and $\{\tau = r\} \in X_0^r$, the event $B = \{X_{r \wedge n} = i_n, \tau = r\}$ is in X_0^r. Therefore, by the Markov property, $P(X_{r+k} = j \mid X_r = i, X_{r \wedge n} = i_n, \tau = r) = P(X_{r+k} = j \mid X_r = i) = p_{ij}(k)$. Finally, expression (2.36) reduces to

$$\sum_{r \geq 0} p_{ij}(k) P(\tau = r, X_{r \wedge n} = i_n, X_r = i) = p_{ij}(k) P(X_{\tau=i}, X_{\tau \wedge n} = i_n).$$

Therefore, the left-hand side of (2.35) is just $p_{ij}(k)$. Similar computations show that the right-hand side of (2.35) is also $p_{ij}(k)$, so that (α) is proven.

(β) We must show that for all states $i, j, k, i_{n-1}, \ldots, i_1$,

$$P(X_{\tau+n+1} = k \mid X_{\tau+n} = j, X_{\tau+n-1} = i_{n-1}, \ldots, X_\tau = i)$$
$$= P(X_{\tau+n+1} = k \mid X_{\tau+n} = j) = p_{jk}.$$

But the first equality follows from the fact proven in (α) that for the stopping time $\tau' = \tau + n$, the processes before and after τ' are independent given $X_{\tau'} = j$. The second equality is obtained by the same calculations as in the proof of (α). $\square$

2.5.2 Regenerative Cycles

Let $N_i := \sum_{n \geq 1} 1_{\{X_n = i\}}$ denote the number of visits to state i strictly after time 0.

Theorem 2.5.9 *The distribution of N_i given $X_0 = j$ is*

$$P_j(N_i = r) = \begin{cases} f_{ji} f_{ii}^{r-1}(1 - f_{ii}) \text{ for } r \geq 1 \\ 1 - f_{ji} \text{ for } r = 0, \end{cases} \tag{2.37}$$

where $f_{ji} := P_j(T_i < \infty)$ and T_i is the return time to i.

Proof. For $r = 0$, this is just the definition of f_{ji}. Now let $r \geq 1$, and assume (2.37) to be true for all k ($1 \leq k \leq r$). In particular,

$$P_j(N_i > r) = 1 - \sum_{k=0}^{r} P_j(N_i = k) = f_{ji} f_{ii}^r.$$

Denoting by τ_r the rth return time to state i,

$$\begin{aligned} P_j(N_i = r + 1) &= P_j(N_i = r + 1, X_{\tau_{r+1}} = i) \\ &= P_j(\tau_{r+2} - \tau_{r+1} = \infty, X_{\tau_{r+1}} = i) \\ &= P_j(\tau_{r+2} - \tau_{r+1} = \infty \mid X_{\tau_{r+1}} = i) P_j(X_{\tau_{r+1}} = i). \end{aligned}$$

But

$$\begin{aligned} P_j(\tau_{r+2} - \tau_{r+1} = \infty \mid X_{\tau_{r+1}} = i) &= P(\tau_{r+2} - \tau_{r+1} = \infty \mid X_{\tau_{r+1}} = i, X_0 = j) \\ &= P(\tau_{r+2} - \tau_{r+1} = \infty \mid X_{\tau_{r+1}} = i) \end{aligned}$$

by the strong Markov property. Since $\tau_{r+2} - \tau_{r+1}$ is the return time to i of the process after τ_{r+1}, the strong Markov property gives

$$P(\tau_{r+2} - \tau_{r+1} = \infty \mid X_{\tau_{r+1}} = i) = P(T_i = \infty \mid X_0 = i).$$

Also,
$$P_j(X_{\tau_{r+1}} = i) = P_j(N_i > r)$$
(if $N_i \le r$, then $X_{\tau_{r+1}} = X_\infty = \Delta \notin E$). Therefore,
$$P_j(N_i = r + 1) = P_i(T_i = \infty)P_j(N_i > r) = (1 - f_{ii})f_{ji}f_{ii}^r.$$

The result is therefore proven by induction.   □

The distribution of N_i given $X_0 = j$ and given $N_i \ge 1$ is geometric (exercise). This has two main consequences. Firstly,
$$P_i(T_i < \infty) = 1 \Leftrightarrow P_i(N_i = \infty) = 1.$$

In words: if starting from i you almost surely return to i, then you will visit i infinitely often. Secondly, we have
$$E_i[N_i] = \sum_{r=1}^\infty rP_i(N_i = r) = \sum_{r-1}^\infty rf_{ii}^r(1 - f_{ii}) = \frac{f_{ii}}{1 - f_{ii}}.$$

In particular,
$$P_i(T_i < \infty) < 1 \Leftrightarrow E_i[N_i] < \infty.$$
We collect these results for future reference.

Theorem 2.5.10 *For any state $i \in E$,*
$$P_i(T_i < \infty) = 1 \Leftrightarrow P_i(N_i = \infty) = 1$$

and
$$P_i(T_i < \infty) < 1 \Leftrightarrow P_i(N_i = \infty) = 0 \Leftrightarrow E_i[N_i] < \infty.$$

In particular, the event $\{N_i = \infty\}$ has P_i-probability 0 or 1.

Consider a Markov chain with a state conventionally denoted by 0 such that $P_0(T_0 < \infty) = 1$. In view of the last theorem, the chain starting from state 0 will return infinitely often to this state. Let $\tau_1 = T_0, \tau_2, \ldots$ be the successive return times to 0, and set $\tau_0 \equiv 0$.

By the strong Markov property, for any $k \ge 1$, the process after τ_k is independent of the process before τ_k (observe that condition $X_{\tau_k} = 0$ is always satisfied), and the process after τ_k is a Markov chain with the same transition matrix as the original chain, and with initial state 0, by construction. Therefore, we have the following

Theorem 2.5.11 *Let $\{X_n\}_{n\geq 0}$ be an* HMC *with an initial state 0 that is almost surely visited infinitely often. Denoting by $\tau_0 = 0, \tau_1, \tau_2, \ldots$ the successive times of visit to 0, the pieces of trajectory*

$$\{X_{\tau_k}, X_{\tau_k+1}, \ldots, X_{\tau_{k+1}-1}\}, k \geq 0,$$

are independent and identically distributed.

Such pieces are called the *regenerative cycles* of the chain between visits to state 0. Each random time τ_k is a *regeneration time*, in the sense that $\{X_{\tau_k+n}\}_{n\geq 0}$ is independent of the past $X_0, \ldots, X_{\tau_k-1}$ and has the same distribution as $\{X_n\}_{n\geq 0}$. In particular, the sequence $\{\tau_k - \tau_{k-1}\}_{k\geq 1}$ is IID.

EXAMPLE 2.5.12: RETURNS TO ZERO OF THE 1-D SYMMETRIC WALK. Let $\tau_1 - \tau_0, \tau_2, \ldots$ be the successive return times to state 0 of the random walk on $\mathbb{Z}$ of Example 2.2.5 with $p = \frac{1}{2}$. We shall admit that $P_0(\tau_0 < \infty) = 1$, a fact that will be proved in the next chapter, and obtain the probability distribution of τ_0 given $X_0 = 0$.

Observe that for $n \geq 1$,

$$P_0(X_{2n} = 0) = \sum_{k\geq 1} P_0(\tau_k = 2n),$$

and therefore, for all $z \in \mathbb{C}$ such that $|z| < 1$,

$$\sum_{n>1} P_0(X_{2n} = 0)z^{2n} = \sum_{k>1}\sum_{n>1} P_0(\tau_k = 2n)z^{2n} = \sum_{k>1} E_0[z^{\tau_k}].$$

But $\tau_k = \tau_1 + (\tau_2 - \tau_1) + \cdots + (\tau_k - \tau_{k-1})$ and therefore, since $\tau_1 = \tau_0$,

$$E_0[z^{\tau_k}] = (E_0[z^{\tau_0}])^k.$$

In particular,

$$\sum_{n\geq 0} P_0(X_{2n} = 0)z^{2n} = \frac{1}{1 - E_0[z^{\tau_0}]}$$

(note that the latter sum includes the term for $n = 0$, that is, 1). Direct evaluation of the left-hand side yields

$$\sum_{n\geq 0} \frac{1}{2^{2n}} \frac{(2n)!}{n!n!} z^{2n} = \frac{1}{\sqrt{1 - z^2}}.$$

Therefore, the generating function of the return time to 0 given $X_0 = 0$ is

$$E_0[z^{T_0}] = 1 - \sqrt{1 - z^2}.$$

Its first derivative

$$\frac{z}{\sqrt{1 - z^2}}$$

tends to ∞ as $z \to 1$ from below via real values. Therefore, by Abel's theorem (see the Appendix),

$$E_0[T_0] = \infty.$$

We see that although given $X_0 = 0$ the return time is almost surely finite, it has an infinite expectation.

_____

2.6   Exercises

Exercise 2.6.1. BACKWARD RECURRENCE TIME
Prove (2.2) in Example 2.1.3.

Exercise 2.6.2. A TRANSITION GRAPH
An HMC $\{X_n\}_{n \geq 0}$ with state space $E = \{0, 1, 2\}$ has the following transition matrix:

$$\mathbf{P} = \begin{array}{c} \\ 0 \\ 1 \\ 2 \end{array} \begin{pmatrix} \begin{array}{ccc} 0 & 1 & 2 \end{array} \\ \begin{array}{ccc} 0.2 & 0.5 & 0.3 \\ 0.1 & 0.1 & 0.8 \\ 0.5 & 0.2 & 0.3 \end{array} \end{pmatrix}.$$

Draw its transition graph and "read" from it $P(X_3 = 1 \mid X_0 = 1)$ and $P(X_7 = 2 \mid X_4 = 0)$.

Exercise 2.6.3. PAST AND FUTURE, TAKE 1
For an HMC $\{X_n\}_{n \geq 0}$ with state space E, prove that for all $n \in \mathbb{N}$, and all states $i_0, i_1, \ldots, i_{n-1}, i, j_1, j_2 \in E$,

$$P(X_{n+2} = j_2, X_{n+1} = j_1 \mid X_n = i, X_{n-1} = i_{n-1}, \ldots, X_0 = i_0)$$

$$= P(X_{n+2} = j_2, X_{n+1} = j_1 \mid X_n = i),$$

whenever both sides are well-defined.

Exercise 2.6.4. PAST AND FUTURE, TAKE 2

The Markov property does not imply that the past and the future are independent given *any* information concerning the present. Find a simple example of an HMC $\{X_n\}_{n\geq 0}$ with state space $E = \{1, 2, 3, 4, 5, 6\}$ such that

$$P(X_2 = 6 \mid X_1 \in \{3, 4\}, X_0 = 2) \neq P(X_2 = 6 \mid X_1 \in \{3, 4\}).$$

Exercise 2.6.5. AN URN

Consider N balls numbered from 1 to N and placed in two urns A and B. Suppose that at stage n, urn A contains X_n balls. One then chooses a ball among the N balls at random (we may suppose that the balls are numbered and that a lottery gives the number of the selected ball, which can be in either of the two urns), and then chooses an urn, A with probability p, B with probability $q = 1 - p$. The selected ball is then placed in the selected urn, and the number of balls in urn A is now X_{n+1}. Admitting that $\{X_n\}_{n\geq 0}$ is an HMC, give its transition matrix.

Exercise 2.6.6. THE FIRST ESCAPE TIME

Let $\{X_n\}_{n\geq 0}$ be an HMC with state space E and transition matrix $\mathbf{P}$. Let τ be the first time n for which $X_n \neq X_0$, where $\tau = +\infty$ if $X_n = X_0$ for all $n \geq 0$. Compute $\mathrm{E}[\tau \mid X_0 = i]$ in terms of p_{ii}.

Exercise 2.6.7. AGGREGATION OF STATES

Let $\{X_n\}_{n>0}$ be an HMC with state space E and transition matrix $\mathbf{P}$, and let $E = \sum_{k=1}^{\infty} A_k$ be a partition of E. Define the process $\{\hat{X}_n\}_{n\geq 0}$ with state space $\hat{E} = \{\hat{1}, \hat{2}, \ldots\}$ by $\hat{X}_n = \hat{k}$ if and only if $X_n \in A_k$. Show that a necessary and sufficient condition for $\{\hat{X}_n\}_{n\geq 0}$ to be an HMC for *any* initial distribution μ of $\{X_n\}_{n\geq 0}$ is that $\sum_{j\in A_\ell} p_{ij}$ be independent of $i \in A_k$ for all k, ℓ, and that in this case, $\hat{p}_{\hat{k}\hat{\ell}} = \sum_{j\subset A_\ell} p_{ij}$ (any $i \in A_k$) is the general entry of the transition matrix $\hat{\mathbf{P}}$ of $\{\hat{X}_n\}_{n\geq 0}$.

Exercise 2.6.8. INVENTORY

Find the transition matrix of the HMC in Example 2.2.7 with the parameters $s = 2, S = 5$, and the following probability distribution for the demand: $P(Z = i) = \frac{1}{5}$ for all $1 \leq i \leq 4$.

Exercise 2.6.9. RECORDS.

Let $\{Z_n\}_{n\geq 1}$ be an IID sequence of geometric random variables: For $k \geq 0$, $P(Z_n = k) = (1 - p)^k p$, where $p \in (0, 1)$. Let $X_n = \max(Z_1, \ldots, Z_n)$ be the *record value* at time n, and suppose X_0 is an $\mathbb{N}$-valued random variable independent of the sequence $\{Z_n\}_{n\geq 1}$. Show that $\{X_n\}_{n\geq 0}$ is an HMC and give its transition matrix.

Exercise 2.6.10. THE SNAKE CHAIN, TAKE 1

Let $\{X_n\}_{n\geq 0}$ be an HMC with state space E and transition matrix $\mathbf{P}$. Define for $L \geq 1, Y_n = (X_n, X_{n+1}, \ldots, X_{n+L})$. The process $\{Y_n\}_{n\geq 0}$ takes its values in $F = \{(i_0, \ldots, i_L) \in E^{L+1}; \ p_{i_0 i_1} p_{i_1 i_2} \cdots p_{i_{L-1} i_L} > 0\}$. Prove that $\{Y_n\}_{n\geq 0}$ is an HMC and give the general entry of its transition matrix. (The chain $\{Y_n\}_{n\geq 0}$ is called the *snake chain* of length $L + 1$ associated with $\{X_n\}_{n\geq 0}$.) This exercise is continued in Exercises 2.6.22 and 2.6.25.

Exercise 2.6.11. A CODING SCHEME

In certain digital communication systems, a sequence of 0's and 1's (bits) is encoded into a sequence of 0's, +1's, and −1's as follows. If the input sequence contains a 0, the output sequence contains a 0 at the same place. If the input sequence contains a 1, then the output sequence will have a −1 or a +1. The choice between −1 and +1 is made in such a way that −1's and +1's alternate in the output sequence. The first 1 is encoded as +1. For instance, 011101 becomes $0, +1, -1, +1, 0, -1$. Find an automaton with four states $+1$, -1, 0_+, and 0_- for which the sequence of visited states, not counting the initial state 0_+, is exactly the encoded sequence (where 0_+ and 0_- are rewritten as 0) when it is fed by the input sequence.

Suppose that the input sequence is IID, with 0 and 1 equiprobable. The sequence of states visited by the automaton is then an HMC. Compute its transition matrix $\mathbf{P}$, its stationary distribution π, and its iterates $\mathbf{P}^n$. Call $\{Y_n\}_{n\geq 0}$ the output sequence (taking its values in $\{0, -1, +1\}$). Compute $\lim_{n\to\infty}\{\mathrm{E}[Y_n Y_{n+k}] - \mathrm{E}[Y_n]\mathrm{E}[Y_{n+k}]\}$ for all $k \geq 0$.

Exercise 2.6.12. PROVE THEOREM 2.2.3

Prove Theorem 2.2.3.

Exercise 2.6.13. RAT AND CAT

Rat and Cat move between two rooms, using different paths. Their motions are independent, governed by their respective transition matrices

$$
\begin{array}{cc}
\begin{array}{cc} 1 & 2 \end{array} & \begin{array}{cc} 1 & 2 \end{array} \\
\begin{array}{c} 1 \\ 2 \end{array}\begin{pmatrix} 0.2 & 0.8 \\ 0.8 & 0.2 \end{pmatrix}, & \begin{array}{c} 1 \\ 2 \end{array}\begin{pmatrix} 0.3 & 0.7 \\ 0.6 & 0.4 \end{pmatrix}.
\end{array}
$$

Cat starts from room 1, Rat from room 2. If they are ever in the same room, Cat eats Rat. How long will Rat survive on the average?

Exercise 2.6.14. TIE-BREAK
Give a Markovian model of a *tie-break* between two tennis players with the respective probabilities α and β of winning a point on their own service. Compute the probability that A wins when he starts serving. How long will the tie-break last on average when A starts serving? (Recall the rules of a tie-break when A starts serving: A has one service then B has two; A has two, etc. The first player with 7 points and at least 2 points ahead wins. If a player reaches 7 points with only a 1-point advantage, the tie-break continues until one of the players makes a break of 2 points.)

Exercise 2.6.15. STREET GANGS
Three characters, $A, B,$ and C, armed with guns, suddenly meet at the corner of a Washington, D.C. street, whereupon they naturally start shooting at one another. Each street-gang kid shoots every tenth second, as long as he is still alive. The probability of a hit for A, B, and C are α, β, and γ respectively. A is the most hated, and therefore, as long as he is alive, B and C ignore each other and shoot at A. For historical reasons not developed here, A cannot stand B, and therefore he shoots only at B while the latter is still alive. Lucky C is shot at if and only if he is in the presence of A alone or B alone. What are the survival probabilities of $A, B,$ and C, respectively?

Exercise 2.6.16. STATE AVOIDANCE
Let $\{X_n\}_{n\geq0}$ be a homogeneous MC with state space $E - \{1, 2, 3, 4\}$ and transition matrix

$$\mathbf{P} = \begin{array}{c} \\ 1 \\ 2 \\ 3 \\ 4 \end{array} \begin{array}{cccc} 1 & 2 & 3 & 4 \\ \left(\begin{array}{cccc} 0.2 & 0.3 & 0.5 & 0 \\ 0 & 0.2 & 0.3 & 0.5 \\ 0.5 & 0 & 0.2 & 0.3 \\ 0.3 & 0.5 & 0 & 0.2 \end{array} \right) \end{array}.$$

What is the probability that when starting from state 1, the chain hits state 3 *before* it hits state 4?

Exercise 2.6.17. MACHINE REPLACEMENT
Consider the chain of Example 2.1.3 and suppose that $P(U < \infty) = 1$. Show that it is irreducible and that its period is $d = \gcd\{n \geq 1; P(U_1 = n) > 0\}$.

Exercise 2.6.18. THE REPAIR SHOP

Show that a necessary and sufficient condition of irreducibility of the chain of Examples 2.2.6 and 2.4.9 is that $P(Z_1 = 0) > 0$ *and* $P(Z_1 \geq 2) > 0$.

Exercise 2.6.19. CYCLIC CLASSES

Show that the transition graph in Figure 2.6.1 is irreducible. Find its period and its cyclic classes.

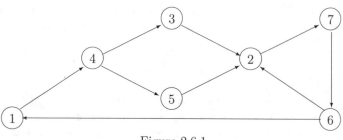

Figure 2.6.1

Exercise 2.6.20. NECESSARY AND SUFFICIENT CONDITION OF APERIODICITY

Let $\mathbf{P}$ be an irreducible transition matrix on the *finite* state space E. Show that a necessary and sufficient condition for $\mathbf{P}$ to be aperiodic is the existence of an integer m such that $\mathbf{P}^m$ has all its entries positive.

Exercise 2.6.21. LOOPS

Show that an irreducible transition matrix $\mathbf{P}$ with at least one state $i \in E$ such that $p_{ii} > 0$ is aperiodic.

Exercise 2.6.22. THE SNAKE CHAIN, TAKE 2

Consider the *snake chain* of Exercise 2.6.10. Show that if $\{X_n\}_{n \geq 0}$ is irreducible, then so is $\{Y_n\}_{n \geq 0}$. Compare the periods of these two chains in the irreducible case.

Exercise 2.6.23. SUCCESSIVE VISITS TO A CYCLIC CLASS

Let $\mathbf{P}$ be an irreducible transition matrix on the countable state space E. Let d be its period. Show that the restriction of $\mathbf{P}^d$ to any cyclic class is aperiodic.

Exercise 2.6.24. JUST AN EXAMPLE

Give the stationary distribution of the HMC with state space $E = \{1, 2, 3\}$ and transition matrix

$$\mathbf{P} = \begin{matrix} & \begin{matrix} 1 & 2 & 3 \end{matrix} \\ \begin{matrix} 1 \\ 2 \\ 3 \end{matrix} & \begin{pmatrix} 1-\alpha & \alpha & 0 \\ 0 & 1-\beta & \beta \\ \gamma & 0 & 1-\gamma \end{pmatrix} \end{matrix},$$

where $\alpha, \beta, \gamma \in (0, 1)$.

Exercise 2.6.25. THE SNAKE CHAIN, TAKE 3

Consider the *snake chain* of Exercises 2.6.10 and 2.6.22. Show that if $\{X_n\}_{n \geq 0}$ has a stationary distribution π, then $\{Y_n\}_{n \geq 0}$ also has a stationary distribution. Which one?

Exercise 2.6.26. NO STATIONARY DISTRIBUTION

Prove that the HMC of Exercise 2.6.9 cannot have a stationary distribution.

Exercise 2.6.27. ACCESS TO THE MEMORY OF THE CPU

Requests for access to the memory of the central processing unit of a computer are modeled as follows. Time is discrete and divided into time slots of equal length. At any time slot one request arrives (with probability θ) or no request at all (probability $1 - \theta$). The events indicating the presence or the absence of a request in the distinct slots are independent. The access times to the memory address of the successive requests are IID, and P(access time $= i$) $= \frac{1}{N}$ ($1 \leq i \leq N$). The requests are processed in first-come-first-served order, one at a time. The requests wait to be processed in a buffer. As soon as a request has accessed its memory address it leaves the buffer, and the next request at the head of the line is processed.

Find a necessary relation between θ and N for a steady state to exist (the state is the number of requests in the buffer). Then give the generating function of the distribution and the mean of the number of requests in the buffer.

Exercise 2.6.28. PAST AND FUTURE, TAKE 3

Let $\{X_n\}_{n\geq 0}$ be an HMC on the state space E with transition matrix $\mathbf{P}$, and suppose that $\mathrm{P}(X_0 = i) = \pi(i) > 0$ for all $i \in E$, where π is a stationary distribution. Define the matrix $\mathbf{Q} = \{q_{ij}\}_{i,j\in E}$ by $\pi(i)q_{ij} = \pi(j)p_{ji}$. Construct $\{X_{-n}\}_{n\geq 1}$ by

$$\mathrm{P}(X_{-1} = i_1, X_{-2} = i_2, \ldots, X_{-k} = i_k \mid X_0 = i, X_1 = j_1, \ldots, X_n = j_n)$$
$$= \mathrm{P}(X_{-1} = i_1, X_{-2} = i_2, \ldots, X_{-k} = i_k \mid X_0 = i) = q_{ii_1}q_{i_1 i_2} \cdots q_{i_{k-1}i_k}$$

for all $k \geq 1, n \geq 1, i, i_1, \ldots, i_k, j_1, \ldots, j_n \in E$. Show that $\{X_n\}_{n\in\mathbb{Z}}$ is an HMC with transition matrix $\mathbf{P}$ and that $\mathrm{P}(X_n = i) = \pi(i)$, for all $i \in E$, all $n \in \mathbb{Z}$.

Exercise 2.6.29. TRUNCATION

Let $\mathbf{P}$ be a transition matrix on the countable state space E, with the positive stationary distribution π. Let A be a subset of the state space, and define the truncation of $\mathbf{P}$ on A to be the transition matrix $\mathbf{Q}$ indexed by A and given by

$$q_{ij} = p_{ij} \text{ if } i, j \in A, \ i \neq j,$$
$$q_{ii} = p_{ii} + \sum_{k\in\bar{A}} p_{ik}.$$

Show that if $(\mathbf{P}, \pi)$ is reversible, then so is $(\mathbf{Q}, \frac{\pi}{\pi(A)})$.

Exercise 2.6.30. CHANGE TIMES

Let $\{X_n\}_{n\geq 0}$ be an HMC on the state space E with transition matrix $\mathbf{P}$. Define the sequence $\{\tau_k\}_{k\geq 0}$ recursively by $\tau_0 = 0$ and for $k \geq 0$,

$$\tau_{k+1} = \inf\{n \geq \tau_k + 1; X_n \neq X(\tau_k)\}$$

$(= +\infty$ if $\tau_k = \infty$ or if $X_n = X(\tau_k)$ for all $n \geq \tau_k + 1)$. Show that for all $k \geq 0$, τ_k is a stopping time of $\{X_n\}_{n\geq 0}$. Define for all $n \geq 0, Y_n = X(\tau_n)$ $(= \Delta \notin E$ if $\tau_n = \infty)$. Show that $\{Y_n\}_{n\geq 0}$ is an HMC, and give its state space and transition matrix.

Exercise 2.6.31. RETURNS TO A GIVEN SET

Let $\{X_n\}_{n\geq 0}$ be an HMC on the state space E with transition matrix $\mathbf{P}$. Let $\{\tau_k\}_{k\geq 1}$ be the successive return times to a given subset $F \subset E$. Assume these times are almost surely finite. Let $X_0 \equiv 0 \in F$, and define $Y_n = X(\tau_n)$. Show that $\{Y_n\}_{n\geq 0}$ is an HMC with state space F.

Exercise 2.6.32. TRANSITION TYPES

Let $\{X_n\}_{n\geq 0}$ be an irreducible HMC with state space E and transition matrix $\mathbf{P}$. Let H be a subset of $E \times E$. One says that a transition of type H is observed at time k if $(X_{k-1}, X_k) \in H$. Let $\tau(0), \tau(1), \ldots$ be the sequence of times of transitions of type H, with $\tau(n) = +\infty$ if the total number of transitions of type H is less than or equal to n. Observe that $\tau(0) \geq 1$, since X_{-1} is not defined. Define for each $n \geq 0, Y_n = X(\tau_n)$ if $\tau_n < \infty$, and $Y_n = \Delta$, an arbitrary element outside E, if $\tau_n = \infty$. Suppose that $\sum_{(i,j)\in H} p_{ij} > 0$.

Show that for some subset $\widetilde{E} \subset E$, to be identified, $\{Y_n\}_{n\geq 0}$ is an irreducible HMC on $\widetilde{E} \cup \{\Delta\}$. Compare with Exercises 2.6.30 and 2.6.31.

Chapter 3

Recurrence and Ergodicity

Consider a Markov chain taking its values in $E = \mathbb{N}$. There is a possibility that for any initial state $i \in \mathbb{N}$ the chain will never visit i after some finite random time. This is often an undesirable feature. For example, if the chain counts the number of customers waiting in line at a service counter (we shall see Markovian models of waiting lines, or *queues*, at different places in this book), such a behavior implies that the waiting line will eventually go beyond the limits of the waiting facility. In a sense, the corresponding system is unstable.

The good notion of stability for an irreducible HMC is that of *positive recurrence*, when any given state is visited infinitely often and in addition the average time between two successive visits to this state is finite. The main issue is that of finding sufficient, and maybe necessary, conditions guaranteeing stability.

3.1 The Potential Matrix Criterion

3.1.1 Recurrent and Transient States

We begin with the *potential matrix* criterion (necessary and sufficient condition), which is of mainly theoretical interest, and the *stationary distribution criterion*. Before this, we give the basic formal definitions of recurrence and transience.

Recall that T_i denotes the *return* time to state i.

Definition 3.1.1 *State $i \in E$ is called* recurrent *if*

$$P_i(T_i < \infty) = 1 \,,$$

and otherwise it is called transient. *A recurrent state $i \in E$ is called* positive recurrent *if*

$$E_i[T_i] < \infty \,,$$

© Springer Nature Switzerland AG 2020
P. Brémaud, *Markov Chains*, Texts in Applied Mathematics 31,
https://doi.org/10.1007/978-3-030-45982-6_3

and otherwise it is called null recurrent.

EXAMPLE 3.1.2: SUCCESS RUNS. The rule of the game is the following: A coin is tossed repeatedly and whenever the result is tails (probability $q = 1 - p$), you climb one step up the ladder, but if the result is heads, you fall all the way down. The sequence $\{X_n\}_{n\geq 0}$ recording your successive positions on the ladder forms a homogeneous Markov chain with state space $E = \mathbb{N}$ and is a special case of the chain with the transition graph in Figure 3.1.1.

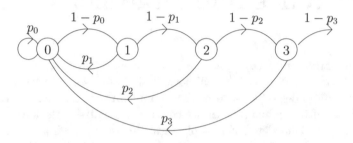

Figure 3.1.1. Transition graph of the success-runs chain

The state space of the chain of Figure 3.1.1 is $E = \mathbb{N}$ and we impose the condition

$$p_i \in (0, 1) \text{ for all } i \in E\,,$$

guaranteeing irreducibility. We shall compute the probability of returning to state 0 and the mean return time to state 0, and from the expressions obtained, we shall deduce the nature of state 0.

There is just one way of going from state 0 back to state 0 in exactly n steps. The corresponding path is $0, 1, 2, \ldots, n - 1, 0$. Therefore $\mathrm{P}_0(T_0 = 1) = p_0$ and

$$\mathrm{P}_0(T_0 = n) = (1 - p_0) \cdots (1 - p_{n-2})p_{n-1} \quad (n \geq 1)\,.$$

Letting $u_0 := 1$ and

$$u_n := (1 - p_0) \cdots (1 - p_{n-1}) \quad (n \geq 1)\,,$$

we obtain

$$\mathrm{P}_0(T_0 = n) = u_{n-1} - u_n \quad (n \geq 1)\,.$$

Since

$$\mathrm{P}_0(T_0 < \infty) = \sum_{n=1}^{\infty} \mathrm{P}_0(T_0 = n) = \lim_{m\uparrow\infty} \sum_{n=1}^{m} \mathrm{P}_0(T_0 = n) = \lim_{m\uparrow\infty} (1 - u_m)\,,$$

we have

$$P_0(T_0 < \infty) = 1 - \lim_{m\uparrow\infty} \prod_{i=0}^{m-1} (1 - p_i).$$

Therefore, in view of a classical result on infinite products (see the Appendix)

$$P_0(T_0 < \infty) = 1 \Leftrightarrow \prod_{i=0}^{\infty} (1 - p_i) = 0 \Leftrightarrow \sum_{i=0}^{\infty} p_i = \infty.$$

In general, it is not easy to check whether a given state is transient or recurrent. One of the goals of the theory of Markov chains is to provide criteria of recurrence. Sometimes, one is happy with just a sufficient condition or a necessary condition.

The problem of finding useful (easy to check) conditions of recurrence is an active area of research. However, the theory has a few conditions that qualify as useful and are applicable to many practical situations. Although the next criterion is of theoretical rather than practical interest, it will be helpful in a few situations, for instance in the study of recurrence of random walks (Examples 3.1.4 and 3.1.5 below).

3.1.2   A Criterion of Recurrence

The *potential matrix* associated with the transition matrix $\mathbf{P}$ is the matrix

$$\mathbf{G} := \sum_{n\geq 0} \mathbf{P}^n.$$

Its general term

$$g_{ij} = \sum_{n=0}^{\infty} p_{ij}(n) = \sum_{n=0}^{\infty} P_i(X_n = j) = \sum_{n=0}^{\infty} E_i[1_{\{X_n=j\}}] = E_i\left[\sum_{n=0}^{\infty} 1_{\{X_n=j\}}\right]$$

is the average number of visits to state j, given that the chain starts from state i.

Theorem 3.1.3 *State $i \in E$ is recurrent if and only if*

$$\sum_{n=0}^{\infty} p_{ii}(n) = \infty.$$

Proof. This is merely a rephrasing of Theorem 2.5.10. □

EXAMPLE 3.1.4: 1-D RANDOM WALK. The corresponding Markov chain was described in Example 2.2.5. The non-null terms of its transition matrix are

$$p_{i,i+1} = p \text{ and } p_{i,i-1} = 1 - p \quad (i \in \mathbb{Z}),$$

where $p \in (0,1)$. We shall study the nature (recurrent or transient) of any one of its states, say, 0. We have $p_{00}(2n+1) = 0$ and

$$p_{00}(2n) = \frac{(2n)!}{n!n!}p^n(1-p)^n.$$

By Stirling's equivalence formula $n! \sim (n/e)^n\sqrt{2\pi n}$, the above quantity is equivalent to

$$\frac{[4p(1-p)]^n}{\sqrt{\pi n}}, \tag{3.1}$$

and the nature of the series $\sum_{n=0}^{\infty} p_{00}(n)$ (convergent or divergent) is that of the series with general term (3.1). If $p \neq \frac{1}{2}$, in which case $4p(1-p) < 1$, the latter series converges, and if $p = \frac{1}{2}$, in which case $4p(1-p) = 1$, it diverges. In summary, the states of the 1-D random walk are transient if $p \neq \frac{1}{2}$, recurrent if $p = \frac{1}{2}$.

Example 2.5.12 shows that for the *symmetric* ($p = \frac{1}{2}$) 1-D random walk, the states are in fact *null recurrent*.

EXAMPLE 3.1.5: 3-D SYMMETRIC RANDOM WALK. The state space of this HMC is $E = \mathbb{Z}^3$. Denoting by e_1, e_2 and e_3 the canonical basis vectors of $\mathbb{R}^3$ (respectively $(1,0,0)$, $(0,1,0)$, and $(0,0,1)$), the non-null terms of the transition matrix of the 3-D symmetric random walk are given by

$$p_{x,x\pm e_i} = \frac{1}{6}.$$

We elucidate the nature of state, say, $0 = (0,0,0)$. Clearly, $p_{00}(2n+1) = 0$ for all $n \geq 0$ and (exercise)

$$p_{00}(2n) = \sum_{0 \leq i+j \leq n} \frac{(2n)!}{(i!j!(n-i-j)!)^2}\left(\frac{1}{6}\right)^{2n}.$$

This can be rewritten as

$$p_{00}(2n) = \sum_{0 \leq i+j \leq n} \frac{1}{2^{2n}}\binom{2n}{n}\left(\frac{n!}{i!j!(n-i-j)!}\right)^2\left(\frac{1}{3}\right)^{2n}.$$

By the *trinomial formula*

$$\sum_{0 \le i+j \le n} \frac{n!}{i!j!(n-i-j)!} \left(\frac{1}{3}\right)^n = 1 \,,$$

and therefore

$$p_{00}(2n) \le K_n \frac{1}{2^{2n}} \binom{2n}{n} \left(\frac{1}{3}\right)^n \,,$$

where

$$K_n := \max_{0 \le i+j \le n} \frac{n!}{i!j!(n-i-j)!} \,.$$

For large values of n, K_n is bounded as follows. Let i_0 and j_0 be the values of i, j that maximize $n!/(i!j!(n-i-j)!)$ in the domain of interest $0 \le i+j \le n$. From the definition of i_0 and j_0, the quantities

$$\frac{n!}{(i_0-1)!j_0!(n-i_0-j_0+1)!} \,,$$

$$\frac{n!}{(i_0+1)!j_0!(n-i_0-j_0-1)!} \,,$$

$$\frac{n!}{i_0!(j_0-1)!(n-i_0-j_0+1)!} \,,$$

$$\frac{n!}{i_0!(j_0+1)!(n-i_0-j_0-1)!} \,,$$

are bounded by

$$\frac{n!}{i_0!j_0!(n-i_0-j_0)!} \,,$$

The corresponding inequalities reduce to

$$n - i_0 - 1 \le 2j_0 \le n - i_0 + 1 \text{ and } n - j_0 - 1 \le 2i_0 \le n - j_0 + 1,$$

and this shows that for large n, $i_0 \sim n/3$ and $j_0 \sim n/3$. Therefore, for large n,

$$p_{00}(2n) \sim \frac{n!}{(n/3)!(n/3)!2^{2n}\,e^n} \binom{2n}{n} \,.$$

By Stirling's equivalence formula, the right-hand side of the latter equivalence is in turn equivalent to $\frac{3\sqrt{3}}{2(\pi n)^{3/2}}$, the general term of a divergent series. State 0 is therefore transient.

Suppose that state $i \in E$ is recurrent, and accessible from state $j \in E$. That is, starting from j, the probability of visiting i at least once is positive (accessibility of i from j), and starting from i, the average number of visits to i is infinite (recurrence of i). Therefore, starting from j the average number of visits to i is infinite:

$$\mathrm{E}_j[N_i] = \sum_{n \geq 1} p_{ji}(n) = \infty \,.$$

Similarly, if i is transient, then for any state $j \in E$,

$$\mathrm{E}_j[N_i] = \sum_{n \geq 1} p_{ji}(n) < \infty \,.$$

3.1.3   Structure of the Transition Matrix

A theoretical application of the potential matrix criterion is to the proof that recurrence is a (communication) class property.

Theorem 3.1.6 *If i and j communicate, they are either both recurrent or both transient.*

Proof. By definition, i and j communicate if and only if there exist integers M and N such that $p_{ij}(M) > 0, p_{ji}(N) > 0$. Going from i to j in M steps, then from j to j in n steps, then from j to i in N steps, is just one way of going from i back to i in $M + n + N$ steps. Therefore, $p_{ii}(M + n + N) \geq p_{ij}(M)p_{jj}(n)p_{ji}(N)$. Similarly, $p_{jj}(N + n + M) \geq p_{ji}(N)p_{ii}(n)p_{ij}(M)$. Therefore, writing $\alpha = p_{ij}(M)p_{ji}(N)$ (a strictly positive quantity), we have that

$$p_{ii}(M + N + n) \geq \alpha p_{jj}(n) \text{ and } p_{jj}(M + N + n) \geq \alpha p_{ii}(n).$$

This implies that the series $\sum_{n=0}^{\infty} p_{ii}(n)$ and $\sum_{n=0}^{\infty} p_{jj}(n)$ either both converge or both diverge. Theorem 3.1.3 concludes the proof.   □

It will be proven later in this chapter that positive recurrence (resp., null recurrence) is also a class property, in the sense that if states i and j communicate and if one of them is positive recurrent (resp., null recurrent), then so is the other.

Therefore, all the states of an irreducible Markov chain are of the same nature: transient, positive recurrent, or null recurrent. We shall therefore call it a transient chain, a positive recurrent chain, or a null recurrent chain, and to determine to which category it belongs, it suffices to study *one* state, selecting the state for which the computations seem easiest (such as state 0 for the chain of Example 3.1.4).

It follows from the above discussion that there are two types of communication classes: the *transient classes* and the *recurrent classes*. Call T the set of all transient states and R the set of all recurrent states. The set R may be composed of several disjoint communication classes R_1, R_2, etc. Any recurrent communication class, R_1 for instance, is closed. Indeed, if the chain can travel from $i \in R_1$ to some $j \in E$, it will have to come back to i, since i is recurrent, and therefore i and j must communicate, so that j must be in R_1.

The communication structure of a transition matrix is therefore as shown in Figure 3.1.2.

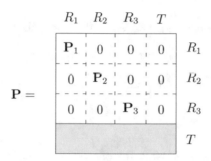

Figure 3.1.2

3.2 Recurrence

3.2.1 Invariant Measures

The notion of invariant measure plays a central role in the recurrence theory of Markov chains. It extends the notion of stationary distribution.

Definition 3.2.1 *A non-trivial (that is, non-null) vector $x = \{x_i\}_{i \in E}$ is called an invariant measure of the stochastic matrix $\mathbf{P} = \{p_{ij}\}_{i,j \in E}$ if for all $i \in E$,*

$$x_i \in [0, \infty)$$

and

$$x_i = \sum_{j \in E} x_j p_{ji}. \tag{3.2}$$

(In abbreviated notation, $0 \le x < \infty$ and $x^T \mathbf{P} = x^T$.)

Theorem 3.2.2 *Let* $\mathbf{P}$ *be the transition matrix of an irreducible recurrent* HMC $\{X_n\}_{n\geq 0}$. *Let* 0 *be an arbitrary state and let* T_0 *be the return time to* 0. *Let for all* $i \in E$

$$x_i := E_0\left[\sum_{n\geq 1} 1_{\{X_n=i\}} 1_{\{n\leq T_0\}}\right] \tag{3.3}$$

(for $i \neq 0$, x_i *is the expected number of visits to state* i *before returning to* 0*). Then, for all* $i \in E$,

$$x_i \in (0, \infty),$$

and x *is an invariant measure of* $\mathbf{P}$.

Observe that if $1 \leq n \leq T_0$, then $X_n = 0$ if and only if $n = T_0$. Therefore,

$$x_0 = 1. \tag{3.4}$$

Also, $\sum_{i\in E}\sum_{n\geq 1} 1_{\{X_n=i\}} 1_{\{n\leq T_0\}} = \sum_{n\geq 1}\left\{\sum_{i\in E} 1_{\{X_n=i\}}\right\} 1_{\{n\leq T_0\}} = \sum_{n\geq 1} 1_{\{n\leq T_0\}} = T_0$, and therefore

$$\sum_{i\in E} x_i = E_0[T_0]. \tag{3.5}$$

For the proof of Theorem 3.2.2, introduce the quantity

$$_0p_{0i}(n) := E_0[1_{\{X_n=i\}} 1_{\{n\leq T_0\}}] = P_0(X_1 \neq 0, \cdots, X_{n-1} \neq 0, X_n = i).$$

This is the probability, starting from state 0, of visiting i at time n before returning to 0. (It is sometimes called a *taboo probability*, 0 being a "taboo", that is, untouchable, state.) From the definition of x,

$$x_i = \sum_{n\geq 1} {}_0p_{0i}(n). \tag{3.6}$$

Proof. We first prove (3.2). Observe that

$$_0p_{0i}(1) = p_{0i}$$

and, by first-step analysis, for all $n \geq 2$,

$$_0p_{0i}(n) = \sum_{j\neq 0} {}_0p_{0j}(n-1)p_{ji}$$

(Exercise 3.4.7). Summing up all the above equalities, and taking (3.6) into account, we obtain

$$x_i = p_{0i} + \sum_{j\neq 0} x_j p_{ji},$$

that is, (3.2), since $x_0 = 1$ (see (3.4)).

Next we show that $x_i > 0$ for all $i \in E$. Indeed, by iterating (3.2), we find that $x^T = x^T \mathbf{P}^n$, that is, since $x_0 = 1$,

$$x_i = \sum_{j \in E} x_j p_{ji}(n) = p_{0i}(n) + \sum_{j \neq 0} x_j p_{ji}(n).$$

If x_i were null for some $i \in E$, $i \neq 0$, the latter equality would imply that $p_{0i}(n) = 0$ for all $n \geq 0$, which means that 0 and i do not communicate, in contradiction to the irreducibility assumption.

It remains to show that $x_i < \infty$ for all $i \in E$. As before, we find that

$$1 = x_0 = \sum_{j \in E} x_j p_{j0}(n)$$

for all $n \geq 1$, and therefore if $x_i = \infty$ for some i, necessarily $p_{i0}(n) = 0$ for all $n \geq 1$, and this also contradicts irreducibility. $\square$

Theorem 3.2.3 *The invariant measure of an irreducible recurrent stochastic matrix is unique up to a multiplicative factor.*

Proof. The proof of Theorem 3.2.2 showed that for an invariant measure y of an irreducible chain, $y_i > 0$ for all $i \in E$, and therefore one can define for all $i, j \in E$ the matrix $\mathbf{Q}$ by

$$q_{ji} := \frac{y_i}{y_j} p_{ij}. \tag{3.7}$$

It is a transition matrix, since $\sum_{i \in E} q_{ji} = \frac{1}{y_j} \sum_{i \in E} y_i p_{ij} = \frac{y_j}{y_j} = 1$. The general term of $\mathbf{Q}^n$ is

$$q_{ji}(n) = \frac{y_i}{y_j} p_{ij}(n). \tag{3.8}$$

Indeed, supposing (3.8) true for n,

$$\begin{aligned}
q_{ji}(n+1) &= \sum_{k \in E} q_{jk} q_{ki}(n) = \sum_{k \in E} \frac{y_k}{y_j} p_{kj} \frac{y_i}{y_k} p_{ik}(n) \\
&= \frac{y_i}{y_j} \sum_{k \in E} p_{ik}(n) p_{kj} = \frac{y_i}{y_j} p_{ij}(n+1),
\end{aligned}$$

and (3.8) follows by induction.

Clearly, $\mathbf{Q}$ is irreducible, since $\mathbf{P}$ is irreducible (just observe that $q_{ji}(n) > 0$ if and only if $p_{ij}(n) > 0$ in view of (3.8)). Also, $p_{ii}(n) = q_{ii}(n)$, and therefore

$\sum_{n\geq 0} q_{ii}(n) = \sum_{n\geq 0} p_{ii}(n)$, and this ensures that $\mathbf{Q}$ is recurrent by the potential matrix criterion. Call $g_{ji}(n)$ the probability, relative to the chain governed by the transition matrix $\mathbf{Q}$, of returning to state i for the first time at step n when starting from j. First-step analysis gives $g_{i0}(n+1) = \sum_{j\neq 0} q_{ij} g_{j0}(n)$ (see Exercise 3.4.7), that is, using (3.7),

$$y_i g_{i0}(n+1) = \sum_{j\neq 0} (y_j g_{j0}(n)) p_{ji} \, .$$

Recall that $_0 p_{0i}(n+1) = \sum_{j\neq 0} \, _0 p_{0j}(n) p_{ji}$ or, equivalently,

$$y_0 \, _0 p_{0i}(n+1) = \sum_{j\neq 0} (y_0 \, _0 p_{0j}(n)) p_{ji} \, .$$

The sequences $\{y_0 \, _0 p_{0i}(n)\}$ and $\{y_i g_{i0}(n)\}$ therefore satisfy the same recurrence equation. Their first terms $(n = 1)$, respectively $y_0 \, _0 p_{0i}(1) = y_0 p_{0i}$ and $y_i g_{i0}(1) = y_i q_{i0}$, are equal in view of (3.7). Therefore, for all $n \geq 1$,

$$_0 p_{0i}(n) = \frac{y_i}{y_0} g_{i0}(n) \, .$$

Summing up with respect to $n \geq 1$ and using the fact that $\sum_{n\geq 1} g_{i0}(n) = 1$ ($\mathbf{Q}$ is recurrent), we obtain the announced result: $x_i = \frac{y_i}{y_0}$.   □

Equality (3.5) and the definition of positive recurrence give the following.

Theorem 3.2.4 *An irreducible recurrent* HMC *is positive recurrent if and only if its invariant measures x satisfy*

$$\sum_{i\in E} x_i < \infty \, .$$

Remark 3.2.5 An HMC may well be irreducible and possess an invariant measure, and yet not be recurrent. The simplest example is the 1-D nonsymmetric random walk, which was shown to be transient Example 3.1.4 and yet admits $x_i \equiv 1$ for invariant measure.

3.2.2   A Positive Recurrence Criterion

In the previous section, the irreducible Markov chain was assumed recurrent and it was shown that it has a unique stationary distribution if it is positive recurrent. It was also observed that the existence of an invariant measure is not sufficient for recurrence. It turns out however that the existence of a stationary probability distribution is necessary and sufficient for an irreducible chain (not a priori assumed recurrent) to be recurrent positive.

Theorem 3.2.6 *An irreducible homogeneous Markov chain is positive recurrent if and only if there exists a stationary distribution. Moreover, the stationary distribution π is, when it exists, unique and $\pi > 0$.*

Proof. The direct part follows from Theorems 3.2.2 and 3.2.4. For the converse part, assume the existence of a stationary distribution π. Iterating $\pi^T = \pi^T \mathbf{P}$, we obtain $\pi^T = \pi^T \mathbf{P}^n$, that is, for all $i \in E$,

$$\pi(i) = \sum_{j \in E} \pi(j) p_{ji}(n) \,.$$

If the chain were transient, then, in view of the potential matrix criterion and the discussion following it, for all states i, j,

$$\lim_{n \uparrow \infty} p_{ji}(n) = 0,$$

and since $p_{ji}(n)$ is bounded by 1 uniformly in j and n, by the dominated convergence theorem for series (see the Appendix)

$$\pi(i) = \lim_{n \uparrow \infty} \sum_{j \in E} \pi(j) p_{ji}(n) = \sum_{j \in E} \pi(j) \left(\lim_{n \uparrow \infty} p_{ji}(n) \right) = 0 \,.$$

This contradicts the assumption that π is a stationary distribution (in particular, $\sum_{i \in E} \pi(i) = 1$). The chain must therefore be recurrent, and by Theorem 3.2.4, it is positive recurrent.

The stationary distribution π of an irreducible positive recurrent chain is unique (use Theorem 3.2.3 and the fact that there is no choice for a multiplicative factor but 1). That $\pi(i) > 0$ for all $i \in E$ follows from Theorem 3.2.2. □

Theorem 3.2.7 *Let π be the unique stationary distribution of an irreducible positive recurrent chain. Then*

$$\pi(i) \mathrm{E}_i[T_i] = 1 \,, \tag{3.9}$$

where T_i is the return time to state i.

Proof. This equality is a direct consequence of expression (3.3) for the invariant measure. Indeed, π is obtained by normalization of x: for all $i \in E$,

$$\pi(i) = \frac{x_i}{\sum_{j \in E} x_j},$$

and in particular, for $i = 0$, using (3.4) and (3.5),

$$\pi(0) = \frac{x_0}{\sum_{j \in E} x_j} = \frac{1}{E_0[T_0]}.$$

Since state 0 does not play a special role in the analysis, (3.9) is true for all $i \in E$.   □

The situation is extremely simple when the state space is finite.

Theorem 3.2.8 *An irreducible* HMC *with finite state space is positive recurrent.*

Proof. We first show recurrence. If the chain were transient, then, from the potential matrix criterion and the observations following it, for all $i, j \in E$,

$$\sum_{n \geq 0} p_{ij}(n) < \infty,$$

and therefore, since the state space is finite

$$\sum_{j \in E} \sum_{n \geq 0} p_{ij}(n) < \infty.$$

But the latter sum is equal to

$$\sum_{n \geq 0} \sum_{j \in E} p_{ij}(n) = \sum_{n \geq 0} 1 = \infty,$$

a contradiction. Therefore, the chain is recurrent. By Theorem 3.2.2 it has an invariant measure x. Since E is finite, $\sum_{i \in E} x_i < \infty$, and therefore the chain is positive recurrent (Theorem 3.2.4).   □

Remark 3.2.9 A "verbal proof" of recurrence in Theorem 3.2.8 is available: The states cannot all be visited only a finite number of times; otherwise, there would exist a finite random time after which no state is visited.

EXAMPLE 3.2.10: RANDOM WALK REFLECTED AT 0. This chain has the state space $E = \mathbb{N}$ and the transition graph of Figure 3.3.1. It is assumed that p_i (and therefore $q_i = 1 - p_i$) are in the open interval $(0, 1)$ for all $i \in E$, so that the chain is irreducible.

In this case, the invariant measure equation $x^T = x^T \mathbf{P}$ takes the form

$$\begin{aligned}
x_0 &= x_1 q_1, \\
x_i &= x_{i-1} p_{i-1} + x_{i+1} q_{i+1}, \ i \geq 1,
\end{aligned}$$

Figure 3.3.1. Reflected random walk

with $p_0 = 1$. The general solution is, for $i \geq 1$,

$$x_i = x_0 \frac{p_0 \cdots p_{i-1}}{q_1 \cdots q_i}.$$

The positive recurrence condition $\sum_{i \in E} x_i < \infty$ is

$$1 + \sum_{i \geq 1} \frac{p_0 \cdots p_{i-1}}{q_1 \cdots q_i} < \infty, \tag{3.10}$$

and if it is satisfied, the stationary distribution π is obtained by normalization of the general solution. This gives

$$\pi(0) = \left(1 + \sum_{i \geq 1} \frac{p_0 \cdots p_{i-1}}{q_1 \cdots q_i} \right)^{-1}, \tag{3.11}$$

and for $i \geq 1$,

$$\pi(i) = \pi(0) \frac{p_0 \cdots p_{i-1}}{q_1 \cdots q_i}. \tag{3.12}$$

In the special case where $p_i = p$, $q_i = q = 1 - p$, the positive recurrence condition becomes $1 + \frac{1}{q} \sum_{j \geq 0} \left(\frac{p}{q} \right)^j < \infty$, that is to say $p < q$ or, equivalently, $p < \frac{1}{2}$.

EXAMPLE 3.2.11: SUCCESS RUNS AND MACHINE REPLACEMENT. The result of Example 3.1.2 will be derived once again, this time via the stationary distribution criterion. Let $q_i := 1 - p_i$. Equality $x^T = x^T \mathbf{P}$ takes the form

$$x_0 = p_0 x_0 + p_1 x_1 + p_2 x_2 + \cdots,$$

and for $i \geq 1$,

$$x_i = q_{i-1} x_{i-1}.$$

Therefore, leaving aside the first equality, for $i \geq 1$,

$$x_i = (q_0 q_1 \cdots q_{i-1}) x_0.$$

Discarding the possibility $x_0 \leq 0$, which would imply that x is negative or null, the first equation is satisfied if and only if

$$1 = p_0 + q_0 p_1 + q_0 q_1 p_2 + \cdots ,$$

that is, since $q_0 q_1 \cdots q_{n-1} p_n = q_0 q_1 \cdots q_{n-1} - q_0 q_1 \cdots q_n,$

$$\prod_{i=0}^{\infty} q_i = 0 . \tag{3.13}$$

Since $q_i \in (0, 1)$, the convergence criterion for infinite products (see the Appendix) tells us that this is in turn equivalent to

$$\sum_{i=0}^{\infty} p_i = \infty .$$

The divergence of the series $\sum_{i=0}^{\infty} p_i$ is therefore a necessary and sufficient condition of existence of an invariant measure.

Under condition (3.13), there exists an invariant measure, and this measure has finite mass ($\sum_{i=0}^{\infty} x_i < \infty$) if and only if

$$1 + \sum_{n=1}^{\infty} \left(\prod_{i=0}^{n-1} q_i \right) < \infty. \tag{3.14}$$

The stationary distribution is then given by

$$\pi(0) = \left(1 + \sum_{n=1}^{\infty} \left(\prod_{i=0}^{n-1} q_i \right) \right)^{-1} \tag{3.15}$$

and for $i \geq 1$,

$$\pi(i) = \left(\prod_{j=0}^{i-1} q_j \right) \pi_0 . \tag{3.16}$$

In Exercise 3.4.2, the reader is invited to verify that the success-runs chain of Example 3.1.2 and the machine replacement chain of Example 2.1.3 are the same if one lets

$$p_i := \frac{P(U = i + 1)}{P(U > i)}.$$

Inequality (3.14) then reads $E[U] < \infty$, and (3.15) and (3.16) give

$$\pi(i) = \frac{P(U > i)}{E[U]}. \tag{3.17}$$

_____

The stationary distribution criterion can also be used to prove instability.

_____

EXAMPLE 3.2.12: INSTABILITY OF ALOHA. A typical situation in a multiple-access satellite communications system is the following. Users—each one identified with a message—contend for access to a single-channel satellite communications link for the purpose of transmitting messages. Two or more messages in the air at the same time jam each other and therefore are not successfully transmitted. The users are somehow able to detect a collision of this sort and will try to retransmit the message involved in a collision later. The difficulty in such communications systems resides mainly in the absence of cooperation among users, who are all unaware of the intention to transmit of competing users.

The _slotted_ ALOHA _protocol_ imposes on the users the following rules (see Figure 3.3.2):

(i) Transmissions and retransmissions of messages can start only at equally spaced moments; the interval between two consecutive (re-)transmission times is called a _slot_; the duration of a slot is always larger than that of any message.

(ii) All _backlogged_ messages, that is, those messages having already tried unsuccessfully, maybe more than once, to get through the link, require retransmission independently of one another with probability $\nu \in (0, 1)$ at each slot. This is the so-called _Bernoulli retransmission policy_.

(iii) The _fresh messages_ (those presenting themselves for the first time) immediately attempt to get through.

Let X_n be the number of backlogged messages at the beginning of slot n. The backlogged messages behave independently, and each one has probability ν of attempting retransmission in slot n. In particular, if there are $X_n = k$ backlogged messages, the probability that i among them attempt to retransmit in slot n is

$$b_i(k) = \binom{k}{i} \nu^i (1 - \nu)^{k-i}. \tag{3.18}$$

Let A_n be the number of fresh requests for transmission in slot n. The sequence $\{A_n\}_{n\geq0}$ is assumed IID with the distribution

$$P(A_n = j) = a_j.$$

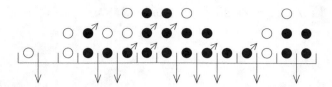

○   fresh message

●   backlogged message, not authorized to attempt retransmission

●↗   backlogged message, authorized to attempt retransmission

↓   successful transmission (or retransmission)

Figure 3.3.2. The ALOHA protocol

The quantity

$$\lambda = \mathrm{E}[A_n] = \sum_{i=1}^{\infty} i a_i$$

is the *traffic intensity*. We suppose that $a_0 + a_1 \in (0,1)$, so that $\{X_n\}_{n\geq 0}$ is an irreducible HMC. Its transition matrix is

$$p_{ij} = \begin{cases} b_1(i)a_0 \text{ if } j = i - 1, \\ [1 - b_1(i)]a_0 + b_0(i)a_1 \text{ if } j = i, \\ [1 - b_0(i)]a_1 \text{ if } j = i + 1, \\ a_{j-i} \text{ if } j \geq i + 2. \end{cases}$$

The proof is by accounting. For instance, the first line corresponds to one among the i backlogged messages having succeeded to retransmit, and for this there should be no fresh arrival (probability a_0) and only one of the i backlogged messages allowed to retransmit (probability $b_1(i)$). The second line corresponds to one of the two events "no fresh arrival and zero or strictly more than two retransmission requests from the backlog" and "zero retransmission request from the backlog and one fresh arrival."

Our objective in this example is to show that the system using the Bernoulli retransmission policy is *not stable*, in the sense that the chain $\{X_n\}_{n\geq 0}$ is *not positive recurrent*. Later on, a remedy to this situation will be proposed. To prove instability, we must contradict the existence of a stationary distribution π.

If such a stationary distribution existed, it should satisfy the balance equations

$$\pi(i) = \pi(i)\{[1 - b_1(i)]a_0 + b_0(i)a_1\} + \pi(i-1)[1 - b_0(i-1)]a_1$$
$$+ \pi(i+1)b_1(i+1)a_0 + \sum_{\ell=2}^{\infty} \pi(i-\ell)a_\ell$$

where $\pi(j) = 0$ if $j < 0$. Writing

$$P_N = \sum_{i=0}^{N} \pi(i)$$

and summing up the balance equations from $i = 0$ to N, we obtain

$$P_N - \pi(N)b_0(N)a_1 + \pi(N+1)b_1(N+1)a_0 + \sum_{\ell=0}^{N} a_\ell P_{N-\ell}.$$

This in turn gives

$$P_N(1 - a_0) = \pi(N)b_0(N)a_1 + \pi(N+1)b_1(N+1)a_0 + \sum_{\ell=1}^{N} a_\ell P_{N-\ell}.$$

But since P_N increases with N and $\sum_{\ell=1}^{N} a_\ell \leq \sum_{\ell=1}^{\infty} a_\ell = 1 - a_0$, we have

$$\sum_{\ell=1}^{N} a_\ell P_{N-\ell} \leq P_{N-1}(1 - a_0),$$

and therefore

$$P_N(1 - a_0) \leq \pi(N)b_0(N)a_1 + \pi(N+1)b_1(N+1)a_0 + P_{N-1}(1 - a_0),$$

from which it follows that

$$\frac{\pi(N+1)}{\pi(N)} \geq \frac{1 - a_0 - b_0(N)a_1}{b_1(N+1)a_0}.$$

Using expression (3.18), we obtain

$$\frac{\pi(N+1)}{\pi(N)} \geq \frac{(1 - a_0) - (1 - \nu)^N a_1}{(N+1)\nu(1 - \nu)^N a_0}.$$

For all values of $\nu \in (0, 1)$, the right-hand side of this inequality eventually becomes infinite, and this contradicts the equality $\sum_{N=1}^{\infty} \pi(N) = 1$ and the inequalities $\pi(N) > 0$ that π should satisfy as the stationary distribution of an irreducible Markov chain.

———

3.3   Empirical Averages

3.3.1   The Ergodic Theorem

The ergodic theorem for Markov chains gives conditions which guarantee that empirical averages of the type

$$\frac{1}{N} \sum_{k=1}^{N} f(X_k, \dots, X_{k+L})$$

converge to probabilistic averages.

Proposition 3.3.1 *Let $\{X_n\}_{n\geq 0}$ be an irreducible recurrent* HMC *and let x denote the canonical invariant measure associated with state $0 \in E$ given by (3.3). Define for $n \geq 1$*

$$\nu(n) = \sum_{k=1}^{n} 1_{\{X_k=0\}} \,. \tag{3.19}$$

Let $f : E \to \mathbb{R}$ be such that

$$\sum_{i\in E} |f(i)| x_i < \infty \,. \tag{3.20}$$

Then, for any initial distribution μ, P_μ-*a.s.,*

$$\lim_{N\uparrow\infty} \frac{1}{\nu(N)} \sum_{k=1}^{N} f(X_k) = \sum_{i\in E} f(i) x_i \,. \tag{3.21}$$

Before the proof, we shall harvest the most interesting consequences.

Theorem 3.3.2 *Let $\{X_n\}_{n\geq 0}$ be an irreducible positive recurrent Markov chain with the stationary distribution π, and let $f : E \to \mathbb{R}$ be such that*

$$\sum_{i\in E} |f(i)| \pi(i) < \infty \,. \tag{3.22}$$

Then for any initial distribution μ, P_μ-*a.s.,*

$$\lim_{n\uparrow\infty} \frac{1}{N} \sum_{k=1}^{N} f(X_k) = \sum_{i\in E} f(i) \pi(i) \,. \tag{3.23}$$

Proof. Apply Proposition 3.3.1 to $f \equiv 1$. Condition (3.20) is satisfied, since in the positive recurrent case, $\sum_{i \in E} x_i < \infty$. Therefore, P_μ-a.s.,

$$\lim_{N \uparrow \infty} \frac{N}{\nu(N)} = \sum_{j \in E} x_j \, .$$

Now, f satisfying (3.22) also satisfies (3.20), since x and π are proportional, and therefore, P_μ-a.s.,

$$\lim_{N \uparrow \infty} \frac{1}{\nu(N)} \sum_{k=1}^{N} f(X_k) = \sum_{i \in E} f(i) x_i \, .$$

Combining the above equalities gives, P_μ-a.s.,

$$\lim_{N \to \infty} \frac{1}{N} \sum_{k=1}^{N} f(X_k) = \lim_{N \to \infty} \frac{\nu(N)}{N} \frac{1}{\nu(N)} \sum_{k=1}^{N} f(X_k) = \frac{\sum_{i \in E} f(i) x_i}{\sum_{j \in E} x_j} \, ,$$

from which (3.23) follows, since π is obtained by normalization of x. $\qquad \square$

Corollary 3.3.3 *Let $\{X_n\}_{n \geq 1}$ be an irreducible positive recurrent Markov chain with the stationary distribution π, and let $g : E^{L+1} \to \mathbb{R}$ be such that*

$$\sum_{i_0, \ldots, i_L} |g(i_0, \ldots, i_L)| \pi(i_0) p_{i_0 i_1} \cdots p_{i_{L-1} i_L} < \infty \, .$$

Then for all initial distributions μ, P_μ-a.s.

$$\lim \frac{1}{N} \sum_{k=1}^{N} g(X_k, X_{k+1}, \ldots, X_{k+L}) = \sum_{i_0, i_1, \ldots, i_L} g(i_0, i_1, \ldots, i_L) \pi(i_0) p_{i_0 i_1} \cdots p_{i_{L-1} i_L} \, .$$

Proof. Apply Theorem 3.3.2 to the "snake chain" $\{(X_n, X_{n+1}, \ldots, X_{n+L})\}_{n \geq 0}$, which is (see Exercises 2.6.10, 2.6.22 and 2.6.25) irreducible recurrent and admits the stationary distribution

$$\pi(i_0) p_{i_0 i_1} \cdots p_{i_{L-1} i_L} \, .$$

$\qquad \square$

Note that

$$\sum_{i_0, i_1, \ldots, i_L} g(i_0, i_1, \ldots, i_L) \pi(i_0) p_{i_0 i_1} \cdots p_{i_{L-1} i_L} = E_\pi[g(X_0, \ldots, X_L)] \, .$$

Proof. (*of Proposition 3.3.1.*) Let $T_0 = \tau_1, \tau_2, \tau_3, \ldots$ be the successive return times to state 0, and define

$$U_p = \sum_{n=\tau_p+1}^{\tau_{p+1}} f(X_n).$$

In view of the regenerative cycle theorem, $\{U_p\}_{p\geq 1}$ is an IID sequence. Moreover, assuming $f \geq 0$ and using the strong Markov property,

$$\mathrm{E}[U_1] = \mathrm{E}_0\left[\sum_{n=1}^{T_0} f(X_n)\right] = \mathrm{E}_0\left[\sum_{n=1}^{T_0} \sum_{i\in E} f(i) 1_{\{X_n=i\}}\right]$$

$$= \sum_{i\in E} f(i)\mathrm{E}_0\left[\sum_{n=1}^{T_0} 1_{\{X_n=i\}}\right] = \sum_{i\in E} f(i) x_i.$$

This quantity is finite by hypothesis and therefore the strong law of large numbers applies to give

$$\lim_{n\uparrow\infty} \frac{1}{n} \sum_{p=1}^{n} U_p = \sum_{i\in E} f(i) x_i,$$

that is,

$$\lim_{n\uparrow\infty} \frac{1}{n} \sum_{k=T_0+1}^{\tau_{n+1}} f(X_k) = \sum_{i\in E} f(i) x_i. \tag{3.24}$$

Observing that

$$\tau_{\nu(n)} \leq n < \tau_{\nu(n)+1},$$

we have

$$\frac{\sum_{k=1}^{\tau_{\nu(n)}} f(X_k)}{\nu(n)} \leq \frac{\sum_{k=1}^{n} f(X_k)}{\nu(n)} \leq \frac{\sum_{k=1}^{\tau_{\nu(n)+1}} f(X_i)}{\nu(n)}.$$

Since the chain is recurrent, $\lim_{n\uparrow\infty} \nu(n) = \infty$, and therefore, from (3.24), the extreme terms of the above chain of inequalities tend to $\sum_{i\in E} f(i) x_i$ as n goes to ∞, and this implies (3.21). The case of a function f of arbitrary sign is obtained by considering (3.21) written separately for $f^+ = \max(0, f)$ and $f^- = \max(0, -f)$, and then taking the difference of the two equalities obtained this way. The difference is not an undetermined form $\infty - \infty$ due to hypothesis (3.20). $\qquad\square$

The version of the ergodic theorem for Markov chains featured in Theorem 3.3.2 is a kind of strong law of large numbers, and it can be used in simulations to compute, when π is unknown, quantities of the type $\mathrm{E}_\pi[f(X_0)]$.

EXAMPLE 3.3.4: FIXED-AGE RETIREMENT. We adopt the machine replacement interpretation of the success-runs Markov chain (Example 3.2.11). Assume positive recurrence. A visit of the chain to state 0 corresponds to a breakdown of a machine, and therefore, in view of the ergodic theorem,

$$\pi(0) = \lim_{N\uparrow\infty} \frac{1}{N} \sum_{k=1}^{N} 1_{\{X_k=0\}}$$

is the empirical frequency of breakdowns. Recall that

$$\pi(0) = \mathrm{E}_0[T_0]^{-1},$$

where T_0 is the return time to 0. Here,

$$\mathrm{E}_0[T_0] = \mathrm{E}[U],$$

and therefore

$$\lim_{N\uparrow\infty} \frac{1}{N} \sum_{k=1}^{N} 1_{\{X_k=0\}} = \frac{1}{\mathrm{E}[U]}. \tag{3.25}$$

Suppose that the cost of a breakdown is so important that it is better to replace a working machine during its lifetime (breakdown implies costly repairs, whereas replacement only implies moderate maintenance costs). The *fixed-age retirement policy* fixes an integer $T \geq 1$ and requires that a machine having reached age T be immediately replaced. We are interested in computing the empirical frequency of breakdowns (not replacements).

The success-runs chain corresponding to this situation is the same as before, except that the times U_n are replaced by $V_n = U_n \wedge T$. Also, a replacement (not breakdown) occurs at time n if and only if $X_n = 0$ and $X_{n-1} = T - 1$. But $X_{n-1} = T - 1$ implies $X_n = 0$, and therefore a replacement occurs at time n if and only if

$$X_{n-1} = T - 1.$$

The empirical frequency of replacements is, therefore, in view of the ergodic theorem,

$$\lim_{N\uparrow\infty} \frac{1}{N} \sum_{k=1}^{N} 1_{\{X_k=T-1\}} = \pi(T - 1).$$

Formula (3.17) gives

$$\pi(T - 1) = \frac{\mathrm{P}(V \geq T)}{\mathrm{E}[V]},$$

and therefore, since $V = U \wedge T$,

$$\pi(T - 1) = \frac{P(U \geq T)}{E[U \wedge T]}.$$

The empirical frequency of visits to state 0 is, by (3.25),

$$\frac{1}{E[U \wedge T]}.$$

The empirical frequency of breakdowns is therefore

$$\frac{1}{E[U \wedge T]} - \frac{P(U \geq T)}{E[U \wedge T]} = \frac{P(U < T)}{E[U \wedge T]}.$$

EXAMPLE 3.3.5: CASH MANAGEMENT. The level of cash in a bank at the beginning of day n is X_n, the number of idling dollars in the safe. For each day a dollar spends idling in the safe, the banker incurs a loss of r dollars. In the absence of control, the cash level fluctuates as a symmetric random walk.

The management policy uses two integers s and S, where $0 < s < S$. The number S is an upper barrier: If at the beginning of day n, the cash level is $X_n = S - 1$ and if during day n one more dollar enters the bank, then during the night from day n to day $n + 1$, $S - s$ dollars are transferred to a more profitable occupation (treasury bills, etc.), and the level of cash at the beginning of day $n+1$ is therefore $X_{n+1} = s$.

Also, if $X_n = 1$ and one more dollar is removed from the safe during day n, the safe is replenished at level s during the night, so that $X_{n+1} = s$.

Therefore, the process $\{X_n\}_{n \geq 0}$ takes its values in $\{1, \ldots, S - 1\}$ if we assume that $1 \leq X_0 \leq S - 1$. A typical trajectory of the cash level is as shown in Figure 3.3.1.

The process $\{X_n\}_{n \geq 0}$ is an HMC with the transition graph of Figure 3.3.1, where all transition probabilities are equal to $\frac{1}{2}$. In addition to the storing cost of \$$r$ per day per dollar, the banker incurs a transaction cost of \$$a$ whenever one dollar is moved into or out of the safe. The long-run cost per day of the (s, S) policy is therefore

$$\lim_{N \uparrow \infty} \frac{1}{N} \sum_{n=1}^{N} \left[\left(\sum_{k=1}^{S-1} rk 1_{\{X_n=k\}} \right) + a(S - s) 1_{\{X_n=S-1, X_{n+1}=s\}} + as 1_{\{X_n=1, X_{n+1}=s\}} \right].$$

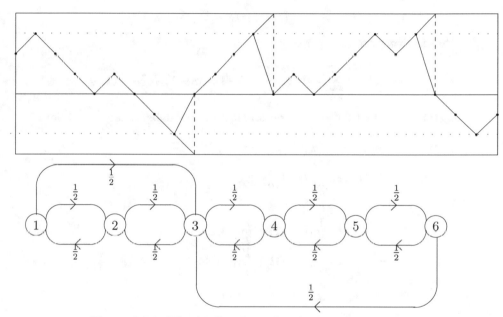

Figure 3.3.1. The (s, S) policy of cash management

The chain $\{X_n\}_{n\geq 0}$ is obviously irreducible, and since the state space is finite, it is positive recurrent. Let π be its stationary distribution.

By the ergodic theorem,

$$\lim_{N\uparrow\infty} \frac{1}{N} \sum_{n=1}^{N} \left(\sum_{k=1}^{S-1} rk1_{\{X_n=k\}} \right) = r \sum_{k=1}^{S-1} k\pi(k)$$

and

$$\lim_{N\uparrow\infty} \frac{1}{N} \ \sum_{n=1}^{N}(\alpha(S-s)1_{\{X_n=S-1,X_{n+1}=s\}} + \alpha s1_{\{X_n=1,X_{n+1}=s\}})$$
$$= \alpha(S-s)\pi(S-1)\tfrac{1}{2} + \alpha s\pi(1)\tfrac{1}{2},$$

so that the long run cost per day is

$$C = r \sum_{k=1}^{S-1} k\pi(k) + \frac{1}{2}\alpha(S-s)\pi(S-1) + \frac{1}{2}\alpha s\pi(1).$$

The stationary distribution π is determined by the global balance equations

$$\pi(1) \ = \ \frac{1}{2}\pi(2),$$
$$\pi(2) \ = \ \frac{1}{2}\pi(1) + \frac{1}{2}\pi(3),$$
$$\vdots$$
$$\pi(s-1) \ = \ \frac{1}{2}\pi(s-2) + \frac{1}{2}\pi(s),$$
$$\pi(s) \ = \ \frac{1}{2}\pi(s-1) + \frac{1}{2}\pi(s+1) + \frac{1}{2}\pi(S-1) + \frac{1}{2}\pi(1),$$
$$\pi(s+1) \ = \ \frac{1}{2}\pi(s) + \frac{1}{2}\pi(s+2),$$
$$\vdots$$
$$\pi(S-2) \ = \ \frac{1}{2}\pi(S-3) + \frac{1}{2}\pi(S-1),$$
$$\pi(S-1) \ = \ \frac{1}{2}\pi(S-2),$$

together with the normalizing condition $\sum_{i=1}^{S-1}\pi(i) = 1$. Since the rank of the transition matrix $\mathbf{P}$ is $|E|-1$, one of the $S-1$ balance equations can be removed, and it can be chosen arbitrarily. We choose the equation relative to s.

From equations 1 to $s-1$, we obtain

$$\pi(i) = i\pi(1) \quad (1 \leq i \leq s).$$

Equations $s + 1$ to S give for $j \in [1, S - s]$,

$$\pi(S - j) = j\pi(S - 1).$$

Taking $i = s$ and $j = S - s$, we have $\pi(s) = s\pi(1) = (S - s)\pi(S - 1)$, so that

$$\pi(S - 1) = \frac{s}{S - s}\pi(1).$$

From the normalization equation we have

$$\pi(1)(1 + \cdots + s) + \pi(S - 1)(1 + \cdots + (S - s - 1)) = 1,$$

that is,

$$\pi(1)\frac{s(s + 1)}{2} + \pi(S - 1)\frac{(S - s - 1)(S - s)}{2} = 1,$$

and taking into account the expression of $\pi(S - 1)$ in terms of $\pi(1)$, we have

$$\pi(1)\left[\frac{s(s + 1)}{2} + \frac{(S - s - 1)s}{2}\right] = 1,$$

that is,

$$\pi(1) = \frac{2}{sS}.$$

Therefore, finally,

$$\pi(i) = \begin{cases} \frac{2i}{sS} & \text{if } i \in [1, s], \\ \frac{2(S - i)}{(S - s)S} & \text{if } i \in [s + 1, S]. \end{cases}$$

The cost C can therefore be computed in terms of s and S and this can be exploited for optimization purposes.

3.3.2   The Renewal Reward Theorem

Let $\{S_n\}_{n \geq 1}$ be an IID sequence of positive random variables such that $E[S_1] < \infty$, and let R_0 be a finite non-negative random variable independent of this sequence. Define for all $n \geq 0$, $R_{n+1} = R_n + S_{n+1}$ and for $t \geq 0$, $N(t) = \sum_{n=1}^{\infty} 1_{\{R_n \leq t\}}$.

Theorem 3.3.6 *Let $\{Y_n\}_{n \geq 1}$ be an IID sequence of random variables such that $E[|Y_1|] < \infty$. Then*

$$\lim_{t \uparrow \infty} \frac{N(t)}{t} = \frac{1}{E[S_1]} \tag{3.26}$$

and

$$\lim_{t \uparrow \infty} \frac{\sum_{n=1}^{N(t)} Y_n}{t} = \frac{E[Y_1]}{E[S_1]}. \tag{3.27}$$

Proof. Since $R_{N(t)} \leq t < R_{N(t)+1}$, we have

$$\frac{N(t)}{R_{N(t)+1}} < \frac{N(t)}{t} \leq \frac{N(t)}{R_{N(t)}}.$$

But the rightmost term is the inverse of

$$\frac{R_{N(t)}}{N(t)} = \frac{R_0 + \sum_{n=1}^{N(t)} S_n}{N(t)}.$$

By the strong law of large numbers and the fact that $\lim_{t \uparrow \infty} N(t) = \infty$ (the S_n's are finite), this quantity tends to $E[S_1]$ and, similarly, $\frac{R_{N(t)+1}}{N(t)} = \frac{R_{N(t)+1}}{N(t)+1} \frac{N(t)+1}{N(t)}$ tends to $E[S_1]$ as $t \to \infty$.

The proof of (3.27) follows from the SLNN and (3.26), since

$$\frac{\sum_{n=1}^{N(t)} Y_n}{t} = \frac{\sum_{n=1}^{N(t)} Y_n}{N(t)} \cdot \frac{N(t)}{t}.$$

$\square$

The result of the next example extends Formula (3.3) giving the invariant measure.

EXAMPLE 3.3.7: REGENERATIVE FORM OF THE STATIONARY DISTRIBUTION. Let $\{X_n\}_{n \geq 0}$ be a positive recurrent HMC with state space E and stationary distribution π. Let S be a stopping time of this chain and let i be a state, such that, P_i-almost surely, $S \in (0, \infty)$ and $X_S = i$. Then for all $j \in E$,

$$E_i \left[\sum_{k=0}^{S-1} 1_{\{X_n=j\}} \right] = E_i [S] \pi(j). \tag{3.28}$$

To prove this, observe that S, being a stopping time of the chain, can be written as

$$S = \varphi (X_0, X_1, \dots)$$

for some functional φ, namely

$$S = \sum_{m=0}^{\infty} m \psi_m(X_0, \dots, X_m),$$

where $\psi_m(X_0, \dots, X_m) = 1_{\{S=m\}}$. Define the sequence $\{S_n\}_{n \geq 1}$ by $S_1 = S$, and for $k \geq 0$,

$$S_{k+1} = \varphi (X_{R_k}, X_{R_k+1}, \dots),$$

where $R_0 = 0$ and, for $k \geq 1$, $R_k = S_1 + \cdots + S_k$. As a matter of fact, since S is a stopping time, S_{k+1} depends only on the cycle $C_k = (X_{R_k}, \ldots, X_{R_k+S_{k+1}-1})$ (recall that $X_{R_k+S_{k+1}} = i$). But by the strong Markov property, the cycles $\{C_k\}_{k \geq 0}$ are IID under P_i and in particular $\{S_n\}_{n \geq 1}$ is IID.

We first suppose that $E_i[S] < \infty$. The renewal reward theorem with

$$Y_n = \sum_{k=R_{n-1}}^{R_n - 1} 1_{\{X_k = j\}}$$

(observe that $E_i[|Y_1|] \leq E_i[S_1] < \infty$) gives

$$\lim_{t \uparrow \infty} \frac{\sum_{n=1}^{N(t)} Y_n}{l} = \frac{E_i\left[\sum_{k=0}^{S-1} 1_{\{X_k = j\}}\right]}{E_i[S]}.$$

But

$$\frac{\sum_{n=1}^{N(t)} Y_n}{t} = \frac{\sum_{k=R_0}^{R_{N(t)}-1} 1_{\{X_k = j\}}}{t} = \frac{\sum_{k=R_0}^{R_{N(t)}-1} 1_{\{X_k = j\}}}{R_{N(t)}} \cdot \frac{R_{N(t)}}{t}.$$

Now, $\lim_{t \uparrow \infty} \frac{R_{N(t)}}{t} = 1$ (exercise), and therefore, by the ergodic theorem,

$$\lim_{t \uparrow \infty} \frac{\sum_{n-1}^{N(t)} Y_n}{t} = \pi(j).$$

This proves the theorem when $E_i[S] < \infty$. If $E_i[S] = \infty$, write (6.13) with S replaced by $S^{(n)} = S \wedge \tau_n(i)$, where $\tau_n(i)$ is the n return time to i, and let $n \to \infty$ to obtain $E_i\left[\sum_{k=0}^{S-1} 1_{\{X_k = j\}}\right] = \infty$.

EXAMPLE 3.3.8: $P_i(T_j < T_i)$[1] Let i and j be two distinct states and let S be the first time of return to i after the first visit to j. Then $E_i[S] = E_i[T_j] + E_j[T_i]$ (use the strong Markov property at T_j). Also,

$$E_i\left[\sum_{n=0}^{S-1} 1_{\{X_n = j\}}\right] = E_i\left[\sum_{n=T_j}^{S-1} 1_{\{X_n = j\}}\right] = E_j\left[\sum_{n=0}^{T_i-1} 1_{\{X_n = j\}}\right],$$

where the last equality is the strong Markov property. Therefore, by (6.13),

$$E_j\left[\sum_{n=0}^{T_i-1} 1_{\{X_n = j\}}\right] = \pi(j)\left(E_i[T_j] + E_j[T_i]\right). \tag{3.29}$$

---

[1][Aldous and Fill, 1998]

Using words, the left-hand side of this equality is

$$E_j[\text{number of visits to } j \text{ before } i].$$

Now, the probability that j is not visited between two successive visits of i is $P_i(T_j > T_i)$. Therefore, the number of visits to i (including time 0) before T_j has a geometric distribution with parameter $p = P_i(T_j > T_i)$, and the average number of such visits is

$$\frac{1}{P_i(T_j < T_i)}.$$

Therefore, by (3.29), after exchanging the roles of i and j,

$$E_i(T_j < T_i) = \frac{1}{\pi(i)(E_i[T_j] + E_j[T_i])}. \tag{3.30}$$

Theorem 6.3.6 gives an algebraic method of computation of $E_i[T_j]$ when the state space is finite.

3.4 Exercises

Exercise 3.4.1. RETURN TO ZERO OF THE TWO-STATE HMC
For an HMC with state space $E = \{0, 1\}$ and transition matrix

$$\mathbf{P} = \begin{array}{c} 0 \\ 1 \end{array}\begin{array}{c} \overset{\displaystyle 0}{} \quad \overset{\displaystyle 1}{} \\ \begin{pmatrix} 1 - \alpha & \alpha \\ \beta & 1 - \beta \end{pmatrix}, \end{array}$$

compute $P_0(T_0 = n)$ and $E_0[T_0]$.

Exercise 3.4.2. SUCCESS RUNS AND MACHINE REPLACEMENT
Verify that the success-runs chain of Example 3.1.2 and the machine-replacement chain of Example 2.1.3, are the same if one lets

$$p_i := \frac{P(U = i + 1)}{P(U > i)}.$$

Exercise 3.4.3. TRANSIENT STATE
Let $\{X_n\}_{n \geq 0}$ be an HMC with state space E and n-step transition matrix $\mathbf{P}^n = \{p_{ij}(n)\}_{i,j \in E}$. Prove directly that if $i \in E$ is a transient state, then $\lim_{n \uparrow \infty} p_{ji}(n) = 0$ for all $j \in E$.

Exercise 3.4.4. THE REPAIR SHOP
Consider the HMC of Example 2.2.6. Recall that $X_n \in E \equiv N$, and for all $n \geq 0$,

$$X_{n+1} = (X_n - 1)^+ + Z_{n+1},$$

where $a^+ = \sup(a, 0)$ and $\{Z_n\}_{n \geq 1}$ is an IID sequence of random variables with values in $\mathbb{N}$. Prove directly, using the strong law of large numbers, that if $E[Z_1] > 1$, state 0 is transient, and that if $E[Z_1] < 1$, state 0 is recurrent.

Exercise 3.4.5. THE POTENTIAL MATRIX
Show that the transition matrix $\mathbf{P}$ and the potential matrix $\mathbf{G}$ are related by

$$\mathbf{G}(I - \mathbf{P}) = I,$$

where I is the identity matrix. Conclude from this that if the state space E is finite, the potential matrix $\mathbf{G}$ cannot have all its entries finite. Conclude that an irreducible HMC with finite state space is recurrent.

Exercise 3.4.6. 2-D SYMMETRIC RANDOM WALK
The symmetric random walk on $\mathbb{Z}^2$ is an HMC $\{X_n\}_{n \geq 0}$ with state space $E = \mathbb{Z}^2$ and with a transition matrix having for non-null entries

$$p_{x, x \pm e_i} = \frac{1}{4}$$

$(x \in \mathbb{Z}^2, i = 1, 2)$, where $e_1 = (1, 0)$ and $e_2 = (0, 1)$.
Compute $P(X_{2n} = (0, 0) \mid X_0 = (0, 0))$ and show that the above HMC is irreducible and recurrent. Show that it is in fact null-recurrent.

Exercise 3.4.7. TABOO PROBABILITIES
Let $\{X_n\}_{n \geq 0}$ be an irreducible recurrent HMC with state space E and transition matrix $\mathbf{P}$. Let $0 \in E$ be some state, and define ${}_0p_{0i}(n) = E_0[1_{\{X_n=i, n \leq T_0\}}]$ and $f_{ij}(n) = P_i(T_j = n)$. Show that for all $n \geq 0, i \in E$,

$$_0p_{0i}(n+1) = \sum_{j \neq 0} {}_0p_{0j}(n)p_{ji}$$

and

$$f_{i0}(n+1) = \sum_{j \neq 0} p_{ij}f_{j0}(n).$$

Exercise 3.4.8. A PROBABILISTIC INTERPRETATION OF THE INVARIANT MEA-SURE

A countable number of particles move independently in the countable space E, each according to a Markov chain with the transition matrix $\mathbf{P}$. Let $A_n(i)$ be the number of particles in state $i \in E$ at time $n \geq 0$, and suppose that the random variables $A_0(i)$, $i \in E$, are independent Poisson random variables with respective means $\mu(i)$, $i \in E$, where $\mu = \{\mu(i)\}_{i \in E}$ is an invariant measure of $\mathbf{P}$. Show that for all $n \geq 1$, the random variables $A_n(i)$, $i \in E$, are independent Poisson random variables with respective means $\mu(i), i \in E$.

Exercise 3.4.9. DOUBLY STOCHASTIC TRANSITION MATRIX

A stochastic matrix $\mathbf{P}$ on the state space E is called *doubly stochastic* if for all states i, $\sum_{j \in E} p_{ji} = 1$. Suppose in addition that $\mathbf{P}$ is irreducible, and that E is *infinite*. Find the invariant measure of $\mathbf{P}$. Show that $\mathbf{P}$ cannot be positive recurrent.

Exercise 3.4.10. SOME MARKOV CHAIN

Consider the transition matrix

$$\mathbf{P} = \begin{pmatrix} 1 - \alpha & \alpha & 0 \\ 0 & 1 - \beta & \beta \\ \gamma & 0 & 1 - \gamma \end{pmatrix},$$

where $\alpha, \beta, \gamma \in (0, 1)$. Show that it is irreducible and compute directly its stationary probability with the help of Formula (3.3).

Exercise 3.4.11. RETURN TO THE INITIAL STATE

Let τ be the first return time to initial state of an irreducible positive recurrent HMC $\{X_n\}_{n \geq 0}$, that is,

$$\tau = \inf\{n \geq 1; X_n = X_0\},$$

with $\tau = +\infty$ if $X_n \neq X_0$ for all $n \geq 1$. Compute the expectation of τ when the initial distribution is the stationary distribution π. Conclude that it is finite if and only if E is finite. When E is infinite, is this in contradiction to positive recurrence?

Exercise 3.4.12. THE SEQUENCE OF NEW VALUES

Let $\{X_n\}_{n\geq0}$ be an HMC with state space E and transition matrix $\mathbf{P}$, and let $\{Y_n\}_{n\geq0}$ be the sequence of *new values* of $\{X_n\}_{n\geq0}$. For instance,

$$
\begin{pmatrix}
n &=& 0 & 1 & 2 & 3 & 4 & 5 & 6 & 7 & 8 & 9 & 10 \\
X_n &=& 1 & 1 & 1 & 2 & 2 & 1 & 3 & 3 & 3 & 2 & 1 \\
& & Y_0 = 1 & & & Y_1 = 2 & & Y_2 = 1 & Y_3 = 3 & & & Y_4 = 2 & Y_5 = 1
\end{pmatrix}.
$$

(i) Show that $\{Y_n\}_{n\geq0}$ is an HMC and give its transition matrix $\mathbf{Q}$ in terms of $\mathbf{P}$.

(ii) Show that $\{X_n\}_{n\geq0}$ is irreducible and recurrent if and only if $\{Y_n\}_{n\geq0}$ is irreducible and recurrent.

(iii) In the irreducible recurrent case, express the invariant measure of $\{Y_n\}_{n\geq0}$ in terms of $(p_{ii}, i \in E)$ and the invariant measure of $\{X_n\}_{n\geq0}$.

(iv) In the irreducible recurrent case, give a necessary and sufficient condition for $\{Y_n\}_{n\geq0}$ to be positive recurrent when $\{X_n\}_{n\geq0}$ is positive recurrent in terms of the stationary distribution π of $\{X_n\}_{n\geq0}$ and $(p_{ii}, i \in E)$.

(v) In the irreducible recurrent case, give a necessary and sufficient condition for $\{Y_n\}_{n\geq0}$ to be positive recurrent when $\{X_n\}_{n\geq0}$ is null recurrent in terms of an invariant measure μ of $\{X_n\}_{n\geq0}$ and p_{ii}.

Exercise 3.4.13. MOVING STONES

Stones $S_1, \ldots, S_M$ are placed in line. At each time n a stone is selected at random, and this stone and the one ahead of it in the line exchange positions. If the selected stone is at the head of the line, nothing is changed. For instance, with $M = 5$, let the current configuration be $S_2 S_3 S_1 S_5 S_4$. If S_5 is selected, the new situation is $S_2 S_3 S_5 S_1 S_4$, whereas if S_1 is selected, the configuration is not altered. At each step, stone S_i is selected with probability $\alpha_i > 0$. Call X_n the situation at time n, for instance $X_n = (S_{i_1}, \ldots, S_{i_M})$, meaning that stone S_{i_j} is in the jth position.

Show that $\{X_n\}_{n\geq0}$ is an irreducible positive recurrent HMC and that its stationary distribution is given by the formula

$$
\pi(S_{i_1}, \ldots, S_{i_M}) = C\alpha_{i_1}^M \alpha_{i_2}^{M-1} \cdots \alpha_{i_M},
$$

for some normalizing constant C.

Exercise 3.4.14. COINCIDENCES

Let A, B, C and D be four points on the unit circle forming a square, and let $\{Y_n\}_{n\geq0}$ and $\{Z_n\}_{n\geq0}$ be two stochastic processes with state space $\{A, B, C, D\}$ denoted by $\{0, 1, 2, 3\}$, and with the dynamics

$$
\begin{aligned}
Y_{n+1} &= Y_n + B_{n+1} \ (\text{mod } 4), \\
Z_{n+1} &= Z_n + C_{n+1} \ (\text{mod } 4),
\end{aligned}
$$

where $\{B_n\}_{n\geq1}$ and $\{C_n\}_{n\geq1}$ are independent IID sequences with $P(B_n = \pm1) = P(C_n = \pm1) = \frac{1}{3} = P(B_n = 0) = P(C_n = 0)$. Suppose, moreover, that Y_0, Z_0, $\{B_n\}_{n\geq1}$, and $\{C_n\}_{n\geq1}$ are mutually independent. What is the average time separating two successive coincidences of the processes $\{Y_n\}_{n\geq1}$ and $\{Z_n\}_{n\geq1}$?

Exercise 3.4.15. THE KNIGHT COMES BACK HOME

A knight moves randomly on a chessboard, making each admissible move with equal probability, and starting from a corner. What is the average time he takes to return to the corner he started from?

Exercise 3.4.16. INSPECTION AT THE PRODUCTION LINE

Consider a production line where each manufactured item may be defective with probability $p \in (0, 1)$. The following inspection plan is proposed with a view to detecting defective items without checking every single one. It has 2 phases: In phase A, the probability of inspecting an article is $r \in (0, 1)$. In phase B, all the articles are inspected. One switches from phase A to phase B as soon as a defective item is detected. One switches from phase B to phase A as soon as a sequence of N successive acceptable items has been found.

Let $\{X_n\}_{n\geq0}$ be the process taking the values $E_0, \ldots, E_N$, where if $j \in [0, N-1]$, E_j means that the inspection plan is in phase B with j successive good items observed, and E_N means that the plan is in phase A.

(a) Prove that $\{X_n\}_{n\geq0}$ is an irreducible HMC, give its transition graph, and show that it is positive recurrent. Find its stationary distribution.

(b) Find the proportion of items inspected and give the *efficiency* of the inspection plan, which is by definition equal to the ratio of the long-run proportion of *detected* defective items over the proportion of defective items.

Exercise 3.4.17. A SEQUENCE OF A'S AND B'S

A sequence of A's and B's is formed as follows. The first item is chosen at random, $P(A) = P(B) = \frac{1}{2}$, as is the second item, independently of the first one. When the first $n \geq 2$ items have been selected, the $(n + 1)$st is chosen, conditionally with respect to the pair at position $n - 1$ and n, as follows:

$$P(A \mid AA) = \frac{1}{2}, P(A \mid AB) = \frac{1}{2}, P(A \mid BA) = \frac{1}{4}, P(A \mid BB) = \frac{1}{4}.$$

What is the proportion of A's and B's in a long chain?

Exercise 3.4.18. RETURN TIMES TO A SET OF STATES

Let $\{X_n\}_{n \geq 0}$ be an irreducible positive recurrent HMC with stationary distribution π. Let A be a subset of the state space E and let $\{\tau(k)\}_{k \geq 1}$ be the sequence of return times to A. Show that

$$\lim_{k \uparrow \infty} \frac{\tau(k)}{k} = \frac{1}{\sum_{i \in A} \pi(i)}.$$

Exercise 3.4.19. AN ABSORPTION FORMULA

Let $\mathbf{P}$ be a transition matrix on $E = \{0, 1, \ldots, N - 1, N\}$ with three communicating classes, $\{0\}$, $\{N\}$, and $\{1, \ldots, N - 1\}$, respectively recurrent, recurrent, and transient (in particular, 0 and N are closed states). Suppose, moreover, that 0 and N are not isolated states, in the sense that there exists at least one $i \in E$ with $p_{i0} > 0$ and one $j \in E$ with $p_{jN} > 0$. In this situation 0 and N are two absorbing states.

We wish to compute u_k, the mean time to absorption in $\{0, N\}$ when starting from k ($1 \leq k \leq N - 1$). In order to do this, consider the transition matrix $\hat{\mathbf{P}}$ on E obtained by modifying the first and last rows of $\mathbf{P}$ in such a way that $\hat{p}_{0k} = 1 = \hat{p}_{Nk}$. The matrix $\hat{\mathbf{P}}$ is easily seen to be irreducible, and since E is finite, it has a unique stationary distribution, denoted by $\hat{\pi}$.

Prove that

$$u_k = \frac{1}{\hat{\pi}(0) + \hat{\pi}(N)} - 1.$$

(Use the result of Exercise 3.4.18.)

Exercise 3.4.20. THE URN OF EHRENFEST
Consider the urn of Ehrenfest with a total number of particles $N = 2M$. Let T_0
and T_M be, respectively, the first time when the compartment A becomes empty
and the first time when the two compartments have an equal number of particles.
Compute the ratio

$$\frac{P_0(T_M < T_0)}{P_M(T_0 < T_M)}.$$

Exercise 3.4.21. NON-OVERLAPPING OCCURRENCES OF A PATTERN
Consider the array of a's and b's in part A of the figure below where the letter in
a given position has been chosen at random in the set $\{a, b\}$, independently of the
other letters, with $P(a) = P(b) = \frac{1}{2}$. Suppose we are interested in counting the
occurrences of the special pattern of part B of the figure below without counting
overlapping patterns. For instance, with the data shown in part A of the figure
below, two such patterns are counted; the third one is not counted because it
overlaps the second one (part C).

$$
\begin{array}{cccccccccc}
b & a & b & b & a & b & a & b & b & b \\
b & a & a & a & b & b & a & a & a & a
\end{array}
\quad A
$$

$$
\begin{array}{ccc}
b & \cdot & b \\
\cdot & a & a
\end{array}
\quad B
$$

YES   YES   NO   C

Find an automaton that successively reads the 2-letter columns from left to right,
and that has a privileged state $*$ with the following property: the automaton enters
(or stays in) state $*$ if and only if it has just discovered a special pattern that is
nonoverlapping with a previously discovered one. What is the long-run proportion
of non-overlapping patterns?

Chapter 4

Long-Run Behavior

Consider an HMC that is irreducible and positive recurrent. In particular, if its initial distribution is the stationary distribution, it keeps the same distribution at all times. The chain is then said to be in the *stationary regime*, or in *equilibrium*, or in *steady state*.

A question arises naturally: What is the long-run behavior of the chain when the initial distribution μ is *arbitrary*? For instance, will it *converge to equilibrium*, and in which sense?

4.1 Coupling

4.1.1 Convergence in Variation

For an ergodic HMC, the type of convergence of interest is not almost-sure convergence but convergence in variation of the distribution at time n to the stationary distribution. This type of convergence is relative to a metric structure that we proceed to define.

Definition 4.1.1 *Let E be a countable space and let α and β be probability distributions on E. The* distance in variation $d_V(\alpha, \beta)$ *between α and β is defined by*

$$d_V(\alpha, \beta) = \frac{1}{2}|\alpha - \beta| = \frac{1}{2}\sum_{i \in E}|\alpha(i) - \beta(i)|. \tag{4.1}$$

The distance in variation *between two random variables X and Y with values in E and respective distributions $\mathcal{L}(X)$ and $\mathcal{L}(Y)$ is $d_V(\mathcal{L}(X), \mathcal{L}(Y))$, and it is denoted with a slight extension of the notation by $d_V(X, Y)$.*

© Springer Nature Switzerland AG 2020
P. Brémaud, *Markov Chains*, Texts in Applied Mathematics 31,
https://doi.org/10.1007/978-3-030-45982-6_4

Lemma 4.1.2 *Let X and Y be two random variables with values in the same countable space E. Then*

$$\sup_{A \subset E} |P(X \in A) - P(Y \in A)| = \sup_{A \subset E} \{P(X \in A) - P(Y \in A)\} = d_V(X, Y).$$

Proof. For the first equality observe that for each A there is a B such that $|P(X \in A) - P(Y \in A)| = P(X \in B) - P(Y \in B)$ (take $B = A$ or $\bar{A}$). For the second equality, write

$$P(X \in A) - P(Y \in A) = \sum_{i \in E} 1_A(i)\{P(X = i) - P(Y = i)\}$$

and observe that the right-hand side is maximal for

$$A = \{i \in E;\ P(X = i) > P(Y = i)\}.$$

Also, for any $A \subset E$,

$$\sum_{i \in E} 1_A(i)\{P(X = i) - P(Y = i)\} + \sum_{i \in E} 1_{\bar{A}}(i)\{P(X = i) - P(Y = i)\} = 0$$

because $\sum_{i \in E}\{P(X = i) - P(Y = i)\} = 0$. For the specific set A above, $P(X = i) - P(Y = i)$ equals $|P(X = i) - P(Y = i)|$ on A, and equals $-|P(X = i) - P(Y = i)|$ on $\bar{A}$. Therefore, for this particular set A,

$$\begin{aligned}
\sum_{i \in E} 1_A(i)\{P(X = i) - P(Y = i)\} &= \sum_{i \in E} 1_A(i)|P(X = i) - P(Y = i)| \\
&= \sum_{i \in E} 1_{\bar{A}}(i)|P(X = i) - P(Y = i)| \\
&= \frac{1}{2} \sum_{i \in E} |P(X = i) - P(Y = i)|.
\end{aligned}$$

$\square$

For two probability distributions α and β on the countable set E, denote by $\mathcal{D}(\alpha, \beta)$ the collection of random vectors (X, Y) taking their values in $E \times E$, and with marginal distributions α and β, that is, $\alpha = \mathcal{L}(X)$, $\beta = \mathcal{L}(Y)$.

Theorem 4.1.3 *For any $(X, Y) \in \mathcal{D}(\alpha, \beta)$,*

$$P(X = Y) \leq 1 - d_V(\alpha, \beta),$$

and equality is attained by some pair $(X, Y) \in \mathcal{D}(\alpha, \beta)$, which is then said to realize maximal coincidence.

Proof. For all $A \subset E$,

$$P(X \neq Y) \geq P(X \in A, Y \in \bar{A}) = P(X \in A) - P(X \in A, Y \in A)$$
$$\geq P(X \in A) - P(Y \in A),$$

and therefore

$$P(X \neq Y) \geq \sup_{A \subset E} \{P(X \in A) - P(Y \in A)\} = d_V(\alpha, \beta).$$

To finish the proof, it suffices to construct $(X, Y) \in \mathcal{D}(\alpha, \beta)$ realizing equality.

We shall need the following observations (Exercise 4.4.1):

$$\frac{1}{2}|\alpha - \beta| = \sum_{i \in E}(\alpha(i) - \beta(i))^+ = \sum_{i \in E}(\beta(i) - \alpha(i))^+ = 1 - \sum_{i \in E}\min(\alpha(i), \beta(i)). \quad (4.2)$$

Let now U, Z, V, and W be independent random variables; U takes its values in $\{0,1\}$, and Z, V, W take their values in E. The distributions of these random variables is given by

$$\begin{aligned}
P(U = 1) &= 1 - d_V(\alpha, \beta), \\
P(Z = i) &= \min(\alpha(i), \beta(i))/(1 - d_V(\alpha, \beta)), \\
P(V = i) &= (\alpha(i) - \beta(i))^+/d_V(\alpha, \beta), \\
P(W = i) &= (\beta(i) - \alpha(i))^+/d_V(\alpha, \beta).
\end{aligned}$$

Letting

$$X := UZ + (1 - U)V \text{ and } Y := UZ + (1 - U)W,$$

we have

$$\begin{aligned}
P(X = i) &= P(U = 1, Z = i) + P(U = 0, V = i) \\
&= P(U = 1)P(Z = i) + P(U = 0)P(V = i) \\
&= \min(\alpha(i), \beta(i)) + (\alpha(i) - \beta(i))^+ = \alpha(i),
\end{aligned}$$

and similarly, $P(Y = i) = \beta(i)$. Therefore, $(X, Y) \in \mathcal{D}(\alpha, \beta)$. Also, $P(X = Y) = P(U = 1) = 1 - d_V(\alpha, \beta)$. $\square$

EXAMPLE 4.1.4: MAXIMAL COINCIDENCE OF $\{0,1\}$-VALUED RANDOM VARI-
ABLES. One seeks a pair of $\{0,1\}$-valued random variables with prescribed marginals

$$P(X = 1) = a \, , \; P(Y = 1) = b \, ,$$

where $a, b \in (0, 1)$, and such that $P(X = Y)$ is maximal. In the notation of the
above theory,

$$\alpha = (1 - a, a) \, , \; \beta = (1 - b, b) \, ,$$

and therefore

$$d_V(\alpha, \beta) = |a - b| \, .$$

Suppose for definiteness that $a \geq b$. The random U, Z, V, W of the construction of
Theorem 4.1.3 have the following distributions.

$$
\begin{aligned}
P(U = 1) &= 1 - a + b \, , \\
P(Z = 1) &= \frac{b}{1 - a + b}, \; P(Z = 0) = \frac{1 - a}{1 - a + b}, \\
V &= 1 \, , \\
W &= 0 \, .
\end{aligned}
$$

Here $X = UZ + 1 - U$, $Y = UZ$.

Definition 4.1.5 *Let $\{\alpha_n\}_{n \geq 0}$ and β be probability distributions on a countable
space E. If $\lim_{n \uparrow \infty} d_V(\alpha_n, \beta) = 0$, the sequence $\{\alpha_n\}_{n \geq 0}$ is said to converge in
variation to the probability distribution β.*

*Let $\{X_n\}_{n \geq 0}$ be an E-valued stochastic process. If for some probability distri-
bution π on E, the distribution $\mathcal{L}(X_n)$ of the random variable X_n converges in
variation to π, that is, if*

$$\lim_{n \uparrow \infty} \sum_{i \in E} |P(X_n = i) - \pi(i)| = 0 \, , \tag{4.3}$$

then $\{X_n\}_{n \geq 0}$ is said to converge in variation *to π.*

There is some abuse of terminology in the above definition (it is the state
random variable, not the process, that converges in variation). However, in this
book, such abuse turns out to be harmless and very convenient.

If the process $\{X_n\}_{n \geq 0}$ converges in variation to π, then

$$\lim_{n \uparrow \infty} E[f(X_n)] = \pi(f)$$

for all bounded functions $f : E \to R$, where

$$\pi(f) = \sum_{i \in E} \pi(i) f(i) \,. \qquad (4.4)$$

Indeed, if M is an upper bound of $|f|$, then

$$|\mathrm{E}[f(X_n)] - \pi(f)| = \left| \sum_{i \in E} f(i) \left(\mathrm{P}(X_n = i) - \pi(i) \right) \right| \leq M \sum_{i \in E} |\mathrm{P}(X_n = i) - \pi(i)| \,.$$

4.1.2 The Coupling Method

Definition 4.1.6 *Two stochastic processes $\{X'_n\}_{n \geq 0}$ and $\{X''_n\}_{n \geq 0}$ taking their values in the same countable state space E are said to* couple *if there exists an almost surely finite random time τ such that*

$$n \geq \tau \Rightarrow X'_n = X''_n \,.$$

The random variable τ is called a coupling time *of the two processes.*

Theorem 4.1.7 *For any coupling time τ of $\{X'_n\}_{n \geq 0}$ and $\{X''_n\}_{n \geq 0}$, we have the fundamental coupling inequality:*

$$d_V(X'_n, X''_n) \leq \mathrm{P}(\tau > n) \,. \qquad (4.5)$$

Proof. For all $A \subset E$,

$$
\begin{aligned}
\mathrm{P}(X'_n \in A) - \mathrm{P}(X''_n \in A) &= \mathrm{P}(X'_n \in A, \ \tau \leq n) + \mathrm{P}(X'_n \in A, \ \tau > n) \\
&\quad - \mathrm{P}(X''_n \in A, \ \tau \leq n) - \mathrm{P}(X''_n \in A, \tau > n) \\
&= \mathrm{P}(X'_n \in A, \ \tau > n) - \mathrm{P}(X''_n \in A, \ \tau > n) \\
&\leq \mathrm{P}(X'_n \in A, \ \tau > n) \\
&\leq \mathrm{P}(\tau > n) \,.
\end{aligned}
$$

Inequality (4.5) then follows from Lemma 4.1.2. □

Observe that Definition 4.1.6 concerns only the marginal distributions of the process, not the process itself. Therefore, if there exists another process $\{X'_n\}_{n \geq 0}$ with $\mathcal{L}(X_n) = \mathcal{L}(X'_n)$ ([1]) for all $n \geq 0$, and if there exists a third process $\{X''_n\}_{n \geq 0}$ such that $\mathcal{L}(X''_n) = \pi$ for all $n \geq 0$, then (4.3) follows from (4.5) if we can show

---

[1] This notation means that the random variables X_n and X'_n have the same probability distribution.

that τ is finite. This trivial observation is useful because of the resulting freedom in the choice of $\{X'_n\}$ and $\{X''_n\}$. In particular, one can use dependent versions.

The framework of the *coupling method*[2] is now in place. It remains to construct $\{X'_n\}_{n\geq 0}$ and $\{X''_n\}_{n\geq 0}$ that couple and that mimic $\{X_n\}_{n\geq 0}$ and π in the sense that $\mathcal{L}(X'_n) = \mathcal{L}(X_n)$ and $\mathcal{L}(X'_n) = \pi$ for all $n \geq 0$.

4.2 Convergence to Steady State

4.2.1 The Positive Recurrent Case

The main result concerns ergodic (that is, irreducible, positive recurrent and *aperiodic*) HMCs.

Theorem 4.2.1 *Let* $\mathbf{P}$ *be an ergodic transition matrix on the countable state space* E. *For all probability distributions* μ *and* ν *on* E,

$$\lim_{n\uparrow\infty} d_V(\mu^T\mathbf{P}^n, \nu^T\mathbf{P}^n) = 0\,.$$

In particular, if ν is the stationary distribution π,

$$\lim_{n\uparrow\infty} |\mu^T\mathbf{P}^n - \pi^T| = 0,$$

and with $\mu = \delta_j$, the probability distribution putting all its mass on j,

$$\lim_{n\uparrow\infty} \sum_{i\in E} |p_{ji}(n) - \pi(i)| = 0\,.$$

From the discussion following Definition 4.1.6, it suffices to construct two coupling chains with initial distributions μ and ν, respectively.

Theorem 4.2.2 *Let* $\{X_n^{(1)}\}_{n\geq 0}$ *and* $\{X_n^{(2)}\}_{n\geq 0}$ *be two independent ergodic* HMCs *with the same transition matrix* $\mathbf{P}$ *and initial distributions* μ *and* ν, *respectively. Let* $\tau = \inf\{n \geq 0;\ X_n^{(1)} = X_n^{(2)}\}$, *with* $\tau = \infty$ *if the chains never intersect. Then* τ *is, in fact, almost surely finite. Moreover, the process* $\{X'_n\}_{n\geq 0}$ *defined by*

$$X'_n = \begin{cases} X_n^{(1)} & \text{if } n \leq \tau\,, \\ X_n^{(2)} & \text{if } n \geq \tau\,, \end{cases} \tag{4.6}$$

is an HMC *with transition matrix* $\mathbf{P}$. *(See Figure 4.2.1.)*

---

[2]The coupling idea comes from [Doeblin, 1938] and has a wide range of applications; see [Lindvall, 2002] who also gives historical comments.

Proof. Consider the product HMC $\{Z_n\}_{n \geq 0}$ defined by $Z_n = (X_n^{(1)}, X_n^{(2)})$. It takes values in $E \times E$, and the probability of transition from (i, k) to (j, ℓ) in n steps is $p_{ij}(n) p_{k\ell}(n)$. This chain is irreducible. Indeed, since **P** is irreducible and *aperiodic*, by Theorem 2.3.10, there exists an m such that for all pairs (i, j) and (k, ℓ), $n \geq m$ implies $p_{ij}(n) p_{k\ell}(n) > 0$. This implies that the period of the product chain is 1, again by Theorem 2.3.10.

Clearly, $\{\pi(i)\pi(j)\}_{(i,j) \in E^2}$ is a stationary distribution for the product chain, where π is the stationary distribution of **P**. Therefore, by the stationary distribution criterion, the product chain is positive recurrent. In particular, it reaches the diagonal of E^2 in finite time, and consequently, $\mathrm{P}(\tau < \infty) = 1$.

It remains to show that $\{X_n'\}_{n \geq 0}$ given by (4.6) is an HMC with transition matrix **P**. This is a consequence of the strong Markov property applied to the product chain. The details are left for the reader (Exercise 4.4.5). $\square$

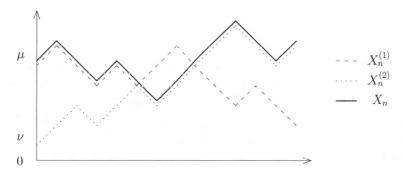

Figure 4.2.1. Independent coupling

Theorem 4.2.1 concerns ergodic chains, and aperiodicity is needed there to guarantee that the product chain is irreducible (Exercise 4.4.6). For periodic chains, the situation is different, but the result follows directly from the ergodic case.

Theorem 4.2.3 *Let* **P** *be an irreducible positive recurrent transition matrix on the countable space E, with period d. Let π be its stationary distribution. If μ is a probability distribution such that $\mu(C_0) = 1$ for some cyclic class C_0, then*

$$\lim_{n \uparrow \infty} \sum_{i \in C_0} |(\mu^T \mathbf{P}^{nd})_i - d\pi(i)| = 0.$$

Proof. Consider the restriction of $\mathbf{P}^d$ to C_0, which is irreducible and aperiodic (see Exercise 2.6.23). It is positive recurrent, since it has an invariant measure with finite mass, namely π restricted to C_0. It remains to show that $d\pi$ restricted to C_0 is a probability distribution, that is, $\pi(C_0) = 1/d$. By the ergodic theorem,

$$\lim_{N\uparrow\infty} \frac{1}{N} \sum_{n=1}^{n} 1_{\{X_n \in C_0\}} = \pi(C_0),$$

and since $X_n \in C_0$ once every d steps, the left-hand side equals $1/d$. □

4.2.2 The Null Recurrent Case

Theorem 4.2.3 concerns the positive recurrent case. The proof of the null recurrent version requires more care:

Theorem 4.2.4 ([3]) *Let* $\mathbf{P}$ *be an irreducible null recurrent transition matrix on* E. *Then for all* $i, j \in E$,

$$\lim_{n\uparrow\infty} p_{ij}(n) = 0. \tag{4.7}$$

Proof. The periodic case follows from the aperiodic case by considering the restriction of $\mathbf{P}^d$ to C_0, an arbitrary cyclic class, and observing that this restriction is also null recurrent. Therefore, $\mathbf{P}$ will be assumed aperiodic.

In this case, we have seen that the product HMC $\{Z_n\}_{n\geq 0} = \{X_n^{(1)}, X_n^{(2)})\}_{n\geq 0}$ is irreducible and aperiodic. However, it cannot be argued that it is recurrent, even if each of its components is recurrent. One must therefore separate the two possible cases.

First, suppose the product chain is transient. Its n-step transition probability from (i, i) to (j, j) is $[p_{ij}(n)]^2$, and it tends to 0 as $n \to \infty$. The result is therefore proved in this particular case.

Suppose now that the product chain is recurrent. The coupling argument used in the aperiodic case applies and yields

$$\lim_{n\uparrow\infty} |\mu^T \mathbf{P}^n - \nu^T \mathbf{P}^n| = 0 \tag{4.8}$$

for arbitrary initial distributions μ and ν. Suppose now that for some $i, j \in E$, (4.7) is not true. One can then find a sequence $\{n_k\}_{k\geq 0}$ of integers strictly increasing to ∞ such that

$$\lim_{k\uparrow\infty} p_{ij}(n_k) = \alpha > 0.$$

---

[3][Orey, 1971].

For fixed $i \in E$ chosen as above, the sequence $(\{p_{is}(n_k), s \in E\})_{k \geq 0}$ of vectors of $[0,1]^E$ is compact in the topology of pointwise convergence. Therefore (Tychonov's theorem; see the Appendix), there exists a subsequence $\{m_\ell\}_{\ell \geq 0}$ of integers strictly increasing to ∞ and a vector $\{x_s, s \in E\} \in [0,1]^E$ such that for all $s \in E$,

$$\lim_{\ell \uparrow \infty} p_{is}(m_\ell) = x_s \,.$$

Now, $x_j = \alpha > 0$, and therefore $\{x_s, s \in E\}$ is non-trivial. Since $\sum_{s \in E} p_{is}(m_\ell) = 1$, it follows from Fatou's lemma that $\sum_{s \in E} x_s \leq 1$. Writing (4.8) with $\mu = \delta_i$ and $\nu = \delta_i \mathbf{P}$, we have that

$$\lim_{n \uparrow \infty} d_V(\delta_i^T \mathbf{P}^n - (\delta_i^T \mathbf{P})\mathbf{P}^n) = \lim_{n \uparrow \infty} d_V(\delta_i^T \mathbf{P}^n - (\delta_i^T \mathbf{P}^n)\mathbf{P}) = 0 \,,$$

and in particular,

$$\lim_{n \uparrow \infty} \left(p_{is}(n) - \sum_{k \in E} p_{ik}(n)p_{ks} \right) = 0 \,,$$

and then,

$$\lim_{\ell \uparrow \infty} \left(p_{is}(m_\ell) - \sum_{k \in E} p_{ik}(m_\ell)p_{ks} \right) = 0 \,.$$

Therefore,

$$x_s = \lim_{\ell \uparrow \infty} \sum_{k \in E} p_{ik}(m_\ell)p_{ks} \,.$$

By Fatou's lemma again,

$$\lim_{\ell \uparrow \infty} \sum_{k \in E} p_{ik}(m_\ell)p_{ks} \geq \sum_{k \in E} \lim_{\ell \uparrow \infty} p_{ik}(m_\ell)p_{ks} = \sum_{k \in E} x_k p_{ks} \,.$$

Therefore

$$x_s \geq \sum_{k \in E} x_k p_{ks}.$$

Summing up with respect to s:

$$\sum_{s \in E} x_s \geq \sum_{s \in E} \sum_{k \in E} x_k p_{ks} = \sum_{k \in E} \left(x_k \sum_{s \in E} x_k p_{ks} \right) = \sum_{k \in E} x_k \,,$$

which shows that the inequality ($\geq$) can only be an equality. In other words, $\{x_s, s \in E\}$ is a non-trivial invariant measure of $\mathbf{P}$ of finite total mass, which implies that $\mathbf{P}$ is positive recurrent, a contradiction. Therefore, (4.7) cannot be contradicted. $\square$

EXAMPLE 4.2.5: THERMODYNAMIC IRREVERSIBILITY. According to the macroscopic theory of thermodynamics, systems progress in an orderly and irreversible manner towards equilibrium. Consider, for instance, a system of N particles in a box divided into two similar compartments A and B by a fictive membrane. If at the origin of time, all particles are in A, they will rather quickly reorganize themselves and "settle to equilibrium", a macroscopic state in which the contents of A and B are thermodynamically equivalent.

Boltzmann claimed that there was an arrow of time in the direction of increasing entropy, and indeed, in the diffusion experiment, equality between the thermodynamic quantities in both compartments corresponds to maximal entropy.

Zermelo, who obviously was not sleeping in the back of the classroom, argued that in view of the time reversibility of the laws of physics, the Boltzmann theory should at least be discussed. Zermelo held a strong position in this controversy. Indeed, there is a famous result of mechanics, Poincaré's recurrence theorem, which implies that in the situation where at time 0 all molecules are in A, then whatever the time T, there will be a subsequent time $t > T$ at which all the molecules will again gather in A. This phenomenon, predicted by irrefutable mathematics, is of course never observed in daily life, where it would imply that the chunk of sugar that one patiently dissolves in one's cup of coffee could escape ingestion by reforming itself at the bottom of the cup.

Boltzmann's theory was hurt by this striking and seemingly inescapable argument. Things had to be clarified. Fortunately, Tatiana and Paul Ehrenfest came up with their Markov chain model, and in a sense saved the edifice that Boltzmann had constructed.

Without going into the details, let us just say that the Ehrenfest model is an approximation of the real diffusion phenomenon that is *qualitatively correct* from the point of view of statistical mechanics. Also, at first sight, it is subject to Zermelo's attack, presenting both features that the latter found incompatible: an irreversible tendency towards equilibrium and recurrence. Here the role of Poincaré's recurrence theorem is played by the Markov chain recurrence theorem, stating that an irreducible chain with a stationary distribution visits any fixed state, say 0, infinitely often. The irreversible tendency towards equilibrium is contained in the theorem of convergence to steady state, according to which the distribution at time n converges to the stationary distribution whatever the initial distribution as n tends to infinity.[4] Thus, according to Markov-chain theory, convergence to

---

[4] *Stricto sensu* this statement is not true, due to the periodicity of the chain. However, such periodicity is an artefact created by the discretization of time, and it would disappear in the continuous-time model, or in a slight modification of the discrete-time model.

statistical equilibrium and recurrence are not antagonistic, and we are here at the epicenter of Zermelo's refutation.

One can show that recurrence is *not observable* for states far from $L = \frac{N}{2}$, assuming that N is even. For instance, the average time to reach 0 from state L is

$$\frac{1}{2L}2^{2L}(1 + O(L)) \tag{4.9}$$

whereas the average time to reach state L from state 0 is less than

$$L + L\log L + O(1). \tag{4.10}$$

See Chapter III, Section 5 of [Bhattacharya and Waymire, 1990] for a derivation of the above estimates. With $L = 10^6$ and one unit of mathematical time equal to 10^{-5} seconds, the return time to equilibrium when compartment A is initially empty is of the order of a second, whereas it would take of the order of

$$\frac{1}{2 \cdot 10^{11}} \times 2^{2^{10^6}} \text{ seconds}$$

to pass from full to empty, which is an astronomical time. These numbers teach us not to spend too much time stirring the coffee, or hurry to swallow it for fear of recrystallization of the chunk of sugar. From a mathematical point of view, being in the steady state at a given time does not prevent the chain from being in a rare state, only it is there rarely. The rarity of the state is equivalent to long recurrence times, so long that when there are more than a few particles in the boxes, it would take an astronomical time to witness the effects of Poincaré's recurrence theorem. Note that Boltzmann rightly argued that the recurrence times in Poincaré's theorem are extremely long, but his heuristic arguments failed to convince.

Here is another manifestation of thermodynamic irreversibility in the Ehrenfest model.

EXAMPLE 4.2.6: NEWTON'S LAW OF COOLING. Let $g(z, n) = E[z^{X_n}]$ be the generating function of X_n. Using the basic rules of conditional expectation, we have

$$
\begin{aligned}
E[z^{X_{n+1}} \mid X_n] &= E[z^{X_{n+1}}1_{\{X_{n+1}=X_n-1\}} \mid X_n] + E[z^{X_{n+1}}1_{\{X_{n+1}=X_n+1\}} \mid X_n] \\
&= E[z^{X_n-1}1_{\{X_{n+1}=X_n-1\}} \mid X_n] + E[z^{X_n+1}1_{\{X_{n+1}=X_n+1\}} \mid X_n] \\
&= z^{X_n-1}P(X_{n+1} = X_n - 1 \mid X_n) + z^{X_n+1}P(X_{n+1} = X_n + 1 \mid X_n),
\end{aligned}
$$

and therefore, taking into account the dynamics of the Ehrenfest model,

$$E[z^{X_{n+1}} \mid X_n] = z^{X_n-1}\frac{X_n}{N} + z^{X_n+1}\left(1 - \frac{X_n}{N}\right)$$

$$= z \cdot z^{X_n} + \frac{1}{N}(1 - z^2)X_n z^{X_n-1}.$$

Taking expectations gives

$$g(z, n+1) = zg(z,n) + \frac{1}{N}(1 - z^2)E[X_n z^{X_n-1}],$$

that is,

$$g(z, n+1) = zg(z,n) + \frac{1}{N}(1 - z^2)g'(z,n).$$

Differentiation of the above identity yields

$$g'(z, n+1) = g(z,n) + zg'(z,n) - \frac{2z}{N}g'(z,n) + \frac{1}{N}(1 - z^2)g''(z,n).$$

Letting $z = 1$, we obtain

$$E[X_{n+1}] = 1 + E[X_n] - \frac{2}{N}E[X_n].$$

Supposing N even $(= 2L)$ and rearranging terms, we have

$$E[X_{n+1} - L] = (1 - \frac{1}{L})E[X_n - L]$$

and therefore

$$E[X_n - L] = E[X_0 - L]\left(1 - \frac{1}{L}\right)^n.$$

Supposing $P(X_0 = i) = 1$, we then have

$$E\left[\frac{X_n - L}{L}\right] = \left(\frac{i - L}{L}\right)\left(1 - \frac{1}{L}\right)^n.$$

In the kinetic theory of heat, $\frac{X_n-L}{L}$ is interpreted as the temperature difference between state X_n and the "equilibrium" state L. To account for a large number of particles, we let L tend to infinity and make i depend on L in such a way that

$$\lim_{L\to\infty} \frac{i(L) - L}{L} = \theta(0).$$

The number $\theta(0)$ is interpreted as the initial deviation from the equilibrium temperature in compartment A. Also, the discrete time unit 1 is now interpreted as

Δ units of real time. If X_n is the state at real time t, we must then have $t = n\Delta$. Interpreting $\frac{1}{2L\Delta}$ as the average proportion of the total energy that passes through the membrane per unit of real time, we require that as L tends to infinity, this quantity remains constant, that is, $\frac{1}{L\Delta} = \gamma$. In particular, Δ tends to zero as L increases to infinity. Therefore, observing that $n = L\gamma t$,

$$\lim_{L\to\infty}\left(1 - \frac{1}{L}\right)^n = \lim_{L\to\infty}\left[\left(1 - \frac{1}{L}\right)^L\right]^{\gamma t} = e^{-\gamma t},$$

and finally, with $\theta(t) = \lim_{L\to\infty} E[\frac{X_n - L}{L}]$,

$$\theta(t) = \theta(0)e^{-\gamma t}.$$

We conclude this section by mentioning a classical reference on thermodynamic irreversibility as probabilists understand it, the article of M. Kac (1947), "Random Walk and the Theory of Brownian Motion", *American Mathematical Monthly, 54*, 369–391.

4.3 Convergence Rates, a First Look

4.3.1 Convergence Rates via Coupling

Knowing that an ergodic Markov chain converges to equilibrium, the next question is, How fast? The first result below is not explicit, but it can be used *in principle* for chains with an infinite number of states.

We recall at this point the meaning of the o (*small o*) symbol. It represents a function defined in a neighborhood of zero such that $\lim_{t\to 0}\frac{|o(t)|}{t} = 0$.

Theorem 4.3.1 *Suppose that the coupling time τ in Theorem 4.2.2 is such that*

$$E[\psi(\tau)] < \infty \tag{4.11}$$

for some non-decreasing function $\psi : \mathbb{N} \to \mathbb{R}_+$ such that $\lim_{n\uparrow\infty}\psi(n) = \infty$. Then for any initial distributions μ and ν

$$|\mu^T \mathbf{P}^n - \nu^T \mathbf{P}^n| = o\left(\frac{1}{\psi(n)}\right). \tag{4.12}$$

Proof. Since ψ is non-decreasing, $\psi(\tau)1_{\{\tau > n\}} \geq \psi(n)1_{\{\tau > n\}}$, and therefore

$$\psi(n)P(\tau > n) \leq E[\psi(\tau)1_{\{\tau > n\}}].$$

Now,

$$\lim_{n \uparrow \infty} \mathrm{E}[\psi(\tau) 1_{\{\tau > n\}}] = 0$$

by dominated convergence (Section A.3), since $\lim_{n \uparrow \infty} \psi(\tau) 1_{\{\tau > n\}} = 0$, by the finiteness of τ, and $\psi(\tau) 1_{\{\tau > n\}}$ is bounded by the integrable random variable $\psi(\tau)$.   □

Remark 4.3.2 Time τ in Theorems 4.2.2 and 4.3.1 is the entrance time of the product chain in the diagonal set of $E \times E$. In principle, the distribution of τ can be explicitly computed. However the actual computations are usually difficult when the state space is infinite.[5]

Theorem 4.3.3 *Let $\mathbf{P}$ be an ergodic transition matrix on the finite state space E. Then for any initial distributions μ and ν, one can construct two HMCs $\{X_n\}_{n \geq 0}$ and $\{Y_n\}_{n \geq 0}$ on E with the same transition matrix $\mathbf{P}$, and the respective initial distributions μ and ν, in such a way that they couple at a finite time τ such that $\mathrm{E}[e^{\alpha \tau}] < \infty$ for some positive α.*

Proof. Exercise 4.4.4.   □

4.3.2   The Perron–Frobenius Theorem

When the state space is finite, we can rely on the standard results of linear algebra to study the asymptotic behavior of homogeneous Markov chains. Indeed, the asymptotic behavior of the distribution at time n of the chain is entirely described by the asymptotic behavior of the n-step transition matrix $\mathbf{P}^n$, and the latter depends on the eigenstructure of $\mathbf{P}$. The *Perron–Frobenius theorem* detailing the eigenstructure of non-negative matrices is therefore all that is needed, at least in theory.

The principal result of Perron and Frobenius is that convergence to steady state of an ergodic finite state space HMC is geometric, with relative speed equal to the second-largest eigenvalue modulus (SLEM). It is true that there are a number of interesting models, especially in biology, where the eigenstructure of the transition matrix can be extracted.[6] Nevertheless, this situation remains exceptional. It is therefore important to find estimates, more precisely, upper and lower bounds, of the SLEM. This will be done in Chapter 9.

---

[5][Lindvall, 2002] has examples of application of Theorem 4.2.2 to infinite state space HMCs.

[6]See for instance [Iosifescu, 1980] or [Karlin and Taylor, 1975].

The basic results of the theory of matrices relative to eigenvalues and eigenvectors are reviewed in the appendix (Subsection A.2.1), from which we extract the following one, relative to a square matrix A of dimension r with *distinct* eigenvalues.

Let $\lambda_1, \ldots, \lambda_r$ be the r distinct eigenvalues and let $u_1, \ldots, u_r$ and $v_1, \ldots, v_r$ be the associated sequences of left and right-eigenvectors, respectively. Then, $u_1, \ldots, u_r$ form an independent collection of vectors, and so do $v_1, \ldots, v_r$. Also, $u_i^T v_j = 0$ if $i \neq j$. Since eigenvectors are determined up to multiplication by an arbitrary non-null scalar, one can choose them in such a way that $u_i^T v_i = 1$ $(1 \leq i \leq r)$. We then have the spectral decomposition

$$A^n = \sum_{i=1}^{r} \lambda_i^n v_i u_i^T . \tag{4.13}$$

EXAMPLE 4.3.4: CONVERGENCE RATE FOR THE TWO-STATE CHAIN. The transition matrix on $E = \{1, 2\}$

$$\mathbf{P} = \begin{pmatrix} 1 - \alpha & \alpha \\ \beta & 1 - \beta \end{pmatrix},$$

where $\alpha, \beta \in (0, 1)$ has for characteristic polynomial $(1 - \alpha - \lambda)(1 - \beta - \lambda) - \alpha\beta$ admits the roots $\lambda_1 = 1$ and

$$\lambda_2 = 1 - \alpha - \beta.$$

Observe at this point that $\lambda = 1$ is always an eigenvalue of a stochastic $r \times r$ matrix $\mathbf{P}$, associated with the right-eigenvector $v = \mathbf{1}$ with all entries equal to 1, since $\mathbf{P}\mathbf{1} = \mathbf{1}$. Also, the stationary distribution

$$\pi^T = \left(\frac{\beta}{\alpha + \beta}, \frac{\alpha}{\alpha + \beta} \right)$$

is the left-eigenvector corresponding to the eigenvalue 1. In this example, the representation (4.13) takes the form

$$\mathbf{P}^n = \frac{1}{\alpha + \beta} \begin{pmatrix} 1 - \alpha & \alpha \\ \beta & 1 - \beta \end{pmatrix} + \frac{(1 - \alpha - \beta)^n}{\alpha + \beta} \begin{pmatrix} \alpha & -\alpha \\ 1 - \beta & -\beta \end{pmatrix}$$

and therefore, since $|1 - \alpha - \beta| < 1$,

$$\lim_{n \uparrow \infty} \mathbf{P}^n = \frac{1}{\alpha + \beta} \begin{pmatrix} 1 - \alpha & \alpha \\ \beta & 1 - \beta \end{pmatrix} .$$

In particular, the result of convergence to steady state,

$$\lim_{n \uparrow \infty} \mathbf{P}^n = \mathbf{1}\pi^T = \mathbf{P}^\infty,$$

was recovered for this special case in a purely algebraic way. In addition, this algebraic method gives the convergence speed, which is exponential and determined by the second-largest eigenvalue absolute value:

$$(\mathbf{P}^n - \mathbf{P}^\infty) = \frac{(1 - \alpha - \beta)^n}{\alpha + \beta} \begin{pmatrix} \alpha & -\alpha \\ -\beta & -\beta \end{pmatrix}.$$

This is a general fact, which follows from the Perron–Frobenius theory of non-negative matrices below.

Definition 4.3.5 *A matrix $A = \{a_{ij}\}_{1 \leq i,j \leq r}$ with real coefficients is called non-negative (resp., positive) if all its entries are non-negative (resp., positive). A non-negative matrix A is called* stochastic *if $\sum_{j=1}^{r} a_{ij} = 1$ for all i, and* substochastic *if $\sum_{j=1}^{r} a_{ij} \leq 1$ for all i, with strict inequality for at least one i.*

Non-negativity (resp., positivity) of A is denoted by $A \geq 0$ (resp., $A > 0$). If A and B are two matrices of the same dimensions with real coefficients, the notation $A \geq B$ (resp., $A > B$) means that $A - B \geq 0$ (resp., $A - B > 0$).

The *communication graph* of a square non-negative matrix A is the oriented graph with the state space $E = \{1, \ldots, r\}$ as its set of vertices and an oriented edge from vertex i to vertex j if and only if $a_{ij} > 0$.

Definition 4.3.6 *A non-negative square matrix A is called* irreducible *(resp., irreducible aperiodic) if it has the same communication graph as an irreducible (resp., irreducible aperiodic) stochastic matrix. It is called* primitive *if there exists an integer k such that $A^k > 0$.*

Remark 4.3.7 A non-negative matrix is primitive if and only if it is irreducible and aperiodic (Exercise 2.6.20).

Theorem 4.3.8 *Let A be a non-negative primitive $r \times r$ matrix. There exists a real eigenvalue λ_1 with algebraic as well as geometric multiplicity one such that $\lambda_1 > 0$, and $\lambda_1 > |\lambda_j|$ for any other eigenvalue λ_j. Moreover, the left-eigenvector u_1 and the right-eigenvector v_1 associated with λ_1 can be chosen positive and such that $u_1^T v_1 = 1$.*

Let $\lambda_2, \lambda_3, \ldots, \lambda_r$ be the eigenvalues of A other than λ_1 ordered in such a way that

$$\lambda_1 > |\lambda_2| \geq \cdots \geq |\lambda_r| \tag{4.14}$$

and if $|\lambda_2| = |\lambda_j|$ for some $j \geq 3$, then $m_2 \geq m_j$, where m_j is the algebraic multiplicity of λ_j. Then

$$A^n = \lambda_1^n v_1 u_1^T + O(n^{m_2-1}|\lambda_2|^n), \qquad (4.15)$$

where $O(f(n))$ represents a function of n such that there exists $\alpha, \beta \in \mathbb{R}, 0 < \alpha \leq \beta < \infty$, such that $\alpha f(n) \leq O(f(n)) \leq \beta f(n)$ for all n sufficiently large.

If in addition, A is stochastic (resp., substochastic), then $\lambda_1 = 1$ (resp., $\lambda_1 < 1$).

If A is stochastic but not irreducible, then the algebraic and geometric multiplicities of the eigenvalue 1 are equal to the number of communication classes.

If A is stochastic and irreducible with period $d > 1$, then there are exactly d distinct eigenvalues of modulus 1, namely the dth roots of unity, and all other eigenvalues have modulus strictly less than 1.

For the proof, see [Seneta, 1981] or [Gantmacher, 1959].

EXAMPLE 4.3.9: RATES OF CONVERGENCE VIA THE PERRON–FROBENIUS THEOREM. If $\mathbf{P}$ is a transition matrix on $E = \{1, \ldots, r\}$ that is irreducible and aperiodic, and therefore primitive, then

$$v_1 = \mathbf{1}, \ u_1 = \pi,$$

where π is the unique stationary distribution. Therefore

$$\mathbf{P}^n = \mathbf{1}\pi^T + O(n^{m_2-1}|\lambda_2|^n), \qquad (4.16)$$

which generalizes Example 4.3.4.

EXAMPLE 4.3.10: JUST ANOTHER EXAMPLE. The characteristic polynomial of the doubly stochastic matrix

$$\mathbf{P} = \frac{1}{12} \begin{pmatrix} 0 & 6 & 6 \\ 4 & 3 & 5 \\ 8 & 3 & 1 \end{pmatrix}$$

is

$$\det(\lambda I - \mathbf{P}) = (\lambda - 1)(\lambda + \frac{1}{6})(\lambda + \frac{1}{2}).$$

Since this matrix is doubly stochastic, $v_1 = \frac{1}{3}(1,1,1)^T$ and $u_1 = (1,1,1)^T$ are a right-eigenvector and a left-eigenvector, respectively, corresponding to the eigenvalue $\lambda_1 = 1$ and such that $u_1^T v_1 = 1$. Elementary computations yield for the right and left-eigenvectors corresponding to $\lambda_2 = -\frac{1}{2}$ and $\lambda_3 = -\frac{1}{6}$, that

$$
u_2 = \frac{1}{12}(2,-1,-1)^T, \quad v_2 = (4,1,-5)^T, \quad u_3 = \frac{1}{4}(-2,3,-1)^T, \quad v_3 = (0,1,-1)^T,
$$

where, here again, $u_2^T v_2 = u_3^T v_3 = 1$. Therefore,

$$
\mathbf{P}^n = \frac{1}{3}\begin{pmatrix} 1 & 1 & 1 \\ 1 & 1 & 1 \\ 1 & 1 & 1 \end{pmatrix} + \left(-\frac{1}{2}\right)^n \frac{1}{12}\begin{pmatrix} 8 & -4 & -4 \\ 2 & -1 & -1 \\ -10 & 5 & 5 \end{pmatrix}
$$
$$
+ \left(-\frac{1}{6}\right)^n \frac{1}{4}\begin{pmatrix} 0 & 0 & 0 \\ -2 & 3 & -1 \\ 2 & -3 & 1 \end{pmatrix}.
$$

The convergence to the steady state is geometric with relative speed $\frac{1}{2}$.

4.3.3   Quasi-stationary Distributions

[7] Let $\{X_n\}_{n\geq 0}$ be an HMC with finite state space E. Suppose that the set of recurrent states R and the set of transient states T are both nonempty. In the block decomposition of $\mathbf{P}$ with respect to the partition $R + T = E$,

$$
\mathbf{P} = \begin{pmatrix} D & 0 \\ B & \mathbf{Q} \end{pmatrix},
$$

the matrix $\mathbf{Q}$ is substochastic, since B is not identically null (otherwise, the transient set would be closed, and therefore recurrent, being finite). We assume, in addition, that $\mathbf{Q}$ is irreducible and aperiodic. Therefore,

$$
\mathbf{Q}^n = \lambda_1^n v_1 u_1^T + O(n^{m_2-1}|\lambda_2|^n), \tag{4.17}
$$

where λ_1, v_1, u_1, m_2, and λ_2 are as in Theorem 4.3.8 with $A = \mathbf{Q}$. In particular, $\lambda_1 \in (0,1)$ and $|\lambda_2| < \lambda_1$.

Let $\nu = \inf\{n \geq 0; X_n \in R\}$ be the entrance time into R. This time is almost surely finite since T is a finite set (this will be proved in Remark 6.1.4). We want

---

[7]Quasi-stationary distributions were introduced in [Bartlett, 1957]. The reader is referred to [Iosifescu, 1980] or [Seneta, 1981] for additional information.

to find the distribution of X_n for large n, conditioned by the fact that X_n is still in T. For this we compute for $i, j \in T$,

$$P_i(X_n = j \mid \nu > n) = \frac{P_i(X_n = j, \nu > n)}{P_i(\nu > n)} = \frac{P_i(X_n = j)}{P_i(X_n \in T)}.$$

Therefore,

$$P_i(X_n = j \mid \nu > n) = \frac{p_{ij}(n)}{\sum_{k \in T} p_{ik}(n)}.$$

In view of (4.17),

$$p_{ik}(n) = \lambda_1^n v_1(i) u_1(k) + O(n^{m_2 - 1} |\lambda_2|^n).$$

Therefore,

$$P_i(X_n = j \mid \nu > n) = \frac{u_1(j)}{\sum_{k \in T} u_1(k)} + O\left(n^{m_2 - 1} \left|\frac{\lambda_2}{\lambda_1}\right|^n\right), \tag{4.18}$$

and in particular,

$$\lim_{n \uparrow \infty} P_i(X_n = j \mid \nu > n) = \frac{u_1(j)}{\sum_{k \in T} u_1(k)}. \tag{4.19}$$

The probability distribution $\{u_1(i)/\sum_{k \in T} u_1(k)\}_{i \in T}$ is called the *quasi-stationary distribution* of the chain relative to T.

4.3.4   Dobrushin's Ergodic Coefficient

When the state space E is finite and the chain is ergodic, *Dobrushin's ergodic coefficient*[8] provides a computable geometric rate of convergence to steady state and will be especially useful in the study of non-homogeneous Markov chains (Chapter 12).

Let E, F, G denote countable sets, finite when so indicated. A stochastic matrix indexed by $F \times E$ is a matrix whose rows (indexed by F) are probability distributions on E.

Definition 4.3.11 *Let $\mathbf{Q}$ be a stochastic matrix indexed by $F \times E$. Its ergodic coefficient is*

$$\delta(\mathbf{Q}) := \frac{1}{2} \sup_{i,j \in F} \sum_{k \in E} |q_{ik} - q_{jk}|$$

$$= \sup_{i,j \in F} d_V(q_{i\cdot}, q_{j\cdot}) = \sup_{i,j \in F} \sup_{A \subseteq E} (q_{iA} - q_{jA}).$$

---

[8][Dobrushin, 1956].

Observe that $0 \leq \delta(\mathbf{Q}) \leq 1$ and that, by the result of Exercise 4.4.12,

$$\delta(\mathbf{Q}) = 1 - \inf_{i,j \in F} \sum_{k \in E} q_{ik} \wedge q_{jk} \,. \tag{4.20}$$

The ergodic coefficient is in general useless if F is infinite. In particular, if the stochastic matrix $\mathbf{Q}$ has two orthogonal rows (that is, rows i, j such that $q_{ik} q_{jk} = 0$ for all $k \in E$), and this is the most frequent case with an infinite state space, then $\delta(\mathbf{Q}) = 1$. However, for finite state spaces, this notion becomes very powerful.

EXAMPLE 4.3.12: TWO-ROW MATRIX. The ergodic coefficient of the stochastic matrix

$$\mathbf{Q} = \begin{pmatrix} \mu^T \\ \nu^T \end{pmatrix} = \begin{pmatrix} \mu_1 & \mu_2 & \mu_3 & \cdots \\ \nu_1 & \nu_2 & \nu_3 & \cdots \end{pmatrix}$$

is the distance in variation between the two rows: $\delta(\mathbf{Q}) = d_V(\mu, \nu)$.

EXAMPLE 4.3.13: COUPLING AND THE ERGODIC COEFFICIENT. We are going to construct two HMCs $\{X_n^{(1)}\}_{n \geq 0}$ and $\{X_n^{(2)}\}_{n \geq 0}$ with the same transition matrix $\mathbf{P}$, assumed irreducible, and with a strictly positive ergodic coefficient, in such a way that they couple at a time τ stochastically smaller than a geometric random variable with parameter $p = 1 - \delta(\mathbf{P})$. The construction is as follows. Let $\alpha_i(j) := p_{ij}$. Suppose that at time n, $X_n^{(1)} = i_1$ and $X_n^{(2)} = i_2$. Then, $X_{n+1}^{(1)}$ and $X_{n+1}^{(2)}$ will be distributed according to the distributions α_{i_1} and α_{i_2}, respectively. This can be done in such a way that

$$P\left(X_{n+1}^{(1)} = X_{n+1}^{(2)} \mid X_n^{(1)} = i_1, X_n^{(2)} = i_2\right) = d_V\left(\alpha_{i_1}, \alpha_{i_2}\right) \,.$$

The latter quantity is $\geq 1 - \delta(\mathbf{P}) > 0$. Therefore the two chains will meet for the first time (and from then on be identical) at a time τ that is stochastically smaller that a geometric random variable with parameter $p = 1 - \delta(\mathbf{P})$.

The following sub-multiplicativity property of the ergodic coefficient (*Dobrushin's inequality*) will play a fundamental role.

Theorem 4.3.14 *Let* $\mathbf{Q}_1$ *and* $\mathbf{Q}_2$ *be two stochastic matrices indexed by* $F \times G$ *and* $G \times E$, *respectively. Then*

$$\delta(\mathbf{Q}_1 \mathbf{Q}_2) \leq \delta(\mathbf{Q}_1) \delta(\mathbf{Q}_2) \,.$$

Proof. From Lemma 4.1.2, for any stochastic matrix $\mathbf{Q} = \{q_{ij}\}$ indexed by $E \times F$,

$$\frac{1}{2} \sum_{k \in E} |q_{ik} - q_{jk}| = \sup_{A \subseteq E} \sum_{k \in A} (q_{ik} - q_{jk}),$$

and in particular, with $\mathbf{Q}_1 = \{a_{ij}\}$ and $\mathbf{Q}_2 = \{b_{ij}\}$,

$$\delta(\mathbf{Q}_1 \mathbf{Q}_2) = \frac{1}{2} \sup_{i,j \in F} \sup_{A \subseteq E} \sum_{k \in A} \left(\sum_{\ell \in G} (a_{i\ell} - a_{j\ell}) b_{\ell k} \right).$$

But

$$\sum_{\ell \in G} (a_{i\ell} - a_{j\ell})^+ = \sum_{\ell \in G} (a_{i\ell} - a_{j\ell})^- = \frac{1}{2} \sum_{\ell \in G} |a_{i\ell} - a_{j\ell}|,$$

and therefore

$$\sum_{k \in A} \sum_{\ell \in G} (a_{i\ell} - a_{j\ell}) b_{\ell k} = \sum_{\ell \in G} (a_{i\ell} - a_{j\ell})^+ \sum_{k \in A} b_{\ell k} - \sum_{\ell \in G} (a_{i\ell} - a_{j\ell})^- \sum_{k \in A} b_{\ell k}$$

$$\leq \left(\frac{1}{2} \sum_{\ell \in G} |a_{i\ell} - a_{j\ell}| \right) \sup_{\ell \in G} \sum_{k \in A} b_{\ell k}$$

$$- \left(\frac{1}{2} \sum_{\ell \in G} |a_{i\ell} - a_{j\ell}| \right) \inf_{\ell \in G} \sum_{k \in A} b_{\ell k}$$

$$\leq \left(\frac{1}{2} \sum_{\ell \in G} |a_{il} - a_{j\ell}| \right) \left(\sup_{\ell, \ell' \in G} \sum_{k \in A} (b_{\ell k} - b_{\ell' k}) \right).$$

The announced inequality then follows from the identity

$$\sup_{\ell, \ell' \in G} \sup_{A \subseteq E} \sum_{k \in A} (b_{\ell k} - b_{\ell' k}) = \frac{1}{2} \sup_{\ell, \ell' \in G} \sum_{k \in E} |b_{\ell k} - b_{\ell' k}| = \delta(\mathbf{Q}_2).$$

$\square$

Theorem 4.3.15 *Let $\mathbf{P}$ be a stochastic matrix indexed by E, and let μ and ν be two probability distributions on E. Then*

$$d_V(\mu^T \mathbf{P}^n, \nu^T \mathbf{P}^n) \leq d_V(\mu, \nu) \delta(\mathbf{P})^n. \tag{4.21}$$

Proof. The proof is by recurrence. Since

$$\begin{pmatrix} \mu^T \mathbf{P}^{n+1} \\ \nu^T \mathbf{P}^{n+1} \end{pmatrix} = \begin{pmatrix} \mu^T \mathbf{P}^n \\ \nu^T \mathbf{P}^n \end{pmatrix} \mathbf{P}$$

and (see Example 4.3.12)

$$d_V(\mu^T \mathbf{P}^n, \nu^T \mathbf{P}^n) = \delta \left(\begin{pmatrix} \mu^T \mathbf{P}^n \\ \nu^T \mathbf{P}^n \end{pmatrix} \right).$$

Therefore, by Lemma 4.3.14,

$$d_V(\mu^T \mathbf{P}^{n+1}, \nu^T \mathbf{P}^{n+1}) \leq d_V(\mu^T \mathbf{P}^n, \nu^T \mathbf{P}^n) \delta(\mathbf{P}),$$

from which (4.21) follows by iteration.   □

Remark 4.3.16 Theorem 4.3.15 implies that Dobrushin's coefficient is an upper bound of the SLEM.

Recall the notations $|v| := \sum_i |v_i|$ for a vector $v = \{v_i\}$ and $|\mathbf{Q}| := \sup_i \left(\sum_k |q_{ik}| \right)$ for a matrix (not necessarily square) $\mathbf{Q} = \{q_{ik}\}$.

Corollary 4.3.17 *Let* $\mathbf{Q}_1$, $\mathbf{Q}_2$ *and* P *be stochastic matrices indexed by* $E \times E$. *Then*

$$|(\mathbf{Q}_1 - \mathbf{Q}_2)\mathrm{P}| \leq |\mathbf{Q}_1 - \mathbf{Q}_2| \delta(\mathrm{P}).$$

Proof. Let μ_k and ν_k be the k-th row of $\mathbf{Q}_1$ and $\mathbf{Q}_2$ respectively. The inequalities to be verified:

$$\sup_{k \in E} |\mu_k \mathrm{P} - \nu_k \mathrm{P}| \leq \sup_{k \in E} |\mu_k - \nu_k| \delta(\mathrm{P}) \quad (k \in E),$$

are direct consequences of Theorem 4.3.15.   □

4.4   Exercises

Exercise 4.4.1. AN ALTERNATIVE EXPRESSION OF THE VARIATION DISTANCE
(a) Prove (4.2).

(b) Show that

$$d_V(\alpha, \beta) = \sup_{i; |y(i)| \leq 1} \left(\sum_{i=1}^{r} \alpha(i) y(i) - \sum_{i=1}^{r} \beta(i) y(i) \right).$$

(c) Show that if $\lambda_i \in [0,1]$ and $\sum_{i=1}^{K} \lambda_i = 1$, then

$$d_V\left(\sum_{i=1}^{K} \lambda_i \alpha_i, \sum_{i=1}^{K} \lambda_i \beta_i\right) \leq \sum_{i=1}^{K} \lambda_i d_V(\alpha_i, \beta_i),$$

where $\{\alpha_i\}_{1\leq i\leq K}$ and $\{\beta_i\}_{1\leq i\leq K}$ are probability distributions on E.

Exercise 4.4.2. PROBABILITY OF COINCIDENCE AND VARIATION DISTANCE
Let f_1 and f_2 be two probability density functions on $\mathbb{R}^k$, and call $\mathcal{D}(f_1, f_2)$ the collection of couples of $\mathbb{R}^k$-valued random variables (X, Y) such that X and Y have the PDF f_1 and f_2 respectively.
Show that if $(X, Y) \in \mathcal{D}(f_1, f_2)$,

$$P(X = Y) \leq 1 - \frac{1}{2} \int_{\mathbf{R}^k} |f_1(x) - f_2(x)| dx$$

and that equality can be attained (X and Y are then said to be *maximally coupled*).

Exhibit (X, Y) achieving equality in the following cases:

(i) $k = 1$, X uniform on $[0, \theta_1]$, Y uniform on $[0, \theta_2]$.

(ii) $k = 2$, X uniform on $[0, \theta_1]^2$, Y uniform on $[0, \theta_2]^2$.

(iii) $k = 1$, X exponential with parameter θ_1, Y exponential with parameter θ_2.

Exercise 4.4.3. DISTANCE IN VARIATION AND BAYESIAN TESTS
Consider the following situation. A real random variable X and another random variable $\Theta \in \{1, 2\}$ have a joint distribution described as follows. For $i = 1, 2$, conditionally to $\Theta = i$, the random variable X has the PDF $f_i(x)$, and $P(\Theta = i) = \frac{1}{2}$. Find a partition $A_1 + A_2 = \mathbb{R}$ such that if we define $\widehat{\Theta} = i$ if $X \in A_i$, then the probability of error, that is, the probability that the *estimate* $\widehat{\Theta}$ is different from the actual value Θ, is minimized. What is then the probability of error $P(\widehat{\Theta} \neq \Theta)$?

Exercise 4.4.4. MARKOV COUPLING
Let $\mathbf{P}$ be a transition matrix on the finite state space E. Define

$$\delta(\mathbf{P}) = \frac{1}{2} \sup_{i,j \in E} \sum_{k \in E} |p_{ik} - p_{jk}|.$$

Let $\{X_n\}_{n\geq 0}$ and $\{Y_n\}_{n\geq 0}$ be two E-valued stochastic processes with the following properties. The initial states are maximally coupled with marginal distributions

μ and ν, respectively. Also, $\{(X_n, Y_n)\}_{n \geq 0}$ is an HMC with transition probabilities $q_{i,j;k,l}$ where for fixed (i, j), the probability distribution $q_{i,j;\cdot,\cdot}$ realizes maximal coincidence of $p_{i,\cdot}$ and $p_{j,\cdot}$ (this type of coupling is called *Markov coupling*.)

(i) Show that $\{X_n\}_{n \geq 0}$ and $\{Y_n\}_{n \geq 0}$ are two HMCs with the transition matrix $\mathbf{P}$.

(ii) Show that if $\delta(\mathbf{P}) < 1$, then $\{X_n\}_{n \geq 0}$ and $\{Y_n\}_{n \geq 0}$ couple in finite time τ and that $\mathrm{E}[e^{\alpha \tau}] < \infty$ for some positive α.

(iii) Show that if $\delta(\mathbf{P})^m < 1$ for some integer $m > 1$, then $\{X_n\}_{n \geq 0}$ and $\{Y_n\}_{n \geq 0}$ couple in finite time τ and that $\mathrm{E}[e^{\alpha \tau}] < \infty$ for some positive α.

(iv) Deduce from this that if $\mathbf{P}$ is ergodic, $\{X_n\}_{n \geq 0}$ and $\{Y_n\}_{n \geq 0}$ couple in finite time τ and that $\mathrm{E}[e^{\alpha \tau}] < \infty$ for some positive α.

Exercise 4.4.5. FINISHING THE PROOF OF THEOREM 4.2.2
Prove in detail the last assertion of Theorem 4.2.2.

Exercise 4.4.6. THE PRODUCT OF TWO INDEPENDENT RANDOM WALKS
Let $\{X_n^1\}_{n \geq 0}$ and $\{X_n^2\}_{n \geq 0}$ be two independent irreducible HMCs with the same transition matrix $\mathbf{P}$. Give a counterexample showing that if $\mathbf{P}$ is not aperiodic, the product chain $\{(X_n^1, X_n^2)\}_{n \geq 0}$ may be reducible.

Exercise 4.4.7. COUPLING TIME OF TWO TWO-STATE HMCS
Let $\{X_n\}_{n \geq 0}$ and $\{Y_n\}_{n \geq 0}$ be two independent HMCs on the state space $E = \{1, 2\}$, with the same transition matrix

$$\mathbf{P} = \begin{pmatrix} 1 - \alpha & \alpha \\ \beta & 1 - \beta \end{pmatrix},$$

where $\alpha, \beta \in (0, 1)$. Let τ be the first time n when $X_n = Y_n$. Compute the probability distribution of τ when $X_0 = 1, Y_0 = 2$.

Exercise 4.4.8. NECESSARY AND SUFFICIENT CONDITION OF REVERSIBILITY
Let $\mathbf{P}$ be an ergodic transition matrix on the countable state space E. Show that a necessary and sufficient condition for the corresponding HMC to be reversible is that for all states $i, i_1, \ldots, i_{k-1}$

$$p_{ii_1} p_{i_1 i_2} \cdots p_{i_{k-1} i} = p_{ii_{k-1}} \cdots p_{i_2 i_1} p_{i_1 i}.$$

Exercise 4.4.9. DO NOT FORGET YOUR UMBRELLA
A man commutes every weekend between his town house and his country house. Each Saturday, he leaves town for the countryside, and he comes back on Sunday. He owns N umbrellas which he takes back and forth between his two houses. He takes one, if available, if it is raining when he leaves town (resp., the countryside). The probability of rain on a given day is p.

Compute the steady state probability of finding exactly i umbrellas in the country house on a given Saturday night.

Exercise 4.4.10. MORE ON QUASI-STATIONARY DISTRIBUTIONS
Using the notation of Section 4.3.3 on quasi-stationary distributions, show that for any transient state j,

$$\lim_{m\uparrow\infty,n\uparrow\infty} P(X_n - j|\nu > m + n) = \frac{u_1(j)v_1(j)}{\sum_{i\in T} u_1(i)v_1(i)}.$$

Exercise 4.4.11. RATE OF CONVERGENCE
For the homogeneous Markov chain with state space $E = \{1, 2, 3\}$ and transition matrix

$$\mathbf{P} = \begin{pmatrix} 1-\alpha & \alpha & 0 \\ 0 & \beta & 1-\beta \\ \gamma & 0 & 1-\gamma \end{pmatrix},$$

where $\alpha, \beta, \gamma \in (0, 1)$, compute $\lim_{n\uparrow\infty} \mathbf{P}^n$ and give the corresponding rate of convergence.

Exercise 4.4.12. PROVE (4.20)
Prove in detail (4.20).

Exercise 4.4.13. $|\delta(\mathbf{P}) - \delta(\mathbf{Q})| \le |\mathbf{P} - \mathbf{Q}|$
Show that for two transition matrices $\mathbf{P}$ and $\mathbf{Q}$ on the same space,

$$|\delta(\mathbf{P}) - \delta(\mathbf{Q})| \le |\mathbf{P} - \mathbf{Q}|.$$

Chapter 5

Discrete-Time Renewal Theory

In the analytic approach to Markov chains, the convergence to steady state of an ergodic HMC is a consequence of a result on power series called the *renewal theorem* by the probabilists. In this chapter however the renewal theorem will not be the essential step in the proof of the convergence theorem but, on the contrary, it will be obtained as a corollary of the latter.

The renewal theorem will be applied to prove convergence to equilibrium of stochastic processes of a more general nature than homogeneous Markov chains, namely the *regenerative processes*.

5.1 The Renewal Process

5.1.1 The Renewal Equation

We start with the basic definitions. Let $\{S_n\}_{n\geq 1}$ be an IID sequence of random variables with values in $\overline{\mathbb{N}} := \{1, 2, \ldots, +\infty\}$ and with the probability distribution

$$P(S_1 = k) = f_k \quad (k \geq 1).$$

Let

$$R_{n+1} := R_n + S_{n+1} \quad (n \geq 0),$$

where R_0 is an arbitrary random variable with values in $\mathbb{N}$ (in particular, $R_0 < \infty$).

Definition 5.1.1 *The sequence* $\{R_n\}_{n\geq 0}$ *is called a* delayed *(by R_0)* renewal sequence *with the* renewal distribution $\{f_k\}_{k\geq 1}$. *If $R_0 \equiv 0$, it is called an* undelayed renewal sequence *or, more simply, a* renewal sequence. *If $P(S_1 = \infty) = 0$, the renewal sequence (delayed or not) is called a* proper renewal sequence *and* $\{f_k\}_{k\geq 1}$ *is called a* proper renewal distribution. *Otherwise, if $P(S_1 = \infty) > 0$, one speaks of a* defective renewal sequence *and of a* defective renewal distribution.

© Springer Nature Switzerland AG 2020
P. Brémaud, *Markov Chains*, Texts in Applied Mathematics 31,
https://doi.org/10.1007/978-3-030-45982-6_5

The quantity

$$\alpha := P(S_1 = \infty)$$

is the *defect* of the renewal distribution. The random time R_k is the k-th *renewal time*, and the sequence $\{S_n\}_{n \geq 1}$ is the *inter-renewal sequence*.

Definition 5.1.2 *With the renewal distribution $\{f_k\}_{k \geq 1}$ is associated the* renewal equation

$$u_n = v_n + \sum_{k=1}^{n} f_k u_{n-k}$$

(for $n = 0$, this reduces to $u_0 = v_0$) where the v_n's are real numbers such that

$$\sum_{k=0}^{\infty} |v_k| < \infty. \tag{5.1}$$

The sequence $\{u_n\}_{n \geq 0}$ is the unknown sequence, *and $\{v_n\}_{n \geq 0}$ is the* data.

Remark 5.1.3 Since u_n can be computed recursively as a function of $u_0, \ldots, u_{n-1}$ and $v_0, \ldots, v_n$, a solution of the renewal equation always exists and is unique.

EXAMPLE 5.1.4: LIFETIME OF A DEFECTIVE RENEWAL SEQUENCE. Define the *lifetime* L of a defective renewal sequence by

$$L := \inf\{R_k; k \geq 0, S_{k+1} = \infty\}.$$

It is the last renewal time at finite distance. We shall see that $u_n := P(L > n)$ satisfies a renewal equation. For this, write

$$1_{\{L>n\}} = 1_{\{L>n\}} 1_{\{S_1>n\}} + 1_{\{L>n\}} 1_{\{S_1 \leq n\}}.$$

Observe that $\{L > n, S_1 > n\} = \{n < S_1 < \infty\}$. Also, denoting by $\hat{L}$ the lifetime associated with the renewal sequence $\{R_{n+1} - R_1\}_{n \geq 0}$, we have the set identity $\{L > n, S_1 \leq n\} = \{\hat{L} > n - S_1, S_1 \leq n\}$. Therefore,

$$P(L > n) = P(n < S_1 < \infty) + P(\hat{L} > n - S_1, S_1 \leq n).$$

Since L and $\hat{L}$ have the same distribution and since $\hat{L}$ is independent of S_1,

$$P(\hat{L} > n - S_1, S_1 \leq n) = \sum_{k=1}^{n} P(\hat{L} > n-k) P(S_1 = k) = \sum_{k=1}^{n} P(L > n-k) P(S_1 = k).$$

Therefore u_n satisfies the renewal equation with data $v_n = P(n < S_1 < \infty)$.

Definition 5.1.5 *Define the* Dirac sequence $\{\delta_n\}_{n\geq 0}$ *by* $\delta_0 = 1, \delta_n = 0$ *for* $n \geq 1$. *When the data is the Dirac sequence, the renewal equation is called the* basic renewal equation *and its solution is called the* fundamental solution.

The fundamental solution will be denoted by $\{h_n\}_{n\geq 0}$ and therefore $h_0 = 1$, and for $n \geq 1$,

$$h_n = \sum_{k=1}^{n} f_k h_{n-k} \, .$$

The fundamental solution has a very simple interpretation. In fact, h_n is the probability that n is a renewal time (we then say, for short, "n is renewal"). It suffices to show that $u_n - P(n$ is renewal) is the unique solution of the basic renewal equation. Clearly, $u_0 = 1$. Also,

$$
\begin{aligned}
P(n \text{ is renewal}) &= \sum_{k=0}^{n-1} P(n \text{ is renewal, last renewal strictly before } n \text{ is } k) \\
&- \sum_{i=0}^{\infty}\sum_{k=0}^{n-1} P(S_{i+1} - n - k, \ k - R_i) \\
&= \sum_{i=0}^{\infty}\sum_{k=0}^{n-1} P(S_{i+1} = n - k)P(k = R_i) \\
&= \sum_{k=0}^{n-1} P(S_1 = n - k)\left(\sum_{i=0}^{\infty} P(k = R_i)\right) \\
&= \sum_{k=0}^{n-1} P(S_1 = n - k)P(k \text{ is renewal}) \\
&= \sum_{k=0}^{n-1} u_k f_{n-k} = \sum_{k=1}^{n} f_k u_{n-k}.
\end{aligned}
$$

Therefore,

$$h_k = P(k \text{ is renewal}) \, .$$

In particular, if ν_n is the number of renewal times R_k in the interval $[0, n]$, then

$$\nu_n = \sum_{k=0}^{n} h_k \, .$$

We now introduce a definition and a convenient notation. The *convolution* of two real sequences $\{x_n\}_{n\geq 0}$ and $\{y_n\}_{n\geq 0}$ is the real sequence

$$z_n := \sum_{k=0}^{n} x_k y_{n-k} \quad (n \geq 0),$$

for short: $z = x * y$.

Theorem 5.1.6 *The renewal equation has a unique solution*

$$u = h * v. \tag{5.2}$$

Proof. Existence and uniqueness have already been noted. To check that the announced solution is correct, write the renewal equation as $u = v + f * u$ (with $f_0 = 0$) and the fundamental equation as $h = \delta + f * h$. Inserting (5.2) into the renewal equation gives $h * v = v + f * (h * v)$ which is indeed true, since the right-hand side is $v + (f * h) * v = v + (h - \delta) * v = v + h * v - \delta * v$, that is, $h * v$, because $\delta * v = v$.   □

EXAMPLE 5.1.7: GEOMETRIC INTER-RENEWAL TIMES. When the distribution of the typical inter-renewal time is geometric, that is,

$$P(S_1 = k) = p(1 - p)^{k-1} \quad (k \geq 1),$$

the fundamental solution is given by $h_0 = 1$ and

$$h_n = p \quad (n \geq 1),$$

as can be readily checked. The solution of the general renewal equation is then

$$u_n = v_n + p(v_0 + \cdots + v_{n-1}).$$

One observes in this particular case that since $\lim_{n\uparrow\infty} v_n = 0$ in view of assumption (5.1),

$$\lim_{n\uparrow\infty} u_n = p \sum_{k=0}^{\infty} v_k = \frac{\sum_{k\geq 0} v_k}{\sum_{k\geq 1} k f_k}.$$

This result will be generalized by the renewal theorem.

5.1.2 The Renewal Theorem

Definition 5.1.8 *The renewal distribution $\{f_k\}_{k\geq 1}$ is called* lattice *(resp., non-lattice) if $d := \gcd\{k \; ; \; k \geq 1, f_k > 0\} > 1$ (resp., $= 1$). The integer d is called the* span *of the renewal distribution.*

The main result can now be stated.

Theorem 5.1.9 *Let $\{f_k\}_{k\geq 1}$ be a non-lattice and proper renewal distribution. For the unique solution of the renewal equation with data satisfying assumption (5.1),*

$$\lim_{n\uparrow\infty} u_n = \frac{\sum_{k\geq 0} v_k}{\sum_{k\geq 1} k f_k}, \tag{5.3}$$

where the ratio on the right-hand side is 0 if $\sum_{k\geq 1} k f_k = \infty$.

Proof. In two parts.

A. Assume the result true for the fundamental solution, that is,

$$\lim_{n\uparrow\infty} h_n = \frac{1}{\sum_{k\geq 1} k f_k} := h_\infty. \tag{5.4}$$

From expression (5.2) of the solution in terms of the fundamental solution, we obtain

$$\sum_{k=0}^{n}(h_{n-k} - h_\infty)v_k = u_n - h_\infty \sum_{k=0}^{n} v_k.$$

The result follows if we can prove that the left-hand side of the above equality converges to 0 as $n \to \infty$. Indeed, with $g(n,k) = (h_{n-k} - h_\infty)v_k 1_{\{k\leq n\}}$, we have for fixed k, $\lim_{n\uparrow\infty} g(n,k) = 0$, and $|g(n,k)| \leq |v_k|$, where $\sum_{k\geq 0} |v_k| < \infty$. Therefore, by dominated convergence for series (see the Appendix), $\lim_{n\uparrow\infty} \sum_{k\geq 0} g(n,k) = \sum_{k\geq 0} \lim_{n\uparrow\infty} g(n,k) = 0$.

B. It remains to prove (5.4). For this, introduce a Markov chain with state space $E = \mathbb{N}$ if the support of $\{f_k\}_{k\geq 1}$ is unbounded and $\{0,\ldots,M-1\}$ if $M < \infty$ is the largest value of S_1. We suppose for definiteness that $E = \mathbb{N}$. The non-null entries of the transition matrix are

$$\begin{aligned} p_{i,i-1} &= 1 \quad (i \geq 1), \\ p_{0i} &= f_{i+1} \quad (i \geq 0). \end{aligned}$$

The corresponding transition graph is depicted in Figure 5.4.1. Note that this is the transition graph of the forward recurrence time HMC $\{X_n\}_{n\geq 0}$ defined by

$$X_n = \inf\{R_k \; ; \; R_k \geq n\} - n.$$

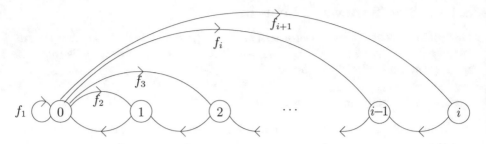

Figure 5.4.1. Transition graph of the forward recurrence chain

This chain is clearly irreducible. The distribution of the return time to state 0 is

$$P_0(T_0 = n) = f_n .$$

Event $\{T_0 = n\}$ implies event $\{X_n = 0\}$ and therefore $P_0(X_n = 0) \geq P_0(T_0 = n)$. Consequently, the set $A = \{n \geq 1; p_{00}(n) > 0\}$ contains the set $B = \{n \geq 1; f_n > 0\}$, and therefore the gcd of A is smaller than or equal to the gcd of B. Therefore, the gcd of A equals 1, that is, the chain is aperiodic.

Since the renewal distribution is assumed proper, we have that $P_0(T_0 < \infty) = \sum_{n \geq 1} f_n = 1$ and therefore the chain is recurrent. If $E_0[T_0] < \infty$, it is ergodic, and then

$$\lim_{n \uparrow \infty} p_{00}(n) = \pi_0 = \frac{1}{E_0[T_0]} .$$

If the chain is not ergodic but only null recurrent, then $\lim_{n \uparrow \infty} p_{00}(n) = 0$ by Orey's theorem (Theorem 4.2.4). In both cases, since $E_0[T_0] = \sum_{k \geq 1} k f_k$,

$$\lim_{n \uparrow \infty} p_{00}(n) = \frac{1}{\sum_{k \geq 1} k f_k} .$$

The proof of (5.3) is complete because $p_{00}(n) = P(n \text{ is renewal}) = h_n$.   □

Theorem 5.1.10 *Under the same conditions as in Theorem 5.1.9, except that the span d of the renewal distribution is now strictly greater than 1, the solution of the renewal equation with data satisfying (5.1) is such that*

$$\lim_{N \uparrow \infty} u_{r+Nd} = d \frac{\sum_{k \geq 0} v_{r+kd}}{\sum_{k \geq 1} k f_k} \quad (0 \leq r \leq d-1). \qquad (5.5)$$

Proof. Observe that when $\{f_k\}_{k\geq 1}$ is proper and lattice with span d, the distribution $\{f_{Nd}\}_{N\geq 1}$ is proper and non-lattice. On the other hand, the renewal equation splits into d renewal equations. The r-th one $(0 \leq r \leq d - 1)$ is

$$u_{r+Nd} = v_{r+Nd} + \sum_{\ell=1}^{N} f_{\ell d} u_{r+Nd-\ell d},$$

where N is the time variable. The renewal theorem can be applied to each one, and we obtain (5.5) after observing that

$$\sum_{N=1}^{\infty} N f_{Nd} = \frac{1}{d} \sum_{N=1}^{\infty} N d f_{Nd} = \frac{1}{d} \sum_{k=1}^{\infty} k f_k.$$

$\square$

5.1.3 Defective Renewal Sequences

Theorem 5.1.11 *Suppose that the renewal distribution is defective and that the data sequence of the renewal equation is non-negative and satisfies instead of (5.1)*

$$\lim_{n\uparrow\infty} v_n = v_\infty < \infty. \tag{5.6}$$

The solution of the renewal equation then satisfies

$$\lim_{n\uparrow\infty} u_n = \frac{v_\infty}{\alpha}, \tag{5.7}$$

where $\alpha = \mathrm{P}(S_1 = \infty)$ is the defect of the renewal distribution.

Proof. The forward recurrence HMC in the proof of Theorem 5.1.9 now has $E = \mathbb{N} \cup \{+\infty\}$ for state space. All states besides $+\infty$ are transient. In particular, the average number of visits to 0 is finite:

$$\nu_\infty = \sum_{k=0}^{\infty} h_k < \infty.$$

From the expression of the solution

$$u_n = \sum_{k=0}^{n} h_k v_{n-k},$$

we therefore obtain by the dominated convergence for series (see the Appendix),

$$\lim_{n\uparrow\infty} u_n = \left(\sum_{k=0}^{\infty} h_k\right) v_\infty = \nu_\infty v_\infty \,.$$

Now, the probability of n visits to 0 is $(1-\alpha)^{n-1}\alpha$, and therefore the average number of visits to 0 is $\nu_\infty = \frac{1}{\alpha}$. $\square$

Theorem 5.1.12 *Suppose that the renewal distribution is non-lattice, defective and such that there exists a $\gamma > 1$ such that*

$$\sum_{k=0}^{\infty} \gamma^n f_n = 1\,. \tag{5.8}$$

Suppose moreover that the data sequence satisfies

$$\sum_{k=0}^{\infty} \gamma^n |v_n| < \infty\,.$$

The solution of the renewal equation then satisfies

$$\lim_{n\uparrow\infty} \gamma^n u_n = \frac{\sum_{k=0}^{\infty} \gamma^k v_k}{\sum_{k=0}^{\infty} k\gamma^k f_k}\,. \tag{5.9}$$

Proof. Observe that if we let $\widetilde{f}_n := \gamma^n f_n$, $\widetilde{v}_n := \gamma^n v_n$ and $\widetilde{u}_n := \gamma^n u_n$, then

$$\widetilde{u}_n = \widetilde{v}_n + \sum_{k=1}^{n} \widetilde{f}_k \widetilde{u}_{n-k}\,.$$

This renewal equation is non-lattice and proper, and therefore the announced result follows from the renewal theorem. $\square$

Remark 5.1.13 A consequence of (5.8) is the exponential decay of the renewal distribution. This shows in particular that (5.8) is an assumption that is *not always satisfied.*

EXAMPLE 5.1.14: CONVERGENCE RATE IN THE DEFECTIVE CASE. The situation is that of Theorem 5.1.12, where in addition the renewal distribution is assumed non-lattice and such that (5.8) holds for some $\gamma > 1$. We seek to understand how u_n tends to u_∞. For this we define $\hat{u}_n = u_n - u_\infty$. Rewriting the renewal equation for u_n as

$$u_n - u_\infty = v_n - u_\infty + \sum_{k=1}^{n} f_k(u_{n-k} - u_\infty) + u_\infty \sum_{k=1}^{n} f_k ,$$

we see that $\hat{u}_n$ satisfies a renewal equation with data

$$\hat{v}_n = v_n - u_\infty P(S_1 > n) .$$

We can therefore apply Theorem 5.1.12 to obtain, after rearrangement,

$$\lim_{n \uparrow \infty} \gamma^n \left(u_n - \frac{v_\infty}{\alpha} \right) = \frac{1}{\gamma} \left\{ \frac{\sum_{k=0}^{\infty} \gamma^k v_k}{\sum_{k=0}^{\infty} \gamma^k P(S_1 > k)} - \frac{v_\infty}{P(S_1 = \infty)} \right\} .$$

An *excessive* renewal equation is one for which $\sum_{k=1}^{\infty} f_n > 1$. Theorem 5.1.12 then has an obvious counterpart. Note however that in the excessive case (5.8) *always* has a solution γ, and of course this solution is in $(0, 1)$.

EXAMPLE 5.1.15: THE LOTKA–VOLTERRA MODEL, TAKE 1. This model concerns (in this example) a population of women. At each time $n \in \mathbb{Z}$, an average number u_n of daughters is born. Each of them gives birth independently of the other women. The average number of daughters of any given woman in the kth year of her life, $k \geq 1$, is f_k. At time 0 the population has $\alpha(i)$ women of age i. Expressing that u_n is the sum of v_n, the average number of daughters born at time n from mothers born at or before time 0, and of r_n, the average number of daughters born at time n from mothers born strictly after time 0 and up to time n, we obtain the renewal equation with data sequence

$$v_n = \sum_{i=0}^{\infty} \alpha(i) f_{n+i} .$$

In this context, the renewal equation is known as the *Lotka–Volterra equation*. Denote by

$$\rho = \sum_{k=1}^{\infty} f_n$$

the average number of daughters of any given woman, and assume that this number is positive, finite and different from 1. Assume that γ defined by (5.8) exists and that the renewal distribution is non-lattice. Denoting by C the right-hand side of (5.9),

$$\lim_{n\uparrow\infty} \gamma^n u_n = C .$$

Note that $\gamma < 1$ if $\rho > 1$, and $\gamma > 1$ if $\rho < 1$. The first case corresponds to exponential explosion, whereas the second case is that of extinction at exponential rate.

5.2   Regenerative Processes

5.2.1   Renewal Equations for Regenerative Processes

The common feature that regenerative processes share with the homogeneous recurrent Markov chains is the existence of regenerative cycles.

Definition 5.2.1 *Let $\{Z_n\}_{n\geq 0}$ be a stochastic process with values in an arbitrary state space E and let $\{R_n\}_{n\geq 0}$ be a delayed renewal sequence. The stochastic process $\{Z_n\}_{n\geq 0}$ is said to be regenerative with respect to the renewal sequence $\{R_n\}_{n\geq 0}$ if for all $k \geq 0$, $\{Z_{n+R_k}\}_{n\geq 0}$ is independent of $R_0, S_1, \ldots, S_k$ and has the same distribution as $\{Z_{n+R_0}\}_{n\geq 0}$.*

Note that the definition does not require that $\{Z_{n+R_k}\}_{n\geq 0}$ be independent of Z_n $(0 \leq n \leq R_k - 1)$, although in many examples this is the case. The freedom resulting from the relaxed conditions of Definition 5.2.1 can be very useful.

EXAMPLE 5.2.2: RECURRENT MARKOV CHAINS. Let $\{X_n\}_{n\geq 0}$ be an irreducible recurrent HMC, with arbitrary initial distribution. Let $\{R_n\}_{n\geq 0}$ be the successive *hitting* times of state 0. The regenerative cycle theorem (Theorem 2.5.11) tells us that $\{X_n\}_{n\geq 0}$ is regenerative with respect to $\{R_n\}_{n\geq 0}$.

EXAMPLE 5.2.3: RELIABILITY. Let $\{U_n\}_{n\geq 1}$ and $\{V_n\}_{n\geq 1}$ be two independent IID sequences of positive integer-valued random variables. Define the sequence $\{S_n\}_{n\geq 1}$ by $S_n = U_n + V_n$, and let $\{R_n\}_{n\geq 0}$ be the associated undelayed renewal sequence. Define a $\{0, 1\}$-valued process $\{Z_n\}_{n\geq 0}$ as in Figure 5.2.1. Clearly, $\{Z_n\}_{n\geq 0}$ is a regenerative process with respect to $\{R_n\}_{n\geq 0}$.

Regenerative processes generate renewal equations and are the main motivation for the study of such equations. For instance, if $f : E \to \mathbb{R}$ is a non-negative

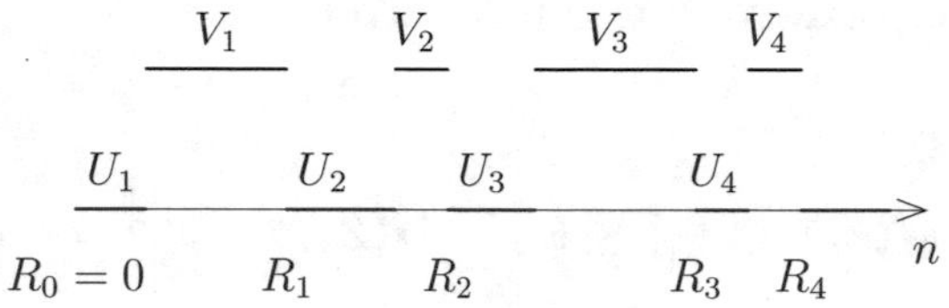

$$\begin{array}{cccc} \underline{V_1} & \underline{V_2} & \underline{V_3} & \underline{V_4} \end{array}$$

Figure 5.2.1. A sample path of the reliability process

function, and if $\{Z_n\}_{n\geq 0}$ is an E-valued process regenerative with respect to the undelayed renewal sequence $\{R_n\}_{n\geq 0}$, then the sequence $\{u_n\}_{n\geq 0}$, where $u_n = \mathrm{E}[f(Z_n)]$, satisfies a renewal equation. In fact,

$$\mathrm{E}[f(Z_n)] = \mathrm{E}[f(Z_n)1_{\{n<S_1\}}] + \mathrm{E}[f(Z_n)1_{\{n\geq S_1\}}]$$

and, with $\widetilde{Z}_n := Z_{n+S_1}$,

$$
\begin{aligned}
\mathrm{E}[f(Z_n)1_{\{n\geq S_1\}}] &= \mathrm{E}[f(\widetilde{Z}_{n-S_1})1_{\{n\geq S_1\}}] \\
&= \sum_{k=1}^{\infty} \mathrm{E}[f(\widetilde{Z}_{n-S_1})1_{\{n\geq S_1\}}1_{\{S_1=k\}}] \\
&= \sum_{k=1}^{n} \mathrm{E}[f(\widetilde{Z}_{n-k})1_{\{S_1=k\}}] \\
&= \sum_{k=1}^{n} \mathrm{E}[f(\widetilde{Z}_{n-k})]\mathrm{P}(S_1=k) \\
&= \sum_{k=1}^{n} \mathrm{E}[f(Z_{n-k})]\mathrm{P}(S_1=k),
\end{aligned}
$$

where the independence of S_1 and $\{\widetilde{Z}_n\}_{n>0}$ and the assumption of equidistribution of $\{\widetilde{Z}_n\}_{n\geq 0}$ and $\{Z_n\}_{n\geq 0}$ have been taken into account. Therefore,

$$\mathrm{E}[f(Z_n)] = \mathrm{E}[f(Z_n)1_{\{n<S_1\}}] + \sum_{k=1}^{n} \mathrm{E}[f(Z_{n-k})]\mathrm{P}(S_1=k),$$

which is precisely the renewal equation with data

$$v_n = \mathrm{E}[f(Z_n)1_{\{n<S_1\}}].$$

5.2.2   The Regenerative Theorem

Note that

$$\sum_{n=0}^{\infty} |v_n| = \sum_{n=0}^{\infty} |E[f(Z_n)1_{\{n<S_1\}}]| \le E\left[\sum_{n=0}^{\infty} |f(Z_n)|1_{\{n<S_1\}}\right] = E\left[\sum_{n=0}^{S_1-1} |f(Z_n)|\right].$$

Therefore, by the renewal theorem:

Theorem 5.2.4 *Let $\{Z_n\}_{n\ge 0}$ be a regenerative process with respect to the underlayed renewal sequence $\{R_n\}_{n\ge 0}$ and let $f : E \to \mathbb{R}$ be such that*

$$E\left[\sum_{n=0}^{S_1-1} |f(Z_n)|\right] < \infty. \tag{5.10}$$

If the distribution of S_1 is proper and non-lattice, then

$$\lim_{n\uparrow\infty} E[f(Z_n)] = \frac{E\left[\sum_{n=0}^{S_1-1} f(Z_n)\right]}{E[S_1]}. \tag{5.11}$$

The last formula is *Smith's formula*.[1]

EXAMPLE 5.2.5: THE CENTRAL FORMULA OF RELIABILITY. This is a continuation of Example 5.2.3. We assume that $S_1 = U_1 + V_1$ is proper and non-lattice. With $f(z) = 1_{\{0\}}(z)$, and assuming $E[U_1] < \infty$, (5.11) gives

$$\lim_{n\uparrow\infty} P(Z_n = 0) = \frac{E[U_1]}{E[U_1] + E[V_1]},$$

since $E[f(Z_n)] = E[1_{\{0\}}(Z_n)] = P(Z_n = 0)$, and $\sum_{n=0}^{S_1-1} 1_{\{0\}}(Z_n) = U_1$.

EXAMPLE 5.2.6: THE BUS PARADOX. Consider the renewal sequence of Definition 5.2.1 with $R_0 = 0$. Define for each $n \ge 0$ the backward recurrence time B_n and the forward recurrence time F_n by

$$B_n := n - L_n, \; F_n := N_n - n,$$

where

$$L_n := \sup\{R_k; \; k \ge 0, \; R_k \le n\}$$

and

$$N_n := \inf\{R_k; k > 0, \; R_k > n\}.$$

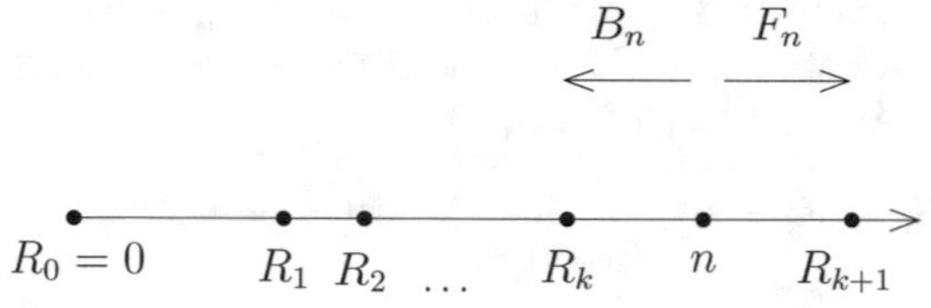

Here, index k is random

Figure 5.2.2. Backward and forward recurrence times

In particular, if $n = R_m$ for some m, then $B_n = 0$ and $F_n = R_{m+1} - R_m = S_{m+1}$. Observe that $F_n \geq 1$ for all $n \geq 0$. Also, if $R_m \leq n < R_{m+1}$, then $B_n + F_n = S_{m+1}$.

Theorem 5.2.4 with $Z_n = (B_n, F_n)$ and $f(Z_n) = 1_{\{B_n=i\}}1_{\{F_n=j\}}$ gives, provided that the distribution of S_1 is proper and non lattice,

$$\lim_{n\uparrow\infty} P(B_n = i, F_n = j) = \frac{P(S_1 = i + j)}{E[S_1]}. \tag{5.12}$$

Indeed the sum $\sum_{n=0}^{S_1-1} 1_{\{B_n=i,F_n=j\}}$ has at most one non-zero term, in which case it is equal to 1. For this term, say corresponding to the index $n = n_0$, $B_{n_0} + F_{n_0} = S_1 = i + j$. Therefore the sum is equal to $1_{\{S_1=i+j\}}$.

Summing up (5.12) from $j = 1$ to ∞ and remembering that $F_n \geq 1$, we obtain

$$\lim_{n\uparrow\infty} P(B_n = i) = \frac{P(S_1 > i)}{E[S_1]}.$$

Similarly, for the forward recurrence time,

$$\lim_{n\uparrow\infty} P(F_n = j) = \frac{P(S_1 \geq j)}{E[S_1]}.$$

The roles of B_n and F_n are not symmetric. To restore symmetry, one must consider B_n and $F_n' = F_n - 1$ (recall that $F_n \geq 1$). Then

$$\lim_{n\uparrow\infty} P(F_n' = j) = \frac{P(S_1 > j)}{E[S_1]}.$$

Since $B_n + F_n = S_m$ for some (random) m determined by the condition $R_m \leq n < R_{m+1}$, one might expect that $P(B_n + F_n = k) = P(S_1 = k)$. But this is in general wrong and constitutes the apparent *paradox of recurrence times*, also called the

---

[1][Smith, 1955].

bus paradox (see Exercise 5.3.7). It is true that $P(B_n + F_n = k) = P(S_m = k)$, but m is random and therefore there is no reason why S_m should have the same distribution as S_1. As a matter of fact,

$$\lim_{n\uparrow\infty} P(B_n + F_n = k) = \lim_{n\uparrow\infty} \sum_{i+j=k} P(B_n = i, F_n = j)$$

$$= \sum_{\substack{i,j \\ i+j=k}} \frac{P(S_1 = i + j)}{E[S_1]} = \frac{kP(S_1 = k)}{E[S_1]}.$$

Theorem 5.2.7 *Let $\{Z_n\}_{n\geq 0}$ be a possibly delayed regenerative process (recall however the current general assumption that $R_0 < \infty$). Let $f : E \to \mathbb{R}$ be such that*

$$\lim_{n\uparrow\infty} E[f(Z_n)1_{\{n<R_0\}}] = 0 \tag{5.13}$$

and

$$E\left[\sum_{k=R_0}^{R_1-1} |f(Z_k)|\right] < \infty. \tag{5.14}$$

Then, if the renewal distribution is proper and non-lattice,

$$\lim_{n\uparrow\infty} E[f(Z_n)] = \frac{E\left[\sum_{k=R_0}^{R_1-1} f(Z_k)\right]}{E[S_1]}. \tag{5.15}$$

Proof. It suffices to show that the limit of $E[f(Z_n)1_{n\geq R_0}]$ equals the right-hand side of (5.15). Introduce $\{\widetilde{Z}_n\}_{n\geq 0} = \{Z_{n+R_0}\}_{n\geq 0}$, and observe that this process is an undelayed regenerative process with respect to $\{R_n - R_0\}_{n\geq 1}$ that is proper and non-lattice. We have

$$E[f(Z_n)1_{\{n\geq R_0\}}] = E[f(\widetilde{Z}_{n-R_0})1_{\{n\geq R_0\}}] = \sum_{k=0}^{n} E[f(\widetilde{Z}_{n-k})P(R_0 = k).$$

By the undelayed version of the regenerative theorem, we have that

$$\lim_{n\uparrow\infty} E[f(\widetilde{Z}_n)] = \frac{E\left[\sum_{k=R_0}^{R_1-1} f(Z_k)\right]}{E[R_1 - R_0]},$$

and therefore, by dominated convergence for series (see the Appendix),

$$\lim_{n\uparrow\infty} \sum_{k=0}^{n} E[f(\widetilde{Z}_{n-k})]P(R_0 = k) = \frac{E\left[\sum_{k=R_0}^{R_1-1} f(Z_k)\right]}{E[R_1 - R_0]}.$$

□

Remark 5.2.8 A useful case where (5.13) is satisfied is that when f is bounded (use dominated convergence).

5.3 Exercises

Exercise 5.3.1. AT A CROSSWALK
At a crosswalk, cars pass on a single lane at times $R_0 = 0, R_1, R_2, \ldots$, where $\{R_n\}_{n \geq 0}$ is a proper renewal sequence. A pedestrian arriving at time 0 crosses the lane as soon as he sees a time interval $x > 0$ between two consecutive cars. How long must he wait, on average?

Exercise 5.3.2. LIFETIME
Let L be the lifetime of a defective renewal sequence. Show that $\lim_{n \uparrow \infty} P(L > n) = 0$ and give the corresponding rate of convergence. Treat in detail the case where the inter-renewal sequence is geometric.

Exercise 5.3.3. AVERAGE NUMBER OF RENEWALS IN AN INTERVAL
Let $\nu((a,b])$ be the average number of renewal epochs in the integer interval $(a, b]$ of a proper non-lattice renewal sequence. What is the limit as $n \uparrow \infty$ of $\nu((n+a, n+b])$?

Exercise 5.3.4. THE CASE $f_0 > 0$
Suppose that the typical inter-renewal time S_1 of a renewal sequence is proper but that $P(S_1 = 0) = f_0 > 0$. Otherwise, suppose that $\gcd\{n \geq 1; f_n > 0\} = 1$. Show that the solution of the *extended* renewal equation

$$u_0 = v_0, \quad u_n = v_n + \sum_{k=0}^{n} f_k u_{n-k} \quad (n \geq 1)$$

(notice the additional term in the sum, corresponding to $k = 0$) satisfies, under the summability condition (5.1),

$$\lim_{n \uparrow \infty} u_n = \frac{\sum_{k \geq 0} v_k}{\sum_{k=1}^{\infty} k f_k}.$$

Exercise 5.3.5. THE LOTKA–VOLTERRA MODEL, TAKE 2

In the Lotka–Volterra population model, what is in the critical case $\rho = 1$ the average number of daughters born at a given large time, when $\alpha(i) = 0$ for all $i > 0$, and $\alpha(0) = 1$? Suppose now that $f_k = e^{-\beta} \frac{\theta^{k-1}}{(k-1)!}$ $(k \geq 1)$. Discuss the asymptotic behavior of u_n, the average number of daughters born at time n, in terms of the positive parameters β and θ (use the same initial conditions as in the first part of the problem).

Exercise 5.3.6. A MAINTENANCE STRATEGY

A given machine can be in either one of three states: G (*good*), M (*in mainte-nance*) or R (*in repair*). Its successive periods where it is in state G (*resp.* M, R) form an independent and identically distributed sequence $\{S_n\}_{n\geq 0}$ (*resp.* $\{U_n\}_{n\geq 0}$, $\{V_n\}_{n\geq 0}$) with finite mean. All these sequences are assumed mutually independent. The maintenance policy uses a number $T > 0$. If the machine has age T and has not failed, it goes to state M. If it fails before it has reached age T, it enters state R. After a period of maintenance or of repair, we consider that the machine starts anew, and enters a G period.

Find the steady state probability that the machine is operational. (Note that "good" does not mean "operational." The machine can be "good" but, due to the operations policy, in maintenance, and therefore not operational.)

Exercise 5.3.7. WAITING FOR THE BUS

(a) In Example 5.2.6, when the typical inter-renewal time is geometric, compute $\lim_{n\uparrow\infty} P(F_n + B_n = k)$, $\lim_{n\uparrow\infty} P(F_n = k)$, and $\lim_{n\uparrow\infty} P(F_n + B_n = k)$.

(b) Under what circumstances do we have $\lim_{n\uparrow\infty} P(F_n + B_n = k) = P(S_1 = k)$ for all $k \geq 1$?

Exercise 5.3.8. RELIABILITY IN THE LATTICE CASE

In Example 5.2.3, we abandon the non-lattice hypothesis. Show that

$$\lim_{t\uparrow\infty} \frac{1}{t} \int_0^t 1_{\{Z(s)=0\}} ds = \frac{E[U_1]}{E[U_1] + E[V_1]},$$

where $Z(s) = 0$ means that the machine is in repair at time s.

Chapter 6

Absorption and Passage Times

The results obtained so far concern *irreducible recurrent* HMCs. They must be completed by the study of chains with *several communication classes*. A typical question is: What is the probability of being absorbed by a given recurrent class when starting from a given transient state? This kind of problem was previously addressed in Chapter 2 in terms of first-step analysis (see, for instance, the tennis Examples 2.2.13 and 2.2.12). This approach leads to a system of linear equations with boundary conditions, for which the solution is unique if the state space is finite. With an infinite state space, the uniqueness issue cannot be overlooked and the absorption problem will therefore be reconsidered with this in mind, and also with the intention of finding general matrix-algebraic expressions for the solutions. This is done in Sections 6.1 and 6.2. We shall then return to the study of irreducible Markov chains in Section 6.3 with the purpose of finding algebraic expressions for the passage times from one state to another.

6.1 Life Before Absorption

6.1.1 Infinite Sojourns

We learned in Subsection 3.1.3 that the state space E can be decomposed as

$$E = T + \sum_j R_j,$$

© Springer Nature Switzerland AG 2020
P. Brémaud, *Markov Chains*, Texts in Applied Mathematics 31,
https://doi.org/10.1007/978-3-030-45982-6_6

where $R_1, R_2, \ldots$ are the disjoint recurrent classes and T is the collection of transient states, and that the transition matrix can be block-partitioned as

$$
\mathbf{P} = \begin{array}{ccccc}
R_1 & R_2 & \cdots & T & \\
\mathbf{P}_1 & 0 & 0 & 0 & R_1 \\
0 & \mathbf{P}_2 & 0 & 0 & R_2 \\
0 & 0 & \ddots & 0 & \vdots \\
B(1) & B(2) & \cdots & \mathbf{Q} & T
\end{array}
$$

or, in condensed notation,

$$
\mathbf{P} = \begin{array}{cc} D & 0 \\ B & \mathbf{Q} \end{array} \tag{6.1}
$$

Note that the number of recurrent classes as well as the number of transient states may be infinite.

There is the possibility, when the set of transient states is infinite, of never being absorbed by the recurrent set. We shall consider this problem first and then proceed to derive the distribution of the time to absorption by the recurrent set, and then finally the probability of being absorbed by a given recurrent class.

Let A be a subset of the state space E (typically the set of transient states, but not necessarily). We seek to obtain for any initial state $i \in A$ the probability of remaining forever in A:

$$
v(i) := P_i \left(\cap_{r \geq 0} \{ X_r \in A \} \right) . \tag{6.2}
$$

Letting

$$
v_n(i) := \mathrm{P}_i(X_1 \in A, \ldots, X_n \in A) ,
$$

we have, by monotone sequential continuity,

$$
\lim_{n \uparrow \infty} \downarrow v_n(i) = v(i) .
$$

But for $j \in A$,

$$
P_i(X_1 \in A, \ldots, X_{n-1} \in A, X_n = j) = \sum_{i_1 \in A} \cdots \sum_{i_{n-1} \in A} p_{i i_1} \cdots p_{i_{n-1} j}
$$

is the general term $q_{ij}(n)$ of the n-th iterate of $\mathbf{Q}$, the restriction of $\mathbf{P}$ to the set A. Therefore,

$$
v_n(i) = \sum_{j \in A} q_{ij}(n) \quad (i \in A) ,
$$

that is,

$$v_n = \mathbf{Q}^n \mathbf{1}_A, \tag{6.3}$$

where $\mathbf{1}_A$ is the column vector indexed by A with all entries equal to 1. From this equality we obtain

$$v_{n+1} = \mathbf{Q} v_n.$$

Letting $n \uparrow \infty$, we have by dominated convergence that $v = \mathbf{Q}v$. Moreover, $\mathbf{0}_A \le v \le \mathbf{1}_A$, where $\mathbf{0}_A$ is the column vector indexed by A with all entries equal to 0. The above result can be refined as follows:

Theorem 6.1.1 *The vector v defined by (6.2) is the maximal solution of*

$$v = \mathbf{Q}v \quad (\mathbf{0}_A \le v \le \mathbf{1}_A).$$

Moreover, either $v = \mathbf{0}_A$ or $\sup_{i \in A} v(i) = 1$.

Proof. Only maximality and the last statement remain to be proven. To prove maximality consider a vector u indexed by A such that $u = \mathbf{Q}u$ and $\mathbf{0}_A \le u \le \mathbf{1}_A$. Iteration of $u = \mathbf{Q}u$ yields $u = \mathbf{Q}^n u$, and $u \le \mathbf{1}_A$ implies that $\mathbf{Q}^n u \le \mathbf{Q}^n \mathbf{1}_A = v_n$. Therefore $u \le v_n$, which gives $u \le v$ by passage to the limit.

To prove the last statement of the theorem, let $c = \sup_{i \in A} v(i)$. From $v \le c\mathbf{1}_A$, we obtain as above $v \le cv_n$ and therefore, at the limit, $v \le cv$. This implies either $v = \mathbf{0}_A$ or $c = 1$. $\qquad \square$

Remark 6.1.2 Equation $v = \mathbf{Q}v$ reads

$$v(i) - \sum_{j \in A} p_{ij} v(j) \quad (i \in A).$$

First-step analysis gives this as a *necessary* condition. However, it does not help to determine which solution to choose in case there are several.

EXAMPLE 6.1.3: RANDOM WALK REFLECTED AT 0. The transition matrix of the random walk on $\mathbb{N}$ with a reflecting barrier at 0,

$$\mathbf{P} = \begin{pmatrix} 0 & 1 & & & \\ q & 0 & p & & \\ & q & 0 & p & \\ & & q & 0 & p \\ & & & \ddots & \ddots & \ddots \end{pmatrix},$$

where $p \in (0,1)$, is clearly irreducible. Intuitively, if $p > q$, there is a drift to the right, and one expects the chain to be transient. This will be proven formally by showing that the probability $v(i)$ of never visiting state 0 when starting from any state $i \geq 1$ is strictly positive. In order to apply Theorem 6.1.1 with $A = \mathbb{N} \backslash \{0\}$, we must find the general solution of $u = \mathbf{Q}u$. This equation reads

$$
\begin{aligned}
u(1) &= pu(2), \\
u(2) &= qu(1) + pu(3), \\
u(3) &= qu(2) + pu(4), \\
&\cdots
\end{aligned}
$$

and its general solution is

$$
u(i) = u(1) \sum_{j=0}^{i-1} \left(\frac{q}{p}\right)^{j}.
$$

The largest value of $u(1)$ respecting the constraint $u(i) \in [0,1]$ is $u(1) = 1 - \left(\frac{q}{p}\right)$. The solution $v(i)$ is obtained by assigning this value, and therefore

$$
v(i) = 1 - \left(\frac{q}{p}\right)^{i}. \tag{6.4}
$$

Remark 6.1.4 If the set of transient states T is *finite*, the probability of infinite sojourn in T is null. This intuitively obvious fact can also be deduced from (6.3) with $A = T$. Indeed, with the notation of (6.1),

$$
\mathbf{P}^{n} = \begin{pmatrix} D^{n} & 0 \\ B_{n} & \mathbf{Q}^{n} \end{pmatrix}, \tag{6.5}
$$

and therefore the general term of $\mathbf{Q}^{n}$ is $p_{ij}(n)$. But we know that for any transient state j, $\lim_{n \uparrow \infty} p_{ij}(n) = 0$ for all $i \in E$, and therefore if T is finite, $v(i) = \lim_{n \uparrow \infty} \sum_{j \in T} p_{ij}(n) = 0$.

EXAMPLE 6.1.5: THE REPAIR SHOP, TAKE 3. Continuation of Example 2.2.6. In this model, the state space is $E = \mathbb{N}$ and the transition matrix is

$$
\mathbf{P} = \begin{pmatrix} a_0 & a_1 & a_2 & a_3 & \cdots \\ a_0 & a_1 & a_2 & a_3 & \cdots \\ & a_0 & a_1 & a_2 & \cdots \\ & & a_0 & a_1 & \cdots \\ & & & & \cdots \end{pmatrix},
$$

where a_i $(i \geq 0)$ is a probability distribution on $\mathbb{N}$ of mean

$$\rho := \sum_{k=0}^{\infty} k a_k.$$

It is assumed that $a_0 > 0$ and $a_0 + a_1 < 1$, so that the chain is irreducible (Exercise 2.6.18).

One can use Theorem 6.1.1 to prove that if $\rho \leq 1$, the chain is recurrent. For this consider the matrix

$$\mathbf{Q} = \begin{pmatrix} a_1 & a_2 & a_3 & \cdots \\ a_0 & a_1 & a_2 & \cdots \\ & a_0 & a_1 & \cdots \\ & & & \cdots \end{pmatrix},$$

which is the restriction of $\mathbf{P}$ to $A_i = \{i+1, i+2, \ldots\}$ for any $i \geq 0$. Thus, the maximal solution of $v = \mathbf{Q}v$, $\mathbf{0}_A \leq v \leq \mathbf{1}_A$ (where $A = A_i$) has, in view of Theorem 6.1.1, the following two interpretations. Firstly, for $i \geq 1$,

$$v(i) = P_i(X_n \geq 1, n \geq 0),$$

that is, $v(i)$ is the probability of never visiting 0 when starting from $i \geq 1$. Secondly, $v(1)$ is the probability of never visiting $\{0, 1, \ldots, i\}$ when starting from $i+1$. Since when starting from $i+1$, the chain visits $\{0, 1, \ldots, i\}$ if and only if it visits i, $v(1)$ is the probability of never visiting i when starting from $i+1$.

We can write the probability of visiting 0 when starting from $i+1$ as

$$1 - v(i+1) = (1 - v(1))(1 - v(i)),$$

since in order to go from $i+1$ to 0 one must first reach i, and then go to 0. Therefore, with

$$v(1) := 1 - \beta,$$

for all $i \geq 1$,

$$v(i) = 1 - \beta^i.$$

This solution depends on a parameter $\beta \in [0, 1]$. To determine β, let us write the first equality of $v = \mathbf{Q}v$:

$$v(1) = a_1 v(1) + a_2 v(2) + \cdots,$$

that is,

$$(1 - \beta) = a_1(1 - \beta) + a_2(1 - \beta^2) + \cdots.$$

Taking $\sum_{i\geq 0} a_i = 1$ into account, this reduces to

$$\beta = g(\beta), \qquad\qquad (\star)$$

where $g(z)$ is the generating function of the distribution a_k $(k \geq 0)$. Also, all other equations of $v = \mathbf{Q}v$ reduce to $(\star)$.

Recall Theorem 1.2.8, which tells us that under the irreducibility assumptions $a_0 > 0$, $a_0 + a_1 < 1$, $(\star)$ has only one solution in $[0, 1]$, namely $\beta = 1$ if $\rho \leq 1$, whereas if $\rho > 1$, it has two solutions in $[0, 1]$, $\beta = 1$ and $\beta = \beta_0 \in (0, 1)$.

Theorem 6.1.1 says that one must take the smallest β satisfying $(\star)$. Therefore, if $\rho > 1$, the probability of visiting state 0 when starting from state $i \geq 1$ is $1 - v(i) = \beta_0^i < 1$, and therefore the chain is transient. If $\rho \leq 1$, the latter probability is $1 - v(i) = 1$, and therefore the chain is recurrent.

We know from Example 2.4.9, that for $\rho = 1$, there is no stationary distribution. Therefore,

If $\rho = 1$, the chain is null recurrent.

If $\rho > 1$, the chain is transient.

It remains to elucidate the situation where $\rho < 1$, and we shall see later on that in this case the chain is positive recurrent.

_____

6.1.2   Time to Absorption

We now turn to the computation of the distribution of the time τ of exit from the transient set T. Theorem 6.1.1 tells us that $v = \{v(i)\}_{i\in T}$, where $v(i) := \mathrm{P}_i(\tau = \infty)$, is the largest solution of $v = \mathbf{Q}v$ subject to the constraints $\mathbf{0}_T \leq v \leq \mathbf{1}_T$, where $\mathbf{Q}$ is the restriction of $\mathbf{P}$ to the transient set T.

The probability distribution of τ when the initial state is $i \in T$ is readily computed starting from the identity

$$\mathrm{P}_i(\tau = n) = \mathrm{P}_i(\tau \geq n) - \mathrm{P}_i(\tau \geq n + 1)$$

and the observation $\{\tau \geq n\} = \{X_{n-1} \in T\}$ for $n \geq 1$, from which we obtain, for $n \geq 1$,

$$\mathrm{P}_i(\tau = n) = \mathrm{P}_i(X_{n-1} \in T) - \mathrm{P}(X_n \in T) = \sum_{j\in T}(p_{ij}(n-1) - p_{ij}(n)).$$

Now, $p_{ij}(n)$ is for $i, j \in T$ the general term of $\mathbf{Q}^n$, and therefore:

Theorem 6.1.6

$$P_i(\tau = n) = \{(\mathbf{Q}^{n-1} - \mathbf{Q}^n)\mathbf{1}_T\}_i.$$

In particular, if $P_i(\tau = \infty) = 0$,

$$P_i(\tau > n) = \{\mathbf{Q}^n\mathbf{1}_T\}_i.$$

In particular, when T *is finite, for any distribution* ν *such that* $\nu(T) = 1$, *we have*

$$P_\nu(\tau > n) = \nu^T \mathbf{Q}^n \mathbf{1}_T.$$

(The reader will perhaps excuse the author for the local conflict of notation in the above formula, where T denotes transposition and a set.)

6.2 Absorption Probabilities

6.2.1 The First Fundamental Matrix

The results of the previous section concern the probability of remaining forever in the transient set or, taking the dual point of view, the probability of never being absorbed by the recurrent set. It remains to compute the probability of being absorbed by a given recurrent class R_j when starting from an initial state $i \in T$. For this, we shall introduce the notion of a *fundamental matrix*.

We shall see that only the case where $i \in T$ and $j \in T$ requires analysis. We are then looking at the number of visits to $j \in T$ before absorption by $R - R_1 \cup R_2 \cup \cdots$ when starting from $i \in T$.

Recall that when starting from any state i, the mean number of visits to a state j is

$$E_i\left[\sum_{n=0}^{\infty} 1_{\{X_n=j\}}\right] = \sum_{n=0}^{\infty} p_{ij}(n), \tag{6.6}$$

and this is the (i,j) element of the potential matrix

$$\mathbf{G} = \sum_{n=0}^{\infty} \mathbf{P}^n. \tag{6.7}$$

Only the pairs (i,j) where i and j are transient require treatment, because in all other situations the quantity (6.6) is either null or infinite. For instance, if i and j are in the same recurrent class, the number of visits to j starting from i is infinite, whereas it is zero if i and j belong to different recurrent classes, or if i is a recurrent state and j is a transient state. If i is transient and j is recurrent, the corresponding term is infinite if j is accessible from i, and it is zero if j is not accessible from j.

To evaluate (6.7) for $i \in T$ and $j \in T$, one uses the representation (6.1) of the transition matrix to obtain

$$\mathbf{G} = \begin{pmatrix} E & 0 \\ 0 & S \end{pmatrix},$$

where

$$S := \sum_{n=0}^{\infty} \mathbf{Q}^n. \tag{6.8}$$

Definition 6.2.1 *An* HMC *with at least one transient state and one recurrent state is called an* absorbing chain, *and the matrix S is its* fundamental matrix.

Remark 6.2.2 There are in fact *two* different types of fundamental matrices, one for absorbing chains that we just introduced and one for ergodic chains that will be defined in Section 6.3.

Denoting by I the identity matrix indexed by T, observing that

$$(I - \mathbf{Q}) \left(\sum_{n=0}^{N} \mathbf{Q}_n \right) = I - \mathbf{Q}^{N+1}$$

and letting N tend to ∞, we find that S is a solution of

$$S(I - \mathbf{Q}) = I.$$

If T is finite, S is therefore the inverse of $I - \mathbf{Q}$. When T is infinite, there may be several solutions for S. However:

Theorem 6.2.3 *The matrix S given by (6.8) is the smallest solution of $S(I-\mathbf{Q}) = I$ subject to the constraint $S \geq 0$.*

Proof. Let S be a solution. Writing $S(I - \mathbf{Q}) = I$ as $S = I + S\mathbf{Q}$ and iterating the latter equality gives

$$S = I + \mathbf{Q} + \cdots + \mathbf{Q}^n + S\mathbf{Q}^{n+1}.$$

But $S \geq 0$ implies $S\mathbf{Q}^{n+1} \geq 0$, and therefore

$$S \geq I + \mathbf{Q} + \cdots + \mathbf{Q}^n.$$

Passing to the limit, we obtain $S \geq \sum_{n=0}^{\infty} \mathbf{Q}^n$. ☐

EXAMPLE 6.2.4: JUST AN EXAMPLE. The transition matrix

$$
\mathbf{P} = \begin{pmatrix}
0.4 & 0.3 & 0.3 & & & & & \\
0 & 0.6 & 0.4 & & & & & \\
0.5 & 0.5 & 0 & & & & & \\
& & & 0 & 1 & & & \\
& & & 0.8 & 0.2 & & & \\
0 & 0 & 0 & & & 0.4 & 0.6 & 0 \\
0.4 & 0.4 & 0 & & & 0 & 0 & 0.2 \\
0.1 & 0 & 0.3 & & & 0.6 & 0 & 0
\end{pmatrix}
$$

with state space $E = \{1, 2, 3, 4, 5, 6, 7, 8\}$ has two recurrent classes $R_1 = \{1, 2, 3\}$ and $R_2 = \{4, 5\}$ and one transient class $T = \{6, 7, 8\}$. Here

$$
\mathbf{Q} = \begin{pmatrix}
0.4 & 0.6 & 0 \\
0 & 0 & 0.2 \\
0.6 & 0 & 0
\end{pmatrix}
$$

and

$$
S = (I - \mathbf{Q})^{-1} = \frac{1}{66} \begin{pmatrix}
125 & 75 & 15 \\
15 & 75 & 15 \\
75 & 45 & 75
\end{pmatrix}.
$$

Also, no state of T leads to R_2, and therefore $g_{ij} = 0$ for $i \in T, j \in R_2$. All states of T lead to R_1, and therefore $g_{ij} = +\infty$ for $i \in T, j \in R_1$. In summary:

$$
\mathbf{G} = \begin{pmatrix}
+\infty & +\infty & +\infty & 0 & 0 & 0 & 0 & 0 \\
+\infty & +\infty & +\infty & 0 & 0 & 0 & 0 & 0 \\
+\infty & +\infty & +\infty & 0 & 0 & 0 & 0 & 0 \\
0 & 0 & 0 & +\infty & +\infty & 0 & 0 & 0 \\
0 & 0 & 0 & +\infty & +\infty & 0 & 0 & 0 \\
+\infty & +\infty & +\infty & 0 & 0 & & & \\
+\infty & +\infty & +\infty & 0 & 0 & & & \\
+\infty & +\infty & +\infty & 0 & 0 & & &
\end{pmatrix},
$$

where the missing part is S.

6.2.2   The Absorption Matrix

We seek to compute the probability of absorption by a given recurrent class when starting from a given transient state. As we shall see later, it suffices for the theory to treat the case where the recurrent classes are singletons. We therefore suppose that the transition matrix has the form

$$\mathbf{P} = \begin{pmatrix} I & 0 \\ B & \mathbf{Q} \end{pmatrix}. \tag{6.9}$$

Let f_{ij} be the probability of absorption by recurrent class $R_j = \{j\}$ when starting from the transient state i. From (6.9),

$$\mathbf{P}^n = \begin{pmatrix} I & 0 \\ L_n & \mathbf{Q}^n \end{pmatrix},$$

where $L_n = (I + \mathbf{Q} + \cdots + \mathbf{Q}^n)B$. Therefore, $\lim_{n\uparrow\infty} L_n = SB$. For $i \in T$, the (i,j) term of L_n is

$$L_n(i,j) = \mathrm{P}(X_n = j | X_0 = i).$$

Now, if T_j is the first time of visit to $R_j = \{j\}$ after time 0, then

$$L_n(i,j) = \mathrm{P}_i(T_j \leq n),$$

since $R_j := \{j\}$ is a closed state. Letting n tend to ∞, we obtain:

Theorem 6.2.5 *Consider an* HMC *with transition matrix* $\mathbf{P}$ *as in (6.9). The probability of absorption by the recurrent class* $R_j = \{j\}$ *starting from the transient state* i *is*

$$\mathrm{P}_i(T_{R_j} < \infty) = (SB)_{ij}. \tag{6.10}$$

The general case, where the recurrence classes are not necessarily singletons, can be reduced to the singleton case as follows. Let $\mathbf{P}^*$ be the matrix obtained from the transition matrix $\mathbf{P}$ given by (6.6), by grouping for each j the states of recurrent class R_j into a single state $\hat{j}$:

$$\mathbf{P}^* = \begin{pmatrix} \hat{1} & \hat{2} & \cdots & T & \\ 1 & 0 & 0 & 0 & \hat{1} \\ 0 & 1 & 0 & 0 & \hat{2} \\ 0 & 0 & \ddots & 0 & \vdots \\ b_{\hat{1}} & b_{\hat{2}} & \cdots & \mathbf{Q} & T \end{pmatrix}, \tag{6.11}$$

where $b_j = B(j)\mathbf{1}_T$ is obtained by summation of the columns of $B(j)$, the matrix consisting of the columns $i \in R_j$ of B.

The probability f_{iR_j} of absorption by class R_j when starting from $i \in T$ equals $\hat{f}_{i\hat{j}}$, the probability of ever visiting $\hat{j}$ when starting from i, computed for the chain with transition matrix $\mathbf{P}^*$.

EXAMPLE 6.2.6: JUST ANOTHER EXAMPLE. Consider the chain with state space $E = \{1, 2, 3, 4, 5, 6, 7\}$ and transition matrix

$$\mathbf{P} = \begin{pmatrix} 0.5 & 0.5 & & & & & \\ 0.8 & 0.2 & & & & & \\ & & 0 & 0.4 & 0.6 & & \\ & & 1 & 0 & 0 & & \\ & & 1 & 0 & 0 & & \\ 0.1 & 0 & 0.2 & 0.1 & 0.2 & 0.3 & 0.1 \\ 0.1 & 0.1 & 0.1 & 0 & 0.1 & 0.2 & 0.4 \end{pmatrix}.$$

It has two recurrent classes $R_1 = \{1, 2\}$, $R_2 = \{3, 4, 5\}$ and one transient class $T = \{6, 7\}$. In the notation of the current subsection,

$$B_1 = \begin{pmatrix} 0.1 & 0 \\ 0.1 & 0.1 \end{pmatrix}, B_2 = \begin{pmatrix} 0.2 & 0.1 & 0.2 \\ 0.1 & 0 & 0.1 \end{pmatrix}, \mathbf{Q} = \begin{pmatrix} 0.3 & 0.1 \\ 0.2 & 0.4 \end{pmatrix},$$

and therefore

$$b_1 = \begin{pmatrix} 0.1 \\ 0.2 \end{pmatrix}, b_2 = \begin{pmatrix} 0.6 \\ 0.2 \end{pmatrix}, \hat{B} = \begin{pmatrix} 0.1 & 0.5 \\ 0.2 & 0.2 \end{pmatrix}.$$

Also,

$$\mathbf{P}^* = \begin{pmatrix} 1 & 0 & 0 & 0 \\ 0 & 1 & 0 & 0 \\ 0.1 & 0.6 & 0.3 & 0.1 \\ 0.2 & 0.2 & 0.2 & 0.4 \end{pmatrix}.$$

Computations yield

$$S = (I - \mathbf{Q})^{-1} = \begin{pmatrix} 1.5 & 0.25 \\ 0.5 & 1.75 \end{pmatrix}, S\hat{B} = \begin{pmatrix} 0.2 & 0.8 \\ 0.4 & 0.6 \end{pmatrix}.$$

In particular the probability of absorption from transient state 6 to class $\{3, 4, 5\}$ is 0.8.

EXAMPLE 6.2.7: SIBMATING. In the diploid organisms a given hereditary character is carried by a pair of *genes*. We consider the situation in which each gene can take two forms, called *alleles*, denoted by a and A. Such was the case in the historical experiments performed around the year 1865 by the monk Gregor Mendel, who studied the hereditary transmission of the nature of the skin in a species of green peas. The two alleles corresponding to the genes or character "nature of the skin" are a for "wrinkled" and A for "smooth." The genes are grouped by pairs, and there are two alleles, and thus three *genotypes* are possible for the character under study: aa, Aa (same as aA), and AA. With each genotype is associated a *phenotype*, which is the external appearance corresponding to the genotype. Genotypes aa, AA have different phenotypes (otherwise, no character could be isolated), and the phenotype of Aa lies somewhere between the phenotypes of aa and AA. Sometimes an allele is *dominant*, say A, and the phenotype of Aa is then the same of the phenotype of AA.

During the reproduction process, each parent contributes to the genetic heritage of the descendant by providing *one* allele of their pair. This is done by intermediaries of the reproduction process, the cells called *gametes* (in the human species, the *spermatozoid* and the *ovula*), which carry only one gene of the pair of genes characteristic of each parent. The gene carried by the gamete is chosen at random from among the pair of genes of the parent. The actual process occurring in the reproduction of diploid cells is called *meiosis* (see Figure 6.2.1).

A given cell possesses two chromosomes. A chromosome can be viewed as a string of genes, each gene being at a specific location in the string.

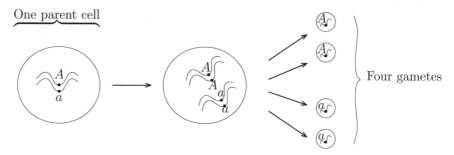

Figure 6.2.1. Meiosis

Let us start from an idealistically infinite population where the genotypes are found in the following proportions:

$$
\begin{array}{ccc}
AA & Aa & aa \\
x & 2z & y
\end{array}
$$

Here x, y, and z are numbers between 0 and 1, and

$$x + 2z + y = 1\,.$$

The two parents are chosen independently (*random mating*), and each of their respective gametes chooses an allele at random independently of each other in the pair carried by the corresponding parent. We leave as an exercise for the reader to prove that the genotypic distribution of all the generations starting with the third one are the same, depending only on the proportions of alleles of type A in the initial population: This is the famous law of Hardy and Weinberg, which holds for random mating.

In *sibmating* (sister-brother mating) two individuals mate and two individuals from their offspring are chosen at random to mate, and this incestuous process goes on through the generations.

We shall denote by X_n the genetic types of the mating pair at the n-th generation. Clearly, $\{X_n\}_{n \geq 0}$ is an HMC with six states representing the different pairs of genotypes $AA \times AA$, $aa \times aa$, $AA \times Aa$, $Aa \times Aa$, $Aa \times aa$, $AA \times aa$, denoted respectively $1, 2, 3, 4, 5, 6$. The following table gives the probabilities of occurrence of the 3 possible genotypes in the descent of a mating pair:

$$
\left.
\begin{array}{c c c c}
 & AA & Aa & aa \\
AA\ AA & 1 & 0 & 0 \\
aa\ aa & 0 & 0 & 1 \\
AA\ Aa & 1/2 & 1/2 & 0 \\
Aa\ Aa & 1/4 & 1/2 & 1/4 \\
Aa\ aa & 0 & 1/2 & 1/2 \\
AA\ aa & 0 & 1 & 0
\end{array}
\right\} \text{parents' genotype}
$$

$$\underbrace{\qquad\qquad\qquad}_{\text{descendant's genotype}}$$

The transition matrix of $\{X_n\}_{n \geq 0}$ is then easily deduced:

$$
\mathbf{P} =
\begin{pmatrix}
1 & & & & & \\
 & 1 & & & & \\
1/4 & & 1/2 & 1/4 & & \\
1/16 & 1/16 & 1/4 & 1/4 & 1/4 & 1/8 \\
 & 1/4 & & 1/4 & 1/2 & \\
 & & & 1 & &
\end{pmatrix}.
$$

The set $R = \{1, 2\}$ is absorbing, and the restriction of the transition matrix to the transient set $T = \{3, 4, 5, 6\}$ is

$$
\mathbf{Q} =
\begin{pmatrix}
1/2 & 1/4 & 0 & 0 \\
1/4 & 1/4 & 1/4 & 1/8 \\
0 & 1/4 & 1/2 & 0 \\
0 & 1 & 0 & 0
\end{pmatrix}.
$$

The fundamental matrix is

$$S = (I - Q)^{-1} = \frac{1}{6} \begin{pmatrix} 16 & 8 & 4 & 1 \\ 8 & 16 & 8 & 2 \\ 4 & 8 & 16 & 1 \\ 8 & 16 & 8 & 8 \end{pmatrix},$$

and the absorption probability matrix is

$$SB = S \begin{pmatrix} 1/4 & 0 \\ 1/16 & 1/16 \\ 0 & 1/4 \\ 0 & 0 \end{pmatrix} = \begin{pmatrix} 3/4 & 1/4 \\ 1/2 & 1/2 \\ 1/4 & 3/4 \\ 1/2 & 1/2 \end{pmatrix}.$$

For instance, the $(3, 2)$ entry, $\frac{3}{4}$, is the probability that when starting from a couple of ancestors of type $Aa \times aa$, the race will end up in the genotype $aa \times aa$.

We shall say a little more about this basic model, in the vein of Subsection 4.3.3 on quasi-stationary distribution. The characteristic equation of Q is

$$(\lambda - \frac{1}{2})(\lambda^3 - \frac{3}{4}\lambda^2 - \frac{1}{8}\lambda + \frac{1}{16}) = 0,$$

and its eigenvalues are, properly ordered in descending order.

$$\lambda_1 = \frac{1 + \sqrt{5}}{4}, \quad \lambda_2 = \frac{1}{2}, \quad \lambda_3 = \frac{1 - \sqrt{5}}{4}, \quad \lambda_4 = \frac{1}{4}.$$

If we define U (resp., V) to be the matrix whose columns are left-eigenvectors (resp., right-eigenvectors) of Q, placed in the same order as the corresponding eigenvalues, one then has

$$U = \begin{pmatrix} 6 + 2\sqrt{5} & 2 & 6 - 2\sqrt{5} & 2 \\ 4 + 4\sqrt{5} & 0 & 4 - 4\sqrt{5} & -2 \\ 6 + 2\sqrt{5} & -2 & 6 - 2\sqrt{5} & 2 \\ 2 & 0 & 2 & -1 \end{pmatrix}$$

and

$$V = \begin{pmatrix} 1 & \sqrt{5} - 1 & 1 & 6 - 2\sqrt{5} \\ 1 & 0 & -1 & 1 \\ 1 & -\sqrt{5} - 1 & 1 & 6 + 2\sqrt{51} \\ 1 & -1 & 1 & -4 \end{pmatrix}.$$

Note that we have not normalized these vectors in order to have $u_i^T v_i = 1$ for all $i \in [1, 4]$. If we do this for the eigenvectors corresponding to λ_1, we then have the new eigenvectors

$$u_1^T = \frac{1}{9 + 4\sqrt{5}}(3 + \sqrt{5}, 2 + 2\sqrt{5}, 3 + \sqrt{5}, 1),$$

$$v_1^T = \frac{9 + 4\sqrt{5}}{20}(1, \sqrt{5} - 1, 1, 6 - 2\sqrt{5}).$$

Note that in addition we have $\sum_{k=1}^{4} u_i(k) = 1$, so that u_1^T is in fact the quasi-stationary distribution. Here the multiplicity of λ_2 is $m_2 = 1$, and therefore the rate of convergence to the quasi-stationary distribution is geometric, with the relative speed

$$\frac{\lambda_2}{\lambda_1} = \frac{2}{1 + \sqrt{5}}.$$

We have seen that the absorption probabilities by the recurrent classes can be obtained as a by-product of the *access matrix*

$$F := \{f_{ij}\},$$

where

$$f_{ij} := P_i(T_j < \infty),$$

because $f_{iR_j} = f_{ik}$ for any $k \in R_j$. For the absorption probabilities, we need only the entries of the access matrix corresponding to the indices $i \in T$, $j \in R$. Note, however, that the other entries of the access matrix can be obtained from the potential matrix. In fact,

(α) if i and j are in the same recurrent class, $f_{ij} = 1$,

(β) if i and j are in two different recurrent classes, $f_{ij} = 0$,

(γ) if i is recurrent and j is transient, $f_{ij} = 0$, and

(δ) if i and j are transient,

$$\frac{f_{ij}}{f_{ii}} = \frac{g_{ij}}{g_{jj}} \text{ if } j \neq i \text{ and } f_{jj} = \frac{g_{jj}}{1 + g_{jj}}.$$

(The last equality follows from Theorem 2.5.9.)

6.2.3   Hitting Times Formula

The next result extends formula (3.3).

Theorem 6.2.8 *Let $\{X_n\}_{n \geq 0}$ be a positive recurrent* HMC *with state space E and stationary distribution π. Let μ be a probability distribution on E and let $S \in \mathbb{N}$ be a stopping time of this chain such that $\mathrm{E}_\mu[S] < \infty$ and $\mathrm{P}_\mu(X_S \in \cdot) = \mu$. Then for all $j \in E$,*

$$\mathrm{E}_\mu\left[\sum_{k=0}^{S-1} 1_{\{X_n=j\}}\right] = \mathrm{E}_\mu[S]\,\pi(j). \tag{6.12}$$

Proof. Let x_j denote the left-hand side of (6.12). If the vector x is an invariant measure of the chain, it must be of the form $x_i = c\pi(i)$ with $c = \sum_i x_i = \mathrm{E}_\mu[S]$, which gives (6.12). For the proof that x is an invariant measure, write (noting that $\mathrm{P}_\mu(X_S = k) = \mathrm{P}_\mu(X_0 = k)$ for the second equality):

$$x_k = \sum_{n=0}^{\infty} \mathrm{P}_\mu(X_n = k, n < S)$$

$$= \sum_{n=0}^{\infty} \mathrm{P}_\mu(X_{n+1} = k, n < S)$$

$$= \sum_{n=0}^{\infty} \sum_j \mathrm{P}_\mu(X_n = j, X_{n+1} = k, n < S)$$

$$= \sum_{n=0}^{\infty} \sum_j \mathrm{P}_\mu(X_n = j, n < S) p_{jk}$$

$$= \sum_j \left(\sum_{n=0}^{\infty} \mathrm{P}_\mu(X_n = j, n < S)\right) p_{jk} = \sum_j x_j p_{jk}\,.$$

$\square$

In fact, (6.12) remains true when $\mathrm{E}_\mu[S] = \infty$ (consider the stopping time $S^{(n)} = S \wedge \tau_n$, where τ_n is the n-return time to i, write (6.12) for $S^{(n)}$, and let $n \to \infty$, to obtain $\mathrm{E}_\mu\left[\sum_{k=0}^{S-1} 1_{\{X_k=j\}}\right] = \infty$).

The particular case that is used in the sequel is for $\mu = \delta_i$. Let S be a stopping time and let i be any state such that $\mathrm{P}_i(X_S = i) = 1$. Then, for all $j \in E$,

$$\mathrm{E}_i\left[\sum_{k=0}^{S-1} 1_{\{X_n=j\}}\right] = \mathrm{E}_i[S]\,\pi(j). \tag{6.13}$$

In the special case where $S = T_i$, the return time to i gives formula (3.3). In the next example, additional information is extracted from (6.12).

EXAMPLE 6.2.9: COMMUTE TIME FORMULAS. ([1]) Let i and j be two distinct states and let S be the first time of return to i after the first visit to j. Then $E_i [S] = E_i [T_j] + E_j [T_i]$ (strong Markov property at T_j). Also,

$$E_i \left[\sum_{n=0}^{S-1} 1_{\{X_n=j\}} \right] = E_i \left[\sum_{n=T_j}^{S-1} 1_{\{X_n=j\}} \right] = E_j \left[\sum_{n=0}^{T_i-1} 1_{\{X_n=j\}} \right],$$

where the last equality is justified by the strong Markov property. Therefore, by (6.13),

$$E_j \left[\sum_{n=0}^{T_i-1} 1_{\{X_n=j\}} \right] = \pi(j) \left(E_i [T_j] + E_j [T_i] \right). \qquad (\star)$$

Using words, the left-hand side of this equality is

$$E_j[\text{number of visits to } j \text{ before } i]. \qquad (6.14)$$

The quantity $E_i [T_j] + E_j [T_i]$ is called the commute time between i and j. It is the average time needed to go from i to j and then return to j, or, in other words the average return time to i with the constraint of visiting j at least once. The quantities $E_i [T_j]$ can be computed, when the state space is finite, via the fundamental matrix (Theorem 6.3.6).

Now, the probability that j is not visited between two successive visits of i is $P_i (T_j > T_i)$. Therefore, the number of visits to i (including time 0) before T_j has a geometric distribution with parameter $p = P_i (T_j > T_i)$, and the average number of such visits is $\frac{1}{P_i(T_j < T_i)}$. Therefore, by $(\star)$, after exchanging the roles of i and j,

$$P_i (T_j < T_i) = \frac{1}{\pi(i) \left(E_i [T_j] + E_j [T_i] \right)}. \qquad (6.15)$$

---

[1][Aldous and Fill, 1998].

6.3   The Second Fundamental Matrix

6.3.1   Definition

The first fundamental matrix concerns absorbing chains. There is another "fundamental matrix", defined this time for ergodic chains, called the *second fundamental matrix*.[2] These two types of fundamental matrices are different objects. The one for absorbing chains is the restriction of the potential matrix to the transient set. For ergodic recurrent chains, this notion is irrelevant since all its entries are infinite. The second fundamental matrix for ergodic chains is interesting because it gives access to a number of quantities such as the mean time $E_i[T_j]$ to return to j from state i, or the variance of the ergodic estimate $\frac{1}{n}\sum_{k=1}^{n} f(X_k)$. For an ergodic HMC with finite state space $E = \{1, 2, \ldots, r\}$ and stationary distribution π, it is the matrix

$$\mathbf{Z} := (I - (\mathbf{P} - \Pi))^{-1}, \tag{6.16}$$

where

$$\Pi := \mathbf{1}\pi^T = \begin{pmatrix} \pi(1) & \cdots & \pi(r) \\ \pi(1) & \cdots & \pi(r) \\ \vdots & & \vdots \\ \pi(1) & \cdots & \pi(r) \end{pmatrix}.$$

Theorem 6.3.1 *For any ergodic transition matrix* $\mathbf{P}$ *on a finite state space, the right-hand side of (6.16) is well defined and*

$$\mathbf{Z} = I + \sum_{n \geq 1}(\mathbf{P}^n - \Pi). \tag{6.17}$$

Proof. First observe that

$$
\begin{aligned}
\Pi\mathbf{P} &= \Pi\,\pi^T\mathbf{P} = \pi^T, \Pi = \mathbf{1}\pi^T, \\
\mathbf{P}\Pi &= \Pi\,\mathbf{P1} = \mathbf{1}, \Pi = \mathbf{1}\pi^T, \\
\Pi^2 &= \Pi\,\Pi = \mathbf{1}\pi^T, \pi^T\mathbf{1} = \mathbf{1}.
\end{aligned}
$$

In particular, for all $k \geq 1$,

$$\mathbf{P}\Pi^k = \Pi = \Pi^k\mathbf{P}.$$

---

[2]The terminology used in this book is perhaps not the standard one. An alternative could be: the fundamental matrix of absorbing chains (the first one), and the fundamental matrix of ergodic chains (the second one).

Therefore,

$$
\begin{aligned}
(\mathbf{P} - \Pi)^n &= \sum_{k=0}^{n} \binom{n}{k} (-1)^{n-k} \mathbf{P}^k \Pi^{n-k} \\
&= \mathbf{P}^n + \left(\sum_{k=0}^{n-1} \binom{n}{k} (-1)^{n-k} \right) \Pi = \mathbf{P}^n - \Pi \, .
\end{aligned}
$$

Therefore, with $A = \mathbf{P} - \Pi$,

$$
(I - A)(I + A + \cdots + A^{n-1}) = I - A^n = I + \mathbf{P}^n - \Pi \, .
$$

Passing to the limit $n \to \infty$,

$$
(I - A)\left(I + \sum_{n \geq 1} A^n\right) = I \, ,
$$

which shows that $I - (\mathbf{P} - \Pi)$ is invertible, with inverse

$$
I + \sum_{n \geq 1} (\mathbf{P} - \Pi)^n = I + \sum_{n \geq 1} (\mathbf{P}^n - \Pi) \, .
$$

$\square$

EXAMPLE 6.3.2: THE DIAGONAL OF THE SECOND FUNDAMENTAL MATRIX.
For fixed $m \geq 1$, let

$$
S := m + \inf \{k > 0 \, ; \, X_{m+k} = i\} \, .
$$

Then by (6.13),

$$
\mathrm{E}_i \left[\sum_{n=0}^{S-1} 1_{X_n = j} \right] = \pi(i) \mathrm{E}_i \left[S \right] \, ,
$$

that is,

$$
\mathrm{E}_i \left[\sum_{n=0}^{m-1} 1_{X_n = j} \right] = \pi(i) \left(m + \mathrm{E}_{\nu_m} \left[S_i \right] \right) \, ,
$$

where S_i is the hitting time of i, and ν_m is the distribution of the chain at time m. Therefore,

$$
\sum_{n=0}^{m-1} (p_{ii}(n) - \pi(i)) = \pi(i) \mathrm{E}_{\nu_m} \left[S_i \right] \, .
$$

Since $\lim_{m \uparrow \infty} \nu_m = \pi$, we have

$$\sum_{n=0}^{\infty} (p_{ii}(n) - \pi(i)) = \pi(i)\mathrm{E}_\pi [S_i] \,,$$

which is easily converted into

$$z_{ii} = \pi(i)\mathrm{E}_\pi [T_i] = \frac{\mathrm{E}_\pi [T_i]}{\mathrm{E}_i [T_i]} \,. \tag{6.18}$$

EXAMPLE 6.3.3: PATTERNS IN COIN TOSSING ([3]) Let $\{Y_n\}_{n \geq 0}$ be an IID sequence of Bernoulli variables, with $\mathrm{P}\,(Y_1 = 1) = \mathrm{P}\,(Y_1 = 0) = \frac{1}{2}$ and let $\{X_n\}_{n \geq 0}$ be the *snake chain* defined by

$$X_n = (Y_n, Y_{n+1}, \dots, Y_{n+L-1})$$

for some $L \geq 1$. Note that both $\{Y_n\}_{n \geq 0}$ and $\{X_n\}_{n \geq 0}$ are irreducible ergodic chains, starting in the steady state. Define

$$\widetilde{z}_{ij} = \sum_{n=0}^{\infty} (p_{ij}(n) - \pi(j)) \tag{6.19}$$

$(= z_{ij} - \pi(j))$, where $\mathbf{P}$ is the transition matrix of $\{X_n\}_{n \geq 0}$ and $\pi(j) = \frac{1}{2^L}$ is its stationary distribution.

For $n \geq L$, X_0 and X_n are independent, and therefore $p_{ij}(n) - \pi(j) = 0$, so that only the first L terms of (6.19) are non-null.

For $n < L$, $p_{ij}(n) > 0$ if and only if the pattern $j = (j_0, \dots, j_{L-1})$ shifted n to the right and the pattern $i = (i_0, \dots, i_{L-1})$ agree where they overlap (see Figure 6.3.1).

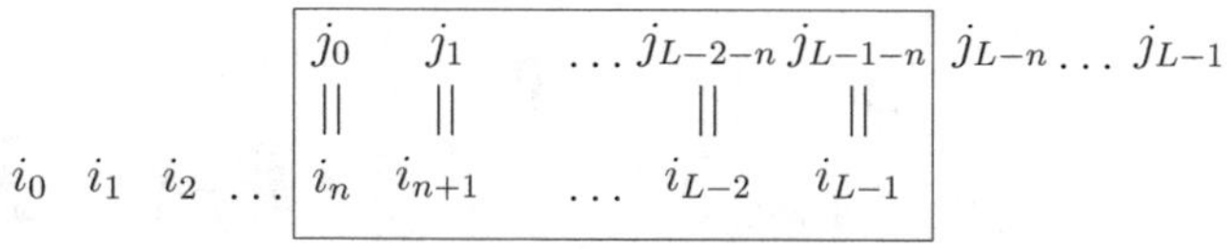

Figure 6.3.1.

---

[3][Aldous and Fill, 1998].

In this case $p_{ij}(n)$ equals $\frac{1}{2^n}$. Therefore, defining

$$c(i,j) = \sum_{n=0}^{L-1} \frac{1}{2^n} \chi(i,j,n), \tag{6.20}$$

where $\chi(i,j,n) = 1$ if and only if the situation depicted in Figure 6.3.1 is realized,

$$\widetilde{z}_{ij} = c(i,j) - L2^{-L}.$$

In view of the result of the previous example,

$$E_\pi[S_i] = 2^L c(i,i) - L.$$

But remember that X_0 is always distributed as π, and that to generate the first pattern, L coin tosses are necessary. Therefore, $2^L c(i,i)$ is the average number of coin tosses needed to see pattern i for the first time.

To illustrate this, consider the pattern $i = $ H'T'T'T'H'T'. We have $c(i,i) = 68$ (see Figure 6.3.2).

H	T	T	T	H	T		2^{-0}	
	H	T	T	T	H		0	
		H	T	T	T		0	$c(\text{HTTTHT}) = 2^6\left(1 + \frac{1}{2^4}\right)$
			H	T	T		0	
				H	T		2^{-4}	
					H		0	

Figure 6.3.2.

An Extension of the Fundamental Matrix

Expression (6.17) is meaningful only if the chain is ergodic. In particular, if the chain is only recurrent positive, but periodic, the series on the right-hand side oscillates. This does not mean, however, that in the periodic case the inverse in (6.16) does not exist. As a matter of fact, it does exist, but it is not given by formula (6.17).

The following is yet another type of matrix related to HMCs, also called a fundamental matrix.[4] This extension does not require, in principle, knowledge of the stationary distribution.

---

[4][Kemeny, 1991]. See [Grinstead and Snell, 1997)] for additional details.

Let b be any vector such that

$$b^T \mathbf{1} \neq 0, \tag{6.21}$$

and define

$$\mathbf{Z} = \left(I - \mathbf{P} + \mathbf{1}b^T\right)^{-1}, \tag{6.22}$$

where $\mathbf{P}$ is an ergodic matrix on the finite space E, with the stationary distribution π. The matrix differs from the usual fundamental matrix in that π is replaced by b.

Theorem 6.3.4 *The inverse matrix in (6.22) exists and*

$$\pi^T = b^T \mathbf{Z}. \tag{6.23}$$

Proof. Since $\pi^T \mathbf{1} = 1$ and $\pi^T (I - \mathbf{P}) = 0$,

$$\pi^T \left(I - \mathbf{P} + \mathbf{1}b^T\right) = \pi^T \mathbf{1}b^T = b^T, \tag{6.24}$$

and therefore, for any vector x such that

$$\left(I - \mathbf{P} + \mathbf{1}b^T\right) x = 0, \tag{6.25}$$

we have

$$b^T x = 0$$

and

$$(I - \mathbf{P}) x = 0.$$

Therefore, x must be a right-eigenvector associated with the eigenvalue $\lambda_1 = 1$, and consequently, x is a multiple of $\mathbf{1}$. But this is compatible with $b^T x = 0$ and $b^T \mathbf{1} \neq 0$ only if $x = 0$. Therefore (6.25) implies $x = 0$, which implies that $\left(I - \mathbf{P} + \mathbf{1}b^T\right)$ is invertible; and (6.24) proves (6.23). $\qquad\square$

6.3.2   The Mutual Time-Distance Matrix

We introduce some more notation: For any square matrix B, $d(B)$ is the diagonal matrix which has the same diagonal as B. In particular, $d(\Pi)^{-1}$ is the diagonal matrix for which the (i,i)-th entry is $\pi(i)^{-1}$. Recall also that $\mathbf{1}\mathbf{1}^T$ is the matrix with all entries equal to 1.

Theorem 6.3.5 *The* mutual time-distance *matrix $M = \{m_{ij}\}_{1 \leq i,j \leq r}$, where $m_{ij} := E_i[T_j]$ $(i, j \in E)$, is given by the formula*

$$M = (I - \mathbf{Z} + \mathbf{1}\mathbf{1}^T d(\mathbf{Z})) d(\Pi)^{-1}. \tag{6.26}$$

Proof. We first note that M has finite entries. Indeed, we already know that $m_{ii} = \mathrm{E}_i[T_i] = 1/\pi(i)$ and that $\pi(i) > 0$. As for m_{ij} when $i \neq j$, it is the mean time to absorption in the modified chain where j is made absorbing. By Remark 6.1.4, it is finite.

By first-step analysis,

$$m_{ij} = 1 + \sum_{k; k \neq j} p_{ik} m_{kj}, \qquad (6.27)$$

that is,

$$M = \mathbf{P}(M - d(M)) + \mathbf{1}\mathbf{1}^T. \qquad (6.28)$$

We now prove that there is but one finite solution of (6.28). To do this, we first show that for any solution M of (6.28), $d(M)$ is necessarily equal to $d(\Pi)^{-1}$. (We know this to be true when M is the mutual distance matrix, but not yet for a general solution of (6.28).) Indeed, premultiplying (6.28) by π^T yields

$$
\begin{aligned}
\pi^T M &= \pi^T \mathbf{P}(M - d(M)) + (\pi^T \mathbf{1})\mathbf{1}^T \\
&= \pi^T (M - d(M)) + \mathbf{1}^T,
\end{aligned}
$$

and therefore $\pi^T d(M) = \mathbf{1}^T$, which implies the announced result.

Now suppose that (6.28) has two finite solutions M_1 and M_2. Since $d(M_1) = d(M_2)$, it follows that

$$M_1 - M_2 = \mathbf{P}(M_1 - M_2).$$

Therefore, any column v of $M_1 - M_2$ is a right-eigenvector of $\mathbf{P}$ corresponding to the eigenvalue 1. We know that the right-eigenspace R_λ and the left-eigenspace L_λ corresponding to any given eigenvalue λ have the same dimension. For $\lambda = 1$, we know that the dimension of L_λ is one, since there is only one stationary distribution for an ergodic chain. Therefore R_λ has dimension 1 for $\lambda = 1$. Thus any right-eigenvector is a scalar multiple of $\mathbf{1}$. Therefore, $M_1 - M_2$ has columns of the type $\alpha \mathbf{1}$ for some α (*a priori* depending on the column). Since $d(M_1) = d(M_2)$, each column contains a zero, and therefore $\alpha = 0$ for all columns, that is, $M_1 - M_2 \equiv 0$.

At this point we have proven that M is the unique finite solution of (6.28). It remains to show that M defined by (6.26) satisfies equation (6.28). Indeed, from (6.26) and $d(M) = d(\Pi)^{-1}$,

$$M - d(\Pi)^{-1} = (-\mathbf{Z} + \mathbf{1}\mathbf{1}^T d(\mathbf{Z}))d(\Pi)^{-1}.$$

Therefore,

$$
\begin{aligned}
\mathbf{P}(M - d(\Pi)^{-1}) &= (-\mathbf{P}\mathbf{Z} + \mathbf{P}\mathbf{1}\mathbf{1}^T d(\mathbf{Z}))d(\Pi)^{-1} \\
&= (-\mathbf{P}\mathbf{Z} + \mathbf{1}\mathbf{1}^T d(\mathbf{Z}))d(\Pi)^{-1} \\
&= M + (-\mathbf{P}\mathbf{Z} - I + \mathbf{Z})d(\Pi)^{-1},
\end{aligned}
$$

where we have used the identity $\mathbf{P1} = \mathbf{1}$ for the second equality and (6.26) again for the third. Using now (6.16) in the form $I - \mathbf{Z} = \Pi - \mathbf{PZ}$, we see that

$$\mathbf{P}(M - d(\Pi)^{-1}) = M - \Pi d(\Pi)^{-1} = M - \mathbf{11}^T,$$

and (6.28) follows, since $d(M) = d(\Pi)^{-1}$. $\square$

Theorem 6.3.6 *Let $\mathbf{Z}$ be the fundamental matrix as in (6.22). Then for all $i \neq j$,*

$$\mathrm{E}_i\left[T_j\right] = \frac{z_{jj} - z_{ij}}{\pi(j)}.$$

Proof. Two preliminary formulas will be needed. First,

$$\mathbf{Z1} = \theta \mathbf{1}, \tag{6.29}$$

where $\theta^{-1} = b^T \mathbf{1}$. Indeed, from the definition of $\mathbf{Z}$,

$$\mathbf{Z}\left(I - \mathbf{P} + \mathbf{1}b^T\right)\mathbf{1} = \mathbf{1}.$$

But $(I - \mathbf{P})\mathbf{1} = 0$, and therefore (6.29) follows.

Next we need the formula

$$\mathbf{Z}\left(I - \mathbf{P}\right) = I - \theta \mathbf{1}b^T, \tag{6.30}$$

which follows from (6.22) and (6.29).

We now proceed to the main part of the proof. Call N the mutual distance matrix M in which the diagonal elements have been replaced by 0's. From (6.28), obtain

$$(I - \mathbf{P})\,N = \mathbf{11}^T - D^{-1},$$

where $D := \operatorname{diag}\{\pi(1), \dots, \pi(n)\}$. Multiplying both sides by $\mathbf{Z}$ and using (6.29), we obtain

$$\mathbf{Z}\left(I - \mathbf{P}\right)N = \theta \mathbf{11}^T - \mathbf{Z}D^{-1}.$$

By (6.30),

$$\mathbf{Z}\left(I - \mathbf{P}\right)N = N - \theta \mathbf{1}b^T N.$$

Therefore,

$$N = \theta \mathbf{11}^T - \mathbf{Z}D^{-1} + \theta \mathbf{1}b^T N.$$

Thus, for all $i, j \in E$,

$$n_{ij} = \theta - \frac{z_{ij}}{\pi(j)} + \theta\left(b^T N\right)_j.$$

For $i = j$, $n_{ij} = \theta - \frac{z_{jj}}{\pi(j)} + \theta\left(b^T N\right)_j = 0$, which gives $\left(b^T N\right)_j$. Finally, for $i \neq j$,

$$n_{ij} = \frac{z_{jj} - z_{ij}}{\pi(j)}.$$

$\square$

6.3.3 Variance of Ergodic Estimates

For a positive recurrent Markov chain, we know from the ergodic theorem that the estimate $\frac{1}{n}\sum_{k=1}^{n} f(X_n)$ of $\langle f\rangle_\pi := \mathrm{E}_\pi[f(X_0)]$ (where $f : E \to \mathbb{R}$ is such that $\langle |f|\rangle_\pi < \infty$, a condition that is always satisfied when the state space is finite) is asymptotically unbiased, in the sense that it converges to $\langle f\rangle_\pi$ as $n \to \infty$.

In the next result, we use the notation $\langle x, y\rangle_\pi$ for $\sum_i \pi(i) x_i y_i$.

Theorem 6.3.7 *Let $\{X_n\}_{n\geq 0}$ be an ergodic Markov chain with finite state space. For any function $f : E \to \mathbb{R}$,*

$$\lim_{n\to\infty} \frac{1}{n} V_{P_\mu}\left(\sum_{k=1}^{n} f(X_k)\right) = 2\langle f, \mathbf{Z}f\rangle_\pi - \langle f, (I + \Pi)f\rangle_\pi \qquad (6.31)$$

for any initial distribution μ.

In (6.31), $f^T = (f(1), \dots, f(r))$, where r is the number of states. The notation V_{P_μ} indicates that the variance is computed with respect to P_μ. The quantity (6.31) will be denoted by $v(f, \mathbf{P}, \pi)$.

Proof. We first treat the case where $\mu = \pi$, the stationary distribution. To simplify, we write V_π for V_{P_π}. Then

$$\frac{1}{n} V_\pi\left(\sum_{k=1}^{n} f(X_k)\right) = \frac{1}{n}\left\{\sum_{k=1}^{n} V_\pi(f(X_k)) + 2\sum_{\substack{k,j=1\\k<j}}^{n} \mathrm{cov}_\pi(f(X_k), f(X_j))\right\}$$

$$= V_\pi(f(X_0)) + \sum_{\ell=1}^{n-1} \frac{n-\ell}{n} \mathrm{cov}_\pi(f(X_0), f(X_\ell))$$

where we have used the fact that when the initial distribution is π, the chain is stationary, and in particular, $\mathrm{cov}_\pi(f(X_k), f(X_j)) = \mathrm{cov}_\pi(f(X_0), f(X_{j-k}))$ for $k < j$. Now,

$$V_\pi(f(X_0)) = \mathrm{E}_\pi[f(X_0)^2] - \mathrm{E}_\pi[f(X_0)]^2$$

$$= \sum_{i\in E} \pi(i)f(i)^2 - \left(\sum_{i\in E} \pi(i)f(i)\right)^2$$

$$= \langle f, f\rangle_\pi - \langle f, \Pi f\rangle_\pi.$$

Also,

$$
\begin{aligned}
\mathrm{cov}_\pi(f(X_0), f(X_\ell)) &= \mathrm{E}_\pi[f(X_0)f(X_k)] - \mathrm{E}_\pi[f(X_0)]^2 \\
&= \sum_{i\in E}\sum_{j\in E}\pi(i)p_{ij}(\ell)f(i)f(j) - \mathrm{E}_\pi[f(X_0)]^2 \\
&= \langle f, \mathbf{P}^\ell f\rangle_\pi - \langle f, \Pi f\rangle_\pi \\
&= \langle f, (\mathbf{P}^\ell - \Pi)f\rangle_\pi .
\end{aligned}
$$

Since $\lim_{n\to\infty}\sum_{\ell=1}^n(\mathbf{P}^\ell - \Pi) = \mathbf{Z} - I$, we have

$$
\lim_{n\to\infty}\sum_{\ell=1}^{n-1}\frac{n-\ell}{n}(\mathbf{P}^\ell - \Pi) = \mathbf{Z} - I.
$$

(This is Cesàro's lemma (Theorem A.1.7): If $A_n = \sum_{\ell=1}^n \alpha_\ell$ tends to A as $n\to\infty$, then $\lim_{n\to\infty}\frac{1}{n}\sum_{\ell=1}^{n-1} A_\ell = A$. But

$$
\frac{1}{n}\sum_{\ell=1}^{n-1} A_\ell = \frac{1}{n}(\alpha_1 + (\alpha_1 + \alpha_2) + \cdots + (\alpha_1 + \cdots + \alpha_{n-1})) = \sum_{\ell=1}^{n-1}\frac{n-\ell}{n}\alpha_\ell.)
$$

Therefore,

$$
\lim_{n\to\infty}\frac{1}{n}V_\pi\left(\sum_{k=1}^n f(X_k)\right) = \langle f, f\rangle_\pi - \langle f, \Pi f\rangle_\pi + 2\langle f, (\mathbf{Z} - I)f\rangle_\pi ,
$$

which is the announced result (for $\mu = \pi$).

To prove the result in the general case where the initial distribution is arbitrary, it suffices to show that for two chains $\{X_n^{(1)}\}_{n\geq 0}$ and $\{X_n^{(2)}\}_{n\geq 0}$ with the same transition matrix $\mathbf{P}$ and arbitrary initial distributions μ and ν, respectively, that couple at a time τ such that $\mathrm{E}[\tau^2] < \infty$ (this is the case here, see Theorem 4.3.3)

$$
\lim_{n\to\infty}\frac{1}{n}V\left(\sum_{k=1}^\infty f(X_k^{(1)})\right) = \lim_{n\to\infty}\frac{1}{n}V\left(\sum_{k=1}^\infty f(X_n^{(2)})\right) .
$$

But with $X_n = X_n^{(1)}$ or $X_n^{(2)}$,

$$
\begin{aligned}
V\left(\sum_{k=1}^{n} f(X_k)\right) &= \mathrm{E}\left[\left(\sum_{k=1}^{n} f(X_k)\right)^2\right] - \mathrm{E}\left[\sum_{k=1}^{n} f(X_k)\right]^2 \\
&= \mathrm{E}\left[\left(\sum_{k=1}^{\tau \wedge n} + \sum_{k=\tau+1}^{n}\right)^2\right] - \left(\mathrm{E}\left[\sum_{k=1}^{\tau \wedge n}\right] + \mathrm{E}\left[\sum_{k=\tau+1}^{n}\right]\right)^2 \\
&= \mathrm{E}\left[\left(\sum_{k=1}^{\tau \wedge n}\right)^2\right] + \mathrm{E}\left[\left(\sum_{k=\tau+1}^{n}\right)^2 + 2\mathrm{E}\left[\left(\sum_{k=1}^{\tau \wedge n}\right)\left(\sum_{k=\tau+1}^{n}\right)\right]\right] \\
&\quad - \mathrm{E}\left[\sum_{k=1}^{\tau \wedge n}\right]^2 - \mathrm{E}\left[\sum_{k=\tau+1}^{n}\right]^2 - 2\mathrm{E}\left[\sum_{k=1}^{\tau \wedge n}\right]\mathrm{E}\left[\sum_{k=\tau+1}^{n}\right].
\end{aligned}
$$

Since $\sum_{k=\tau+1}^{n} f(X_k^{(1)}) = \sum_{k=\tau+1}^{n} f(X_k^{(2)})$, it follows (with obvious shorthand notations) that

$$
\frac{1}{n}\left\{V\left(\sum_{k=1}^{n} f(X_k^{(1)})\right) - \frac{1}{n}V\left(\sum_{k=1}^{n} f(X_k^{(2)})\right)\right\} = \frac{1}{n}A_n + \frac{2}{n}B_n - \frac{2}{n}C_n,
$$

where

$$
\begin{aligned}
A_n &= \left\{\mathrm{E}\left[\left(\sum_{k=1}^{\tau \wedge n}(1)\right)^2\right] - \mathrm{E}\left[\left(\sum_{k=1}^{\tau \wedge n}(2)\right)^2\right] - \mathrm{E}\left[\sum_{k=1}^{\tau \wedge n}(1)\right]^2 + \mathrm{E}\left[\sum_{k=1}^{\tau \wedge n}(2)\right]^2\right\}, \\
B_n &= \left\{\mathrm{E}\left[\left(\sum_{k=\tau+1}^{n}(1,2)\right)\left(\sum_{k=1}^{\tau \wedge n}(1) - \sum_{k=1}^{\tau \wedge n}(2)\right)\right]\right\}, \\
C_n &= \left\{\mathrm{E}\left[\sum_{k=\tau+1}^{n}(1,2)\right]\mathrm{E}\left[\sum_{k=1}^{\tau \wedge n}(1) - \sum_{k=1}^{\tau \wedge n}(2)\right]\right\}.
\end{aligned}
$$

Write

$$
\frac{2}{n}B_n = 2\mathrm{E}\left[\frac{\sum_{k=\tau+1}^{n}(1,2)}{n}\left(\sum_{k=1}^{\tau \wedge n}(1) - \sum_{k=1}^{\tau \wedge n}(2)\right)\right]
$$

and observe that the quantity under the expectation converges, as $n \to \infty$, towards $\mathrm{E}_\pi[f(X_0)]\left(\sum_{k=1}^{\tau}(f(X_k^{(1)}) - f(X_k^{(2)}))\right)$ and is for fixed n bounded in absolute value by $2(\sup|f|)\tau$, an integrable random variable. Therefore, by dominated convergence,

$$
\lim_{n \to \infty} \frac{2}{n}B_n = 2\mathrm{E}_\pi[f(X_0)]\mathrm{E}\left[\sum_{k=1}^{\tau}(f(X_k^{(1)}) - f(X_k^{(2)}))\right].
$$

A similar argument shows that $\frac{2}{n}C_n$ has the same limit. Therefore, $\lim_{n\to\infty}\frac{2}{n}(B_n - C_n) = 0$. As for A_n, it is bounded by $4(\sup|f|)^2 E[\tau^2] < \infty$, and therefore $\lim_{n\to\infty}\frac{1}{n}A_n = 0$. $\square$

We shall now give an expression of the asymptotic variance in terms of the eigenvalues, when $\mathbf{P}$ has r distinct eigenvalues. We have, in view of (4.13),

$$(\mathbf{P}^n - \Pi) = \sum_{i=2}^{r}\lambda_i^n v_i u_i^T,$$

and therefore

$$\mathbf{Z} = I + \sum_{n\geq 1}(\mathbf{P}^n - \Pi) = I + \sum_{i=2}^{r}\frac{\lambda_i}{1 - \lambda_i}v_i u_i^T. \tag{6.32}$$

Also, from (6.31),

$$v(f, \mathbf{P}, \pi) = V_\pi(f(X_0)) + 2\sum_{i=2}^{r}\frac{\lambda_i}{1 - \lambda_i}\langle f, v_i\rangle_\pi (f^T u_i). \tag{6.33}$$

For a reversible pair $(\mathbf{P}, \pi)$, we have $u_i = Dv_i$, and therefore $f^T u_i = \langle f, v_i\rangle_\pi$. Using this observation and (9.5), we obtain from (6.33),

$$v(f, \mathbf{P}, \pi) = \sum_{i=2}^{r}\frac{1 + \lambda_i}{1 - \lambda_i}|\langle f, v_i\rangle_\pi|^2. \tag{6.34}$$

Remark 6.3.8 When one is interested in the speed of convergence to equilibrium, it is the second-largest eigenvalue modulus that is important. If one is interested in simulation, that is, in the computation of $E_\pi[f(X_0)]$ as the ergodic mean $\lim_{n\to\infty}\frac{1}{n}\sum_{k=1}^{n}f(X_k)$, all eigenvalues play a role if we measure the quality of the ergodic estimator by the asymptotic variance, as the above formulas show.

6.4   The Branching Process

This section also concerns an absorption problem, but a specific one with a specific approach. It features what is perhaps the first absorption result in Markov chain theory.

6.4.1   The Galton–Watson Model

A cousin of Darwin, Sir Francis Galton posed in 1873, in the *Educational Times*, the question of evaluating the survival probability of a given line of English peerage.

In the same year and in the same journal, Reverend Watson proposed the method of solution that has now become a textbook classic, and thereby initiated an important branch of probability with applications in nuclear science, in chemistry and in biology.

The Standard Description

The recurrence equation

$$X_{n+1} = \sum_{k=1}^{X_n} Z_{n+1}^{(k)} \tag{6.35}$$

($X_{n+1} = 0$ if $X_n = 0$), where $\{Z_n^{(j)}\}_{n \geq 1, j \geq 1}$ is an IID collection of integer-valued random variables with common generating function

$$g(z) := E\left[z^Z\right] = \sum_{n \geq 0} a_n z^n$$

and independent of the integer-valued random variable X_0, defines a stochastic process $\{X_n\}_{n \geq 0}$ called a *branching process*. It may be interpreted as follows: X_n is the number of individuals in the n-th generation of a given population (humans, particles, etc.). Individual number k of the n-th generation gives birth to $Z_{n+1}^{(k)}$ descendants, and this accounts for Eqn. (6.35). The random variable X_0 is the number of *ancestors*. The appellation "branching process" refers to Francis Galton's original preoccupation with genealogical trees (Figure 6.4.1). This process, also called the *Galton–Watson process*, is in view of the recurrence equation (6.35) and the independence assumptions a homogeneous Markov chain (Theorem 2.2.1).

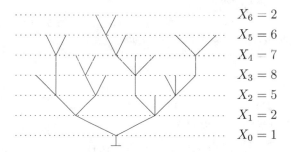

$$X_6 = 2$$
$$X_5 = 6$$
$$X_4 = 7$$
$$X_3 = 8$$
$$X_2 = 5$$
$$X_1 = 2$$
$$X_0 = 1$$

Figure 6.4.1. Sample tree of a branching process

EXAMPLE 6.4.1: THE REPAIR SHOP, TAKE 4. Continuation of Examples 2.2.6 and 2.4.9. This model has an interesting connection with branching process. The first busy cycle length is the first time n at which there is no machine left in the facility. We may suppose that $Z_1 > 0$, by which we mean that we are starting our observation at time 1. It takes $X_1 = Z_1$ units of time before one can start the service of the X_2 machines arriving during the time these X_1 machines are repaired. Then it takes X_3 units of time before one can start the service of the machines arriving during the time these X_2 machines are repaired, and so on, thus defining a sequence $\{X_n\}_{n \geq 1}$ satisfying the relation (6.35) as long as $X_n > 0$; with $X_0 := 1$ and where the $\{Z_n^{(j)}\}_{n \geq 1, j \geq 1}$ are IID with the same distribution as Z_1. Thus, defining the first time $\tau > 1$ at which $X_n = 0$, the first time the repair service facility is empty is $\sum_{i=1}^{\tau-1} X_i$. The probability of eventually having at least one "day off" for the mechanics is therefore the probability of extinction of a branching process $\{X_n\}_{n \geq 1}$ whose typical offspring has the same distribution as Z_1.

Probability of Extinction

The primary quantity of interest is the extinction probability $P(\mathcal{E})$, that is, the probability of absorption of the branching process in state 0.

The following trivial cases are excluded from the analysis: $P(Z = 0) = 0$, $P(Z = 0) = 1$ and $P(Z \geq 2) = 0$.

Theorem 6.4.2 *When there is just one ancestor* $(X_0 = 1)$,

   (a) $P(X_{n+1} = 0) = g(P(X_n = 0))$,

   (b) $P(\mathcal{E}) = g(P(\mathcal{E}))$, and

   (c) *if* $m := E[Z] < 1$ *the probability of extinction is* 1; *and if* $m > 1$, *the probability of extinction is* < 1 *but non-null.*

Proof.

   (a) Let ψ_n be the generating function of X_n. Since X_n is independent of the $Z_{n+1}^{(k)}$'s, by the result of Example 1.2.4,

$$\psi_{n+1}(z) = \psi_n(g(z)).$$

Iterating this equality, we obtain $\psi_{n+1}(z) = \psi_0(g^{(n+1)}(z))$, where $g^{(n)}$ is the n-th iterate of g. Since there is only *one* ancestor, $\psi_0(z) = z$, and therefore $\psi_{n+1}(z) =$

$g^{(n+1)}(z) = g(g^{(n)}(z))$, that is,

$$\psi_{n+1}(z) = g(\psi_n(z)).$$

In particular, since $\psi_n(0) = P(X_n = 0)$, (a) is proven.

(b) An extinction occurs if and only if at some time n (and then for all subsequent times) $X_n = 0$. Therefore

$$\mathcal{E} = \cup_{n=1}^{\infty}\{X_n = 0\}.$$

Since $X_n = 0$ implies $X_{n+1} = 0$, the sequence of events $\{X_n = 0\}_{n \geq 1}$ is non-decreasing, and therefore, by monotone sequential continuity,

$$P(\mathcal{E}) = \lim_{n \uparrow \infty} P(X_n = 0).$$

The generating function g is continuous, and therefore from (a) and the last equation, the probability of extinction satisfies (b).

(c) By Theorem 1.2.8, recalling that the trivial cases where $P(Z = 0) = 1$ or $P(Z \geq 2) = 0$ have been eliminated.

(α) If $E[Z] \leq 1$, the only solution of $x = g(x)$ in $[0, 1]$ is 1, and therefore $P(\mathcal{E}) = 1$. The branching process eventually becomes extinct.

(β) If $E[Z] > 1$, there are two solutions of $x = g(x)$ in $[0, 1]$, 1 and x_0 such that $0 < x_0 < 1$. From the strict convexity and monotonicity of $g : [0, 1] \to [0, 1]$, it follows that the sequence $y_n = P(X_n = 0)$ that satisfies $y_0 = 0$ and $y_{n+1} = g(y_n)$ converges increasingly to x_0. In particular, when the mean number of descendants $E[Z]$ is strictly larger than 1, $P(\mathcal{E}) \in (0, 1)$. □

EXAMPLE 6.4.3: EXTINCTION PROBABILITY FOR A POISSON OFFSPRING. Take for offspring distribution the Poisson distribution with mean $\lambda > 0$, whose generating function is $g(x) = e^{\lambda(x-1)}$. Suppose that $\lambda > 1$ (the supercritical case). The probability of extinction $P(\mathcal{E})$ is the unique solution in $(0, 1)$ of

$$x = e^{\lambda(x-1)}.$$

EXAMPLE 6.4.4: EXTINCTION PROBABILITY FOR A BINOMIAL OFFSPRING. Take for offspring distribution the binomial distribution $\mathcal{B}(N, p)$, with $0 < p < 1$. Recall that its mean is $m = Np$ and that its generating function is $g(x) = (px + (1-p))^N$. Suppose that $Np > 1$ (the supercritical case). The probability of extinction $P(\mathcal{E})$ is the unique solution in $(0, 1)$ of

$$x = (px + (1 - p))^N.$$

EXAMPLE 6.4.5: POISSON BRANCHING AS A LIMIT OF BINOMIAL BRANCHING. Suppose now that $p = \frac{\lambda}{N}$ with $\lambda > 1$ (therefore we are in the supercritical case) and the probability of extinction is given by the unique solution in $(0, 1)$ of

$$x = \left(\frac{\lambda}{N}x + (1 - \frac{\lambda}{N})\right)^N = \left(1 - \frac{\lambda}{N}(1 - x)\right)^N.$$

Letting $N \uparrow \infty$, we see that the right-hand side tends *from below* $(1 - x \le e^{-x})$ to the generating function of a Poisson variable of mean λ. Using this fact and the concavity of the generating functions, it follows that the probability of extinction also tends to the probability of extinction relative to the Poisson distribution.

6.4.2   Tail Distributions

Tail of the Extinction Time Distribution

Let T be the extinction time of the Galton–Watson branching process. The distribution of T is fully described by

$$P(T \le n) = P(X_n = 0) = \psi_n(0) \qquad (n \ge 0)$$

and $P(T = \infty) = 1 - P(\mathcal{E})$. In particular,

$$\lim_{n \uparrow \infty} P(T \le n) = P(\mathcal{E}). \qquad (\star)$$

Theorem 6.4.6 *In the supercritical case* $(m > 1$ *and therefore* $0 < P(\mathcal{E}) < 1)$,

$$P(\mathcal{E}) - P(T \le n) \le g'(P(\mathcal{E}))^n. \qquad (6.36)$$

Proof. The probability of extinction $P(\mathcal{E})$ is the limit of the sequence $x_n = P(X_n = 0)$ satisfying the recurrence equation $x_{n+1} = g(x_n)$ with initial value $x_0 = 0$. We have that

$$0 \le P(\mathcal{E}) - x_{n+1} = P(\mathcal{E}) - g(x_n) = g(P(\mathcal{E})) - g(x_n),$$

that is,

$$\frac{P(\mathcal{E}) - x_{n+1}}{P(\mathcal{E}) - x_n} = \frac{g(P(\mathcal{E})) - g(x_n)}{P(\mathcal{E}) - x_n} \le g'(P(\mathcal{E})),$$

where we have taken the convexity of g into account and $x_n < P(\mathcal{E})$. □

EXAMPLE 6.4.7: CONVERGENCE RATE FOR THE POISSON OFFSPRING DISTRIBUTION. For a Poisson offspring with mean $m = \lambda > 1$, $g'(x) = \lambda g(x)$ and therefore $g'(P(\mathcal{E})) = \lambda P(\mathcal{E})$. Therefore

$$P(\mathcal{E}) - P(T \le n) \le (\lambda P(\mathcal{E}))^n .$$

EXAMPLE 6.4.8: CONVERGENCE RATE FOR THE BINOMIAL OFFSPRING DISTRIBUTION. For a $\mathcal{B}(N,p)$ offspring with mean $m = Np > 1$, $g'(x) = Np\frac{g(x)}{1-p(1-x)}$ and therefore

$$g'(\Gamma(\mathcal{E})) - Np\frac{P(\mathcal{E})}{1 - p(1 - P(\mathcal{E}))}.$$

Taking $p = \frac{\lambda}{N}$,

$$g'(P_N(\mathcal{E})) = \lambda \frac{P_N(\mathcal{E})}{1 - \frac{\lambda}{N}(1 - P_N(\mathcal{E}))},$$

where the notation stresses the dependence of the extinction probability on N.

Tail of the Total Population Distribution

The random tree corresponding to a branching process can be explored in several ways. One way is generation by generation and corresponds to the classical construction of the Galton–Watson process given above. There is an alternative way that will be useful in a few lines. At step n of the exploration, we have a set of *active vertices* $\mathcal{A}_n$ and a set of *explored vertices* $\mathcal{B}_n$. At time 0 there is one active vertex, the root of the branching tree, so that $\mathcal{A}_0 = \{root\}$, and no vertex has been explored yet: $\mathcal{B}_0 = \varnothing$. At step $n \ge 1$, one chooses a vertex v_{n-1} among the vertices active at time n (those in $\mathcal{A}_{n-1}$), and this vertex is added to the set of explored vertices, that is $\mathcal{B}_n = \mathcal{B}_{n-1} \cup \{v_{n-1}\}$, and it is deactivated, whereas its children become active. Therefore, calling ξ_n the number of children of v_{n-1} and denoting by A_n the cardinal of $\mathcal{A}_n$, $A_0 = 1$ and

$$A_n = A_{n-1} - 1 + \xi_n$$

as long as $A_{n-1} > 0$. The exploration stops when there are no active vertices left, at time $Y = \inf\{n > 0; A_n = 0\}$, which is the size of the branching tree. By induction, as long as $A_{n-1} > 0$,

$$A_n = 1 - n + \sum_{i=1}^{n} \xi_i \,.$$

The one-by-one exploration procedure is summarized by the *history of the branching process*, that is, the random string

$$H = (\xi_1, \dots, \xi_Y)$$

taking its values in the subset F of $\mathbb{N}^* := \left(\cup_{k=1} \mathbb{N}^k \right) \cup \mathbb{N}^\infty$, determined by the following constraints: If $x = (x_1, x_2, \dots, x_k) \in \mathbb{N}^k$, $1 - \sum_{i=1}^{n} x_i - n > 0$ for all $n \le k$ and $1 - \sum_{i=1}^{k} x_i - k = 0$; and if $x = (x_1, x_2, \dots) \in \mathbb{N}^\infty$, $1 - \sum_{i=1}^{n} x_i - n > 0$ for all $n \ge 1$. Finite k's correspond to histories with extinction, whereas $x \in \mathbb{N}^\infty$ represents a history without extinction.

For any sequence $(x_1, \dots, x_k) \in F \cap \mathbb{N}^k$,

$$P(H = (x_1, \dots, x_k)) = \prod_{i=1}^{k} a_{x_i} \,.$$

Theorem 6.4.9 *Consider a branching process with a single ancestor in the subcritical case $m := \mathrm{E}[Z] < 1$ (in particular, the total population size Y is finite). We have for the total population size the bound*

$$P(Y > n) \le e^{-nh(1)}$$

where $h(a) = \sup_{t \ge 0}\{at - \log \mathrm{E}[e^{tZ}]\}$.

Proof. Consider the *random walk* $\{W_n\}_{n \ge 0}$ defined by $W_0 = 1$ and

$$W_n = W_{n-1} - 1 + \xi_n \,,$$

where the ξ_i's are IID random variables with the same distribution as Z. Then as long as they are strictly positive, the distributions of $\{W_n\}_{n \geq 0}$ and $\{A_n\}_{n \geq 0}$ are the same. Therefore

$$P(Y > n) = P(W_1 > 0, \dots, W_n > 0) \leq P(W_n > 0)$$

$$= P\left(1 + \sum_{i=1}^{n} \xi_i > n\right) = P\left(\sum_{i=1}^{n} \xi_i \geq n\right).$$

The announced result then follows from the Chernoff bound of Theorem 1.4.6 (here we take $a = 1$ and therefore, from the discussion following the statement of the theorem and the assumption $E[Z] < 1$ for the subcritical case, $h(1) > 0$). □

6.4.3 Conditioning by Extinction

We shall now determine the probability distribution of the history of a supercritical branching conditioned on the event that extinction occurs.

Theorem 6.4.10 *Let $\{a_k\}_{k>0}$ be a supercritical offspring distribution, that is, such that $\sum_{k \geq 0} k a_k > 1$. Let g_a be its generating function and $P(\mathcal{E})$ be the corresponding probability of extinction, the unique solution in $(0, 1)$ of $P(\mathcal{E}) = g_a(P(\mathcal{E}))$. The distribution of the branching process conditioned on extinction is the same as the distribution of a subcritical branching process with offspring distribution*

$$b_k = a_k P(\mathcal{E})^{k-1} \quad (k \geq 0).$$

Proof. We start by checking that this is a probability distribution on the set of non-negative integers and that this distribution is subcritical. In fact,

$$P(\mathcal{E}) = \sum_{k \geq 0} a_k P(\mathcal{E})^k = P(\mathcal{E}) \sum_{k \geq 0} b_k.$$

Let the generating function of $\{b_k\}_{k \geq 0}$ be denoted by g_b. A simple computation reveals that $g_b(x) = P(\mathcal{E})^{-1} g_a(P(\mathcal{E})x)$ and therefore $g_b'(x) = g_a'(P(\mathcal{E})x)$, so that

$$\sum_{k \geq 0} k b_k = g_b'(1) = g_a'(P(\mathcal{E})) < 1$$

(g_a' is a strictly increasing function).

It remains to compute $P(H = (x_1, \dots, x_k) \mid \text{extinction})$ when the underlying off-spring distribution is $\{a_k\}_{k \geq 0}$. For all $k \in \mathbb{N}$ and all $(x_1, \dots, x_k) \in F$

$$P(H = (x_1, \dots, x_k) \mid \text{extinction}) = \frac{P(H = (x_1, \dots, x_k), \text{ extinction})}{P(\text{extinction})}$$
$$= \frac{P(H = (x_1, \dots, x_k))}{P(\text{extinction})},$$

(since the condition $(x_1, \dots, x_k) \in F$ implies extinction at exactly time k for the history $(x_1, \dots, x_k)$). Therefore

$$P(H = (x_1, \dots, x_k) \mid \text{extinction}) = \frac{1}{P(\mathcal{E})} P(H = (x_1, \dots, x_k))$$
$$= \frac{1}{P(\mathcal{E})} \prod_{i=1}^{k} a_{x_i} = \frac{1}{P(\mathcal{E})} \prod_{i=1}^{k} b_{x_i} P(\mathcal{E})^{-(x_i - 1)}$$
$$= P(\mathcal{E})^{k-1-\sum_{i=1}^{k} x_i} \prod_{i=1}^{k} b_{x_i} = \prod_{i=1}^{k} b_{x_i}.$$

(The fourth equality makes use of the relation $\sum_{i=1}^{k} x_i = k - 1$ when $(x_1, \dots, x_k) \in F$.)   □

EXAMPLE 6.4.11: THE POISSON CASE. For a Poisson offspring supercritical distribution with mean $\lambda > 1$,

$$b_k = e^{-\lambda} \frac{\lambda^k}{k!} P(\mathcal{E})^{k-1} = \frac{1}{P(\mathcal{E})} e^{-\lambda} \frac{(\lambda P(\mathcal{E}))^k}{k!}.$$

But in this case $P(\mathcal{E}) = g_a(P(\mathcal{E})) = e^{\lambda(P(\mathcal{E}) - 1)}$, or equivalently

$$\frac{1}{P(\mathcal{E})} e^{-\lambda} = e^{-\lambda P(\mathcal{E})}.$$

Therefore

$$b_k = e^{-\lambda \mathcal{P} - e} \frac{(\lambda P(\mathcal{E}))^k}{k!},$$

which corresponds to a Poisson distribution with mean $\mu = \lambda P(\mathcal{E})$.

6.5 Exercises

Exercise 6.5.1. $P_i(T_j > n)$
Let $\{X_n\}_{n \geq 0}$ be an irreducible HMC with finite state space $E = \{1, 2, \ldots, r\}$ and transition matrix $\mathbf{P}$. Let T_j be the return time to $j \in E$. Show that

$$P_i(T_j > n) = \{\mathbf{Q}_j^n \mathbf{1}_{E \backslash \{j\}}\}_i,$$

where $\mathbf{Q}_j$ is obtained from $\mathbf{P}$ by deleting the jth column and the jth row, and $\mathbf{1}_{E \backslash \{j\}}$ is a column vector of dimension $r - 1$ with all its entries equal to 1.

Exercise 6.5.2. FIRST MEETING TIME
Find the distribution of the first meeting time of two independent HMCs with state space $E = \{1, 2\}$ and transition matrix

$$\mathbf{P} = \begin{pmatrix} 1 - \alpha & \alpha \\ \beta & 1 - \beta \end{pmatrix},$$

where $\alpha, \beta \in (0, 1)$, when their initial states are different.

Exercise 6.5.3. ABSORPTION BY A RECURRENT CLASS
Let $\{X_n\}_{n \geq 0}$ be an HMC on the state space $E = \{1, 2, 3, 4, 5\}$ with transition matrix

$$\mathbf{P} = \begin{pmatrix} 1 & & & & \\ & 0.4 & 0.6 & & \\ & 0.5 & 0.5 & & \\ 0.4 & 0.2 & 0.2 & 0.1 & 0.1 \\ 0.5 & 0 & 0.1 & 0.2 & 0.2 \end{pmatrix}.$$

Compute the probability of absorption in the recurrent class $\{2, 3\}$ when starting from the transient state 5.

Exercise 6.5.4. PROOF OF REMARK 6.1.4
Give a direct proof of the result in Remark 6.1.4 stating that infinite sojourn in a *finite* transient set is almost surely impossible.

Exercise 6.5.5. ABSORPTION WITH A TABOO STATE
Consider an HMC with a non-empty recurrent set R and a transient set T with at least two distinct elements i and j. Give a general method to compute the probability of never visiting j and of being absorbed by a given recurrent class R_k

when starting from i. Apply this to the HMC of Example 6.2.6 with $i = 6$, $j = 7$ and $k = 2$.

Exercise 6.5.6. BORROWING MONEY FROM THE MAFIA
You have 1 dollar and you owe the mafia 5 dollars. Being a probabilist, you decide to make a series of bets with a fair coin. You have the choice between two strategies. Bet 1 dollar each time (the timid strategy), or bet as much as you can but no more than necessary to reach a fortune of 5 dollars (the bold strategy). Find for each strategy your probability of staying alive.

Exercise 6.5.7. A LIMIT THEOREM
Consider an HMC with a non-empty set of transient states T and with at least one positive recurrent class R that is aperiodic. Let π be the unique stationary distribution of the restriction of the chain to R. Prove that for any $i \in T$, $j \in R$, $\lim_{n \uparrow \infty} p_{ij}(n) = f_{iR} \pi(j)$, where f_{iR} is the probability of absorption in R when starting from i.

Exercise 6.5.8. DIAGONALIZABILITY
Let $\mathbf{P}$ be a stochastic matrix on the finite state space E, and let T and R be the sets of transient states and recurrent states, respectively. Suppose, moreover, that the recurrent classes consist of only one state; that is, the block decomposition of the transition matrix with respect to the partition $R + T = E$ is

$$\mathbf{P} = \begin{pmatrix} I & 0 \\ B & \mathbf{Q} \end{pmatrix},$$

where I is the $|R| \times |R|$ identity matrix. Show that $\mathbf{P}$ is diagonalizable if and only if $\mathbf{Q}$ is diagonalizable.

Exercise 6.5.9. 0, 1 OR 2 CHILDREN
Compute the probability of extinction of a Galton–Watson branching process with one ancestor when the probabilities of having $0, 1$, or 2 sons are respectively $\frac{1}{4}, \frac{1}{4}$, and $\frac{1}{2}$.

Exercise 6.5.10. THE BRANCHING PROCESS TRANSITION MATRIX
Show that the Galton–Watson branching process is an HMC. Show that the (i, j)th entry p_{ij} of the transition matrix of this chain is the coefficient of z^j in $(g(z))^i$, where $g(z)$ is the generating function of the number of descendants of a given individual.

Exercise 6.5.11. SEVERAL ANCESTORS
Give the survival probability of the Galton–Watson branching process with $k > 1$ ancestors.

Exercise 6.5.12. MEAN AND VARIANCE OF THE POPULATION SIZE SEQUENCE
In the Galton–Watson branching process model with $k > 1$ ancestors, give the mean and variance of X_n.

Exercise 6.5.13. SIZE OF THE BRANCHING TREE
When the probability of extinction is 1 ($m < 1$), call Y the size of the Galton–Watson branching tree ($Y = \sum_{n \geq 0} X_n$). Prove that

$$g_Y(z) = z\, g_Z\left(g_Y(z)\right).$$

Exercise 6.5.14. CONJUGATE OFFSPRING DISTRIBUTIONS
Fix a probability distribution $\{b_k\}_{k \in \mathbb{N}}$ that is critical ($\sum_{k \in \mathbb{N}} k\, b_k = 1$) and such that $b_0 > 0$. For any $\lambda > 0$, define the *exponentially tilted distribution* $\{a_k(\lambda)\}_{k \in \mathbb{N}}$ by

$$a_k(\lambda) = b_k \frac{\lambda^k}{g_b(\lambda)} = b_k \frac{\lambda^k}{\sum_{k \geq 0} b_k \lambda^k}.$$

(a) Verify that this is indeed a probability distribution on $\mathbb{N}$ that is supercritical if $\lambda > 1$ and subcritical if $\lambda < 1$.

(b) Take for $\{b_k\}_{k \in \mathbb{N}}$ the Poisson distribution with mean 1. What is the conjugate distribution?

(c) A parameter μ is said to be a *conjugate parameter* of $\lambda > 0$ if

$$\frac{\lambda}{g_b(\lambda)} = \frac{\mu}{g_b(\mu)}.$$

Let $\lambda > 1$. Prove that there exists a unique *conjugate parameter* μ of λ such that $\mu \neq \lambda$ which satisfies $\mu < 1$ and is given by

$$\mu = \lambda P(\mathcal{E})(\lambda),$$

where $P(\mathcal{E})(\lambda)$ is the probability of extinction of a Galton–Watson branching process with the offspring distribution $\{a_k(\lambda)\}_{k \in \mathbb{N}}$.

(d) Let $\lambda > 1$. Show that the distribution of the supercritical branching process history with offspring distribution $\{a_k(\lambda)\}_{k \in \mathbb{N}}$ conditioned on extinction is identical to that of the subcritical branching process history with offstring distribution $\{a_k(\mu)\}_{k \in \mathbb{N}}$ where $\mu = \lambda P(\mathcal{E})(\lambda)$ is the conjugate parameter of λ.

Chapter 7

Lyapunov Functions and Martingales

The present chapter is an introduction to a few topics from potential theory and martingale theory, with the purpose of demonstrating the power of martingale theory and of giving examples of the rich interplay between probability and analysis. An important aspect of this chapter concerns the various results complementing the study of recurrence of Chapter 3. In this respect, the single most important result is *Foster's theorem* below.

7.1 Lyapunov Functions

7.1.1 Foster's Theorem

The stationary distribution criterion of positive recurrence of an irreducible chain requires solving the balance equation, an often hopeless enterprise. The following *sufficient* condition is more tractable and indeed quite powerful.

Theorem 7.1.1 [1] *Let* $\mathbf{P}$ *be an irreducible transition matrix on the countable state space* E. *Suppose that there exists a function* $h : E \to \mathbb{R}$ *such that* $\inf_i h(i) > -\infty$,

$$\sum_{k \in E} p_{ik} h(k) < \infty \quad (i \in F), \tag{7.1}$$

and

$$\sum_{k \in E} p_{ik} h(k) \le h(i) - \epsilon \quad (i \notin F), \tag{7.2}$$

---

[1] [Foster, 1953].

© Springer Nature Switzerland AG 2020
P. Brémaud, *Markov Chains*, Texts in Applied Mathematics 31,
https://doi.org/10.1007/978-3-030-45982-6_7

for some finite set F and some $\epsilon > 0$. Then the corresponding HMC *is positive recurrent.*

Proof. This proof does not explicitly use martingale theory, but Foster's theorem is very much related to results in Section 7.3 that are proved via the martingale convergence theorem (Theorem 7.3.1). Recall the notation X_0^n for $(X_0, \ldots, X_n)$. Since $\inf_i h(i) > -\infty$, one may assume without loss of generality that $h \geq 0$, by adding a constant if necessary. Call τ the return time to F and let $Y_n := h(X_n)1_{\{n < \tau\}}$. Equality (7.2) implies that $E[h(X_{n+1}) \mid X_n = i] \leq h(i) - \epsilon$ for all $i \notin F$. For $i \notin F$,

$$
\begin{aligned}
E_i[Y_{n+1} \mid X_0^n] &= E_i[Y_{n+1}1_{\{n<\tau\}} \mid X_0^n] + E_i(Y_{n+1}1_{\{n\geq\tau\}} \mid X_0^n]\\
&= E_i[Y_{n+1}1_{\{n<\tau\}} \mid X_0^n] \leq E_i[h(X_{n+1})1_{\{n<\tau\}} \mid X_0^n]\\
&= 1_{\{n<\tau\}}E_i[h(X_{n+1}) \mid X_0^n] = 1_{\{n<\tau\}}E_i[h(X_{n+1}) \mid X_n]\\
&\leq 1_{\{n<\tau\}}h(X_n) - \epsilon1_{\{n<\tau\}}\,,
\end{aligned}
$$

where the third *equality* comes from the fact that $1_{\{n<\tau\}}$ is a function of X_0^n (Theorem 1.1.65), the fourth *equality* is the Markov property and the last *inequality* is true because P_i-a.s., $X_n \notin F$ on $n < \tau$. Therefore, P_i-a.s.,

$$
E_i[Y_{n+1} \mid X_0^n] \leq Y_n - \epsilon1_{\{n<\tau\}}
$$

and, taking expectations,

$$
0 \leq E_i[Y_{n+1}] \leq E_i[Y_n] - \epsilon P_i(\tau > n)\,.
$$

Iterating the above equality and taking into account the fact that Y_n is non-negative, we obtain

$$
0 \leq E_i[Y_0] - \epsilon \sum_{k=0}^n P_i(\tau > k)\,.
$$

But $Y_0 = h(i)$, P_i-a.s., and $\sum_{k=0}^\infty P_i(\tau > k) = E_i[\tau]$. Therefore, for all $i \notin F$,

$$
E_i[\tau] \leq \epsilon^{-1}h(i).
$$

For $j \in F$, by first-step analysis

$$
E_j[\tau] = 1 + \sum_{i\notin F} p_{ji}E_i[\tau]\,.
$$

Therefore $E_j[\tau] \leq 1 + \epsilon^{-1}\sum_{i\notin F} p_{ji}h(i)$, a finite quantity in view of assumption (7.1): the return time to F starting anywhere in F has finite expectation. Since F is a finite set, this implies positive recurrence in view of the following lemma. $\square$

Lemma 7.1.2 Let $\{X_n\}_{n\geq 0}$ be an irreducible HMC, let F be a finite subset of the state space E and let $\tau(F)$ be the return time to F. If $\mathrm{E}_j[\tau(F)] < \infty$ for all $j \in F$, the chain is positive recurrent.

Proof. Select $i \in F$, and let T_i be the return time of $\{X_n\}$ to i. Let $\tau_1 = \tau(F), \tau_2, \tau_3, \ldots$ be the successive return times to F. It follows from the strong Markov property that $\{Y_n\}_{n\geq 0}$ defined by $Y_0 = X_0 = i$ and $Y_n = X_{\tau_n}$ for $n \geq 1$ is an HMC with state space F (Exercise 2.6.31). Since $\{X_n\}_{n\geq 0}$ is irreducible, so is $\{Y_n\}_{n\geq 0}$. Since F is finite, $\{Y_n\}$ is positive recurrent and in particular $\mathrm{E}_i[\widetilde{T}_i] < \infty$, where $\widetilde{T}_i$ is the return time to i of $\{Y_n\}$. Letting $S_0 := \tau_1$ and $S_k := \tau_{k+1} - \tau_k$ for $k \geq 1$,

$$T_i = \sum_{k=0}^{\infty} S_k 1_{\{k < \widetilde{T}_i\}},$$

and therefore

$$\mathrm{E}_i[T_i] = \sum_{k=0}^{\infty} \mathrm{E}_i[S_k 1_{\{k<\widetilde{T}_i\}}].$$

Now,

$$\mathrm{E}_i[S_k 1_{\{k<\widetilde{T}_i\}}] = \sum_{\ell \in F} \mathrm{E}_i[S_k 1_{\{k<\widetilde{T}_i\}} 1_{\{X_{\tau_k}=\ell\}}],$$

and applying the strong Markov property to $\{X_n\}_{n\geq 0}$ and the stopping time τ_k, and observing that $\{k < \widetilde{T}_i\}$ is an event of the past at time τ_k,

$$\mathrm{E}_i[S_k 1_{\{k<\widetilde{T}_i\}} 1_{\{X_{\tau_k}=\ell\}}] = \mathrm{E}_i[S_k \mid k < \widetilde{T}_i, X_{\tau_k} = \ell] \mathrm{P}_i(k < \widetilde{T}_i, X_{\tau_k} = \ell)$$
$$= \mathrm{E}_i[S_k \mid X_{\tau_k} = \ell] \mathrm{P}_i(k < \widetilde{T}_i, X_{\tau_k} = \ell).$$

Since $\mathrm{E}_i[S_k \mid X_{\tau_k} = \ell] = \mathrm{E}_\ell[\tau(F)]$, the latter expression is bounded by

$$\max_{\ell \in F} \mathrm{E}_\ell[\tau(F)] \times \mathrm{P}_i(k < \widetilde{T}_i, X_{\tau_k} = \ell)$$

and therefore

$$\mathrm{E}_i[T_i] \leq \left(\max_{\ell \in F} \mathrm{E}_\ell(\tau(F))\right) \sum_{k=0}^{\infty} \mathrm{P}_i(\widetilde{T}_i > k) = \left(\max_{\ell \in F} \mathrm{E}_\ell(\tau(F))\right) \mathrm{E}_i[\widetilde{T}_i] < \infty.$$

$\square$

Remark 7.1.3 The function h in Foster's theorem is called a *Lyapunov function* because it plays a role similar to the Lyapunov functions in the stability theory of ordinary differential equations. It has a tendency to decrease along the trajectories of the process, at least outside a finite set of states, called the *refuge*. Since it is non-negative, it cannot decrease forever and therefore it eventually enters the refuge.

   The following corollary of Foster's theorem is sometimes referred to as *Pakes' lemma*.

Corollary 7.1.4 ([2]) *Let $\{X_n\}_{n\geq0}$ be an irreducible* HMC *on $E = \mathbb{N}$ such that for all $n \geq 0$ and all $i \in E$,*

$$E[X_{n+1} \mid X_n = i] < \infty \tag{7.3}$$

and

$$\limsup_{i\uparrow\infty} E[X_{n+1} - X_n \mid X_n = i] < 0. \tag{7.4}$$

Such an HMC *is positive recurrent.*

Proof. Let -2ϵ be the left-hand side of (7.4). In particular, $\epsilon > 0$. By (7.4), for i sufficiently large, say $i > i_0$, $E[X_{n+1} - X_n \mid X_n = i] < -\epsilon$, and therefore the conditions of Foster's theorem are satisfied with $h(i) = i$ and $F = \{i; i \leq i_0\}$. □

EXAMPLE 7.1.5: A RANDOM WALK ON $\mathbb{N}$. Let $\{Z_n\}_{n\geq1}$ be an IID sequence of integrable random variables with values in $\mathbb{Z}$ such that

$$E[Z_1] < 0,$$

and define $\{X_n\}_{n\geq0}$, an HMC with state space $E = \mathbb{N}$, by

$$X_{n+1} = (X_n + Z_{n+1})^+,$$

where X_0 is independent of $\{Z_n\}_{n\geq1}$. Assume irreducibility (the reader is invited to find the necessary and sufficient condition for this). Here

$$E[X_{n+1} - i \mid X_n = i] = E[(i + Z_{n+1})^+ - i]$$
$$= E[-i1_{\{Z_{n+1}\leq-i\}} + Z_{n+1}1_{\{Z_{n+1}>-i\}}] \leq E[Z_11_{\{Z_1>-i\}}].$$

By dominated convergence, the limit of $E[Z_11_{\{Z_1>-i\}}]$ as i tends to ∞ is $E[Z_1] < 0$ and therefore, by Pakes' lemma, the HMC is positive recurrent.

---

[2][Pakes, 1969].

EXAMPLE 7.1.6: THE REPAIR SHOP, TAKE 5. Continuation of Example 6.4.1. Arguments very similar to those of the previous example show that in the repair shop HMC (assumed irreducible; see Exercise 2.6.18), condition $E[Z_1] < 1$ implies positive recurrence. Actually, in view of the results of Example 2.4.9, condition $E[Z_1] < 1$ is in fact *necessary and sufficient* for positive recurrence.

7.1.2 Queueing Applications

EXAMPLE 7.1.7: STABILIZATION OF ALOHA. ([3]) It was proven in Example 3.2.12 that the ALOHA protocol with a fixed retransmission probability ν is unstable. It seems natural to try a retransmission probability $\nu = \nu(k)$ depending on the number k of backlogged messages. In fact, there is a choice of the function $\nu(k)$ that achieves stability of the protocol.[4]

The probability that i among the k backlogged messages at the beginning of slot n retransmit in slot n is as in Example 3.2.12, except that ν is replaced by $\nu(k)$. The same is true for the transition probabilities.

An elementary computation yields

$$E[X_{n+1} - X_n \mid X_n = i] = \lambda - b_1(i)a_0 - b_0(i)a_1. \tag{7.5}$$

Note that $b_1(i)a_0 + b_0(i)a_1$ is the probability of one successful (re-)transmission in a slot given that the backlog at the beginning of the slot is i. Equivalently, since there is at most one successful (re-)transmission in any slot, this is the average number of successful (re-)transmissions in a slot given the backlog i at the start of the slot.

According to Pakes' lemma, it suffices to find a function $\nu(k)$ guaranteeing that

$$\lambda \leq \lim_{i \uparrow \infty} (b_1(i)a_0 + b_0(i)a_1) - \epsilon, \tag{7.6}$$

for some $\epsilon > 0$. We shall therefore study the function

$$g_k(\nu) = (1 - \nu)^k a_1 + k\nu(1 - \nu)^{k-1} a_0,$$

since condition (7.6) is just $\lambda \leq g_i(\nu(i)) - \epsilon$. The derivative of $g_k(\nu)$ is, for $k \geq 2$,

$$g_k'(\nu) = k(1 - \nu)^{k-2}[(a_0 - a_1) - \nu(ka_0 - a_1)].$$

---

[3][Fayolle, 1976].
[4][Fayolle, 1976].

We first assume that $a_0 > a_1$. In this case, for $k \geq 2$, the derivative is null at

$$\nu = \nu(k) = \frac{a_0 - a_1}{ka_0 - a_1},$$

and the corresponding value of $g_k(\nu)$ is a maximum equal to

$$g_k(\nu(k)) = a_0 \left(\frac{k-1}{k - a_1/a_0} \right)^{k-1}.$$

Therefore, $\lim_{k \uparrow \infty} g_k(\nu(k)) = a_0 \exp\left\{ \frac{a_1}{a_0} - 1 \right\}$, and we see that

$$\lambda < a_0 \exp\left\{ \frac{a_1}{a_0} - 1 \right\} \tag{7.7}$$

is a sufficient condition for stability of the protocol. For instance, with a Poisson distribution of arrivals

$$a_i = \mathrm{e}^{-\lambda} \frac{\lambda^i}{i!},$$

condition (7.7) reads

$$\lambda < \mathrm{e}^{-1}$$

(in particular, the condition $a_0 > a_1$ is satisfied a posteriori).

If $a_0 \leq a_1$, the protocol can be shown to be unstable, whatever retransmission policy $\nu(k)$ is adopted (the reader is invited to check this).

———

EXAMPLE 7.1.8: THE STACK ALGORITHM. The slotted ALOHA protocol with constant retransmission probability in Example 3.2.12 was proved unstable and it was shown in Example 7.1.7 that a backlog-dependent retransmission probability could restore stability. The problem then resides in the necessity for each user to know the size of the backlog in order to implement the retransmission policy. This is not practically feasible and one must therefore devise policies based on the actual information available by just listening to the link: collision, no transmission, or successful transmission. Such policies, which in a sense estimate the backlog, have been found that yield stability. However, we shall not discuss them here, and instead we shall consider another type of *collision resolution protocol*, namely, the *stack algorithm*.[5]

In the general form of a collision resolution protocol, when a collision occurs, all new requests are buffered until all the messages involved in the collision have

———

[5][Tsybahov and Mihailov, 1980], [Capetanakis, 1979]. See [Rom and Sidi, 1990] for bibliographical details.

found their way through the link. When these messages have resolved their collision problem, the buffered messages then try to retransmit, maybe enter a collision, and then resolve their collision. Time is therefore divided into successive periods, called *collision-resolution intervals* (CRI).

We shall now be more specific and describe the stack algorithm. Let us examine the fate of the messages arriving in the first slot, which are the messages that arrived during the previous CRI. They all try to retransmit in the first slot of the CRI and therefore, if there are two or more messages, a collision occurs (in the other case, the CRI has lasted just one slot and a new CRI begins in the next slot). An unbiased coin is tossed for each colliding message. If it shows heads, the message joins *layer* 0 of a *stack* whereas if it shows tails, it is placed in layer 1. In the next slot, all messages of layer 0 try the link. If there is no collision (because layer 0 was empty or just contained one message), layer 0 is eliminated, and layer 1 below pops up to become layer 0. If on the contrary there is a collision because layer 0 formed after the first slot contained two or more messages, the colliding messages again flip a coin; those with heads form the new layer 0, those with tails form the new layer 1, and the former layer 1 is pushed bottomwards to form layer 2.

In general, at each step, only layer 0 tries to retransmit. If there is no collision, layer 0 disappears, and the layers $1, 2, 3, \ldots$ become layers $0, 1, 2, \ldots$ If there is a collision, layer 0 splits into layer 0 and layer 1, and layers $1, 2, 3, \ldots$ become layer $2, 3, 4, \ldots$ It should be noted that in this protocol, each user (message) knows at every instant in which layer he or she is, just by listening to the channel that gives the information: collision or no collision. In that sense, the protocol is *distributed*, because there is no central operator broadcasting nonlocally available information, such as the size of the backlog, to all users.

Once a collision is resolved, that is, when all layers have disappeared, a new CRI begins. The number of customers that are starting this CRI are those that have arrived in the CRI that just ended. Figure 7.1.1 gives an example of what happens in a CRI.

If we assume that the fresh requests are as in Example 3.2.12, that is, $\{A_n\}_{n \geq 1}$ is IID, where A_n is the number of new requests in slot n, then the sequence $\{X_n\}_{n \geq 0}$, where X_n is the *length* of the n-th CRI, forms an irreducible HMC. Stability of the protocol is naturally identified with positive recurrence of this chain, which will now be proved with the help of Foster's theorem.

According to Pakes' lemma, it suffices to show that

$$\limsup_{i \uparrow \infty} E[X_{n+1} - X_n \mid X_n = i] < 0 \qquad (7.8)$$

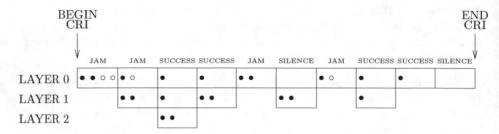

Figure 7.1.1. The stack algorithm

and for all i,

$$\mathrm{E}[X_{n+1} \mid X_n = i] < \infty. \tag{7.9}$$

For this, let Z_n be the number of fresh arrivals in the n-th CRI. We have

$$\begin{aligned}
\mathrm{E}[X_{n+1} \mid X_n = i] &= \sum_{k=0}^{\infty} \mathrm{E}[X_{n+1} \mid X_n = i, Z_n = k] \mathrm{P}(Z_n = k \mid X_n = i) \\
&= \sum_{k=0}^{\infty} \mathrm{E}[X_{n+1} \mid Z_n = k] \mathrm{P}(Z_n = k \mid X_n = i).
\end{aligned}$$

It will be shown that for all $n \geq 0$,

$$\mathrm{E}[X_{n+1} \mid Z_n = k] \leq \alpha k + 1, \tag{7.10}$$

where $\alpha = 2.886$, and therefore

$$\mathrm{E}[X_{n+1} \mid X_n = i] \leq \sum_{k=0}^{\infty} (\alpha k + 1) \mathrm{P}(Z_n = k \mid X_n = i) = \alpha \mathrm{E}[Z_n \mid X_n = i] + 1 \,.$$

By Wald's lemma (Theorem 1.1.43),

$$\mathrm{E}[Z_n \mid X_n = i] = \lambda i \,,$$

where λ is the *traffic intensity*, and therefore

$$\mathrm{E}[X_{n+1} - X_n \mid X_n = i] \leq 1 + i(\lambda \alpha - 1).$$

We see that condition (7.9) is always satisfied and that (7.8) is satisfied, provided that

$$\lambda < \frac{1}{\alpha} = 0.346 \,. \tag{7.11}$$

It remains to prove (7.10). Let $E[X_{n+1} \mid Z_n = k] = L_k$ (it is indeed a quantity independent of n). Clearly,

$$L_0 = L_1 = 1,$$

since with zero or one packet at the beginning of a CRI, there is no collision. When $k \geq 2$, there is a collision, and the k users toss a coin, and depending on the result they split into two sets, layer 0 and layer 1. Among these k users, i obtain heads with probability

$$q_i(k) = \binom{k}{i} \left(\frac{1}{2}\right)^k.$$

The average length of the CRI given that there are $k \geq 2$ customers at the start, and given that the first layer 0 contains i messages, is

$$L_{k,i} = 1 + L_i + L_{k-i}.$$

Indeed, the first slot saw a collision; the i customers in the first layer 0 will take on the average L_i slots to resolve their collision, and L_{k-i} more slots will be needed for the $k - i$ customers in the first-formed layer 1 (these customers are always at the bottom of the stack, in a layer traveling up and down until it becomes layer 0, at which time they start resolving their collision). Since

$$L_k = \sum_{i=0}^{k} q_i(k) L_{k,i},$$

we have

$$L_k = 1 + \sum_{i=0}^{k} q_i(k)(L_i + L_{k-i}).$$

Solving for L_k, we obtain

$$L_k = \frac{1 + \sum_{i=0}^{k-1}[q_i(k) + q_{k-i}(k)]L_i}{1 - q_0(k) - q_k(k)}. \tag{7.12}$$

Suppose that for some $m \geq 2$, and α_m satisfying

$$\alpha_m \geq \sup_{j>m} \frac{\sum_{i=0}^{m-1}(L_i + 1)(q_i(j) + q_{j-i}(j))}{\sum_{i=0}^{m-1} i(q_i(j) + q_{j-i}(j))}, \tag{7.13}$$

it holds that $L_m \leq \alpha_m m - 1$. Then we shall prove that for all $n \geq m$,

$$L_n \leq \alpha_m n - 1. \tag{7.14}$$

We do this by induction, supposing that (7.14) holds true for $n = m, m+1, \ldots,$
$j-1$, and proving that it holds true for $n = j$. Equality (7.12) gives

$$
L_j(1 - q_0(j) - q_j(j)) = 1 + \sum_{i=0}^{j-1}(q_i(j) + q_{j-i}(j))L_i
$$

$$
= 1 + \sum_{i=0}^{m-1} + \sum_{i=m}^{j-1}
$$

$$
\leq 1 + \sum_{i=0}^{m-1} + \sum_{i=m}^{j-1}(q_i(j) + q_{j-i}(j))(\alpha_m i - 1),
$$

where we used the induction hypothesis. The latter term equals

$$
1 + \sum_{i=0}^{m-1}(q_i(j) + q_{j-i}(j))(L_i - \alpha_m i + 1)
$$

$$
+ \sum_{i=0}^{j}(q_i(j) + q_{j-i}(j))(\alpha_m i - 1) - (q_0(j) + q_j(j))(\alpha_m j - 1)
$$

$$
= 1 + \sum_{i=0}^{m-1}(q_i(j) + q_{j-i}(j))(L_i - \alpha_m i + 1) + \alpha_m j - 2 - (q_0(j) + q_j(j))(\alpha_m j - 1),
$$

where we used the identities

$$
\sum_{i=0}^{j} q_i(j) = 1, \quad \sum_{i=0}^{j} i q_i(j) = jp, \quad \sum_{i=0}^{j} i q_{j-i}(j) = j(1-p).
$$

Therefore,

$$
L_j \leq (\alpha_m j - 1) + \frac{\sum_{i=0}^{m-1}(q_i(j) + q_{j-i}(j))(L_i - \alpha_m i + 1)}{1 - q_0(j) - q_j(j)}.
$$

Therefore, for $L_j \leq \alpha_m j - 1$ to hold, it suffices to have

$$
\sum_{i=0}^{m-1}(q_i(j) + q_{j-i}(j))(L_i - \alpha_m i + 1) \leq 0.
$$

We require this to be true for all $j > m$, and (7.13) guarantees this. It can be
checked *numerically* that for $m = 6$ and $\alpha_6 = 2.886$, (7.13) is satisfied, $L_6 \leq$
$2.886 \times 6 - 1$ and equality (7.14) is true for $n = 2, 3, 4, 5$, and this completes the
proof.

7.2 Martingales and Potentials

7.2.1 Harmonic Functions and Martingales

The concept of martingale is a central one in the theory of stochastic processes. One of the objectives of the present section is to introduce the reader to martingale theory and its applications to Markov chains. These will be developed in the next section. Closely connected with martingales are the harmonic functions and potentials.

Let $\{X_n\}_{n\geq 0}$ be an HMC on the countable space E with transition matrix $\mathbf{P}$. In the study of recurrence based on invariant measures or stationary distributions, the principal role is played by the equation $x^T = x^T\mathbf{P}$, where x^T is a row vector. The recurrence/transience criteria of the present chapter are based on the "dual" equation (*resp.*, inequations)

$$\mathbf{P}h = h \quad (\textit{resp.} \ \geq h, \leq h)\,, \tag{7.15}$$

where h is a column vector. In developed form, for all $i \in E$,

$$\sum_{j\in E} p_{ij}h(j) = h(i) \quad (\textit{resp.} \ \geq h(i), \leq h(i))\,.$$

Equation (7.15) is equivalent to

$$\mathrm{E}[h(X_{n+1}) \mid X_n = i] = h(i) \quad (\textit{resp.} \ \geq h(i), \leq h(i))\,,$$

for all $i \in E$. In view of the Markov property, the left-hand side of the above equality is also equal to

$$\mathrm{E}[h(X_{n+1}) \mid X_n = i, X_{n-1} = i_{n-1}, \ldots, X_0 = i_0],$$

and therefore (7.15) is equivalent to

$$\mathrm{E}[h(X_{n+1} \mid X_0^n] = h(X_n) \quad (\textit{resp.} \ \leq h(X_n), \geq h(X_n))\,.$$

This motivates the two following definitions.

Definition 7.2.1 *A function* $h : E \to \mathbb{R}$ *is called* harmonic (*resp.*, subharmonic, superharmonic) *iff*

$$\mathbf{P}h = h, \ \textit{resp.}, \ \geq h, \leq h\,. \tag{7.16}$$

Superharmonic functions are also called excessive *functions. More generally, let* $D \subset E$ *be a subset of the state space, called the* domain, *and denote its complement in* E, *called the* boundary, *by* ∂D. *If* $h : E \to \mathbb{R}$ *satisfies (7.16) on* D, *then it is called* harmonic (*resp.*, subharmonic, superharmonic) *on* D.

Definition 7.2.2 *A real-valued stochastic process $\{Y_n\}_{n\geq 0}$ such that for $n \geq 0$*

(i) Y_n is a function of $X_0, \ldots, X_n$, and

(ii) $\mathrm{E}[|Y_n|] < \infty$ or $Y_n \geq 0$

is called a martingale *(resp.,* submartingale, supermartingale) *with respect to* $\{X_n\}_{n\geq 0}$ *if, moreover,*

$$\mathrm{E}[Y_{n+1} \mid X_0^n] = Y_n, \ resp., \ \geq Y_n, \leq Y_n \,. \tag{7.17}$$

In the above definition, $\{X_n\}_{n\geq 0}$ can be any stochastic process, not necessarily a Markov chain. Also, note that a martingale is a submartingale *and* a supermartingale.

EXAMPLE 7.2.3: HARMONIC FUNCTIONS PRODUCE MARTINGALES. The discussion at the beginning of the current subsection shows that if $h : E \to \mathbb{R}$ is either a function such that $\mathrm{E}[|h(X_n)|] < \infty$ for all $n \geq 0$ or a non-negative function, and if it is harmonic (resp., subharmonic, superharmonic), the process $\{h(X_n)\}_{n\geq 0}$ is, with respect to $\{X_n\}_{n\geq 0}$, a martingale (resp., submartingale, supermartingale).

Condition (7.17) implies that for all $k \geq 1$,

$$\mathrm{E}[Y_{n+k} \mid X_0^n] = Y_n \quad (resp. \ \geq Y_n, \leq Y_n) \,.$$

For instance, in the martingale case, with $k = 2$,

$$\mathrm{E}[Y_{n+2} \mid X_0^n] = \mathrm{E}[\mathrm{E}[Y_{n+2} \mid X_0^{n+1}] \mid X_0^n] = \mathrm{E}[Y_{n+1} \mid X_0^n] = Y_n \,.$$

EXAMPLE 7.2.4: THE LÉVY MARTINGALE. Let $\{X_n\}_{n\geq 0}$ be an HMC with transition matrix $\mathbf{P}$, and let $f : E \to \mathbb{R}$ be a bounded function. Then, the sequence

$$M_n^f := f(X_n) - f(X_0) - \sum_{k=0}^{n-1}(\mathbf{P} - I)f(X_k) \quad (n \geq 0) \tag{7.18}$$

is a martingale with respect to $\{X_n\}_{n\geq 0}$. Indeed, if $|f| \leq K < \infty$,

$$|(\mathbf{P}f)(i)| = \left| \sum_{j \in E} p_{ij} f(j) \right| \leq K \,.$$

Therefore, $\left| M_n^f \right| \leq 2(n+1)K < \infty$ and M_n is integrable. Also,

$$M_{n+1}^f - M_n^f = f(X_{n+1}) - \mathbf{P}f(X_n) \,,$$

and therefore, since

$$E[f(X_{n+1}) \mid X_0^n] = E[f(X_{n+1}) \mid X_n] = \mathbf{P}f(X_n),$$

we have the martingale equality

$$E[M_{n+1}^f - M_n^f \mid X_0^n] = 0.$$

7.2.2 The Maximum Principle

The *maximum principle* is an important result of potential theory, and we give below one of its avatars when the state space is discrete. It is a good opportunity to show the deep and productive links between probability and analysis.

Let $\{X_n\}_{n \geq 0}$ be an HMC with countable state space E and transition matrix $\mathbf{P}$. Let D be an arbitrary subset of E, called the *domain*, and denote by ∂D the complement of D in E, which is called the *boundary*. Let $c : D \to \mathbb{R}$ and $\varphi : \partial D \to \mathbb{R}$ be non-negative functions called the *unit time cost* and the *final cost*, respectively. Let T be the hitting time of ∂D.

For each state $i \in E$, let

$$h(i) := E_i \left[\sum_{0 \leq k < T} c(X_k) + \varphi(X_T) 1_{\{T < \infty\}} \right]. \tag{7.19}$$

The function $h : E \to \overline{\mathbb{R}}$ so defined is non-negative and possibly infinite. It is called the *average cost*. Note that T is not required to be finite and that ∂D may be empty.

Theorem 7.2.5 *Let $h : E \to \overline{\mathbb{R}}_+$ be defined by (7.19). Then:*

(i) The function h is non-negative and such that

$$h = \begin{cases} \mathbf{P}h + c & \text{on } D \\ \varphi & \text{on } \partial D \end{cases} \tag{7.20}$$

(ii) Any non-negative function $u : E \to \overline{\mathbb{R}}$ such that

$$u \geq \begin{cases} \mathbf{P}u + c & \text{on } D \\ \varphi & \text{on } \partial D \end{cases} \tag{7.21}$$

is a majorant of h, that is, $u \geq h$.

(iii) If for all $i \in E$,

$$P_i(T < \infty) = 1,$$

then (7.20) has at most one non-negative bounded solution.

Proof.

(i) Properties $h \geq 0$ and $h = \varphi$ on ∂D are satisfied by definition. First-step analysis gives for $i \in D$,

$$h(i) = c(i) + \sum_{j \in E} p_{ij} h(j) \qquad (\star)$$

(rely on intuitive arguments or see the details after the proof).

(ii) Define for $n \geq 0$ the non-negative function $h_n : E \to \mathbb{R}$ by

$$h_n(i) = \mathrm{E}_i \left[\sum_{k=0}^{n-1} c(X_k) 1_{\{k < T\}} + \varphi(X_T) 1_{\{T < n\}} \right]. \qquad (7.22)$$

Observe that $h_0 \equiv 0$ and that $\lim_{n \uparrow \infty} \uparrow h_n = h$, by monotone convergence. Also, with a proof similar to that of (i),

$$h_{n+1} = \begin{cases} \mathbf{P} h_n + c & \text{on } D, \\ \varphi & \text{on } \partial D. \end{cases} \qquad (7.23)$$

With u as in (7.21), we have $u \geq h_0$. By induction, $u \geq h_n$ (this is true for $n = 0$, and if this true for some n, it is true for $n+1$. Indeed $u \geq \mathbf{P}u + c \geq \mathbf{P}h_n + c = h_{n+1}$ on D, and $u \geq \varphi = h_{n+1}$ on ∂D). Therefore, $u \geq \lim_{n \to \infty} h_n = h$.

(iii) If u is bounded and non-negative, then by Example 7.2.4,

$$M_n = u(X_n) - u(X_0) - \sum_{k=0}^{n-1} (\mathbf{P} - I) u(X_k)$$

is a martingale with respect to $\{X_n\}_{n \geq 0}$. By the optional sampling theorem, for all integers $K \geq 0$, $\mathrm{E}_i[M_{T \wedge K}] = \mathrm{E}_i[M_0] = 0$ and therefore

$$u(i) = \mathrm{E}_i \left[u(X_{T \wedge K}) \right] - \mathrm{E}_i \left[\sum_{k=0}^{T \wedge K - 1} (\mathbf{P} - I) u(X_k) \right] = \mathrm{E}_i \left[u(X_{T \wedge K}) + \sum_{k=0}^{T \wedge K - 1} c(X_k) \right]$$

since by hypothesis $(I - \mathbf{P})u = c$ on D. Since $P_i(T < \infty) = 1$, $\lim_{K \uparrow \infty} \mathrm{E}_i[u(X_{T \wedge K}] = \mathrm{E}_i[u(X_T)]$ by dominated convergence. But $u(X_T) = \varphi(X_T)$ because $u = \varphi$ on ∂D.

Also, $\lim_{K\uparrow\infty} E_i[\sum_{k=0}^{T\wedge K-1} c(X_k)] = E_i[\sum_{k=0}^{T-1} c(X_k)]$ by monotone convergence. Finally,

$$u(i) = E_i \left[\sum_{k=0}^{T-1} c(X_k) + \varphi(X_T) \right] = h(i).$$

Proof of (⋆). Write for $i \in D$,

$$
\begin{aligned}
v(i) &= E_i[c(X_0) + \sum_{1 \le n < T} c(X_n) + \varphi(X_T)1_{\{T<\infty\}}] \\
&= c(i) + E_i \left[\sum_{1 \le n < T} c(X_n) + \varphi(X_T)1_{\{T<\infty\}} \right],
\end{aligned}
$$

that is,

$$v(i) = c(i) + \sum_{j\in E} E_i[Z1_{\{X_1=j\}}],$$

where

$$Z = \sum_{1 \le n < T} c(X_n) + \varphi(X_T)1_{\{T<\infty\}}.$$

Since $X_0 = i \in D$ implies that $T \ge 1$ on $\{X_0 = i\}$, the random variable Z is a function of $X_1, X_2, \ldots$, and therefore, by the Markov property,

$$E_i[Z1_{\{X_1=j\}}] = E[Z \mid X_1 = j]p_{ij}.$$

Now, since $T \ge 1$ on $\{X_0 = i\}$ when $i \in D$, the quantity Z in the above calculations can be rewritten as

$$Z = \sum_{0 \le n < T-1} c(Y_n) + \varphi(Y_{T-1})1_{\{T-1<\infty\}},$$

where $Y_n = X_{n+1}$. Also, for the HMC $\{Y_n\}_{n\ge0}$, $T' = T - 1$ is the hitting time of ∂D, and therefore

$$E[Z \mid X_1 = j] = E \left[\sum_{0 \le n < T'} c(Y_n) + \varphi(Y_{T'})1_{\{T'<\infty\}} \,\Big|\, Y_0 = j \right],$$

and this quantity is just $v(j)$, since $\{X_n\}_{n\ge0}$ and $\{Y_n\}_{n\ge0}$ have the same transition matrix and in particular have the same distribution when their initial states are the same. $\square$

Remark 7.2.6 Theorem 7.2.5 can be rephrased as follows: The function h given by (7.19) is a minorant of all non-negative solutions of (7.21), and for $u = h$, the inequalities in (7.21) become equalities. Moreover, if h is bounded and $P_i(T < \infty) = 1$ for all $i \in E$, then h is the *unique* solution of (7.20).

EXAMPLE 7.2.7: THE DIRICHLET PROBLEM. Let $h : E \to \mathbb{R}$ be a non-negative bounded function that is *harmonic on* $D \subset E$, that is,

$$h = \mathbf{P}h \text{ on } D.$$

If $P_i(T < \infty) = 1$ for all $i \in E$, then h is entirely determined by its value on the boundary ∂D. To see this, call φ the restriction of h to ∂D. Then

$$h = \begin{cases} \mathbf{P}h & \text{on } D, \\ \varphi & \text{on } \partial D. \end{cases}$$

Since φ is bounded, the function $i \mapsto E_i[\varphi(X_T)]$ is bounded, and therefore since $P_i(T < \infty) = 1$ for all $i \in E$, it is the unique non-negative bounded solution of (7.20) where $c = 0$. Hence

$$h(i) = E_i[\varphi(X_T)]. \tag{7.24}$$

As an illustration, let $\mathbf{P}$ be the transition matrix corresponding to a symmetric random walk on $E = \mathbb{Z}^2$: the only transitions allowed are from $i = (i_1, i_2)$ to the four nearest states and are equiprobable. Then, with $e_1 := (1, 0)$, $e_2 := (0, 1)$,

$$4(\mathbf{P} - I)f(i) = \quad (f(i + e_1) - f(i)) - (f(i) - f(i - e_1))$$
$$+ \quad (f(i + e_2) - f(i)) - (f(i) - f(i - e_2)) .$$

Call this quantity $\Delta f(i)$. The function Δf is the (discrete) *Laplacian* of f. With $c = 0$, (7.20) becomes

$$\begin{cases} \Delta^2 h = 0 & \text{on } D, \\ h = \varphi & \text{on } \partial D \end{cases} \tag{7.25}$$

and the reader will recognize here the *Dirichlet problem* of potential theory. In two dimensions, the symmetric random walk is irreducible and recurrent, and therefore, the hitting time T of ∂D is finite if ∂D is not empty. Therefore, in this case, (7.20) has at most one bounded solution. If φ is bounded, the solution is given by (7.24). More generally, if ∂D is not empty and $\sup_i E_i[\varphi(X_T)] < \infty$, then (7.24) gives the *unique* non-negative bounded solution of the Dirichlet problem.

EXAMPLE 7.2.8: INFINITE SOJOURN. Let $c \equiv 0$ in (7.20) and let $\varphi(k) = 1$ for all $k \in \partial D$. Then for $i \in D$,

$$v(i) = 1 - h(i) = P_i(T = \infty)$$

is the probability of infinite sojourn in D. By application of the maximum principle, we retrieve the results of Subsection 6.1.1.

EXAMPLE 7.2.9: OPTIMAL CONTROL. [6] This example is meant to show the link between optimal control and potential theory. A stochastic process $\{X_n\}_{n \geq 0}$ with values in E is "controlled" in the following way. Let $\{\mathbf{P}(a); a \in A\}$, where A is some set called the set of *actions*, be a family of transition matrices on E, with the interpretation that if at time n the controlled process is in state i and if the controller takes action a, then at time $n + 1$ the state will be j with probability $p_{ij}(a)$. A *control strategy* u is a function $u : E \to A$ which prescribes to take action $u(i)$ when the process is in state i. Therefore, under the strategy u, the controlled process is an HMC with transition matrix $\mathbf{P}^u$, where

$$p_{ij}^u := p_{ij}(u(i)).$$

There is a *cost* $V^u(i)$ associated with each strategy u and each initial state i, of the form

$$V^u(i) = E_i^u[\sum_{0 \leq k < T} c^u(X_k) + \varphi^u(X_T)1_{\{T < \infty\}}],$$

where c^u, φ^u and T are as in Theorem 7.2.5, with D fixed, and moreover, $c^u(i) = c(i, u(i))$ and $\varphi^u(i) = \varphi(i, u(i))$, for appropriate functions c and φ. The problem of *optimal control* is that of finding, if it exists, an *optimal strategy* u^*, such that

$$V^{u^*}(i) \geq V^u(i),$$

for all states i and all strategies u.

We have the following result:

Theorem 7.2.10 *Suppose that there exists a function $V : E \to \mathbb{R}$ such that*

$$V(i) = \sup_{a \in A} \left\{ \sum_{j \in E} p_{ij}(a)V(j) + c(i, a) \right\} \qquad (i \in D)$$

---

[6]Optimal control of Markov chains is treated for instance in [Puterman, 1994].

and

$$V(i) = \sup_{a \in A} \varphi(i, a) \quad (i \in \partial D),$$

and that the above supremums are attained for $a = u^*(i)$ *and some function* u^* : $E \to A$. *Then,* u^* *is an optimal control and* $V = V^{u^*}$.

Proof. Since for all controls u,

$$V \geq \mathbf{P}^u V + c^u \text{ on } D,$$

and

$$V \geq \varphi^u \text{ on } \partial D,$$

it follows from the maximum principle that

$$V \geq V^u$$

for all controls u. Also, $V = V^{u^*}$ and therefore u^* is an optimal control. $\square$

7.3   Martingales and HMCs

7.3.1   The Two Pillars of Martingale Theory

One of the main results of martingale theory, which is the key to the recurrence (resp., transience) criteria of the next subsection, is the probabilistic counterpart of the convergence of a bounded non-decreasing sequence of real numbers to a finite limit.

Theorem 7.3.1 *Let* $\{Y_n\}_{n \geq 0}$ *be either a non-negative supermartingale, or a bounded submartingale, with respect to* $\{X_n\}_{n \geq 0}$. *Then, almost surely,* $\lim_{n \uparrow \infty} Y_n$ *exists and is finite.*

The proof is omitted.

A first application of this result to Markov chain theory is the following:

Theorem 7.3.2 *An irreducible recurrent* HMC *has no non-negative superharmonic or bounded subharmonic functions besides the constant functions.*

Proof. If h is non-negative superharmonic (resp., bounded subharmonic), then the stochastic sequence $\{h(X_n)\}_{n\geq 0}$ is a non-negative supermartingale (resp., bounded submartingale) and therefore, by the martingale convergence theorem, it converges to a finite limit Y. Since $\{X_n\}_{n\geq 0}$ visits any state $i \in E$ infinitely often, one must have $Y = h(i)$ almost surely for all $i \in E$. In particular, h is a constant. □

The next result is another pillar of martingale theory. We give a weak version of it that is sufficient for our purpose.

Theorem 7.3.3 *Let $\{M_n\}_{n\geq 0}$ be a martingale with respect to some process $\{X_n\}_{n\geq 0}$, and let T be a stopping time of $\{X_n\}_{n\geq 0}$. Suppose that at least one of the following condition holds:*

(α) *P-a.s, $T \leq n_0$ for some $n_0 \geq 0$, or*

(β) *P-a.s, $T < \infty$ and $|M_n| \leq K < \infty$ when $n < T$.*

Then

$$E[M_T] = E[M_0].\tag{7.26}$$

Proof.

(α) Write

$$M_T - M_0 = \sum_{k=0}^{n_0-1}(M_{k+1} - M_k)1_{\{k<T\}}.$$

Since T is a stopping time of $\{X_n\}_{n\geq 0}$,

$$1_{\{k<T\}} = \varphi(X_0^k)$$

for some function φ, and therefore

$$E[(M_{k+1} - M_k)1_{\{k<T\}}] = E[(M_{k+1} - M_k)\varphi(X_0^k)] = 0.$$

Therefore,

$$E[M_T - M_0] = \sum_{k=0}^{n_0-1} E[(M_{k+1} - M_k)1_{\{k<T\}}] = 0.$$

(β) Apply the result of (α) to the stopping time $T \wedge n_0$ to obtain

$$E[M_{T\wedge n_0}] = E[M_0].$$

Therefore,

$$|E[M_T] - E[M_0]| = |E[M_T] - E[M_{T\wedge n_0}]| \leq 2KP(T > n_0).$$

Since T is finite, $\lim_{n_0\uparrow\infty} P(T > n_0) = 0$, and therefore $E[M_T] = E[M_0]$. □

7.3.2   Transience and Recurrence via Martingales

Theorem 7.3.4 *A necessary and sufficient condition for an irreducible* HMC *to be transient is the existence of some state conventionally called 0 and of a bounded function* $h : E \to \mathbb{R}$, *not identically null and satisfying*

$$h(j) = \sum_{k \neq 0} p_{jk} h(k) \quad (j \neq 0). \tag{7.27}$$

Proof. Let T_0 be the return time to state 0. First-step analysis shows that the bounded function h defined by

$$h(j) = \mathrm{P}_j(T_0 = \infty)$$

satisfies (7.27). If the chain is transient, h is non-trivial. This proves necessity.

Conversely, suppose that (7.27) holds for a not identically null bounded function. Define

$$\widetilde{h}(j) = \begin{cases} h(j) \text{ if } j \neq 0 \\ 0 \text{ if } j = 0 \end{cases}$$

and let $\alpha := \sum_{k \in E} p_{0k} \widetilde{h}(k)$. Changing signs if necessary, α can be assumed ≥ 0. Then $\widetilde{h}$ is subharmonic. If the chain were recurrent, then by Theorem 7.3.2, $\widetilde{h}$ would be a constant. This constant would be equal to $\widetilde{h}(0) = 0$, and this contradicts the assumed non-triviality of h. $\square$

EXAMPLE 7.3.5: THE REPAIR SHOP, TAKE 6. Continuation of Example 7.1.6. We shall show that if $\mathrm{E}[Z_1] > 1$, the system of equations (7.27) admits a bounded non-trivial solution, and therefore, by Theorem 7.3.4, the chain is transient. In fact, trying $y_j = 1 - \zeta^j$ for a solution, we can check that equations (7.27) reduce to a single equation in ζ,

$$\sum_{k \geq 0} \mathrm{P}(Z_1 = k) \zeta^k = \zeta, \tag{7.28}$$

for which (Theorem 1.2.8) there is, under condition $\mathrm{E}[Z_1] > 1$ and the irreducibility condition, a solution $\zeta \in (0, 1)$. Therefore, $h(i) = \zeta^i$ is a solution of (7.27) that is non-trivial and bounded.

The next result is to be compared with Foster's theorem.

Theorem 7.3.6 *Let the* HMC *with transition matrix* $\mathbf{P}$ *be irreducible. Suppose that there exist* (α)*: a function* $h : E \to \mathbb{R}$ *such that* $\{i \; ; \; h(i) < K\}$ *is finite for all finite* K *and* (β)*: a finite subset* F *of* E*, such that*

$$\sum_{k \in E} p_{ik} h(k) \leq h(i), \;\; \text{for all } i \notin F \,.$$

Then the chain is recurrent.

Proof. Since $\{i \; ; \; h(i) < 0\}$ is finite, $\inf h(i) > -\infty$, and therefore, adding a constant if necessary, one may assume without loss of generality that $h \geq 0$. Let $\tau = \tau(F)$ be the return time to F, and define $Y_n = h(X_n)1_{\{n < \tau\}}$. The arguments in the proof of Foster's theorem show that for $i \notin F$, P_i-a.s,

$$\mathrm{E}_i[Y_{n+1}|X_0^n] \leq Y_n \,.$$

Therefore, $\{Y_n\}_{n \geq 0}$ is, under P_i, a non-negative supermartingale with respect to $\{X_n\}_{n \geq 0}$. By the martingale convergence theorem, $\lim_{n \uparrow \infty} Y_n = Y_\infty$ exists and is finite, P_i-a.s.

Suppose, in view of contradiction, that the chain is transient. It must then visit any finite subset of the state space only a finite number of times. In particular, for arbitrary K, we can have $h(X_n) < K$ only for a finite (random) number of indices n. This implies that $\lim_{n \to \infty} h(X_n) = +\infty$, P_j-a.s. (for any $j \in E$). For this to be compatible with the fact that $\{1_{\{n < \tau\}} h(X_n)\}$ has P_i-a.s. a finite limit for $i \notin F$, we must have $\mathrm{P}_i(\tau < \infty) = 1$.

In summary, $\mathrm{P}_i(\tau < \infty) = 1$ for all $i \notin F$. Since F is finite, some state in F must be recurrent, hence the announced contradiction. □

EXAMPLE 7.3.7: REPAIR SHOP, CONCLUSION. We know from the previous examples on the Repair Shop model that the corresponding chain is positive recurrent if and only if $\mathrm{E}[Z_1] < 1$, and that it is transient if $\mathrm{E}[Z_1] > 1$. It remains to examine the case $\mathrm{E}[Z_1] = 1$, for which there are only two possibilities left: transient or null recurrent. It turns out that the chain is null recurrent in this case. Indeed, one easily verifies that Theorem 7.3.6 applies with $h(i) = i$ and $F = \{0\}$. Therefore, the chain is recurrent. Since it is not positive recurrent, it is null-recurrent.

We are now through with the Repair Shop HMC. We have found that

$$\mathrm{P}(Z_1 = 0) > 0 \text{ and } \mathrm{P}(Z_1 \geq 2) > 0$$

is a necessary and sufficient condition for irreducibility, and that in this case if $E[Z_1] < 1$, the chain is positive recurrent,

if $E[Z_1] = 1$, the chain is null recurrent,

if $E[Z_1] > 1$, the chain is transient.

Here is another application of the martingale convergence theorem in the vein of the previous results.

Theorem 7.3.8 *Let the* HMC $\{X_n\}_{n\geq0}$ *with transition matrix* $\mathbf{P}$ *be irreducible and let* $h : E \to \mathbb{R}$ *be a bounded function such that*

$$\sum_{k\in E} p_{ik}h(k) \leq h(i) \quad (i \notin F),$$

for some subset F *(not assumed finite) of the state space. Suppose, moreover, that there exists an* $i \notin F$ *such that*

$$h(i) < h(j) \quad (j \in F). \tag{7.29}$$

Then, the chain is transient.

Proof. Let τ be the return time in F and let $i \notin F$ satisfy (7.29). Defining $Y_n = h(X_{n\wedge\tau})$, we have

$$E_i[Y_{n+1}|X_0^n] = E_i[1_{\{n<\tau\}}h(X_{n+1})|X_0^n] + E_i[1_{\{n\geq\tau\}}h(X_\tau)|X_0^n].$$

The second term on the right-hand side of the above equality is $1_{\{n\geq\tau\}}h(X_\tau) = 1_{\{n\geq\tau\}}Y_n$ (observe that $1_{\{n\geq\tau\}}h(X_\tau)$ is a function of X_0^n), whereas the first term is, in view of calculations already performed in the proof of Foster's theorem, less than or equal to $1_{\{n<\tau\}}h(X_n) = 1_{\{n<\tau\}}Y_n$. Therefore, under P_i, $\{Y_n\}_{n\geq0}$ is a (bounded) supermartingale with respect to $\{X_n\}_{n\geq0}$. By the martingale convergence theorem, the limit Y of $Y_n = h(X_{n\wedge\tau})$ exists and is finite, P_i-almost surely. By bounded convergence, $E_i[Y] = \lim_{n\uparrow\infty} E_i[Y_n]$, and since $E_i[Y_n] \leq E_i[Y_0] = h(i)$ (supermartingale property), we have $E_i[Y] \leq h(i)$.

If τ were P_i-a.s. finite, then Y_n would eventually be frozen at a value $h(j)$ for $j \in F$, and therefore by (7.29), $E_i[Y] > h(i)$, a contradiction with the last inequality.

Therefore, $P_i(\tau < \infty) < 1$, which means that with a strictly positive probability, the chain starting from $i \notin F$ will not return to F. This is incompatible with irreducibility and recurrence. $\square$

7.3.3 Absorption via Martingales

The method of computation of absorption probabilities given in Chapter 6 is essentially algebraic. Other cases may require ingenuity (see Watson's solution of the branching process, Chapter 6.4). Martingale theory, and in particular the martingale convergence theorem, can also be usefully applied, as the next two examples demonstrate.

EXAMPLE 7.3.9: BRANCHING PROCESSES VIA MARTINGALES. It is assumed as in Chapter 6.4 that $P(Z = 0) < 1$ and $P(Z \geq 2) > 0$. The stochastic process

$$Y_n = \frac{X_n}{m^n} \quad (n \geq 0),$$

where m is the average number of descendants of a given individual, is a martingale with respect to $\{X_n\}_{n \geq 0}$. Indeed, each member of the n-th generation gives on the average m sons, and all the individuals of all the generations do this independently. Therefore, $E[X_{n+1}|X_n] = mX_n$, and

$$E\left[\frac{X_{n+1}}{m^{n+1}}|X_0^n\right] = E\left[\frac{X_{n+1}}{m^{n+1}}|X_n\right] = \frac{X_n}{m^n}.$$

By the martingale convergence theorem, almost surely

$$\lim_{n \uparrow \infty} \frac{X_n}{m^n} = Y < \infty.$$

In particular, if $m < 1$, then $\lim_{n \uparrow \infty} X_n = 0$ almost surely. Since X_n takes integer values, this implies that the branching process eventually becomes extinct.

If $m = 1$, then $\lim_{n \uparrow \infty} X_n = X_\infty < \infty$, and it is easily argued that this limit must be 0. Therefore, in this case as well the process eventually becomes extinct.

For the case $m > 1$, we consider the unique solution in $(0, 1)$ of $x = g(x)$ (see Theorem 1.2.8). Suppose we can show that $Z_n = x^{X_n}$ is a martingale. Then, by the martingale convergence theorem, it converges to a finite limit, and therefore X_n has a limit X_∞, which, however, can be infinite. One can easily argue that this limit cannot be other than 0 (extinction) or ∞ (non-extinction). Since $\{Z_n\}_{n \geq 0}$ is a martingale, $x = E[Z_0] = E[Z_n]$, and therefore, by dominated convergence, $x = E[Z_\infty] = E[x^{X_\infty}] = P(X_\infty = 0)$. Therefore, x is the probability of extinction.

It remains to show that $\{Z_n\}_{n \geq 0}$ is a martingale. We have

$$E[x^{X_{n+1}}|X_n = i] = x^i \quad (i \geq 0).$$

This is obvious for $i = 0$. For $i > 0$, X_{n+1} is the sum of i independent random variables with the same generating function g, and therefore, $E[x^{X_{n+1}}|X_n = i] = g(x)^i = x^i$. From this last result and the Markov property,

$$E[x^{X_{n+1}}|X_0^n] = E[x^{X_{n+1}}|X_n] = x^{X_n}.$$

EXAMPLE 7.3.10: A CELLULAR AUTOMATON. Consider a chessboard of size $N \times N$, on which are placed stones, exactly one on each square. Each stone has one among k possible colors. The state X_n of the process at time n is the $N \times N$ matrix with elements in $\{1, \ldots, k\}$ describing the chessboard and the color of the stone in each square. The evolution of $\{X_n\}_{n \geq 0}$ is that of a homogeneous Markov chain, where the transition from X_n to X_{n+1} is as follows. Select one case of the chessboard at random, and change the color of the stone there, the new color being the color of a stone chosen at random among the 4 neighboring stones. To avoid boundary effects, we shall consider that the chessboard is a bi-torus in the sense of Figure 7.3.1. Any monochromatic state is of course closed.

Figure 7.3.1. Neighbors in the cellular automaton model

Denote by Y_n the proportion of red stones at stage n. The process $\{Y_n\}_{n \geq 0}$ is a martingale with respect to $\{X_n\}_{n \geq 0}$. Indeed, Y_n is a function of X_n and is integrable, since it is bounded by 1. Also, $E[Y_{n+1}|X_0^n] = Y_n$, as the following exchange argument shows.

Let α_{n+1} be the box selected at time $n+1$ and let β_{n+1} be the selected neighbor of α_{n+1}. Then, for any pair (α, β) of boxes, $P(\alpha_{n+1} = \alpha, \beta_{n+1} = \beta|X_0^n) = P(\alpha_{n+1} = \beta, \beta_{n+1} = \alpha|X_0^n) = \frac{1}{8N^2}$. Clearly, if the choice $\alpha_{n+1} = \alpha, \beta_{n+1} = \beta$ changes Y_n to $Y_{n+1} = Y_n + \Delta Y_{n+1}$, the choice $\alpha_{n+1} = \beta$, $\beta_{n+1} = \alpha$ changes Y_n to $Y_{n+1} = Y_n - \Delta Y_{n+1}$. Since these two situations are equiprobable, the martingale property easily follows.

By the martingale convergence theorem, $\lim_{n \uparrow \infty} Y_n = Y$ exists, and by dominated convergence $E[Y] = \lim_{n \uparrow \infty} E[Y_n]$. Therefore, since $E[Y_n] = E[Y_0]$, we have

$E[Y] = E[Y_0] = y_0$, where y_0 is the initial proportion of red stones. Because $|\Delta Y_n| = 0$ or $\frac{1}{N^2}$ for all n, $\{Y_n\}$ can converge only if it remains constant after some (random) time, and this constant is either 0 or 1. Since the limit 1 corresponds to absorption by the "all-red" state, we see that the probability of being absorbed by the "all-red" state is equal to the initial proportion of red stones. This analysis being true for all colors, we see that the monochromatic states are the only absorbing states.

The next example gives an idea of how the optional sampling theorem can be used to compute absorption probabilities.

EXAMPLE 7.3.11: THE GAMBLER'S RUIN REVISITED. Consider the symmetric random walk $\{X_n\}_{n \geq 0}$ on $\mathbb{Z}$ with $X_0 = 0$. It is a martingale (with respect to itself). Let T be the first time n for which $X_n = -a$ or $+b$, where $a, b > 0$. This is a stopping time, and moreover $T < \infty$ (use Remark 6.1.4, for instance). We can apply the optional sampling theorem, part (β), with $K = \sup(a, b)$, to obtain

$$0 = E[X_0] = E[X_T].$$

Writing

$$v = P(-a \text{ is hit before } b),$$

we have

$$E[X_T] = -av + b(1 - v),$$

and therefore

$$v = \frac{b}{a + b}.$$

EXAMPLE 7.3.12: A COUNTEREXAMPLE. Consider the symmetric random walk with $X_0 = 0$, as in the previous example, but now define T to be the hitting time of $b > 0$. We know that $T < \infty$, since the symmetric walk on $\mathbb{Z}$ is recurrent. If the optional sampling theorem applied, we would have

$$0 = E[X_0] = E[X_T] = b,$$

an obvious contradiction. The optional sampling theorem does not apply because neither condition (α) nor (β) thereof is satisfied.

7.4   Exercises

Exercise 7.4.1. A WAITING LINE

Consider the following model of a waiting line. At the beginning of each time period, exactly one customer enters the system. In period $n \geq 1$, the service capacity is Z_{n+1}, that is to say, up to Z_{n+1} customers can be served. This means that if there are X_n customers at the beginning of period n, there will be $(X_n - Z_{n+1})^+ + 1$ at the beginning of period $n + 1$. It is assumed that the sequence $\{Z_n\}_{n\geq 1}$ is IID and independent of the initial number of customers, so that $\{X_n\}_{n\geq 0}$ is an HMC. Give a necessary and sufficient condition of irreducibility. Assuming irreducibility, show that the HMC is positive recurrent if $E[Z_1] > 1$.

Exercise 7.4.2. AN ERGODIC HMC

Consider the HMC with integer values and transition matrix

$$
\begin{aligned}
p_{01} &= 1\,, \\
p_{i,i+k} &= \frac{1}{2^{i+1}} \frac{i!}{k!(i-k)!} \quad (i \geq 1,\ 0 \leq k \leq i), \\
p_{i0} &= \frac{1}{2} \quad (i \geq 1)\,.
\end{aligned}
$$

Show that it is ergodic (give *two* proofs). Compute the mean time between two successive visits to 0.

Exercise 7.4.3. ON THE SECOND CONDITION OF FOSTER'S THEOREM

The purpose of this exercise is to show that condition (7.1) is important. Apply Foster's theorem to the *forward recurrence chain* (see the proof of Theorem 5.1.9). Comment.

Exercise 7.4.4. AVERAGE TIME OF RETURN TO THE REFUGE

Show that in Example 7.1.5,

$$
E_i[\tau] \leq \frac{i}{-E[Z_1]}\,,
$$

where τ is, as in the proof of Theorem 7.1.1, the return time to the refuge F. Give similar estimates for Example 7.1.6 and Exercise 7.4.1.

Exercise 7.4.5. ANOTHER LÉVY MARTINGALE

Let $\{X_n\}_{n\geq 0}$ be an HMC with state space E and transition matrix $\mathbf{P}$. Let $f : \mathbb{N} \times E \to \mathbb{R}$ be a function such that for all $n \geq 0$ and $i \in E$,

$$E[|f(n, X_n)|] < \infty$$

and

$$(\mathbf{P}f)(n+1, i) := \sum_{j \in E} p_{ij} f(n+1, j) = f(n, i).$$

Show that

$$M_n = f(n, X_n)$$

defines a martingale $\{M_n\}_{n\geq 0}$ with respect to $\{X_n\}_{n\geq 0}$.

Exercise 7.4.6. TWO MARTINGALES WITH RESPECT TO THE SYMMETRIC RANDOM WALK

Let $\{X_n\}_{n\geq 0}$ be a symmetric random walk on $\mathbb{Z}$. Show that $\{X_n\}_{n\geq 0}$ and $\{X_n^2 - n\}_{n\geq 0}$ are martingales with respect to $\{X_n\}_{n\geq 0}$.

Exercise 7.4.7. $P_i(T < \infty)$ IS A HARMONIC FUNCTION

Let $\{X_n\}_{n\geq 0}$ be an HMC with state space E, and let B be a closed subset of states, that is,

$$i \in B \Rightarrow \sum_{j \in B} p_{ij} = 1.$$

Let T be the hitting time of B, and let

$$h(i) :- P_i(T < \infty) \quad (i \in E).$$

Show that $\{h(X_n)\}_{n\geq 0}$ is a martingale with respect to $\{X_n\}_{n\geq 0}$.

Exercise 7.4.8. ABSORPTION BY 0

Consider the homogeneous Markov chain with state space $E = \{0, 1, \ldots, m\}$ and transition probabilities

$$p_{ij} = \binom{m}{j} \left(\frac{i}{m}\right)^j \left(1 - \frac{i}{m}\right)^{m-j}.$$

In particular, 0 and m are absorbing states. Compute the probability of absorption by state 0.

Exercise 7.4.9. TIME OUT THE STRIP
In Example 7.3.11, compute $E[T]$ using the result of Exercise 7.4.6.

Exercise 7.4.10. THE LÉVY MARTINGALE CHARACTERIZES THE HMC
Let $\{X_n\}_{n\geq 0}$ be a stochastic process with values in E, which is not assumed to
be an HMC. Let $\mathbf{P}$ be some transition matrix on E. Prove that if for all bounded
$f : E \to \mathbb{R}$, $\{M_n^f\}_{n\geq 0}$ defined by (7.18) is a martingale with respect to $\{X_n\}_{n\geq 0}$,
then $\{X_n\}_{n\geq 0}$ is an HMC with transition matrix $\mathbf{P}$.

Exercise 7.4.11. A GENERAL REPRESENTATION OF HARMONIC FUNCTIONS
Let $\{X_n\}_{n\geq 1}$ be an HMC on E, with transition matrix $\mathbf{P}$. Let $D \subset E$, and suppose
that the boundary ∂D is accessible in finite time T from all states. Show that any
non-negative bounded function $h : \partial D \to \mathbb{R}$ that is harmonic on D is of the form

$$h(i) = \sum_{k \in \partial D} h(k)\mathrm{P}_i(\text{the chain enters } \partial D \text{ through } k).$$

Exercise 7.4.12. THE RUIN PROBABILITY VIA MARTINGALES
Show that the function $h(i) = \left(\frac{q}{p}\right)^i$ is harmonic for the random walk on $\mathbb{Z}$ with
$p_{i,i+1} = p$ and $p_{i,i-1} = q = 1 - p$, where $p \in (0,1), p \neq \frac{1}{2}$. Apply the martingale
convergence theorem to obtain the ruin probability of Example 2.2.10.

Chapter 8

Random Walks on Graphs

8.1 Pure Random Walks

A pure random walk is the motion on a graph of a particle that is not allowed to rest and that chooses the next move equiprobably among all possible ones available. This section opens with the classical random walks on $\mathbb{Z}$ and $\mathbb{Z}^3$ and the pure random walk on a graph of which the one-dimensional symmetric random walk on $\mathbb{Z}$ is the simplest example.

8.1.1 The Symmetric Random Walks on $\mathbb{Z}$ and $\mathbb{Z}^3$

In the one-dimensional pure random walk on $\mathbb{Z}$, the nodes are the relative integers and the edges are all the unordered pairs of the type $(i, i+1)$. Example 3.1.4 showed that this HMC is recurrent and from Example 2.5.12 we know that is in fact null recurrent.

Null recurrence of the symmetric random walk on $\mathbb{Z}$ implies that the time required to reach state 0 from a given state k has infinite mean. This suggests that the probability given the initial state k that T_0 takes large values is large. The following result gives a bound on how large it is. More precisely:

Theorem 8.1.1 *For a symmetric random walk on $\mathbb{Z}$,*

$$P_k(T_0 > r) \leq \frac{12|k|}{\sqrt{r}} . \tag{8.1}$$

The proof, given right after Example 8.1.4, relies on the following result of independent interest, called the *reflection principle*.

© Springer Nature Switzerland AG 2020
P. Brémaud, *Markov Chains*, Texts in Applied Mathematics 31,
https://doi.org/10.1007/978-3-030-45982-6_8

Theorem 8.1.2 *For all positive integers j, k and n,*

$$\mathrm{P}_k(T_0 < n,\, X_n = j) = \mathrm{P}_k(X_n = -j)\,,$$

and therefore, summing over $j > 0$,

$$\mathrm{P}_k(T_0 < n,\, X_n > 0) = \mathrm{P}_k(X_n < 0)\,.$$

Proof. By the strong Markov property, for $m < n$,

$$\mathrm{P}_k(T_0 = m,\, X_n = j) = \mathrm{P}_k(T_0 = m)\mathrm{P}_0(X_{n-m} = j)\,.$$

Since the distribution of X_n is symmetric when the initial position is 0, the right-hand side is

$$\mathrm{P}_k(T_0 = m)\mathrm{P}_0(X_{n-m} = -j) = \mathrm{P}_k(T_0 = m,\, X_n = -j)\,,$$

and therefore

$$\mathrm{P}_k(T_0 = m,\, X_n = +j) = \mathrm{P}_k(T_0 = m,\, X_n = -j)\,.$$

Summation over $m < n$ yields

$$\mathrm{P}_k(T_0 < n,\, X_n = j) = \mathrm{P}_k(T_0 < n,\, X_n = -j) = \mathrm{P}_k(X_n = -j),$$

where, for the last equality, it was observed that starting from a positive position and reaching a negative position at time n implies that position 0 has been reached for the first time strictly before time n. $\qquad\square$

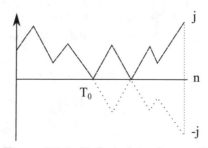

Figure 8.1.1. Reflected random walk

Remark 8.1.3 A combinatorial interpretation of the proof is the following. There is a one-to-one correspondence between the paths that hit 0 before time n and reach position $j > 0$ at time n, and the paths that reach position $-j$ at time n. In fact, given a path that hits 0 before time n and reaches position $j > 0$, associate to it the path that is reflected with respect to position 0 after time T_0 (Figure 8.1.1). This is the reflection principle.

EXAMPLE 8.1.4: THE GAMBLER'S RUIN, TAKE 2. A gambler with initial fortune 1 plays a heads and tails fair coin game with a one dollar stake at each toss. What is the distribution of the duration of the game until he is broke? In other terms, what is the distribution of the return time to 0 of a symmetric random walk starting from position 1? Note that in this case T_0 is necessarily odd. We have by the strong Markov property and the reflection principle (Theorem 8.1.2)

$$
\begin{aligned}
P_1(T_0 = 2m + 1) &= P_1(T_0 > 2m, X_{2m} = 1, X_{2m+1} = 0) \\
&= P_1(T_0 > 2m, X_{2m} = 1)P_1(X_{2m+1} = 0 \mid X_{2m} = 1) \\
&= P_1(T_0 > 2m, X_{2m} = 1)\frac{1}{2} \\
&\quad - \frac{1}{2}\{P_1(X_{2m} - 1) - P_1(T_0 \le 2m, X_{2m} - 1)\} \\
&= \frac{1}{2}\{P_1(X_{2m} = 1) - P_1(T_0 < 2m, X_{2m} = 1)\} \\
&= \frac{1}{2}\{P_1(X_{2m} = 1) - P_1(X_{2m} = -1)\} \\
&= \frac{1}{2}\left\{\binom{2m}{m}2^{-2m} - \binom{2m}{m-1}2^{-2m}\right\} = \frac{\binom{2m}{m}}{m+1}2^{-2m-1}.
\end{aligned}
$$

The way is now clear for the proof of Theorem 8.1.1.

Proof. By symmetry, it suffices to consider the case $k > 0$. The bound is an immediate consequence of the following two results:

$$
P_k(T_0 > r) = P_0(-k < X_r \le +k) \tag{$\star$}
$$

and

$$
P_0(X_r = k) < \frac{3}{\sqrt{r}}. \tag{$\dagger$}
$$

We start with $(\star)$:

$$
\begin{aligned}
P_k(X_r > 0) &= P_k(X_r > 0, T_0 \le r) + P_k(X_r > 0, T_0 > r) \\
&= P_k(X_r > 0, T_0 \le r) + P_k(T_0 > r) \\
&= P_k(X_r > 0, T_0 < r) + P_k(T_0 > r) \\
&= P_k(X_r < 0) + P_k(T_0 > r),
\end{aligned}
$$

where the last equality is the reflection principle. But by symmetry of the random walk, $P_k(X_r < 0) = P_k(X_r > 2k)$. Therefore

$$
\begin{aligned}
P_k(T_0 > r) &= P_k(X_r > 0) - P_k(X_r > 2k) \\
&= P_k(0 < X_r \le 2k) = P_0(-k < X_r \le +k).
\end{aligned}
$$

We now turn to the proof of $(\dagger)$. Let $k = 0, 1, \ldots, r$. Starting from state 0, the event $X_{2r} = 2k$ occurs if and only if there are $r + k$ upward moves and $r - k$ downward moves of the random walks. Therefore

$$
P(X_{2r} = 2k) = \binom{2r}{r+k} 2^{-2r}.
$$

The right-hand side is maximized for $k = 0$, and therefore

$$
P(X_{2r} = 2k) \le \binom{2r}{r} 2^{-2r} \le \sqrt{\frac{8}{\pi}} \frac{1}{\sqrt{2r}},
$$

by Stirling's approximation. To obtain a bound for $P(X_{2r+1} = 2k+1)$, condition on the first move of the random walk and use the previous bound to obtain (Exercise 8.4.4)

$$
P(X_{2r+1} = 2k + 1) \le \frac{4}{\sqrt{\pi}} \frac{1}{\sqrt{2r + 1}}.
$$

$\square$

The Symmetric Random Walk on $\mathbb{Z}^3$ on $\mathbb{Z}^2$

Example 3.1.5 showed that the random walk on $\mathbb{Z}^3$ is transient. One might wonder at this point about the symmetric random walk on $\mathbb{Z}^2$, which moves at each step northward, southward, eastward and westward equiprobably. It is in fact recurrent (Exercise 8.4.5). The symmetric random walk on $\mathbb{Z}^p$ ($p \ge 4$) is transient (Exercise 8.4.6).

8.1.2   The Pure Random Walk on a Graph

Consider a finite non-directed connected graph $G = (V, \mathcal{E})$ where V is the set of *vertices*, or *nodes*, and $\mathcal{E}$ is the set of *edges*. Let d_i be the *index* of vertex i (the number of edges "adjacent" to it). Since there are no isolated nodes (a consequence of the connectedness assumption), $d_i > 0$ for all $i \in V$. Transform this graph into a directed graph by splitting each edge into two directed edges of opposite directions, and make it a transition graph by associating to the directed edge from i to j the transition probability $\frac{1}{d_i}$ (Figure 8.1.2). Note that $\sum_{i \in V} d_i = 2|\mathcal{E}|$.

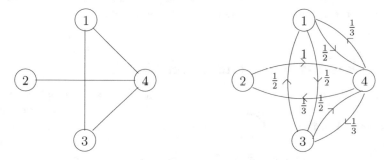

Figure 8.1.2. A random walk on a graph

The corresponding HMC with state space $E \equiv V$ is irreducible (G is connected). It therefore admits a unique stationary distribution π. We make the educated guess that it is given by the formula $\pi(i) = Kd_i$, where K is obtained by normalization: $K = \left(\sum_{j \in E} d_j \right)^{-1} = (2|\mathcal{E}|)^{-1}$. Therefore

$$\pi(i) = \frac{d_i}{2|\mathcal{E}|}.$$

To prove that this is the correct answer, it suffices by Corollary 2.4.15 to show that for all pairs of distinct vertices i and j connected by an edge, the corresponding detailed balance equation is satisfied. This is the case since $p_{ij} = \frac{1}{d_i}$ and $p_{ji} = \frac{1}{d_j}$, so that

$$\pi(i)\frac{1}{d_i} = \pi(j)\frac{1}{d_j}.$$

EXAMPLE 8.1.5: RANDOM WALK ON THE HYPERCUBE, TAKE 2. The random walk on the (n-dimensional) hypercube is the random walk on the graph with set of vertices $E = \{0, 1\}^n$ and edges between vertices x and y that differ in just one coordinate. For instance, in three dimensions, the only possible motions of a particle performing the random walk on the cube is along its edges in both directions. Clearly, whatever the dimension $n \geq 2$, $d_i = \frac{1}{n}$, and the stationary distribution is the uniform distribution.

EXAMPLE 8.1.6: THE LAZY RANDOM WALK. The lazy random walk on the finite graph $(V, \mathcal{E})$ is, by definition, the Markov chain on V with the transition probabilities $p_{ii} = \frac{1}{2}$ and for $i, j \in V$ such that i and j are connected by an edge of the graph, $p_{i,i} = \frac{1}{2d_i}$. This modified chain admits the same stationary distribution as the original random walk. The difference is that the lazy version is always aperiodic, whereas the original version may be periodic.

8.1.3 Spanning Trees and Cover Times

Let $\{X_n\}_{n \in \mathbb{Z}}$ be an irreducible stationary HMC with finite state space E, transition matrix P and stationary distribution π. Let $G = (E, \mathcal{A})$ be the associated directed graph, where $\mathcal{A}$ is the set of *directed edges* (or *arcs*), that is, of ordered pairs of states (i, j) such that $p_{ij} > 0$. The weight of an arc (i, j) is p_{ij}. A *rooted spanning tree* of G is a directed subgraph of G with the following properties:

(i) As an undirected graph, it is a connected graph with E as its set of vertices.

(ii) As an undirected graph, it has no cycles.

(iii) As a directed graph, each of its vertices has out degree 1, except one vertex, the root, which has out degree 0.

Denote by $\mathcal{S}$ the set of spanning trees of G, and by $\mathcal{S}_i$ the subset of $\mathcal{S}$ consisting of rooted spanning trees with summit $i \in V$. The weight $w(S)$ of a rooted spanning tree of $S \in \mathcal{S}$ is the product of the weights of all the directed edges in S.

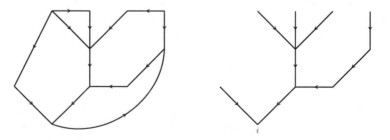

Figure 8.1.3. A directed graph and one of its directed spanning trees

Theorem 8.1.7 *The stationary distribution π of* P *is given by*

$$\pi(i) = \frac{\sum_{S \in \mathcal{S}_i} w(S)}{\sum_{S \in \mathcal{S}} w(S)}.\tag{8.2}$$

Proof. ([1]) Define a stochastic process $\{Y_n\}_{n \in \mathbb{Z}}$ taking its values in $\mathcal{S}$ as follows. The root of Y_n is X_n, say $X_n = i$. Now, by screening the past values $X_{n-1}, X_{n-2}, \dots$ in this order, let $X_{n-\ell_1}$ be the first value different from X_n, let $X_{n-\ell_2}$, $\ell_2 > \ell_1$, be the first value different from X_n and $X_{n-\ell_1}$, let $X_{n-\ell_3}$, $\ell_3 > \ell_2$, be the first value different from X_n, $X_{n-\ell_1}$ and $X_{n-\ell_2}$. Continue this procedure until you have exhausted the (finite) state space E. The spanning tree Y_n is the one with directed edges $(X_{n-\ell_1}, X_{n-\ell_1+1} = X_n)$, $(X_{n-\ell_2}, X_{n-\ell_2+1})$, $(X_{n-\ell_3}, X_{n-\ell_3+1})\dots$

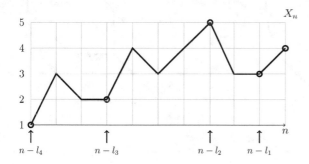
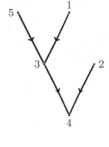

Figure 8.1.4

Since the chain $\{X_n\}_{n \in \mathbb{Z}}$ is stationary, so is the stochastic process $\{Y_n\}_{n \in \mathbb{Z}}$. It is moreover an HMC. We denote by Q_{ST} the transition probability from S to T.

The transition from $Y_n = S \in \mathcal{S}$ with root i to $Y_{n+1} = T \in \mathcal{S}$ with root j in one step is the following:

(a) Add to S the directed (i, j), thus creating a directed spanning graph with a unique directed loop that contains i and j (this may be a self-loop at i).

(b) Delete the unique directed edge of S out of j, say (j, k), thus breaking the loop and producing a rooted spanning tree $T \in \mathcal{S}$ with root j.

(c) Such transition occurs with probability p_{ij}.

---

[1][Anantharam and Tsoucas, 1989].

We now describe the rooted spanning trees S that can lead to the rooted spanning tree T with root j according to the transition matrix Q. T with root j can be obtained from the spanning tree S if and only if S can be constructed from T by the following reverse procedure based on a suitable vertex k:

(α) Add to T the directed edge (j, k), thus creating a directed spanning graph with unique directed loop containing j and k (possibly a self-loop at j).

(β) Delete the unique directed edge (i, j) that lies in the loop, thus breaking the loop and producing a rooted spanning tree $T \in S$ with root i.

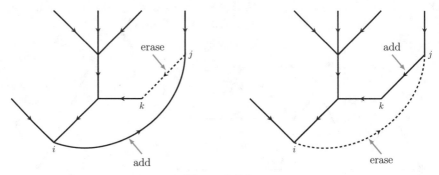

Figure 8.1.5

Let k be the unique vertex used in the reverse procedure. Observing that to pass from T to S, we first added the edge (i, j) and then deleted the unique directed edge (j, k), and that to pass from S to T, we added the directed edge (j, k) and then deleted the edge (j, i). Therefore

$$w(S)Q_{ST} = w(T)R_{TS}$$

where $R_{TS} := p_{jk}$. It follows that

$$\sum_S w(S)Q_{ST} = \sum_S w(T)R_{TS} = w(T)\,.$$

Therefore, the stationary distribution $\{\rho(S)\}_{S \in \mathcal{S}}$ of the chain is

$$\rho(S) = \frac{w(S)}{\sum_{S'} w(S')}\,,$$

and therefore,

$$\pi(i) = \sum_{T \in \mathcal{S}_i} \rho(T) = \frac{\sum_{T \in \mathcal{S}_i} w(T)}{\sum_{T \in \mathcal{S}} w(T)}.$$

$\square$

Corollary 8.1.8 *Let $\{X_n\}_{n \in \mathbb{Z}}$ be the stationary random walk on the complete graph built on the finite state space E. (In particular $p_{ij} = \frac{1}{|E|-1}$ for all $j \neq i$ and the stationary distribution is the uniform distribution on E.) Let for all i ($i \in E$) $\tau_i := \inf\{n \geq 0 \,; X_n = i\}$. The directed graph with directed edges*

$$(X_{\tau_i}, X_{\tau_i - 1}), \qquad i \neq X_0$$

is uniformly distributed over $\mathcal{S}$.

Proof. Use the proof of Theorem 8.1.7 and the time-reversibility of the random walk. $\square$

Definition 8.1.9 *The cover time of an* HMC *is the number of steps it takes to visit all the states.*

Theorem 8.1.10 *The maximum (with respect to the initial state) average cover time of the random walk on a finite graph $(V, \mathcal{E})$ is bounded above by $(|V| - 1) \times (2|\mathcal{E}| - 1) \leq 2|V| \times |\mathcal{E}|$.*

Proof. First observe that the average return time to a given state $i \in E = V$ is $\mathrm{E}_i[T_i] = \frac{1}{\pi(i)} = \frac{2|\mathcal{E}|}{d_i}$. By first-step analysis, denoting by N_i the set of states (vertices) adjacent to i,

$$\frac{2|\mathcal{E}|}{d_i} = \mathrm{E}_i[T_i] = \frac{1}{d_i} \sum_{j \in N_i} (1 + \mathrm{E}_j[T_i])$$

and therefore

$$2|\mathcal{E}| = \sum_{j \in N_i} (1 + \mathrm{E}_j[T_i]) \geq 1 + \mathrm{E}_j[T_i],$$

from which we obtain the bound

$$\mathrm{E}_j[T_i] \leq 2|\mathcal{E}| - 1$$

for any pair (i, j) of states. Let now i_0 be an arbitrary state and consider the spanning circuit obtained by a depth-first census of the vertices of the graph (see Figure 8.1.6), say $i_0, i_1, \ldots i_{2|V|-2} = i_0$.

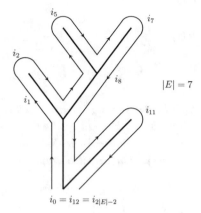

$$i_0 = i_{12} = i_{2|E|-2}$$

Figure 8.1.6

From any vertex i_0 it is possible to traverse the entire spanning tree exactly twice and end up in i_0. Clearly, the average cover time from i_0 is smaller than or equal to the time needed to visit all the vertices of the tree in the order $i_0, i_1, i_{2|V|-2} = i_0$. The (average) time required to go from i_0 to i_1, plus the time needed to go from i_1 to i_0, is less that the return time to i_0, which is in turn bounded by $2|\mathcal{E}| - 1$. The time required to go from i_1 to i_2, plus the time needed to go from i_2 to i_1, is less that the return time to i_1, which is less than $2|\mathcal{E}| - 1$, and so on. Therefore the cover time is bounded by $(|V| - 1) \times (2|\mathcal{E}| - 1) \leq 2|V| \times |\mathcal{E}|$. $\square$

EXAMPLE 8.1.11: COVER TIME OF THE CYCLIC RANDOM WALK. The vertices are n points uniformly distributed on the unit circle, and the n edges are those linking the neighboring vertices. Let c_n denote the cover time for a pure random walk on this n-cycle graph. This average time does not depend on the starting vertex, say 0. Let τ be the first time at which $n - 1$ vertices have been visited. Clearly $\mathrm{E}[\tau] = c_{n-1}$. Also at time τ, the position of the random walker is of the form $i - 1$ or $i + 1$, where i is the vertex that has not been visited yet, say $i - 1$. The random walker will visit i either by walking through the vertices $i - 2, i - 3, \ldots$ or by going directly from $i - 1$ to i. He is in the same situation as the symmetric gambler whose initial fortune is 1 and plays against a gambler whose initial fortune is $n - 1$. The average time before a gambler is broke is $1(n - 1) = n - 1$. Therefore $c_n = c_{n-1} + n - 1$. Since $c_1 = 0$, $c_n = \frac{1}{2}n(n - 1)$. Theorem 8.1.10 would have given the bound $2n^2$.

EXAMPLE 8.1.12: COVER TIME OF THE RANDOM WALK ON THE COMPLETE
GRAPH. The complete graph K_n has n vertices and all possible edges. Therefore
the probability of moving in one step from a given edge to another edge is $\frac{1}{n-1}$.
Consider now the modified walk with loops. From a given edge the probability of
moving to another edge or of staying still is the same: $\frac{1}{n}$. Clearly the cover time in
this modified model is greater than in the original model. For the modified model,
the cover time is the same as the time to complete the collection of the coupon
collector of n objects. Therefore the cover time of the complete graph random
walk is smaller than $(1 + o(1))n \log n$. The bound of Theorem 8.1.10 $2|V| \times |\mathcal{E}|$ is
equal to $2n^2(n-1)$.

8.2   Symmetric Walks on a Graph

8.2.1   Reversible Chains as Symmetric Walks

Let $G = (V, \mathcal{E})$ be a finite graph, that is, V is a finite collection of vertices, or nodes,
and $\mathcal{E}$ is a subset of (unordered) pairs of vertices, denoted by $e = \langle i, j \rangle$. One then
denotes by $i \sim j$ the fact that $\langle i, j \rangle \in \mathcal{E}$. This graph is assumed connected. The
edge, or branch, $e - \langle i, j \rangle$ has a positive number $c_e = c_{ij}$ attached to it such that

$$c_{ij} = c_{ji}. \tag{8.3}$$

In preparation for the electrical network analogy, call c_e the *conductance* of edge e,
and call its reciprocal $R_e = \frac{1}{c_e}$ the *resistance* of e. Denote by C the family $\{c_e\}_{e \in \mathcal{E}}$
and call it the conductance (parameter set) of the network. If $\langle i, j \rangle \notin \mathcal{E}$, let $c_{ij} = 0$
by convention.

Define an HMC on $E := V$ with transition matrix $\mathbf{P}$ given by

$$p_{ij} = \frac{c_{ij}}{C_i},$$

where $C_i = \sum_{j \in V} c_{ij}$ is the normalizing factor. The homogeneous Markov chain
introduced in this way is called the *random walk* on the graph G with conductance
C, or the (G, C)-random walk. The state X_n at time n is interpreted as the
position on the set of vertices of a particle at time n. When on vertex i the
particle chooses to move to an adjacent vertex j with a probability proportional
to the conductance of the corresponding edge, that is, with probability $p_{ij} = \frac{c_{ij}}{C_i}$.
Note that this HMC is irreducible since the graph G is assumed connected and the
conductances are positive. Moreover, if $\sum_{j \in V} C_j < \infty$ (for instance if the graph is

finite), its stationary probability is

$$\pi(i) = \frac{C_i}{\sum_{j \in V} C_j} \tag{8.4}$$

and moreover, it is reversible. To prove this, it suffices to check the reversibility equations

$$\pi(i)\frac{c_{ij}}{C_i} = \pi(j)\frac{c_{ji}}{C_j},$$

using hypothesis (8.3).

EXAMPLE 8.2.1: ILLUSTRATION, TAKE 1. ([2]) The left-hand side of Figure 8.2.1 describes the network in terms of resistances, whereas the right-hand side is in terms of conductances.

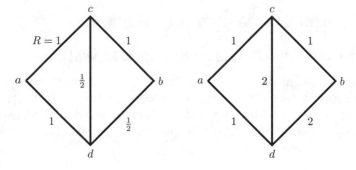

Figure 8.2.1

Figure 8.2.2 shows the transition graph of the associated reversible HMC

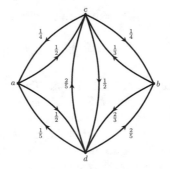

Figure 8.2.2

[2][Doyle and Snell, 2000].

By (8.4), the stationary distribution is $\pi^T = \frac{1}{14}(2,3,4,5)$.

EXAMPLE 8.2.2: THE PURE RANDOM WALK. A symmetric random walk on the graph G is a particular (G, C)-random walk for which $c_e \equiv 1$ $(e \in \mathcal{E})$. In this case, at any given time, the particle being in a given site chooses at random the adjacent site where it will move. We have already seen that the corresponding stationary probability is

$$\pi(i) = \frac{d_i}{2|\mathcal{E}|} \quad (i \in E),$$

where d_i is the degree of node i (the number of nodes to which it is connected) and $|\mathcal{E}|$ is the number of edges.

The connection between random walks and reversible HMCs is in fact both ways. Given a reversible irreducible positive recurrent transition matrix $\mathbf{P} = \{p_{ij}\}_{i,j \in V}$ on V with stationary probability π, we may define the conductance of edge $e = \langle i, j \rangle$ by $c_{ij} = \pi(i)p_{ij}$ $(= c_{ji}$ by reversibility) and define in this way a random walk with the same transition matrix. In particular $C_i = \pi(i)$ and $p_{ij} = \frac{c_{ij}}{C_i}$.

The following result, called the essential edge lemma, is a useful trick for obtaining average passage times. Consider a (G, C)-random walk on a connected graph with the following property. There exist an edge $e = \langle v, x \rangle$ (the essential edge) such that its removal results in two disjoint components. The first (containing v) has for set of vertices $A(v, x)$ and for set of edges $\mathcal{E}(v, x)$, the second (containing x) has for set of vertices $A(x, v)$ and for set of edges $\mathcal{E}(x, v)$.

Lemma 8.2.3 *Under the above condition,*

$$E_v[T_x] = \frac{2\sum_{e \in \mathcal{E}(v,x)} c_e}{c_{vx}} + 1, \tag{8.5}$$

and

$$E_v[T_x] + E_x[T_v] = \frac{2\sum_e c_e}{c_{vx}}. \tag{8.6}$$

Proof. Consider the symmetric random walk with vertices $A(v, x) \cup \{x\}$ and edges $\mathcal{E}(v, x) \cup \{\langle v, x \rangle\}$. For each edge of the modified random walk, the conductance is that of the original random walk. The average time to reach x from v in this restricted graph is obviously equal to that of the original graph. Now, in the restricted random walk, by first-step analysis,

$$E_x[T_x] = 1 + E_v[T_x]$$

and $E_x[T_x]$ is the inverse of the stationary probability of x, that is

$$E_x[T_x] = \frac{2c_{vx} + 2\sum_{e \in \mathcal{E}(v,x)} c_e}{c_{vx}},$$

which gives (8.5). Exchanging the roles of x and v, and combining the two results gives (8.6). $\square$

8.2.2   The Electrical Network Analogy

For finite reversible HMCs, a quantity such as $P_i(T_a < T_b)$ can sometimes be obtained rather simply using an analogy with electrical networks.[3] Once the chain is identified, in a way that will be explained below, with a network of resistances whose nodes are its states, the above quantity is seen to be the effective resistance between nodes a and b. This effective resistance is then computed by successive reductions of the network to a single branch between these nodes. The theory will then be applied to study recurrence in reversible chains with a countable state space.

The setting and notation are those of Subsection 8.2.1. The pair (G, C) will now be interpreted as an electrical network where electricity flows along the edges of the graph (the "branches" of the electrical network). By convention, if $i \not\sim j$, $c_{ij} = 0$. To each directed pair (i, j) there is associated a potential difference Φ_{ij} and a current I_{ij} which are real numbers and satisfy the antisymmetry conditions

$$I_{ji} = -I_{ij} \text{ and } \Phi_{ji} = -\Phi_{ij}$$

for all edges $\langle i, j \rangle$. Two distinct nodes will play a particular role: the source a and the sink b. The currents and the potential differences are linked by the following fundamental laws of electricity:

Kirchoff's potential law: For any sequence of vertices $i_1, i_2, \ldots, i_{n+1}$ such that $i_{n+1} = i_1$ and $i_k \sim i_{k+1}$ for all $1 \le k \le n$,

$$\sum_{\ell=1}^{n} \Phi_{i_\ell, i_{\ell+1}} = 0.$$

Kirchoff's current law: For all nodes $i \in V$, $i \ne a, b$,

$$\sum_{j \in V} I_{ij} = 0.$$

---

[3][Kakutani, 1945], [Kemeny, Snell and Knapp, 1960].

Ohm's law: For all edges $e = \langle i, j \rangle$

$$I_{ij} = c_e \Phi_{ij}.$$

It readily follows from Kirchoff's potential law that there exists a function $\Phi : V \to \mathbb{R}$ determined up to an additive constant such that

$$\Phi_{ij} = \Phi(j) - \Phi(i).$$

Note that, by Ohm's law, the current I_{ij} and the potential difference $\Phi(j) - \Phi(i)$ have the same sign ("currents flow in the direction of increasing potential"). Denote by $I = \{I_{ij}\}_{i,j \in V}$ the current matrix. When the three fundamental laws are satisfied, we say that (Φ, I) is a realization of the electrical network (G, C).

From Kirchoff's current law and Ohm's law, we have that for all $i \neq a, b$,

$$\sum_{i \in V} c_{ij}(\Phi(j) - \Phi(i)) = 0,$$

or equivalently

$$\Phi(i) = \sum_{j \in V} \frac{c_{ij}}{C_i} \Phi(j).$$

Therefore,

Theorem 8.2.4 *The potential function Φ is harmonic on $V \backslash \{a, b\}$ (see the definition in Example 7.2.7) with respect to the (G, C)-random walk.*

In particular, by Example 7.2.7, it is uniquely determined by its boundary values $\Phi(a)$ and $\Phi(b) = 0$.

The Probabilistic Interpretation of Voltage

We shall now interpret a given realization (Φ, I) of the electrical network (G, C) in terms of the associated (G, C)-random walk. We start with the voltage.

Theorem 8.2.5 *Call Φ_1 the solution corresponding to a unit voltage at source a and a null voltage at sink b:*

$$\Phi_1(a) = 1, \qquad \Phi_1(b) = 0.$$

Then, for all $i \in V$,

$$\Phi_1(i) = \mathrm{P}_i \left(T_a < T_b \right).$$

Proof. Using the one-step forward method, one shows that the function h given by $h(i) = \mathrm{P}_i\,(T_a < T_b)$ (the probability that starting from i, a is reached before b) is harmonic on $D = V\backslash\{a,b\}$ and that $h(a) = 1$ and $h(b) = 0$. Recall that a function harmonic on $D = V\backslash\{a,b\}$ is uniquely determined by its values on $\{a,b\}$. Therefore, $\Phi_1 \equiv h$.   □

EXAMPLE 8.2.6: ILLUSTRATION, TAKE 2. Modify the HMC of the running example so as to make states (nodes) a and b absorbing. For $i \in \{a,b,c,d\}$, $h(i)$ defined above is the probability that, starting from i, this HMC is absorbed in a. (Compare with the gambler's ruin problem.)

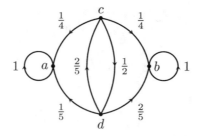

Figure 8.2.3

The recurrence equations for h are

$$h(a) = 1$$
$$h(b) = 0$$
$$h(c) = \frac{1}{4} + \frac{1}{2}h(d)$$
$$h(d) = \frac{1}{5} + \frac{2}{5}h(c).$$

The solution is represented in the figure below, where $\Phi_1 = h$ is the voltage map corresponding to a 1 Volt battery.

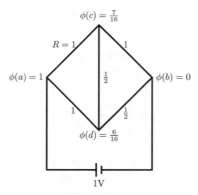

Figure 8.2.4

The Probabilistic Interpretation of Current

We now interpret the current. A particle performs the (G, C)-random walk starting from a, except that now it is supposed to leave the network once it has reached b. We show that the current I_{ij} from i to j is proportional to the expected number of passages of this particle from i to j minus the expected number of passages in the opposite direction, from j to i.

Proof. Let $u(i)$ be the expected number of visits to node i before it reaches b and leaves the network. Clearly $u(b) = 0$. Also for $i \neq a, b$, $u(i) = \sum_{j \in V} u(j) p_{ji}$. But $C_i p_{ij} = C_j p_{ji}$ so that $u(i) = \sum_{j \subset V} u(j) p_{ij} \frac{C_i}{C_j}$ and finally

$$\frac{u(i)}{C_i} = \sum_{j \in V} p_{ij} \frac{u(j)}{C_j}.$$

Therefore the function

$$\Phi(i) := \frac{u(i)}{C_i} \quad (i \in V)$$

is harmonic on $D = V \backslash \{a, b\}$. It is the unique such function whose values at a and at b are specified by

$$\Phi(a) = \frac{u(a)}{C_a}, \qquad \Phi(b) = 0. \tag{$\star$}$$

With such a voltage function,

$$
\begin{aligned}
I_{ij} &= (\Phi(i) - \Phi(j))c_{ij} \\
&= \left(\frac{u(i)}{C_i} - \frac{u(j)}{C_j} \right) c_{ij} \\
&= u(i)\frac{c_{ij}}{C_i} - u(j)\frac{c_{ji}}{C_j} = u(i)p_{ij} - u(j)p_{ji} \,.
\end{aligned}
$$

But $u(i)p_{ij}$ is the expected number of crossings from i to j and $u(j)p_{ji}$ is the expected number of crossings in the opposite direction.   □

Under voltage Φ determined by $(\star)$,

$$
I_a := \sum_{j \in V} I_{aj} = 1
$$

because, in view of the probabilistic interpretation of current in this case, the sum is equal to the expected value of the difference between the number of times the particle leaves a and the number of times it enters a, that is 1 (each time the particle enters a it leaves it immediately, except for the one time when it leaves a forever to be eventually absorbed in b).

Similarly, let $I_{1,a}$ be the current out of a when the unit voltage is applied to a ($\Phi_1(a) = 1$). Since multiplication of the voltage by a factor implies multiplication of the current by the same factor, we have that

$$
\frac{\Phi(a)}{I_a} = \frac{\Phi_1(a)}{I_{1,a}},
$$

that is,

$$
\Phi(a) = \frac{1}{I_{1,a}} \,. \tag{8.7}
$$

———————

EXAMPLE 8.2.7: ILLUSTRATION, TAKE 3. The figure below gives the currents, as given from the voltages by Ohm's law. The current out of a is $I_{1,a} = I_{1,ac} + I_{1,ad} = \frac{9}{16} + \frac{10}{16} = \frac{19}{16}$, by Kirchoff's law.

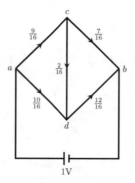

Figure 8.2.5

8.3 Effective Resistance and Escape Probability

8.3.1 Computation of the Effective Resistance

Definition 8.3.1 *The* effective resistance *between a and b is, by definition, the quantity*

$$R_{eff}(a \leftrightarrow b) := \frac{\Phi(a)}{I_a}. \tag{8.8}$$

As we saw before, this quantity does not depend on the value $\Phi(a)$. When $\Phi(a) = \Phi_1(a) = 1$, the effective conductance $C_a = (R_{eff}(a \leftrightarrow b))^{-1}$ therefore equals the current $I_{1,a}$ flowing out of a. In this case

$$I_{1,a} = \sum_{j \in V} (\Phi_1(a) - \Phi_1(j)) c_{aj} = \sum_{j \in V} (1 - \Phi_1(j)) c_{aj}$$

$$= C_a \left(1 - \sum_{j \in V} \Phi_1(j) \frac{c_{aj}}{C_a} \right) = C_a \left(1 - \sum_{j \in V} p_{aj} \Phi_1(j) \right).$$

But the quantity $\left(1 - \sum_{j \in V} p_{aj} \Phi_1(j) \right)$ is the *escape probability*

$$P_{esc} := P_a(T_b < T_a),$$

that is, the probability that the particle starting from a reaches b before returning to a. Therefore

$$P_{esc} = \frac{1}{C_a R_{eff}(a \leftrightarrow b)}.$$

EXAMPLE 8.3.2: ILLUSTRATION, TAKE 4. The effective resistance is

$$R_{eff}(a \leftrightarrow b) = \frac{1}{I_{1,a}} = \frac{1}{\frac{19}{16}} = \frac{16}{19}.$$

In particular, the probability – starting from a – of returning to a before hitting b is $P_{esc} = \frac{1}{C_a R_{eff}(a \leftrightarrow b)} = \frac{1}{2 \times \frac{16}{19}} = \frac{19}{32}.$

EXAMPLE 8.3.3: COMMUTE TIME AND EFFECTIVE RESISTANCE. Recall formula (6.15):

$$P_a(T_b < T_a) = \frac{1}{\pi(a)(E_a[T_b] + E_b[T_a])}$$

and the expression

$$\pi(a) = \frac{C_a}{C}.$$

We obtain the following formula:

$$E_a[T_b] + E_b[T_a] = C R_{eff}(a \leftrightarrow b). \tag{8.9}$$

(The left-hand side is the commute time between a and b.)

In order to compute the effective resistance, we have at our disposition the procedure used in the simplification of resistance networks, such as the following two basic rules.

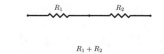

Figure 8.3.1. Series configuration

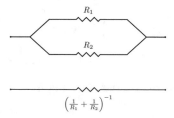

Figure 8.3.2. Parallel configuration

We can also merge nodes with the same voltage.

EXAMPLE 8.3.4: THE CUBIC NETWORK. All the resistances in the network below
are unit resistance.

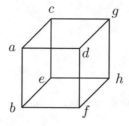

Figure 8.3.3

By symmetry, the nodes c and d have the same voltage, and can therefore be
merged. Similarly for the nodes e and f.

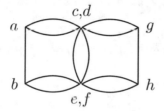

Figure 8.3.4

One can then use the rule for resistances in parallel to further simplify the network:

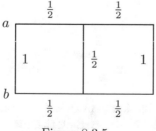

Figure 8.3.5

Alternating series and parallel simplifications, we have:

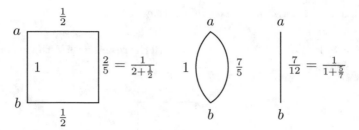

Figure 8.3.6

Therefore the effective resistance between a and b is $R_{eff}(a \leftrightarrow b) = \frac{7}{12}$.

EXAMPLE 8.3.5: THE BINARY TREE. Consider the pure random walk on the full binary tree of depth k. Select two nodes a and b. Let $\mathcal{P}(a \leftrightarrow b)$ be the shortest path linking a and b. In view of computing $P_a(T_b < T_a)$, we make the preliminary observation that this quantity does not change if one cuts all edges that are not in $\mathcal{P}(a \leftrightarrow b)$ and have an endpoint in $\mathcal{P}(a \leftrightarrow b)\backslash\{a\}$. We are therefore left with the graph $\mathcal{P}(a \leftrightarrow b)$ plus, when a is not a leaf of the tree, the edges leading to a that do not belong to $\mathcal{P}(a \leftrightarrow b)$. Therefore $R_{eff}(a \leftrightarrow b) = d(a,b)$, the graph distance between a and b, and therefore $P_a(T_b < T_a) = \frac{1}{3d(a,b)}$ if a is not a leaf, $= \frac{1}{d(a,b)}$ if a is a leaf.

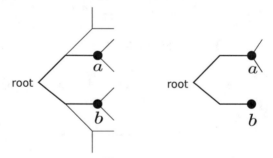

Figure 8.3.7

Another basic rule of reduction of electrical networks is the star-triangle transformation (Exercise 8.4.9). It states that the two following electrical network configurations are equivalent if and only if for $i = 1, 2, 3$,

$$R_i \widetilde{R}_i = \delta$$

where
$$\delta = R_1 R_2 R_3 \left(R_1^{-1} + R_2^{-1} + R_3^{-1} \right) = \frac{\widetilde{R}_1 \widetilde{R}_2 \widetilde{R}_3}{\widetilde{R}_1 + \widetilde{R}_2 + \widetilde{R}_3}.$$

("Equivalence" means that if one network supports the currents i_1, i_2 and i_3 entering the triangle at nodes 1, 2 and 3 respectively, so does the other network.)

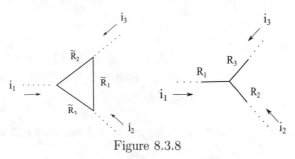

Figure 8.3.8

See Exercise 8.4.15 for an example of application.

8.3.2 Thompson's and Rayleigh's Principles

By definition, a flow on the graph G with source a and sink b is a collection of real numbers $J = \{J_{ij}\}_{i,j \in V}$ such that

(a) $J_{ij} = -J_{ji}$,

(b) $J_{ij} = 0$ if $i \not\sim j$, and

(c) $\sum_{j \in V} J_{ij} = 0$ for all $i \neq a, b$.

Denote by $J_i = \sum_{j \in V} J_{ij}$ the flow out of i. A unit flow J is one for which $J_a = 1$. In general,
$$J_a = -J_b$$
because, since $J_i = 0$ for all $i \neq a, b$,
$$J_a + J_b = \sum_{i \in V} J_i$$
$$= \sum_{i,j \in V} J_{ij} = \frac{1}{2} \sum_{i,j \in V} (J_{ij} + J_{ji}) = 0.$$

Also, for any function $w : V \to \mathbb{R}$,
$$(w(a) - w(b)) J_a = \frac{1}{2} \sum_{i,j \in V} (w(j) - w(i)) J_{ij}. \tag{8.10}$$

Indeed, from the properties of flows,

$$
\sum_{i,j\in V}(w(i)-w(j))J_{ij} = \sum_{i,j\in V}w(i)J_{ij} - \sum_{i,j\in V}w(j)J_{ij}
$$

$$
= \sum_{i,j\in V}w(i)J_{ij} + \sum_{i,j\in V}w(j)J_{ji}
$$

$$
= \sum_{i\in V}w(i)J_i + \sum_{j\in V}w(j)J_j
$$

$$
= w(a)J_a + w(b)J(b) + w(a)J_a + w(b)J(b)
$$

$$
= w(a)J_a - w(b)J(a) + w(a)J_a - w(b)J(a)
$$

$$
= 2(w(a)-w(b))J_a .
$$

The energy dissipated in the network by the flow J is by definition the quantity

$$
E(J) := \frac{1}{2}\sum_{i,j\in V}J_{ij}^2 R_{ij}.
$$

This is a meaningful electrical quantity for the special case where the flow is a current I corresponding to a potential Φ, in which case, by Ohm's law:

$$
E(I) = \frac{1}{2}\sum_{i,j\in V}I_{ij}^2 R_{ij} = \frac{1}{2}\sum_{i,j\in V}I_{ij}(\Phi(j)-\Phi(i)).
$$

Theorem 8.3.6 *The effective resistance between the source a and the sink b is equal to the energy dissipated in the network when the current I_a out of a is the unit current.*

Proof. By (8.10),

$$
E(I) = (\Phi(a)-\Phi(b))I_a = \Phi(a)I_a ,
$$

and by definition (8.8) of the effective resistance $R_{eff}(a\leftrightarrow b)$ between a and b,

$$
E(I) = I_a^2 R_{eff}(a\leftrightarrow b) .
$$

$\square$

The following result is known as *Thomson's principle*.

Theorem 8.3.7 *The energy dissipation $E(J)$ is minimized among all unit flows J by the unit current flow I.*

Proof. Let J be a unit flow from a to b and let I be a unit current flow from a to b. Define $D = J - I$. This is a flow from a to b with $D_a = 0$. We have that

$$\sum_{i,j\in V} J_{ij}^2 R_{ij} = \sum_{i,j\in V} (I_{ij} + D_{ij})^2 R_{ij}$$

$$= \sum_{i,j\in V} I_{ij}^2 R_{ij} + 2 \sum_{i,j\in V} I_{ij} D_{ij} R_{ij} + \sum_{i,j\in V} D_{ij}^2 R_{ij}$$

$$= \sum_{i,j\in V} I_{ij}^2 R_{ij} + 2 \sum_{i,j\in V} (\Phi(j) - \Phi(i)) D_{ij} + \sum_{i,j\in V} D_{ij}^2 R_{ij}.$$

From (8.10) with $w = \Phi$ and $J = D$, the middle term equals $4(\Phi(a) - \Phi(b))D_a = 0$, so that

$$\sum_{i,j\in V} J_{ij}^2 R_{ij} = \sum_{i,j\in V} I_{ij}^2 R_{ij} + \sum_{i,j\in V} D_{ij}^2 R_{ij} \geq \sum_{i,j\in V} I_{ij}^2 R_{ij} .$$

$\square$

We now state and prove *Rayleigh's principle*.

Theorem 8.3.8 *The effective resistance between two points can only increase as any resistance in the circuit increases.*

Proof. Change the resistances R_{ij} to $\overline{R}_{ij} \geq R_{ij}$ and let I and $\overline{I}$ be the corresponding unit current flows. Then

$$\overline{R}_{eff} = \frac{1}{2} \sum_{i,j\subset V} \overline{I}_{ij}^2 \overline{R}_{ij} \geq \frac{1}{2} \sum_{i,j\subset V} \overline{I}_{ij}^2 R_{ij} .$$

But by Thomson's principle,

$$\frac{1}{2} \sum_{i,j\in V} \overline{I}_{ij}^2 R_{ij} \geq \frac{1}{2} \sum_{i,j\in V} I_{ij}^2 R_{ij} = R_{eff}(a \leftrightarrow b) .$$

$\square$

EXAMPLE 8.3.9: SHORTING AND CUTTING. Shorting consists in making some resistances null and therefore decreases the effective resistance. On the contrary, cutting (an edge), which consists in making the corresponding resistance infinite, increases the effective resistance.

8.3.3   Infinite Networks

Consider a (G, C)-random walk where now $G = (V, \mathcal{E})$ is an infinite connected graph with finite-degree vertices. Since the graph is infinite, this HMC may be transient. This subsection gives a method that is sometimes useful in assessing the recurrence or transience of this random walk. Note that once recurrence is proved, we have an invariant measure x, namely $x_i = C_i$ (a finite quantity since each vertex has finite degree). Positive recurrence is then granted if and only if $\sum_{i \in V} C_i < \infty$. (The latter condition alone guarantees the existence of an invariant measure, but remember that existence of an invariant measure does not imply recurrence.)

Some arbitrary vertex will be distinguished, henceforth denoted by 0. Recall that the graph distance $d(i, j)$ between two vertices is the smallest number of edges to be crossed when going from i to j. For $N \geq 0$, let

$$K_N = \{i \in V; \, d(0, i) \leq N\}$$

and

$$\partial K_N = K_N - K_{N-1} = \{i \in V; \, d(0, i) = N\}.$$

Let G_N be the restriction of G to K_N. A graph $\overline{G}_N$ is obtained from G_N by merging the vertices of ∂K_N into a single vertex named b_N. Let $R_{eff}(N) := R_{eff}(0 \leftrightarrow b_N)$ be the effective resistance between 0 and b_N of the network $\overline{G}_N$. Since $\overline{G}_N$ is obtained from $\overline{G}_{N+1}$ by merging the vertices of $\partial K_N \cup \{b_{N+1}\}$, $R_{eff}(N) \leq R_{eff}(N+1)$. In particular, the limit

$$R_{eff}(0 \leftrightarrow \infty) := \lim_{N \uparrow \infty} R_{eff}(N)$$

exists. It may be finite or infinite.

Theorem 8.3.10 *The probability of return to 0 of the (G, C)-random walk is*

$$P_0(X_n = 0 \text{ for some } n \geq 1) = 1 - \frac{1}{C_0 R_{eff}(0 \leftrightarrow \infty)}.$$

In particular, this chain is recurrent if and only if $R_{eff}(0 \leftrightarrow \infty) = \infty$.

Proof. The function h_N defined by

$$h_N(i) := P(X_n \text{ hits } K_N \text{ before } 0)$$

is harmonic on $V_N \backslash \{\{0\} \cup K_N\}$ with boundary conditions $h_N(0) = 0$ and $h_N(i) = 1$ for all $i \in K_N$. Therefore, the function g_N defined by

$$g_N(i) = h_N(i) \quad (i \in K_{N-1} \cup \{b_N\})$$

and $g_N(b_N) = 1$ is a potential function for the network $\overline{G}_N$ with source 0 and sink b_N. Therefore

$$P_0\left(X_n \text{ returns to } 0 \text{ before reaching } \partial K_N\right) = 1 - \sum_{j \sim 0} p_{0j} g_N(j)$$

$$= 1 - \sum_{j \sim 0} \frac{c_{0j}}{C_0}(g_N(j) - g_N(0)).$$

By Ohm's law, $\sum_{j \sim 0} c_{0j}(g_N(j) - g_N(0))$ is the total current $I_N(0)$ out of 0, and therefore since the potential difference between b_N and 0 is 1, $I_N(0) = \frac{1}{R_{eff}(N)}$. Therefore

$$P_0\left(X_n \text{ returns to } 0 \text{ before reaching } \partial K_N\right) = 1 - \frac{1}{C_0 R_{eff}(N)}$$

and the result follows since, by the sequential continuity property of probability,

$$P(X_n = 0 \text{ for some } n \geq 1) = \lim_{N \uparrow \infty} P_0\left(X_n \text{ returns to } 0 \text{ before reaching } \partial K_N\right).$$

$\square$

Theorem 8.3.11 *Consider two sets of conductances C and $\overline{C}$ on the same connected graph $G = (V, \mathcal{E})$ such that for each edge e,*

$$u c_e \leq \overline{c}_e \leq v c_e$$

for some constants u and v, $0 < u \leq v < \infty$. Then the random walks (G, C) and $(G, \overline{C})$ are of the same type (either both recurrent, or both transient).

Proof. Let C^u be the set of conductances on G defined by $c_e^u = u c_e$, and define similarly the set of conductances C^v. Observe that the random walks (G, C^u), (G, C^v) and (G, C) are the same. The rest of the proof follows from Rayleigh's monotonicity law, because (G, C) and $(G, \overline{C})$ then have effective resistances that are both finite or both infinite. $\square$

EXAMPLE 8.3.12: THE SYMMETRIC RANDOM WALK ON $\mathbb{Z}^2$. The symmetric random walk on $\mathbb{Z}^2$ corresponds in the electrical network analogy to the infinite grid where all resistances are unit resistances. The grid is actually infinite and in the figure below only "levels" up to the third one are shown. (Level 0 is the center node, level 1 consists of the 4 nodes at distance 1 from level 0, and more generally, level $i + 1$ consists of the $8i + 4$ nodes at distance 1 from level i.)

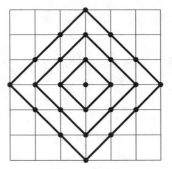

Figure 8.3.9

By Rayleigh's monotonicity law, if one shorts a set of nodes (that is, if the resistances directly linking pairs of nodes in this set are set to 0, in which case the nodes thereof have the same potential), the effective resistance between two nodes is decreased.

Shorting successively the nodes of each $i \geq 1$, we obtain for the effective resistance between node 0 and level $i + 1$ in the shorted network (Figure 8.3.10),

$$\overline{R}_{eff}(i+1) = \sum_{n=0}^{i} \frac{1}{8i+4}.$$

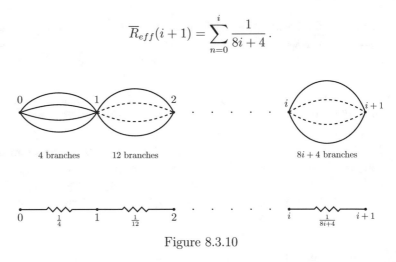

Figure 8.3.10

$\overline{R}_{eff}(i + 1)$ is, by Rayleigh's monotonicity principle, smaller than the actual effective resistance in the full grid between node 0 and any node at level $i + 1$. Therefore, since $\lim_{N \uparrow \infty} \overline{R}_{eff}(N) = \infty$, the two-dimensional symmetric random walk on $\mathbb{Z}^2$ is recurrent.

8.4 Exercises

Exercise 8.4.1. PASSAGE TIMES FOR BIRTH-AND-DEATH PROCESSES
Consider the symmetric random walk on the graph $G = (V, \mathcal{E})$, where $V = \{0, \ldots, N-1\}$ and $\mathcal{E} = \{\langle i-1, i \rangle \, ; \, 1 \leq i \leq N-1\}$. Call w_i the conductance of edge $\langle i-1, i \rangle$. Define $w := \sum_{i=1}^{N-1} w_i$. Let $a, b, c \in V$ be such that $a < b < c$. Show that

(α) $P_b(T_c < T_a) = \frac{\sum_{i=a+1}^{b} w_i^{-1}}{\sum_{i=a+1}^{c} w_i^{-1}}$,

(β) $E_b[T_c] = c - b \sum_{j=b+1}^{c} \sum_{i=1}^{j-1} w_i w_j^{-1}$,

(γ) $E_b[T_c] + E_c[T_b] = w \sum_{i=b+1}^{c} w_i^{-1}$.

Exercise 8.4.2. ON THE CIRCLE
Consider the random walk on the circle. More precisely, there are n points labeled $0, 1, 2, \ldots, n-1$ orderly and equidistantly placed on a circle. A particle moves from one point to an adjacent point in the manner of a random walk on $\mathbb{Z}$. This gives rise to an HMC with the transition probabilities $p_{i,i+1} = p \in (0, 1)$, $p_{i,i-1} = 1 - p$, where, by the "modulo convention", $p_{0,-1} := p_{0,n-1}$ and $p_{n-1,n} := p_{n-1,0}$. Compute the average time it takes to go back to 0 when initially at 0.

Exercise 8.4.3. STREAKS OF 1'S IN A WINDOW OF FAIR COIN TOSSES
Let $\{U_n\}_{n \in \mathbb{Z}}$ be an IID sequence of equiprobable 0's and 1. Define $X_n \in \{0, 1, \ldots, N\}$ by $X_n - 0$ if $U_n = 0$ and

$$X_n = k \text{ if } U_n = 1, U_{n-1} = 1, \ldots, U_{n-k+1} = 1, U_{n-k} = 0.$$

In words, we look at the window of length N just observed at the n-th toss of a sequence of fair coin tosses, and set $X_n = k$ if the length of the last streak of 1's is k. For instance, with $N = 5$ and

$$(U_{-4}, U_{-3}, \ldots, U_5) = (0110110111)$$

we have $X_0 = 1$ (the first window of size 5 is 01101 and the rightmost streak of 1's has length 1), $X_1 = 2$ (the next window of size 5 is 11011 and the rightmost streak of 1's has length 2), $X_2 = 0$ (the next window of size 5 is 1010110 and the rightmost streak of 1's has length 0), $X_3 = 1$ (the next window of size 5 is 0101101 and the rightmost streak of 1's has length 1). The next sliding windows are 11011 and 10111 give respectively $X_3 = 2$ and $X_4 = 3$.

(a) Give the transition matrix of this HMC and its stationary distribution π.

(b) Assuming the chain stationary, give the transition matrix $\widetilde{\mathbf{P}}$ of the time reversed HMC.

(c) Show that, whatever the initial state, the distribution of the reversed chain is already the stationary distribution at the N-th step.

Exercise 8.4.4. FINISHING THE PROOF OF THEOREM 8.1.1
Prove that
$$P(X_{2r+1} = 2k + 1) \leq \frac{4}{\sqrt{\pi}} \frac{1}{\sqrt{2r+1}}.$$

Exercise 8.4.5. NULL RECURRENCE OF THE 2-D SYMMETRIC RANDOM WALK
Show that the 2-D symmetric random walk on $\mathbb{Z}^2$ is null recurrent.

Exercise 8.4.6. TRANSIENCE OF THE 4-D SYMMETRIC RANDOM WALK
Show that the projection of the 4-D symmetric random walk on $\mathbb{Z}^3$ is a lazy symmetric random walk on $\mathbb{Z}^3$. Deduce from this that the 4-D symmetric random walk is transient. More generally, show that the symmetric random walk on $\mathbb{Z}^p$, $p \geq 5$, is transient.

Exercise 8.4.7. THE LINEAR WALK
Consider the pure random walk on the linear graph with vertices $0, 1, 2, \ldots, n$ and edges $\langle i, i+1 \rangle$ $(0 \leq i \leq n-1)$. Compute the cover time. Compare to the rough bound.

Exercise 8.4.8. EHRENFEST VIA THE SPANNING TREE FORMULA
Apply formula (8.2) to the Ehrenfest HMC.

Exercise 8.4.9. THE STAR-TRIANGLE EQUIVALENCE
Show that the electrical network configurations in Figure 8.3.8 are equivalent if and only if for $i = 1, 2, 3$, $R_i \widetilde{R}_i = \delta$, where

$$\delta = R_1 R_2 R_3 \left(R_1^{-1} + R_2^{-1} + R_3^{-1} \right) = \frac{\widetilde{R}_1 \widetilde{R}_2 \widetilde{R}_3}{\widetilde{R}_1 + \widetilde{R}_2 + \widetilde{R}_3}.$$

(Equivalence: if one network supports the currents i_1, i_2 and i_3 entering the triangle at nodes 1, 2 and 3 respectively, so does the other network.)

Exercise 8.4.10. THE URN OF EHRENFEST AS AN ELECTRICAL NETWORK
Describe the Ehrenfest HMC with $N = 2M$ particles in stationary state in terms of electrical networks. Let state M be the source and state N the sink. Compute

the voltage $\Phi_1(i)$ at any state $i \in \{0, 1, \cdots, N\}$ constrained by $\Phi_1(M) = 1$ and $\Phi_1(N) = 0$. Compute $P(T_M < T_N)$.

Exercise 8.4.11. THE SPHERICAL SYMMETRIC TREE
Consider the full spherical tree of degree three (see the figure) and define Γ_N to be the set of nodes at distance N from the root, called 0. Consider the symmetric random walk on the full symmetric spherical tree, where all the edges from Γ_{i-1} to Γ_i have the same resistance $R_i > 0$.

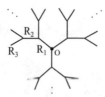

Figure 8.4.1

Show that a necessary and sufficient condition of recurrence of the corresponding symmetric random walk is

$$\sum_{i-1}^{\infty} \frac{R_i}{|\Gamma_i|} = \infty .$$

Exercise 8.4.12. THE LADDER
Find the effective resistance between a and b of the following infinite network of unit resistances.

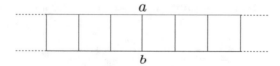

Figure 8.4.2

Exercise 8.4.13. THE BARRIER
Find the effective resistance between a and b of the following infinite network of unit resistances.

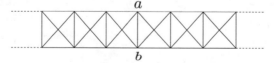

Figure 8.4.3

Exercise 8.4.14. $G_b(a) = C_a R_{eff}(a \leftrightarrow b)$

Consider a (G, C)-random walk, and let a and b be two distinct vertices of G. Let T_b be the first return time to b. Define

$$G_b(a) := \mathrm{E}_a\left[\sum_{n=0}^{\infty} 1_{\{X_n=a\}} 1_{\{n<T_b\}}\right]$$

(the average number of visits to a before b is hit, given that the initial state is a). Show that

$$G_b(a) = C_a R_{eff}(a \leftrightarrow b).$$

Exercise 8.4.15. REDUCING THE FOUR-SQUARE NETWORK

(^{4}) The goal is to find the effective resistance of the following electrical network between nodes a and b. (A resistance not appearing on a branch is conventionally taken equal to 1.)

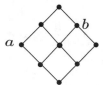

Figure 8.4.4

The successive reduction operations are recorded in the sequence of figures below.

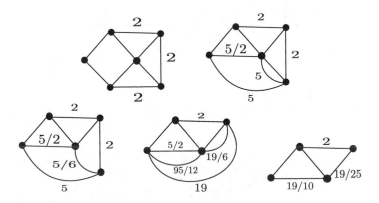

---

^{4}From [Klenke, 2014].

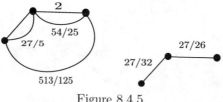

Figure 8.4.5

Finally the effective resistance between a and b is $\frac{27}{32} + \frac{27}{26} = \frac{629}{416}$.

(1) Give the details.

(2) What would be the result if we append to the rightmost node a "square" formed by 4 unit resistances?

Exercise 8.4.16. SYMMETRIC WALK ON THE BINARY TREE
Consider the full binary tree whose root is denoted by 0. Show that

$$R_{eff}(0 \leftrightarrow \infty) = \frac{1}{2}(R_{eff}(0 \leftrightarrow \infty) + 1)$$

and

$$R_{eff}(N+1) = \frac{1}{2}(R_{eff}(N) + 1).$$

Deduce from this that the symmetric random walk on the full binary tree is transient.

Chapter 9

Convergence Rates

9.1 Reversible Transition Matrices

9.1.1 A Characterization of Reversibility

In view of the convergence in variation of an ergodic Markov chain to its stationary distribution, one may consider that the value at a "large" time n of the state is "approximately" distributed according to the stationary distribution. In other terms, the state at time n is an *approximate sample* of the stationary distribution. The quality of this sample is measured in terms of the distance in variation between its distribution and the stationary distribution. This chapter is devoted to the obtention of *convergence speeds* of an ergodic HMC to its stationary distribution and in particular of bounds of the *second largest eigenvalue modulus* of its transition matrix. This is done primarily for reversible HMCs, since most Monte Carlo Markov chains (see Chapter 11) are of this type.

The main result of Perron and Frobenius is that convergence to steady state of an ergodic finite state space HMC is geometric with relative speed equal to the second-largest eigenvalue modulus (SLEM). Even if there are a few interesting models, especially in biology, where the eigenstructure of the transition matrix can be extracted,[1] this situation nevertheless remains exceptional. The added structure of reversible transition matrices allows us to push the analysis further and avoids recourse to the Perron–Frobenius theorem. For convenience, we recall the definition of reversibility, and introduce a slight change in the terminology.

Definition 9.1.1 *Let* $\mathbf{P}$ *be a transition matrix and* π *a strictly positive probability vector on* E. *The pair* $(\mathbf{P}, \pi)$ *is called reversible if the detailed balance equations (2.30) are satisfied.*

---

[1]See for instance [Karlin and Taylor, 1975 and 1981].

© Springer Nature Switzerland AG 2020
P. Brémaud, *Markov Chains*, Texts in Applied Mathematics 31,
https://doi.org/10.1007/978-3-030-45982-6_9

It will also be assumed that the state space is finite, say $E := \{1, 2, \ldots, r\}$, and that $\mathbf{P}$ is irreducible. This implies in particular that π is the unique stationary distribution, that $\pi > 0$ and that $\mathbf{P}$ is positive recurrent. For short, we shall sometimes say: "$\mathbf{P}$ is reversible".

Let $\ell^2(\pi)$ be the real vector space $\mathbb{R}^r$ endowed with the scalar product

$$\langle x, y \rangle_\pi := \sum_{i \in E} x(i) y(i) \pi(i)$$

and the corresponding norm $\|x\|_\pi := \left(\sum_{i \in E} x(i)^2 \pi(i) \right)^{\frac{1}{2}}$. We shall write

$$\langle x \rangle_\pi := \sum_i \pi(i) x(i) = \langle x, 1 \rangle_\pi$$

for the *mean* of x with respect to π. The variance of x with respect to π is

$$\mathrm{Var}_\pi(x) := \sum_i \pi(i) x(i)^2 - \left(\sum_i \pi(i) x(i) \right)^2 = \|x\|_\pi^2 - \langle x \rangle_\pi^2 .$$

Similarly to $\ell^2(\pi)$, $\ell^2(\frac{1}{\pi})$ is defined as the real vector space $\mathbb{R}^r$ endowed with the scalar product

$$\langle x, y \rangle_{\frac{1}{\pi}} := \sum_{i \in E} x(i) y(i) \frac{1}{\pi(i)} .$$

Both $\ell^2(\pi)$ and $\ell^2(\frac{1}{\pi})$ are real Hilbert spaces.

Theorem 9.1.2 *The pair $(\mathbf{P}, \pi)$ is reversible if and only if $\mathbf{P}$ is self-adjoint in $\ell^2(\pi)$, that is, if and only if*

$$\langle \mathbf{P}x, y \rangle_\pi = \langle x, \mathbf{P}y \rangle_\pi \quad (x, y \in \ell^2(\pi)). \tag{9.1}$$

Proof. Suppose $(\mathbf{P}, \pi)$ reversible. Then

$$\langle \mathbf{P}x, y \rangle_\pi = \sum_{i \in E} \left\{ \left(\sum_{j \in E} p_{ij} x(j) \right) y(i) \pi(i) \right\}$$

$$= \sum_{i,j \in E} \pi(i) p_{ij} \, x(j) y(i) = \sum_{i,j \in E} \pi(j) p_{ji} \, y(i) x(j)$$

$$= \sum_{j \in E} \left\{ x(j) \left(\sum_{i \in E} p_{ji} y(i) \right) \pi(j) \right\} = \langle x, \mathbf{P}y \rangle_\pi .$$

Conversely, suppose $\mathbf{P}$ is self-adjoint in $\ell^2(\pi)$. Let δ_k be the k-th vector of the canonical basis of $\mathbb{R}^r$ (the only non-null entry, 1, is in the k-th position). Then the detailed balance equation (2.30) follows from (9.1) with the choice $x = \delta_i$, $y = \delta_j$. $\square$

Let

$$D := D(\pi) := \operatorname{diag}\{\pi(1), \ldots, \pi(r)\} \qquad (9.2)$$

and

$$\mathbf{P}^* := D^{\frac{1}{2}} \mathbf{P} D^{-\frac{1}{2}}.$$

More explicitly,

$$p^*_{ij} = p_{ij} \frac{\sqrt{\pi(i)}}{\sqrt{\pi(j)}} \quad (i, j \in E).$$

Reversibility of $(\mathbf{P}, \pi)$ is equivalent to $\mathbf{P}^*$ being a symmetric matrix.

Note that

$$x^T D y = \langle x, y \rangle_\pi . \qquad (9.3)$$

Since $\mathbf{P}^*$ is symmetric, its eigenvalues are real, it is diagonalizable and the sets of right- and left-eigenvectors are the same.

Choose an orthonormal basis of $\mathbb{R}^r$ formed of right-eigenvectors $w_1, \ldots, w_r$ associated with the eigenvalues $\lambda_1, \ldots, \lambda_r$ respectively. Define u and v by

$$w = D^{-\frac{1}{2}} u, \quad w = D^{\frac{1}{2}} v,$$

where w is a right- (and therefore left-) eigenvector of $\mathbf{P}^*$, corresponding to the eigenvalue λ. In particular,

$$u = Dv . \qquad (9.4)$$

The matrices $\mathbf{P}$ and $\mathbf{P}^*$ have the same eigenvalues, and moreover, v (resp., u) is a right-eigenvector (resp., left-eigenvector) of $\mathbf{P}$ corresponding to the eigenvalue λ.

Orthonormality with respect to the usual Euclidean scalar product of the collection $\{w_1, \ldots, w_r\}$ is equivalent to orthonormality in $\ell^2(\pi)$ of $\{v_1, \ldots, v_r\}$, that is,

$$\langle v_i, v_j \rangle_\pi = \delta_{ij} \quad (1 \le i, j \le r).$$

Similarly, $\{u_1, \ldots, u_r\}$ is an orthonormal collection in $\ell^2(\frac{1}{\pi})$:

$$\langle u_i, u_j \rangle_{\frac{1}{\pi}} = \delta_{ij} \quad (1 \le i, j \le r).$$

The eigenvectors u_1 and v_1 may always be chosen as follows

$$u_1 = \pi, \qquad v_1 = \mathbf{1} .$$

Since $\{v_1, \ldots, v_r\}$ is also a basis of $\mathbb{R}^r$, any vector $x \in \mathbb{R}^r$ can be expressed as $x = \sum_{i \in E} \alpha_i v_i$. In particular, $\langle x, v_j \rangle_\pi = \alpha_j$, and therefore

$$x = \sum_{j=1}^{r} \langle x, v_j \rangle_\pi v_j. \tag{9.5}$$

Similarly,

$$x^T = \sum_{j=1}^{r} \langle x, u_j \rangle_{\frac{1}{\pi}} u_j^T. \tag{9.6}$$

The variance of x with respect to π is

$$\text{Var}_\pi(x) = \sum_{j=2}^{r} |\langle x, v_j \rangle_\pi|^2. \tag{9.7}$$

For all n, $\mathbf{P}^n v_j = \lambda_j^n v_j$, and therefore

$$\mathbf{P}^n x = \sum_{j=1}^{r} \lambda_j^n \langle x, v_j \rangle_\pi v_j. \tag{9.8}$$

Similarly,

$$x^T \mathbf{P}^n = \sum_{j=1}^{r} \lambda_j^n \langle x, u_j \rangle_{\frac{1}{\pi}} u_j^T. \tag{9.9}$$

From (9.8), (9.3), and (9.4), we obtain $\mathbf{P}^n x = \sum_{j=1}^{r} \lambda_j^n v_j u_j^T x$, and we therefore retrieve the representation (4.13) for $A = \mathbf{P}$. For future reference, we give the following form of (9.8):

$$\mathbf{P}^n x - \langle x \rangle_\pi \mathbf{1} = \sum_{j=2}^{r} \lambda_j^n \langle x, v_j \rangle_\pi v_j. \tag{9.10}$$

9.1.2   Convergence Rates in Terms of the SLEM

Theorem 9.1.3 *Letting $\pi_{min} := \min_{k \in E} \pi(k)$, we have the bound*

$$\max_{i \in E} d_V(p_i.(n), \pi) \le \frac{\rho^n}{2\pi_{min}}.$$

Proof. From (9.10), for all $i \in E$,

$$p_{ik}(n) - \pi(k) = \sum_{j=2}^{r} \lambda_j^n v_j(i) v_j(k) \pi(k). \tag{9.11}$$

Therefore,

$$d_V(p_{i\cdot}(n), \pi) \leq \frac{1}{2} \sum_{k=1}^r \left| \sum_{j=2}^r \lambda_j^n v_j(i) v_j(k) \pi(k) \right|$$

$$\leq \frac{1}{2} \sum_{k=1}^r \max_{\ell \in E} \left(\sum_{j=2}^r \lambda_j^n |v_j(i)| |v_j(\ell)| \right) \pi(k)$$

$$= \frac{1}{2} \max_{\ell \in E} \left(\sum_{j=2}^r \lambda_j^n |v_j(i)| |v_j(\ell)| \right).$$

Therefore, denoting by ρ the SLEM of $\mathbf{P}$

$$d_V(p_{i\cdot}(n), \pi) \leq \frac{1}{2} \max_{\ell \in E} \left(\sum_{j=2}^r |v_j(i)| |v_j(\ell)| \right) \rho^n.$$

By Schwarz's inequality,

$$\sum_{j=2}^r |v_j(i)| |v_j(\ell)| \leq \left(\sum_{j=2}^r v_j(i)^2 \right)^{\frac{1}{2}} \left(\sum_{j=2}^r v_j(\ell)^2 \right)^{\frac{1}{2}}$$

$$\leq \left(\sum_{j=1}^r v_j(i)^2 \right)^{\frac{1}{2}} \left(\sum_{j=1}^r v_j(\ell)^2 \right)^{\frac{1}{2}}.$$

Now, from (9.5) with $x = \delta_i$, we have that $\delta_i = \sum_{j=1}^r v_j(i) \pi(i) v_j$. Writing this equality for the i-th coordinate gives $1 - \sum_{j=1}^r v_j(i)^2 \pi(i)$, and therefore

$$\left(\sum_{j=1}^r v_j(i)^2 \right)^{\frac{1}{2}} \left(\sum_{j=1}^r v_j(\ell)^2 \right)^{\frac{1}{2}} \leq (\pi(i) \pi(\ell))^{-\frac{1}{2}}.$$

$\square$

The constant before ρ^n is often too large. The bound may be improved if we start from a specific state i as the next result suggests.

Theorem 9.1.4 *Let $\mathbf{P}$ be a reversible irreducible transition matrix on the finite state space $E = \{1, \ldots, r\}$ with the stationary distribution π. Then for all $n \geq 1$ and all $i \in E$,*

$$d_V(\delta_i^T \mathbf{P}^n, \pi)^2 \leq \frac{p_{ii}(2)}{2\pi(i)} \rho^{2n-2}, \tag{9.12}$$

where ρ is the SLEM of $\mathbf{P}$.

Proof. From (9.10) and (9.7), we have that

$$\|\mathbf{P}^n x - \langle x \rangle_\pi \mathbf{1}\|_\pi^2 = \sum_{j=2}^r |\lambda_j|^{2n} |\langle x, v_j \rangle_\pi|^2 \le \rho^{2n} \mathrm{Var}_\pi(x). \tag{9.13}$$

Now, by reversibility and Schwarz's inequality in $\ell^2(\pi)$,

$$
\begin{aligned}
\left| \sum_{j \in E} p_{ij} x(j) \right|^2 &= \left| \sum_{j \in E} p_{ji} \frac{\pi(j)}{\pi(i)} x(j) \right|^2 \le \left(\sum_{j \in E} \frac{p_{ji}}{\pi(i)} |x(j)| \pi(j) \right)^2 \\
&\le \left(\sum_{j \in E} x(j)^2 \pi(j) \right) \left(\sum_{j \in E} \left(\frac{p_{ji}}{\pi(i)} \right)^2 \pi(j) \right) \\
&= \left(\sum_{j \in E} x(j)^2 \pi(j) \right) \left(\sum_{j \in E} (p_{ji} p_{ij}) \frac{1}{\pi(i)} \right) = \left(\sum_{j \in E} x(j)^2 \pi(j) \right) \frac{p_{ii}(2)}{\pi(i)},
\end{aligned}
$$

that is,

$$\left| \sum_{j \in E} p_{ij} x(j) \right|^2 \le \frac{p_{ii}(2)}{\pi(i)} \|x\|_\pi^2.$$

With $x = \mathbf{P}^{n-1} y - \langle y \rangle_\pi \mathbf{1}$, this gives, in view of (9.13),

$$
\begin{aligned}
\left| \sum_{j=1}^r p_{ij}(n) y(j) - \sum_{i=1}^r \pi(j) y(j) \right|^2 &\le \frac{p_{ii}(2)}{\pi(i)} \|\mathbf{P}^{n-1} y - \langle y \rangle_\pi \mathbf{1}\|_\pi^2 \\
&\le \frac{p_{ii}(2)}{\pi(i)} \mathrm{Var}_\pi(y) \rho^{2n-2}.
\end{aligned}
$$

The result then follows from the following alternative expression of the distance in variation (Exercise 4.4.1):

$$d_V(\alpha, \beta) = \frac{1}{2} \sup \left(\sum_{i=1}^r \alpha(i) y(i) - \sum_{i=1}^r \beta(i) y(i); \; \sup |y(i)| = 1 \right),$$

and the observation that if y is such that $\sup |y(i)| \le 1$, then $\mathrm{Var}_\pi(y) \le 1$. $\qquad \square$

Remark 9.1.5 One may wonder at the power of the bound (9.17) when $p_{ii}(2) = 0$! See however Exercise 9.4.1.

For the next estimate, we need a definition:

Definition 9.1.6 *The χ^2-contrast of distribution α with respect to distribution β is the quantity*

$$\chi^2(\alpha; \beta) := \sum_{i \in E} \frac{(\alpha(i) - \beta(i))^2}{\beta(i)} .$$

Note that

$$\chi^2(\alpha; \pi) = \|\alpha - \pi\|_{\frac{1}{\pi}}^2 . \tag{9.14}$$

Also

$$4 d_V(\alpha, \beta)^2 \le \chi^2(\alpha; \beta) , \tag{9.15}$$

as follows from Schwarz's inequality:

$$\left(\sum_{i \in E} |\alpha(i) - \beta(i)| \right)^2 = \left(\sum_{i \in E} \left| \frac{\alpha(i)}{\beta(i)} - 1 \right| \beta(i)^{\frac{1}{2}} \beta(i)^{\frac{1}{2}} \right)^2$$

$$\le \sum_{i \in E} \left(\frac{\alpha(i)}{\beta(i)} - 1 \right)^2 \beta(i) = \sum_{i \in E} \frac{1}{\beta(i)} (\alpha(i) - \beta(i))^2 .$$

Theorem 9.1.7 *Let $\mathbf{P}$ be a reversible irreducible transition matrix on the finite state space $E = \{1, \ldots, r\}$ with the stationary distribution π. Then for any probability distribution μ on E, and for all $n \ge 1$,*

$$\|\mu^T \mathbf{P}^n - \pi^T\|_{\frac{1}{\pi}} \le \rho^n \|\mu - \pi\|_{\frac{1}{\pi}} . \tag{9.16}$$

Also, for $n \ge 1$, all $i \in E$, and all $A \subseteq E$,

$$|\delta_i^T \mathbf{P}^n(A) - \pi^T(A)| \le \left(\frac{1 - \pi(i)}{\pi(i)} \right)^{\frac{1}{2}} \min \left(\pi(A)^{\frac{1}{2}}, \frac{1}{2} \right) \rho^n , \tag{9.17}$$

where ρ is the SLEM *of $\mathbf{P}$. In particular,*

$$4 d_V(\delta_i^T \mathbf{P}^n, \pi)^2 \le \frac{1 - \pi(i)}{\pi(i)} \rho^{2n} . \tag{9.18}$$

Proof. Recall that $u_1 = \pi$, and therefore $\langle \mu - \pi, u_1 \rangle_{\frac{1}{\pi}} = \sum_{i \in E} (\mu(i) - \pi(i)) = 0$. Therefore, by (9.9), and letting $\alpha_j := \langle \mu - \pi, u_j \rangle_{\frac{1}{\pi}}$, we obtain

$$\|(\mu - \pi)^T \mathbf{P}^n\|_{\frac{1}{\pi}}^2 = \sum_{j=2}^r \alpha_j^2 \lambda_j^{2n} \|u_j\|_{\frac{1}{\pi}}^2 = \sum_{j=2}^r \alpha_j^2 \lambda_j^{2n}$$

$$\le \rho^{2n} \sum_{j=2}^r \alpha_j^2 = \rho^{2n} \|\mu - \pi\|_{\frac{1}{\pi}}^2 ,$$

and (9.16) follows, since $\pi^T \mathbf{P}^n = \pi^T$.

Let $\mu_n^T := \delta_i^T \mathbf{P}^n$. By Schwarz's inequality,

$$
\begin{aligned}
|\mu_n(A) - \pi(A)|^2 &= \left| \sum_{\ell \in A} \left(\frac{\mu_n(\ell)}{\pi(\ell)} - 1 \right) \pi(\ell) \right|^2 \\
&\leq \left(\sum_{\ell \in A} \left(\frac{\mu_n(\ell)}{\pi(\ell)} - 1 \right)^2 \pi(\ell) \right) \pi(A) \\
&\leq \left(\sum_{\ell \in E} \left(\frac{\mu_n(\ell)}{\pi(\ell)} - 1 \right)^2 \pi(\ell) \right) \pi(A) \\
&= \| \delta_i^T \mathbf{P}^n - \pi^T \|_{\frac{1}{\pi}}^2 \pi(A) \\
&\leq \rho^{2n} \| \delta_i - \pi \|_{\frac{1}{\pi}}^2 \pi(A),
\end{aligned}
$$

where the last inequality uses (9.16). But, as simple calculations reveal,

$$
\| \delta_i - \pi \|_{\frac{1}{\pi}}^2 = \frac{1 - \pi(i)}{\pi(i)}, \tag{9.19}
$$

and therefore

$$
|\delta_i^T \mathbf{P}^n(A) - \pi^T(A)| \leq \left(\frac{1 - \pi(i)}{\pi(i)} \right)^{\frac{1}{2}} \pi(A)^{\frac{1}{2}} \rho^n. \tag{9.20}
$$

Now,

$$
|\mu_n(A) - \pi(A)|^2 \leq d_V(\mu_n, \pi)^2 \leq \frac{1}{4} \chi^2(\mu_n; \pi).
$$

But, by (9.16), (9.14), and (9.19)

$$
\chi^2(\mu_n; \pi) = \| \delta_i^T \mathbf{P}^n - \pi^T \|_{\frac{1}{\pi}}^2 \leq \rho^{2n} \| \delta_i - \pi \|_{\frac{1}{\pi}}^2 = \rho^{2n} \frac{1 - \pi(i)}{\pi(i)}.
$$

Therefore,

$$
|\delta_i^T \mathbf{P}^n(A) - \pi^T(A)| \leq \left(\frac{1 - \pi(i)}{\pi(i)} \right)^{\frac{1}{2}} \frac{1}{2} \rho^n. \tag{9.21}
$$

Combining (9.20) and (9.21) gives (9.17). Inequality (9.18) then follows since $d_V(\alpha, \beta) = \sup_{A \subset E} |\alpha(A) - \beta(A)|$. $\square$

9.1.3 Rayleigh's Spectral Theorem

The results of the previous section are useful once a bound for the SLEM is available. Coming back to the eigenvalues of $\mathbf{P}$, we know that $\lambda_1 = 1$ is one of them, with multiplicity 1. It corresponds to the unique right-eigenvector v_1 such that $\|v_1\|_\pi = 1$, namely $v_1 = \mathbf{1}$. Moreover, the eigenvalues of $\mathbf{P}$ all belong to the closed unit disk of $\mathbb{C}$, and in the reversible case of interest in this section, they are real. Therefore, with proper ordering,

$$1 = \lambda_1 > \lambda_2 \geq \cdots \geq \lambda_r \geq -1. \tag{9.22}$$

Remark 9.1.8 Note that this order is different from the one adopted in (4.14) for the statement of the Perron–Frobenius theorem. In (9.22), λ_2 is the second-largest eigenvalue (SLE), whereas in (4.14) it was the eigenvalue with the second-largest *modulus*. The strict inequality $\lambda_1 > \lambda_2$ expresses the fact that λ_1 is the unique eigenvalue equal to 1. We also know from the Perron–Frobenius theorem that the only eigenvalue of modulus 1 and not equal to 1, in this case -1, occurs if and only if the chain is periodic of period $d = 2$. In particular, in the reversible case, the period cannot exceed 2.

It will be convenient to consider the matrix $I - \mathbf{P}$, called the Laplacian of the HMC. Its eigenvalues are $\beta_i = 1 - \lambda_i$ ($1 \leq i \leq r$) and therefore

$$0 = \beta_1 < \beta_2 \leq \cdots \leq \beta_r \leq 2.$$

Clearly, a right-eigenvector of $I - \mathbf{P}$ corresponding to $\beta_i = 1 - \lambda_i$ is v_i, a right-eigenvector of $\mathbf{P}$ corresponding to λ_i.

The Dirichlet form $\mathcal{E}_\pi(x, x)$ associated with a reversible pair $(\mathbf{P}, \pi)$ is defined by

$$\mathcal{E}_\pi(x, x) := \langle (I - \mathbf{P})x, x \rangle_\pi.$$

We shall keep in mind that $\mathcal{E}_\pi(x, x)$ also depends on $\mathbf{P}$. We have

$$\mathcal{E}_\pi(x, x) = \frac{1}{2} \sum_{i,j \in E} \pi(i) p_{ij} (x(j) - x(i))^2 \tag{9.23}$$

$$= \sum_{i<j} \pi(i) p_{ij} (x(j) - x(i))^2. \tag{9.24}$$

Indeed,

$$\langle (I - \mathbf{P})x, x \rangle_\pi = \sum_{i,j \in E} \pi(i) p_{ij} x(i) (x(i) - x(j))$$

$$= \sum_{i,j \in E} \pi(j) p_{ji} x(j) (x(j) - x(i))$$

$$= \sum_{i,j \in E} \pi(i) p_{ij} x(j) (x(j) - x(i)),$$

where the second equality is obtained by an obvious change of indexation, and
the third uses the reversibility of $(\mathbf{P}, \pi)$. Expressing $\mathcal{E}_\pi(x, x)$ as the half-sum of
the second and last terms in the above chain of equalities yields (9.23). Equality
(9.25) then follows from the detailed balance equations $\pi(i)p_{ij} = \pi(j)p_{ji}$.

An analogous (and simpler) computation gives

$$\mathrm{Var}_\pi(x) = \frac{1}{2} \sum_{i,j \in E} \pi(i)\pi(j)(x(j) - x(i))^2 \,. \tag{9.25}$$

The next result gives a characterization of the *second-largest eigenvalue* (SLE)
λ_2, or equivalently of $\beta_2 = 1 - \lambda_2$.

Theorem 9.1.9 *Let* $\mathbf{P}$ *be an irreducible transition matrix on the finite state space*
$E = \{1, 2, \ldots, r\}$ *with stationary distribution* π. *If* $(\mathbf{P}, \pi)$ *is reversible,*

$$\beta_2 = \inf \left\{ \frac{\mathcal{E}_\pi(x, x)}{\mathrm{Var}_\pi(x)} \,;\, \mathrm{Var}_\pi(x) \neq 0 \right\} \,.$$

Remark 9.1.10 Condition $\mathrm{Var}_\pi(x) \neq 0$ just says that x is not, as a function, a
constant. In other words, it is not of the form $x = c\mathbf{1}$ for some $c \in \mathbb{R}$.

Proof. First observe from (9.23) that the ratio $\frac{\mathcal{E}_\pi(x,x)}{\mathrm{Var}_\pi(x)}$ is invariant by translation
since

$$\mathcal{E}_\pi(x, x) = \mathcal{E}_\pi(x - c\mathbf{1}, x - c\mathbf{1}) \text{ and } \mathrm{Var}_\pi(x - c\mathbf{1}) = \mathrm{Var}_\pi(x) \tag{9.26}$$

for any real number c, and invariant by scaling (when replacing x by cx where c
is a non-null real number). Therefore, we may restrict attention to the case where
the variance is 1 and the mean is null. From (9.8), $(I - \mathbf{P})x = \sum_{j=1}^r \beta_j \langle x, v_j \rangle_\pi v_j$,
and therefore

$$\mathcal{E}_\pi(x, x) = \sum_{j=1}^r \beta_j |\langle x, v_j \rangle_\pi|^2 \,.$$

Also from (9.7),

$$\langle x, x \rangle_\pi = \sum_{j=1}^r |\langle x, v_j \rangle_\pi|^2 = 1$$

and

$$\langle x, v_1 \rangle_\pi = \langle x, \mathbf{1} \rangle_\pi = \langle x \rangle_\pi = 0 \,.$$

Therefore

$$\mathcal{E}_\pi(x,x) = \sum_{i=2}^r \sum_{j=2}^r \beta_j |\langle x, v_j \rangle_\pi|^2$$

$$\leq \beta_2 \sum_{j=2}^r |\langle x, v_j \rangle_\pi|^2 = \beta_2.$$

The inequality becomes an equality when $x = v_2$ since $\mathcal{E}_\pi(v_2, v_2) = \beta_2$. $\qquad\square$

Remark 9.1.11 The second largest eigenvalue (SLE) is not in general the second largest eigenvalue modulus (SLEM). Both an upper bound of λ_2 (the SLE) and a lower bound of the smallest eigenvalue λ_r are needed in order to obtain a bound for the SLEM. Note that the lazy Markov chain (see Example 2.4.8) associated with a reversible Markov chain has all its eigenvalues $1 + \lambda_i$ ($1 \leq i \leq r$) non-negative (see (9.22)), and the SLEM is then equal to the SLE.

9.2 Bounds for the SLEM

9.2.1 Bounds via Rayleigh's Characterization

The next theorem gives a method based on Rayleigh's theorem to obtain an upper bound of λ_2 and a lower bound of λ_r.

Theorem 9.2.1 *(a) If $A > 0$ is such that for all $x \in \mathbb{R}^r$,*

$$\mathrm{Var}_\pi(x) \leq A\mathcal{E}_\pi(x,x), \tag{9.27}$$

then, denoting by λ_2 the SLE of $\mathbf{P}$, $\lambda_2 \leq 1 - \frac{1}{A}$.

(b) If there exists a $B > 0$ such that for all $x \in \mathbb{R}^r$,

$$\langle \mathbf{P}x, x \rangle_\pi + \|x\|_\pi^2 \geq B\|x\|_\pi^2, \tag{9.28}$$

then $\lambda_j \geq -1 + B$ ($1 \leq j \leq r$).

Proof. (a) It follows from Theorem 9.1.9 that $\beta_2 \geq 1/A$.

(b) Taking $x = v_j$ in (9.28) and using the fact that $\mathbf{P}v_j = \lambda_j v_j$ gives $\lambda_j + 1 \geq B$.
$\qquad\square$

The following is a useful consequence of Rayleigh's characterization of the second largest eigenvalue.

Theorem 9.2.2 *Consider two reversible* HMCs *on the same finite state space* $E = \{1, 2, \ldots, r\}$ *and let* $(\mathbf{P}, \pi)$ *and* $(\widetilde{\mathbf{P}}\widetilde{\pi})$ *be their respective transition matrices and stationary distributions. Suppose that there exist two positive constants A and B such that*

$$\pi(i) \le A\widetilde{\pi}(i) \text{ and } \mathcal{E}_{\widetilde{\mathbf{P}}}(x, x) \le B\mathcal{E}_{\mathbf{P}}(x, x) \quad (i \in E, x \in E^r).$$

Then, $\widetilde{\beta}_2 \le AB\beta_2$.

Proof. The quantity $\|x - c\mathbf{1}\|^2$ is minimized for $c = \langle x \rangle_\pi$ and is then equal to $\mathrm{Var}_\pi(x)$. In particular, for $c = \langle x \rangle_{\widetilde{\pi}}$,

$$\begin{aligned}
\mathrm{Var}_\pi(x) &\le \|x - \langle x \rangle_{\widetilde{\pi}}\|^2 \\
&= \sum_i \pi(i)(x(i) - \langle x \rangle_{\widetilde{\pi}})^2 \\
&\le A \sum_i \widetilde{\pi}(i)(x(i) - \langle x \rangle_{\widetilde{\pi}})^2 \\
&= A\mathrm{Var}_{\widetilde{\pi}}(x).
\end{aligned}$$

Therefore

$$\frac{1}{\mathrm{Var}_{\widetilde{\pi}}(x)} \le A\frac{1}{\mathrm{Var}_\pi(x)} \text{ and } \mathcal{E}_{\widetilde{\pi}}(x, x) \le B\mathcal{E}_\pi(x, x),$$

so that

$$\frac{\mathcal{E}_{\widetilde{\pi}}(x, x)}{\mathrm{Var}_{\widetilde{\pi}}(x)} \le AB\frac{\mathcal{E}_\pi(x, x)}{\mathrm{Var}_\pi(x)}.$$

Minimizing over the non-null x's yields the announced inequality. $\square$

Weighted Paths

The next two results give an upper bound and a lower bound in terms of the geometry of the transition graph.

In the transition graph associated with $\mathbf{P}$, denote a directed edge $i \to j$ by e, and call $e^- = i$ and $e^+ = j$ its initial vertex and end vertex respectively. Let for any such directed edge e

$$Q(e) := \pi(i)p_{ij}. \tag{9.29}$$

For each ordered pair of *distinct* states (i, j), select arbitrarily one and only one path from i to j (that is, a sequence $i, i_1, \ldots, i_m, j$ such that $p_{ii_1}p_{i_1i_2}\cdots p_{i_mj} > 0$)

which does not use the same edge twice. Let Γ be the collection of paths so selected. For a path $\gamma_{ij} \in \Gamma$, let

$$|\gamma_{ij}|_Q := \sum_{e \in \gamma_{ij}} \frac{1}{Q(e)} = \frac{1}{\pi(i)p_{ii_1}} + \frac{1}{\pi(i_1)p_{i_1 i_2}} + \cdots + \frac{1}{\pi(i_m)p_{i_m j}}.$$

Definition 9.2.3 *The quantity*

$$\kappa = \kappa(\Gamma) := \max_e \sum_{\gamma_{ij} \ni e} |\gamma_{ij}|_Q \pi(i)\pi(j)$$

is the Poincaré coefficient *of* $\mathbf{P}$.

Theorem 9.2.4 ([2]) *Let* $\mathbf{P}$ *be an irreducible transition matrix on the finite state space* E, *with stationary distribution* π, *and assume* $(\mathbf{P}, \pi)$ *to be reversible. Denoting by* λ_2 *its* SLE,

$$\lambda_2 \leq 1 - \frac{1}{\kappa}.$$

Proof. If suffices to show that (9.27) holds for $A = \kappa$. For this, write

$$\text{Var}_\pi(x) = \frac{1}{2} \sum_{i,j \in E} (x(i) - x(j))^2 \pi(i)\pi(j)$$

$$= \frac{1}{2} \sum_{i,j \in E} \left\{ \sum_{e \in \gamma_{ij}} \frac{1}{Q(e)^{\frac{1}{2}}} Q(e)^{\frac{1}{2}} (x(e^-) - x(e^+)) \right\}^2 \pi(i)\pi(j).$$

By Schwarz's inequality, this quantity is bounded above by

$$\frac{1}{2} \sum_{i,j \in E} \left\{ |\gamma_{ij}|_Q \sum_{e \in \gamma_{ij}} Q(e)(x(e^-) - x(e^+))^2 \right\} \pi(i)\pi(j)$$

$$= \frac{1}{2} \sum_e \left\{ Q(e)(x(e^-) - x(e^+))^2 \left[\sum_{\gamma_{ij} \ni e} |\gamma_{ij}|_Q \pi(i)\pi(j) \right] \right\} \leq \mathcal{E}_\pi(x,x)\kappa(\Gamma).$$

$\square$

We now proceed to obtain a lower bound. For each state i, select exactly one closed path σ_i from i to i that does not pass twice through the same edge, and with an

---

[2][Diaconis and Strook, 1991].

odd number of edges (for this to be possible, we assume that $\mathbf{P}$ is aperiodic), and let Σ be the collection of paths so selected. For a path $\sigma_i \in \Sigma$, let

$$|\sigma_i|_Q = \sum_{e \in \sigma_i} \frac{1}{Q(e)} \, .$$

Define

$$\alpha = \alpha(\Sigma) = \max_e \sum_{\sigma_i \ni e} |\sigma_i|_Q \pi(i) \, .$$

Theorem 9.2.5 ([3]) *Let $\mathbf{P}$ be an irreducible and aperiodic transition matrix on the finite state space E with stationary distribution π. If $(\mathbf{P}, \pi)$ is reversible, then*

$$\lambda_r \geq -1 + \frac{2}{\alpha} \, .$$

Proof. It suffices to prove (9.28) with $B = \frac{2}{\alpha}$. For this, we use the easily established identity

$$\frac{1}{2} \sum_{i,j \in E} (x(i) + x(j))^2 \pi(i) p_{ij} = \langle \mathbf{P}x, x \rangle_\pi + \|x\|_\pi^2 \, . \tag{$\star$}$$

If σ_i is a path from i to i with an odd number of edges, of the form $\sigma_i = (i_0 = i, i_1, i_2, \ldots, i_{2m}, i)$, then

$$
\begin{aligned}
x(i) &= \frac{1}{2}\{(x(i_0) + x(i_1)) - (x(i_1) + x(i_2)) + \cdots + (x(i_{2m}) + x(i))\} \\
&= \frac{1}{2} \sum_{e \in \sigma_i} (-1)^{n(e)} (x(e^+) + x(e^-)),
\end{aligned}
$$

where $n(e) = k$ if $e = (i_k, i_{k+1}) \in \sigma_i$. Therefore,

$$\|x\|_\pi^2 = \sum_{i \in E} \frac{\pi(i)}{4} \left\{ \sum_{e \in \sigma_i} \frac{1}{Q(e)^{\frac{1}{2}}} Q(e)^{\frac{1}{2}} (-1)^{n(e)} (x(e^+) + x(e^-)) \right\}^2 ,$$

and by Schwarz's inequality, this quantity is smaller than or equal to

$$\sum_{i \in E} \left\{ \frac{\pi(i)}{4} |\sigma_i|_Q \sum_{e \in \sigma_i} (x(e^+) + x(e^-))^2 Q(e) \right\}$$

$$= \frac{1}{4} \sum_e \left\{ (x(e^+) + x(e^-))^2 Q(e) \sum_{\sigma_i \ni e} |\sigma_i|_Q \pi(i) \right\}$$

$$\leq \frac{\alpha}{4} \sum_e (x(e^-) + x(e^+))^2 Q(e).$$

---

[3][Diaconis and Strook, 1991].

Therefore, in view of $(\star)$,

$$\|x\|_\pi^2 \le \frac{\alpha}{2}\left\{\|x\|_\pi^2 + \langle \mathbf{P}x, x\rangle_\pi\right\},$$

and this is the announced inequality. $\square$

EXAMPLE 9.2.6: RANDOM WALK ON A GRAPH. Recall that the stationary distribution π of the random walk on a finite graph $G = (V, \mathcal{E})$ is given by $\pi(i) = \frac{d_i}{2|\mathcal{E}|}$, where d_i is the degree of vertex i and $|\mathcal{E}|$ is the number of edges. We first apply the bound of Theorem 9.2.4. For any edge e, $Q(e) = \frac{1}{2|\mathcal{E}|}$. Denoting by $|\gamma|$ the length of a path γ,

$$|\gamma_{ij}|_Q = |\gamma_{ij}| \times 2|\mathcal{E}|.$$

Therefore

$$\kappa = \max_e \sum_{\gamma_{ij} \ni e} |\gamma_{ij}| \times 2|\mathcal{E}| \frac{d_i d_j}{4|\mathcal{E}|^2}$$

$$\le \max_e \sum_{\gamma_{ij} \ni e} |\gamma_{ij}| \times \frac{d_{max}^2}{2|\mathcal{E}|}$$

where d_{max} is the maximum degree of a vertex. Therefore

$$\kappa(\Gamma) \le \frac{1}{2|\mathcal{E}|} d_{max}^2 K,$$

where

$$K := \max_e |\{\gamma \in \Gamma; e \in \gamma\}| \times \max\{|\gamma|; \gamma \in \Gamma\}.$$

Finally

$$\lambda_2 \le 1 - \frac{2|\mathcal{E}|}{K d^2}.   \tag{9.30}$$

Similar calculations give for the bound in Theorem 9.2.5

$$\lambda_r \ge -1 + \frac{2}{d_{max}|\sigma|b},   \tag{9.31}$$

where $|\sigma| = \max |\sigma_i|$, and $b := \max_e |\{\sigma \in \Sigma; e \in \sigma\}|$.

The Bottleneck Bound

This bound concerns finite state space irreducible reversible transition matrices $\mathbf{P}$. It is in terms of flows on the transition graph.

For a non-empty set $B \subset E$, define the *capacity* of B,

$$\pi(B) := \sum_{i \in B} \pi(i) \, ,$$

and the *edge flow* out of B,

$$Q\left(B, \overline{B}\right) := \sum_{i \in B, j \in \overline{B}} \pi(i) \, p_{ij} \, .$$

Note that $Q\left(B, \overline{B}\right) = Q\left(\overline{B}, B\right)$ and that $0 \leq Q\left(B, \overline{B}\right) \leq \pi(B) \leq 1$. For non-empty B, define the *bottleneck ratio* of B:

$$\Phi(B) := \frac{Q\left(B, \overline{B}\right)}{\pi(B)} \, .$$

Definition 9.2.7 *The bottleneck ratio of the pair* $(\mathbf{P}, \pi)$ *is*

$$\Phi^* := \inf\left(\Phi(B); 0 < |B| < |E|, \pi(B) \leq \frac{1}{2}\right) . \tag{9.32}$$

EXAMPLE 9.2.8: BOTTLENECK OF THE PURE RANDOM WALK. For the pure random walk on $G = (V, \mathcal{E})$,

$$\pi(i) p_{ij} = \frac{d_i}{2|\mathcal{E}|} \quad d_i = \frac{1}{2|\mathcal{E}|}$$

if $\langle i, j \rangle$ is an edge, $= 0$ otherwise. In this case, defining the internal boundary ∂B to be the set of states $i \in B$ that are connected to an element of $\overline{B}$ by an edge,

$$\Phi(B) = \frac{|\partial B|}{\sum_{i \in B} d_i} \, .$$

Theorem 9.2.9 Cheeger's inequality:

$$1 - 2\Phi^* \leq \lambda_2 \leq 1 - \frac{1}{2}(\Phi^*)^2 \, .$$

Proof. ([4])

(a) Apply Rayleigh's spectral theorem,

$$1 - \lambda_2 \leq \frac{\mathcal{E}_\pi(x, x)}{\|x\|_\pi^2}$$

for any non-trivial vector x such that $\langle x \rangle_\pi = 0$. Select $B \subset E$ such that $\pi(B) \leq \frac{1}{2}$, and define

$$x(i) = \begin{cases} 1 - \pi(B) & \text{if } i \in B, \\ -\pi(B) & \text{if } i \notin B. \end{cases}$$

Then $\langle x \rangle_\pi = 0$ and $\|x\|_\pi^2 = \pi(B)(1 - \pi(B))$. Also,

$$\begin{aligned}
\mathcal{E}_\pi(x, x) &= \frac{1}{2} \sum_{i,j} (x(i) - x(j))^2 \pi(i) p_{ij} \\
&= \frac{1}{2} \sum_{i \in B} (\cdots) \sum_{j \notin B} (\cdots) + \frac{1}{2} \sum_{i \notin B} (\cdots) \sum_{j \in B} (\cdots) \\
&= \frac{1}{2} Q(\overline{B}, B) + \frac{1}{2} Q(B, \overline{B}) = Q(\overline{B}, B).
\end{aligned}$$

Therefore,

$$1 - \lambda_2 \leq \frac{Q(\overline{B}, B)}{\pi(B)(1 - \pi(B))} \leq 2 \frac{Q(\overline{B}, B)}{\pi(B)}.$$

This being true for all B such that $\pi(B) \leq \frac{1}{2}$, we have, by definition of Φ^*,

$$1 - \lambda_2 < 2 \, \Phi^*.$$

(b) Let u be a left-eigenvector of $\mathbf{P}$ associated with an eigenvalue $\lambda \neq 1$. In particular, u is orthogonal to π, the left-eigenvector associated with the eigenvalue $\lambda_1 = 1$, and therefore u has positive as well as negative entries. The same is true for x defined by

$$x(i) = \frac{u(i)}{\pi(i)}.$$

Assume without loss of generality that for some k $(1 \leq k \leq r)$

$$x(1) \geq \cdots \geq x(k) > 0 \geq x(k+1) \geq \cdots \geq x(r),$$

and that $\pi(B) \leq \frac{1}{2}$ for $B := \{1, \ldots, k\}$ (if necessary, change the order of the states, and for the last assumption, change u into $-u$). Let

$$y(i) := \frac{u(i)}{\pi(i)} \mathbb{1}_{\{u(i) > 0\}}.$$

---

[4][Jerrum and Sinclair, 1989].

We have $u^T (I - \mathbf{P}) = u^T (1 - \lambda)$, and therefore

$$u^T (I - \mathbf{P}) y = (1 - \lambda) u^T y = (1 - \lambda) \sum_{i \in B} \pi (i) y (i)^2. \tag{9.33}$$

Also,

$$
\begin{aligned}
u^T (I - \mathbf{P}) y &= \sum_{i \in B} \sum_{j=1}^{r} (\delta_{ji} - p_{ji}) u (j) y (i) \\
&\geq \sum_{i \in B} \sum_{j \in B} (\delta_{ji} - p_{ji}) u (j) y (i) ,
\end{aligned}
$$

since the missing terms $-p_{ji} u (j) y (i)$ corresponding to $i \in B$ and $j \notin B$ are positive or null. Therefore,

$$u^T (I - \mathbf{P}) y \geq \langle y, (I - \mathbf{P}) y \rangle_{\pi} ,$$

and by (9.33), (9.23) and reversibility (Theorem 9.1.2),

$$1 - \lambda \geq \dfrac{\displaystyle\sum_{i<j} \pi (i) p_{ij} (y (i) - y (j))^2}{\displaystyle\sum_{i \in B} \pi (i) y (i)^2}.$$

From $(a + b)^2 \leq 2 (a^2 + b^2)$, we obtain

$$\sum_{i<j} \pi (i) p_{ij} (y (i) + y (j))^2 \leq 2 \sum_{i<j} \pi (i) p_{ij} \left(y (i)^2 + y (j)^2 \right),$$

and, by reversibility,

$$
\begin{aligned}
\sum_{i<j} \pi (i) p_{ij} \left(y (i)^2 + y (j)^2 \right) &= \sum_{i<j} \pi (i) p_{ij} y (i)^2 + \sum_{i<j} \pi (j) p_{ji} y (j)^2 \\
&= \sum_{i \neq j} \pi (i) p_{ij} y (i)^2 \leq \sum_{i \in B} \pi (i) y (i)^2 .
\end{aligned}
$$

Therefore

$$1 - \lambda \geq \dfrac{\displaystyle\sum_{i<j} \pi (i) p_{ij} (y (i) - y (j))^2}{\displaystyle\sum_{i \in B} \pi (i) y (i)^2} \dfrac{\displaystyle\sum_{i<j} \pi (i) p_{ij} (y (i) + y (j))^2}{2 \displaystyle\sum_{i \in B} \pi (i) y (i)^2}.$$

By Schwarz's inequality and the identity $a^2 - b^2 = (a-b)(a+b)$,

$$\left(\sum_{i<j} \pi(i) \, p_{ij} \left(y(i)^2 - y(j)^2 \right) \right)^2$$

$$\leq \left(\sum_{i<j} \pi(i) \, p_{ij} \left(y(i) - y(j) \right)^2 \right) \left(\sum_{i<j} \pi(i) \, p_{ij} \left(y(i) + y(j) \right)^2 \right),$$

and therefore

$$1 - \lambda \geq \frac{1}{2} \left(\frac{\displaystyle\sum_{i<j} \pi(i) \, p_{ij} \left(y(i)^2 - y(j)^2 \right)}{\displaystyle\sum_{i \in B} \pi(i) \, y(i)^2} \right)^2 . \tag{$\dagger$}$$

Define $B_\ell = \{1, \ldots, \ell\}$. We have

$$\sum_{i<j} \pi(i) \, p_{ij} \left(y(i)^2 - y(j)^2 \right) = \sum_{i<j} \pi(i) \, p_{ij} \left(\sum_{i \leq l < j} \left(y(\ell)^2 - y(\ell+1)^2 \right) \right)$$

$$= \sum_{\ell=1}^{k} \left(y(\ell)^2 - y(\ell+1)^2 \right) \sum_{i \in B_\ell, j \notin B_\ell} \pi(i) \, p_{ij}$$

$$= \sum_{\ell=1}^{k} \left(y(\ell)^2 - y(\ell+1)^2 \right) F(B_\ell).$$

Since for $1 \leq \ell < k$, $\pi(B_\ell) \leq \pi(B) \leq \dfrac{1}{2}$, we have $F(B_\ell) \geq \Phi^* \, \pi(B_\ell)$. Therefore,

$$\sum_{i<j} \pi(i) \, p_{ij} \left(y(i)^2 - y(j)^2 \right) \geq \Phi^* \sum_{\ell=1}^{k} \left(y(\ell)^2 - y(\ell+1)^2 \right) \pi(B_\ell)$$

$$= \Phi^* \sum_{\ell=1}^{k} \left\{ \left(y(\ell)^2 - y(\ell+1)^2 \right) \sum_{i=1}^{\ell} \pi(i) \right\}$$

$$= \Phi^* \sum_{i=1}^{k} \left\{ \pi(i) \left(\sum_{\ell=i}^{k} \left(y(\ell)^2 - y(\ell+1)^2 \right) \right) \right\}$$

$$= \Phi^* \sum_{i \in B} \pi(i) \, y(i)^2 .$$

Therefore, from ($\dagger$)

$$1 - \lambda \geq \frac{(\Phi^*)^2}{2} .$$

$\square$

EXAMPLE 9.2.10: THE CYCLIC GRAPH. The vertices are n points uniformly distributed on the unit circle, and the n edges are those linking the neighboring vertices. Take n odd. For any B, $Q(B, \overline{B}) = \frac{1}{n}$ and one may easily check that Φ^* is achieved by any set B of $\frac{n-1}{2}$ consecutive vertices, and then $\Phi^* = \frac{2}{n-1}$. Therefore

$$\lambda_2 \leq 1 - \frac{2}{(n-1)^2} \, .$$

It can be verified that the bound in Example 9.2.6 gives in this special case

$$\lambda_2 \leq 1 - \frac{8n}{(n-1)^2(n+1)}$$

and is therefore of the same order but with a better constant. It turns out that in this case the exact eigenvalues are available,[5] namely $\cos(2\pi \frac{j}{n})$ $(0 \leq j \leq n - 1)$, and therefore $\lambda_2 = 1 - \frac{2\pi^2}{n^2} + O(\frac{1}{n^4})$. The Poincaré bound is therefore comparable, up to a factor π^2, to the actual spectral gap.

9.2.2   Strong Stationary Times

Strong stationary times, by which exact sampling for the stationary distribution of a positive recurrent HMC can be achieved, will be defined right after the next examples.

EXAMPLE 9.2.11: TOP TO RANDOM CARD SHUFFLING, TAKE 1. ([6]) The title refers to a method of shuffling a deck of N cards whereby the top card of the deck is removed and placed at random in the deck, and the procedure is repeated *ad infinitum*.

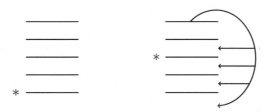

Figure 9.2.1. Top to random shuffling

[5][Diaconis, 1988].
[6][Aldous and Diaconis, 1987].

This defines an irreducible HMC $\{X_n\}_{n\geq 0}$, where a state is a permutation of the deck. In other words, it is a random walk on the group $\mathcal{S}_N$ of permutations on the set of N cards. Its stationary distribution is the uniform distribution (Example 2.4.20). (Alternatively, use symmetry, and to make symmetry more apparent, arrange the cards in a circle rather than in a deck.) Denote by $\star$ the card originally at the bottom. If there are j cards below $\star$ at time n, the $j!$ possible arrangements of these cards are equally likely, as the following inductive argument shows. The statement is true for $n = 0$. Suppose it is true for some $n \geq 0$, then it is true for $n + 1$. In fact two events can take place at time $n + 1$. Either the top card is placed above $\star$ in which case the claim is trivially true, or it is placed under $\star$ and it is also true since inserting at a random position an element in a random permutation of j elements results in a random permutation of $j + 1$ elements.

Let τ_j be the jth time a card is inserted below $\star$. If there are N cards, then, at time τ_{N-1}, card $\star$ has reached the top. Let $\tau = \tau_{N-1} + 1$. Since for $j \leq N - 1$, at time τ_j all the $j!$ arrangements of the j cards below $\star$ are equally likely, the distribution of X_τ is uniform. Note that τ is a X_0^n-stopping time.

EXAMPLE 9.2.12: LAZY WALK ON THE HYPERCUBE, TAKE 2. In Example 2.4.19, the lazy random walk on the N-dimensional hypercube $E = \{0, 1\}^N$ was described distributionwise by the recurrence equation $X_{n+1} = f(X_n, Z_{n+1})$ where $\{Z_n\}_{n\geq 1}$ is an IID sequence of random variables uniformly distributed on $\{1, \ldots, N\}$ independent of the initial state X_0. More precisely, $Z_n = (U_n, B_n)$ where the sequence $\{(U_n, B_n)\}_{n\geq 1}$ is IID and uniformly distributed on $\{1, 2, \ldots, N\} \times \{0, 1\}$. The position at time $n + 1$ is that of X_n except that the bit in position U_{n+1} is replaced by B_{n+1}.

Define a random time τ to be the first time for which the set $\{U_1, U_2, \ldots, U_n\}$ contains all the elements of $\{1, 2, \ldots, N\}$. Because at this time all the coordinates have been replaced by independent fair bits, the distribution at time τ is the uniform distribution, that is, the stationary distribution.

This time, however, τ is not a X_0^n-stopping time. It is a randomized X_0^n-stopping time in the sense of the next definition.

Definition 9.2.13 *Let $\{X_n\}_{n\geq 0}$ be an HMC with the representation as in Theorem 2.2.1. A random time τ with values in $\overline{\mathbb{N}}$ is called a* randomized X_0^n-*stopping time if, for all $k \in \mathbb{N}$, the event $\{\tau = k\}$ is expressible in terms of $X_0, Z_1, \ldots, Z_k$.*

The times τ of Examples 9.2.11 and 9.2.12 are randomized stopping times

(Exercise 9.4.14).

If τ is a randomized X_0^n-stopping, for all $m, n \geq 0$ and for all $i, j \in E$,

$$P\left(X_{m+n} = j \mid X_n = i, \tau \leq n\right) = p_{ij}\left(m\right).$$

Indeed, $\{\tau \leq n\}$ is expressible in terms of $X_0, Z_1, \ldots, Z_n$, and is therefore independent of X_{m+n} given $X_n = j$. Similar formulas, formally identical to the case where τ is a usual, non-randomized, X_0^n-stopping time, hold true and will be used in the calculations below.

Definition 9.2.14 ([7]) *A randomized X_0^n-stopping time τ with respect to the* HMC *$\{X_n\}_{n \geq 0}$ admitting a unique stationary distribution π is called a strong stationary time of this* HMC *iff it is almost surely finite and*

(α) X_τ is distributed according to π and is independent of τ.

If the requirement of independence of X_τ and τ is dropped, τ is simply called a stationary time. The times τ of Examples 9.2.11 and 9.2.12 are strong stationary times (Exercise 9.4.14).

In the above definition, condition (α) is equivalent to either one of the following two conditions:

(β) For all $i \in E$ and all $n \geq 0$,

$$P\left(X_n = i \mid \tau = n\right) = \pi\left(i\right).$$

(γ) For all $i \in E$ and all $n \geq 0$,

$$P\left(X_n = i \mid \tau \leq n\right) = \pi\left(i\right).$$

The reader is invited to provide the proof (Exercise 9.4.16).

Also, if either (α), or (β), or (γ), holds, then $\{X_{\tau+n}\}_{n \geq 0}$ is a stationary HMC with the transition matrix $\mathbf{P}$ and is independent of τ. To check this, just write

$$
\begin{aligned}
P(\tau = k, &X_\tau = i_0, X_{\tau+1} = i_1, \ldots, X_{\tau+n} = i_n) \\
&= P(\tau = k, X_k = i_0, X_{k+1} = i_1, \ldots, X_{k+n} = i_n) \\
&= P(\tau = k, X_k = i_0)P(X_{k+1} = i_1, \ldots, X_{k+n} = i_n \mid \tau = k, X_k = i_0) \\
&= P(\tau = k)\pi(i_0)P(X_{k+1} = i_1, \ldots, X_{k+n} = i_n \mid X_k = i_0) \\
&= P(\tau = k)P_\pi(X_k = i_0, X_{k+1} = i_1, \ldots, X_{k+n} = i_n) \\
&= P(\tau = k)P_\pi(X_0 = i_0, X_1 = i_1, \ldots, X_n = i_n).
\end{aligned}
$$

The announced result then follows.

---

[7][Fill, 1991], [Aldous and Diaconis, 1987], [Diaconis and Fill, 1991].

Convergence Rates via Strong Stationary Times

EXAMPLE 9.2.15: LAZY WALK ON THE CIRCLE, TAKE 2. [8] Let $\{X_n\}_{n\geq 0}$ be a symmetric random walk on $E = \mathbb{Z}_d$, the integers modulo d, identified with d points on the circle (Figure 9.2.2). It moves one step in either direction or remains still, each motion with probability $\frac{1}{3}$.

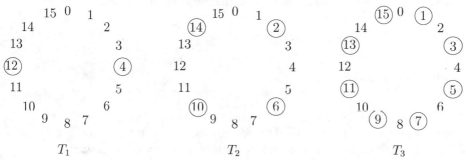

$$T_1 \qquad\qquad T_2 \qquad\qquad T_3$$

Figure 9.2.2. Strong stationary time for the symmetric walk on the circle

This chain is clearly ergodic with the uniform probability on E. A strong stationary time can be constructed as follows in the case $d = 2^a$. We treat the case $d = 2^4 = 16$ for definiteness.

Starting from 0, let T_1 be the first time either state 4 or 12 is visited. Clearly, X_{T_1} is uniformly distributed on $\{4, 12\}$ and is independent of T_1. Next, let T_2 be the first time after T_1 when the chain visits the states at distance 2 from X_{T_1}. Then X_{T_2} is uniformly distributed on $\{2, 6, 10, 14\}$ and is independent of T_2. Time T_3 is now the first time after T_2 when the chain hits a state at distance 1 from X_{T_2}. Then X_{T_3} is uniformly distributed on the odd numbers $\{1, 3, 5, 7, 9, 11, 13, 15\}$ and is independent of T_3. Finally, let T be the first time after T_3 where the chain makes a clockwise move or stays still. We can take T as the desired strong stationary time, since it is independent of X_T, and X_T is uniform on E.

For $d = 2^a$, the distances successively traveled are $\pm 2^{a-2}, \pm 2^{a-3}, \ldots, \pm 1 = \pm 2^{a-a}$. The mean time to travel at distance b of this symmetric walk is $\frac{3}{2}b^2$. The last step from T_{a-1} to $T_a = T$ takes $\frac{3}{2}$ time units on average. Therefore

$$\mathrm{E}_0[T_a] = \frac{3}{2}(2^{2a-4} + \ldots + 4 + 1) = \frac{3}{2}2^{2a}(2^{-4} + 2^{-6} + \ldots + 2^{-2(a-1)} + 2^{-2a}).$$

---

[8][Diaconis and Fill, 1991].

Therefore, for $a \geq 2$,

$$\mathrm{E}_0[T_a] \leq \frac{3}{16} 2^{2a} = \frac{3}{16} d^2.$$

By Markov's inequality,

$$\mathrm{P}_0(T_a > n) \leq \frac{\mathrm{E}_0[T_a]}{n} \leq \frac{3}{16} \frac{d^2}{n},$$

and therefore, since the result would be the same for any state,

$$d_V(\mu^T \mathbf{P}^n, \pi^T) \leq \frac{3}{16} \frac{d^2}{n}.$$

In this case, $t_{mix}(\varepsilon) \leq \frac{3}{16} \frac{d^2}{\varepsilon}$.

The tail of the distribution of a strong stationary time gives a bound for the rate of convergence in variation of an ergodic HMC. This is the content of Theorem 9.2.18 below. For this, it is convenient to use the notion of separation.

Definition 9.2.16 *Let α and β be two probability distributions on the denumerable space E. The separation of α from β is the quantity*

$$s(\alpha; \beta) := \max_{i \in E} \left(1 - \frac{\alpha(i)}{\beta(i)} \right).$$

Note that $0 \leq s(\alpha; \beta) \leq 1$. (For the lower bound, observe that one cannot have $\alpha(i) > \beta(i)$ for all i.)

Theorem 9.2.17 *Let α and β be two probability distributions on the denumerable space E. Then*

$$d_V(\alpha; \beta) \leq s(\alpha; \beta).$$

Proof. Recall that $d_V(\alpha; \beta) = \sum_{i\,;\,\beta(i)>\alpha(i)} (\beta(i) - \alpha(i))$. But the latter sum equals

$$\sum_{i\,;\,\beta(i)>\alpha(i)} \beta(i) \left(1 - \frac{\alpha(i)}{\beta(i)} \right) \leq \left(\sum_{i\,;\,\beta(i)>\alpha(i)} \beta(i) \right) s(\alpha; \beta) \leq s(\alpha; \beta).$$

$\square$

Theorem 9.2.18 *Let* $\mathbf{P}$ *be transition matrix of an irreducible positive recurrent* HMC $\{X_n\}_{n\geq 0}$ *with the stationary distribution* π. *If* τ *is a strong stationary time of the chain with initial distribution* μ, *then*

$$s\left(\mu^T \mathbf{P}^n; \pi^T\right) \leq P(\tau > n).$$

Proof. By (γ) after Definition 9.2.14,

$$P\left(X_n = i\right) \geq P\left(X_n = i, \tau \leq n\right) = \left(1 - P\left(\tau > n\right)\right) \pi\left(i\right).$$

Therefore, for all i,

$$P\left(\tau > n\right) \geq 1 - \frac{P\left(X_n = i\right)}{\pi\left(i\right)}.$$

$\square$

Theorem 9.2.19 *Let* τ *be a strong stationary time of the* HMC $\{X_n\}_{n\geq 0}$. *Then* τ *is also a stationary coupling time of the same chain.*

Proof. For each $m \geq 0$, define on $\{\tau = m\}$ the process $\left\{Y_n^{(m)}\right\}_{n\geq m}$ by

$$Y_n^{(m)} = X_n \text{ if } n \geq m.$$

Since for $n \geq m$, by definition of a strong stationary time, $P\left(X_n = i, \tau = m\right) = \pi\left(i\right) P\left(\tau = m\right)$, we see that, conditionally on $\{\tau = m\}$, $\left\{Y_n^{(m)}\right\}_{n\geq m}$ is a stationary HMC. It can be extended to a stationary HMC $\left\{Y_n^{(m)}\right\}_{n\geq 0}$. Letting $Y_n := Y_n^{(m)}$ on $\{\tau = m\}$, we obtain an HMC $\{Y_n\}_{n\geq 0}$ that is stationary and such that $X_n = Y_n$ for $n \geq \tau$. $\square$

9.2.3 Reversibilization

Suppose $\mathbf{P}$ is an ergodic transition matrix on the finite state space $E = \{1, 2, \ldots, r\}$, with stationary distribution π. This time, $(\mathbf{P}, \pi)$ is not assumed reversible. What can be done to catch up with the results obtained above for the reversible case?

Consider the transition matrix $\tilde{\mathbf{P}}$ of the time-reversed chain, defined by

$$\tilde{p}_{ij} := \frac{\pi(j)p_{ji}}{\pi(i)},$$

or, in compact form, with $D = D(\pi)$ as in (9.2),

$$\tilde{\mathbf{P}} := D^{-1}\mathbf{P}^T D.$$

The matrix $M = M(\mathbf{P}) := \mathbf{P}\tilde{\mathbf{P}}$, that is,

$$M = \mathbf{P}D^{-1}\mathbf{P}^T D, \tag{9.34}$$

is reversible with respect to its stationary distribution π. To prove this assertion, we have to verify that $M^* := D^{\frac{1}{2}}MD^{-\frac{1}{2}}$ is symmetric, that is,

$$D^{\frac{1}{2}}MD^{-\frac{1}{2}} = \left(D^{\frac{1}{2}}MD^{-\frac{1}{2}}\right)^T.$$

(The right-hand side is

$$D^{-\frac{1}{2}}M^T D^{\frac{1}{2}} = D^{-\frac{1}{2}}D\mathbf{P}D^{-1}\mathbf{P}^T D^{\frac{1}{2}} = D^{\frac{1}{2}}\mathbf{P}D^{-1}\mathbf{P}^T D^{\frac{1}{2}}$$

whereas the left-hand side is $D^{\frac{1}{2}}\mathbf{P}D^{-1}\mathbf{P}^T DD^{-\frac{1}{2}} = D^{\frac{1}{2}}\mathbf{P}D^{-1}\mathbf{P}^T D^{\frac{1}{2}}$.)

The eigenvalues of M are real and all belong to $[-1, +1]$. In fact, they all belong to the interval $[0, 1]$. To see this, observe that M has the same eigenvalues as $D^{\frac{1}{2}}MD^{-\frac{1}{2}}$ and that the latter matrix is $(D^{\frac{1}{2}}\mathbf{P}D^{-\frac{1}{2}})(D^{\frac{1}{2}}\mathbf{P}D^{-\frac{1}{2}})^T$, a symmetric definite non-negative matrix. In particular, the SLE is equal to the SLEM.

The matrix M given by (9.34) is the *multiplicative reversibilization* of $\mathbf{P}$. See Exercise 9.4.13 for another type of reversibilization.

Theorem 9.2.20 ([9]) *Let $\gamma_1 = \gamma_1(M)$ be the second-largest eigenvalue of $M = \mathbf{P}\tilde{\mathbf{P}}$, where $\mathbf{P}$ is an ergodic transition matrix on the finite state space E. Then for any probability distribution ν on E,*

$$|\nu^T\mathbf{P}^n - \pi^T|^2 \le \gamma_1(M)^n\chi^2(\nu; \pi). \tag{9.35}$$

Inequality (9.35) is called the χ^2-contrast bound.

Proof. The following identity ([10]) will be needed:

$$\text{Var}_\pi(x) = \text{Var}_\pi(\tilde{\mathbf{P}}x) + \langle(1 - M)x, x\rangle_\pi. \tag{9.36}$$

It is proven as follows. First, from (9.26), if we let $\hat{x} = x - \langle x\rangle_\pi \mathbf{1}$, then

$$\begin{aligned}
\langle(I - M)x, x\rangle_\pi &= \langle(I - M)\hat{x}, \hat{x}\rangle_\pi = \|\hat{x}\|_\pi^2 - \langle M\hat{x}, \hat{x}\rangle_\pi \\
&= \|\hat{x}\|_\pi^2 - \langle\mathbf{P}\tilde{\mathbf{P}}\hat{x}, \hat{x}\rangle_\pi = \|\hat{x}\|_\pi^2 - \|\tilde{\mathbf{P}}\hat{x}\|_\pi^2,
\end{aligned}$$

---

[9][Fill, 1991].
[10][Mihaïl, 1989].

where the fact that $\tilde{\mathbf{P}}$ is the adjoint of $\mathbf{P}$ in $\ell^2(\pi)$ was taken into account. The identity (9.36) follows since $\|\hat{x}\|_\pi^2 = \text{Var}_\pi(x)$ and $\|\tilde{\mathbf{P}}\hat{x}\|_\pi^2 = \text{Var}_\pi(\tilde{\mathbf{P}}x)$.

Now let $\chi_n^2 := \chi^2(\nu^T \mathbf{P}^n; \pi)$ and $\rho_n(i) := \frac{(\nu^T \mathbf{P}^n)(i)}{\pi(i)}$. One verifies by inspection that $\text{Var}_\pi(\rho_n) = \chi_n^2$ and $\tilde{\mathbf{P}}\rho_n = \rho_{n+1}$. Therefore, from (9.36),

$$\chi_n^2 = \chi_{n+1}^2 + \langle (1 - M)\rho_n, \rho_n \rangle_\pi .$$

By Rayleigh's spectral theorem,

$$\langle (1 - M)\rho_n, \rho_n \rangle_\pi \geq (1 - \gamma_1(M)) \text{Var}_\pi(\rho_n) = (1 - \gamma_1(M))\chi_n^2 ,$$

and therefore $\chi_{n+1}^2 \leq \gamma_1 \chi_n^2$, from which it follows that $\chi_n^2 \leq \gamma_1^n \chi_0^2$. But by (9.15), $d_V(\nu, \pi)^2 \leq \chi^2(\nu; \pi)$, and this finishes the proof. $\qquad\square$

9.3 Mixing Times

9.3.1 Basic Definitions

For a positive recurrent HMC with transition matrix $\mathbf{P}$ and stationary distribution π, define for all $n \geq 0$

$$d(n) := \max_{i \in E} d_V(\delta_i \mathbf{P}^n, \pi), \quad \bar{d}(n) := \max_{i,j \in E} d_V(\delta_i \mathbf{P}^n, \delta_j \mathbf{P}^n) . \qquad (9.37)$$

These quantities are equivalent in the following sense:

Theorem 9.3.1

$$d(n) \leq \bar{d}(n) \leq 2d(n) .$$

Proof. The rightmost inequality follows from the triangle inequality

$$d_V\left(\delta_i^T \mathbf{P}^n, \delta_j^T \mathbf{P}^n\right) \leq d_V\left(\delta_i^T \mathbf{P}^n, \pi\right) + d_V\left(\delta_j^T \mathbf{P}^n, \pi\right) .$$

Now, for all $k \geq 0$, $\pi(k) = \sum_j \pi(j)p_{jk}(n)$, and therefore

$$
\begin{aligned}
\pi(A) &= \sum_{k \in A} \sum_j \pi(j)p_{jk}(n) = \sum_j \pi(j) \left(\sum_{k \in A} p_{jk}(n) \right) \\
&= \sum_j \pi(j) \mathbf{P}_j(X_n \in A) = \sum_j \pi(j)\delta_j^T \mathbf{P}^n(A) .
\end{aligned}
$$

Therefore

$$d_V\left(\delta_i^T \mathbf{P}^n, \pi\right) = \sup_{A \subseteq E}\left(\delta_i^T \mathbf{P}^n(A) - \pi(A)\right)$$

$$= \sup_{A \subseteq E}\left|\sum_j \pi(j)\left(\delta_i^T \mathbf{P}^n(A) - \delta_j^T \mathbf{P}^n(A)\right)\right|$$

$$\le \sup_{A \subseteq E}\sum_j \pi(j)\left|\delta_i^T \mathbf{P}^n(A) - \delta_j^T \mathbf{P}^n(A)\right|$$

$$= \sum_j \pi(j)\sup_{A \subseteq E}\left|\delta_i^T \mathbf{P}^n(A) - \delta_j^T \mathbf{P}^n(A)\right|$$

$$= \sum_j \pi(j)d_V\left(\delta_i^T \mathbf{P}^n, \delta_j^T \mathbf{P}^n\right) \le d_V\left(\delta_k^T \mathbf{P}^n, \delta_j^T \mathbf{P}^n\right),$$

for any $k \in E$. Hence the leftmost inequality. $\square$

Define the following mixing times. For $\varepsilon > 0$,

$$t_{mix}(\varepsilon) := \inf\{n \ge 0\,;\, d(n) \le \varepsilon\}, \tag{9.38}$$

and

$$t_{mix} := t_{mix}(1/4). \tag{9.39}$$

Lemma 9.3.2 *The function $\overline{d}$ is sub-multiplicative, that is, for all integers m, n:*

$$\overline{d}(n + m) \le \overline{d}(n) \times \overline{d}(m).$$

Proof. Let Y and Z be two random variables with respective distributions $\delta_i^T \mathbf{P}^n$ and $\delta_j^T \mathbf{P}^n$, and realizing maximal coupling for these distributions, that is,

$$d_V(\delta_i^T \mathbf{P}^n, \delta_j^T \mathbf{P}^n) = \mathrm{P}(Y \ne Z).$$

Observe that

$$p_{i,k}(n + m) = \sum_\ell p_{i,\ell}(n)p_{\ell,k}(m) = \sum_\ell \mathrm{P}(Y = \ell)p_{\ell,k}(m) = \mathrm{E}\left[p_{Y,k}(m)\right]$$

and similarly, $p_{j,k}(n + m) = \mathrm{E}\left[p_{Z,k}(m)\right]$. Therefore,

$$p_{i,k}(n + m) - p_{j,k}(n + m) = \mathrm{E}\left[p_{Y,k}(m) - p_{Z,k}(m)\right]$$

and

$$d_V(\delta_i^T \mathbf{P}^{n+m}, \delta_j^T \mathbf{P}^{n+m})$$

$$= \frac{1}{2} \sum_k |p_{i,k}(n+m) - p_{j,k}(n+m)| = \frac{1}{2} \sum_k |\mathrm{E}\left[p_{Y,k}(m) - p_{Z,k}(m)\right]|$$

$$\leq \mathrm{E}\left[\frac{1}{2} \sum_k |p_{Y,k}(m) - p_{Z,k}(m)|\right] = \mathrm{E}\left[d_V\left(p_{Y,\cdot}(m), p_{Z,\cdot}(m)\right)\right].$$

The quantity under expectation is null if $Y = Z$ and is in any case bounded by $\bar{d}(n)$. Therefore

$$d_V(\delta_i^T \mathbf{P}^{n+m}, \delta_j^T \mathbf{P}^{n+m}) \leq \mathrm{E}\left[\bar{d}(n) 1_{Y \neq Z}\right]$$

$$= \bar{d}(n) \mathrm{P}(Y \neq Z) = \bar{d}(n) d_V(\delta_i^T \mathbf{P}^n, \delta_j^T \mathbf{P}^n).$$

Maximizing over i, j yields the announced result. $\square$

When N is an integer, by Lemma 9.3.2,

$$d(N t_{mix}(\varepsilon)) \leq \bar{d}(N t_{mix}(\varepsilon)) \leq \bar{d}(t_{mix}(\varepsilon))^N \leq (2\varepsilon)^N.$$

In particular, with $\varepsilon = \frac{1}{4}$,

$$d(N t_{mix}) \leq 2^{-N}.$$

With $N = N(\varepsilon) := \lceil \log_2 \varepsilon^{-1} \rceil$, $2^{-N} \leq \varepsilon$, and therefore, $d(N t_{mix}) \leq \varepsilon$, which implies that

$$t_{mix}(\varepsilon) \leq \lceil \log_2 \varepsilon^{-1} \rceil t_{mix}. \tag{9.40}$$

9.3.2 Upper Bounds via Coupling

We now show how to compute mixing times via coupling. Recall that two random sequences $\{X_n\}_{n \geq 0}$ and $\{Y_n\}_{n \geq 0}$ with values in the same set E are said to couple at time τ if $n \geq \tau$ implies that $X_n = Y_n$. By Theorem 4.1.7, $d_V(X_n, Y_n) \leq \mathrm{P}(\tau \geq n)$. Applying this inequality to the coupling time of two HMCs with the same transition matrix $\mathbf{P}$ with initial states i and j respectively, we have that

$$d_V(\delta_i \mathbf{P}^n, \delta_j \mathbf{P}^n) \leq \mathrm{P}_{i,j}(\tau \geq n).$$

By Markov's inequality,

$$\mathrm{P}_{i,j}(\tau \geq n) \leq \frac{\mathrm{E}_{i,j}[\tau]}{n}.$$

Therefore

Theorem 9.3.3

$$d(n) \leq \max_{i,j \in E} P_{i,j}(\tau \geq n) \leq \max_{i,j \in E} \frac{E_{i,j}[\tau]}{n}.$$

EXAMPLE 9.3.4: LAZY WALK ON THE CIRCLE, TAKE 1. This is by definition a lazy random walk on the graph consisting of N points regularly placed on a circle with an edge between each pair of adjacent vertices. The stationary distribution is the uniform distribution. We construct a Markovian coupling $\{X_n, Y_n\}_{n \geq 0}$ in the following way. At time n, supposing $X_n \neq Y_n$, a fair coin is tossed. If heads, the first particle moves one step in the direction chosen at random by means of another fair coin tossed independently and the other particle stays still. If tails, the second particle moves one step in the direction chosen at random by means of another fair coin tossed independently and the other particle stays still. The two particles make identical moves as soon as they collide for the first time. Calling D_n the distance between the two particles, $\{D_n\}_{n \geq 0}$ is a symmetric random walk on $\{0, 1, \ldots, N\}$ with absorbing states 0 and N. The coupling time is the first time τ where the symmetric random walk is absorbed at 0 or N. Therefore $E_{i,j}[\tau] = k(N-k)$ where k is the distance between i and j, and $d(n) \leq \max_{i,j \in E} \frac{E_{i,j}[\tau]}{n} \leq \frac{N^2}{4n}$. The right-hand side equals $\frac{1}{4}$ for $n = N^2$, therefore $t_{mix} \leq N^2$.

EXAMPLE 9.3.5: TOP TO RANDOM CARD SHUFFLING, TAKE 2. ([11]) In the notation of Example 9.2.11,

$$\tau = \tau_1 + (\tau_2 - \tau_1) + \cdots + (\tau - \tau_{N-1}),$$

where $\tau - \tau_{N-1} = 1$. At time τ_i, card $\star$ has i cards below it, and the probability that the current top card is inserted below $\star$ is therefore $\frac{i+1}{N}$. Therefore, $\tau_{i+1} - \tau_i$ is geometric:

$$P(\tau_{i+1} - \tau_i = k) = \frac{i+1}{N} \left(1 - \frac{i+1}{N}\right)^{k-1}.$$

Consider now the following problem: Sample uniformly with replacement an urn containing N balls, and denote by V the number of draws until each ball has been sampled at least once. Let V_i be the number of draws until i distinct balls have been sampled at least once. We have the identity

$$V = (V - V_{N-1}) + \cdots + (V_2 - V_1) + V_1.$$

---

[11][Aldous and Diaconis, 1987].

Once i distinct balls have been drawn at least once, there is a probability $\frac{N-i}{N}$ of sampling a ball not previously sampled. Therefore, $V_i - V_{i-1}$ is geometric:

$$P(V_i - V_{i-1} = k) = \frac{N-i}{N}\left(1 - \frac{N-i}{N}\right)^{k-1}.$$

In particular, τ and V have the same distribution. For each ball b, let A_b be the event that ball b was not drawn in the first $m = N \log(N) + cN$ draws, $c \geq 0$. We have

$$P(V > m) = P(\cup_b A_b) = N(1 - \frac{1}{N})^m \leq N\mathrm{e}^{-\frac{m}{N}} = N\mathrm{e}^{-\log(N)-c} = \mathrm{e}^{-c}.$$

Therefore

$$d(N \log(N) + cN) \leq (P(\tau > N \log(N) + cN) \leq \mathrm{e}^{-c},$$

where $d(k) = d_V(\mu^T \mathbf{P}^k, \pi^T)$. In particular, $t_{mix}(\varepsilon) \leq N \log N - \log(\varepsilon) N$.

9.3.3 Lower Bounds

Consider an irreducible ergodic HMC on the finite state space E, with transition matrix $\mathbf{P}$, and with a uniform stationary distribution π. Let $d_+(i)$ be the out-degree of state i, that is, the number of directed edges in the transition graph out of vertex i: $d_+(i) := |\{j \in E; p_{ij} > 0\}|$, and let $d_{+,max} := \max_{i \in E} d_+(i)$. Therefore, starting from any state, the maximum number of states accessible in n steps is at most $d_{+,max}^n$. The distribution of X_n is therefore concentrated on a subset of E with at most $d_{+,max}^n$ elements. In particular, for any state i,

$$d_V(\delta_i^T \mathbf{P}^n, \pi) \geq \frac{1}{|E|}(|E| - d_{+,max}^n).$$

In particular, if $d_{+,max}^n \leq (1-\varepsilon)|E|$, that is, if $n \leq \frac{\log((1-\varepsilon)|E|)}{\log d_{+,max}}$, then $d(n) \geq \varepsilon$. This implies that

$$t_{mix}(\varepsilon) \geq \frac{\log((1-\varepsilon)|E|)}{\log d_{+,max}}.$$

EXAMPLE 9.3.6: RANDOM WALK ON A GRAPH. Let d_{max} be the maximal degree of the graph $G = (V, \mathcal{E})$. For the random walk on this graph, $d_{+,max} = d_{max}$ and therefore

$$t_{mix}(\varepsilon) \geq \frac{\log((1-\varepsilon)|E|)}{\log d_{max}}.$$

From the directed transition graph of an irreducible ergodic HMC on the finite state space E, with transition matrix $\mathbf{P}$, construct a graph whose vertex set is E and with an edge linking i and j if and only if $p_{ij} + p_{ji} > 0$. The diameter D of the chain is by definition the diameter of this graph, that is, the maximal graph distance between two vertices. If i_0 and j_0 are two states at the maximal graph distance D, then $\delta_{i_0} \mathbf{P}^{\lfloor (D-1)/2 \rfloor}$ and $\delta_{i_0} \mathbf{P}^{\lfloor (D-1)/2 \rfloor}$ have disjoint support, and therefore $\overline{d}(\lfloor (D-1)/2 \rfloor) = 1$. In particular, for any $\varepsilon < \frac{1}{2}$,

$$t_{mix}(\varepsilon) \geq \left\lfloor \frac{D-1}{2} \right\rfloor.$$

For the next result, recall definition (9.32) of the bottleneck ratio.

Theorem 9.3.7 *For an ergodic* HMC *with transition matrix* $\mathbf{P}$ *and bottleneck ratio* Φ^*,

$$t_{mix} \geq \frac{1}{4\Phi^*}.$$

Proof. Denote by π_B the restriction of π to the set $B \subset E$, and by ρ_B the probability π conditioned by B:

$$\pi_B(A) = \pi(A \cap B),\ A \subseteq B, \text{ and } \rho_B(A) = \frac{\pi(A \cap B)}{\pi(B)},\ A \subseteq E.$$

We have

$$\pi(B)d_V(\rho_B\mathbf{P}, \rho_B) = \pi(B) \sum_{i;\rho_B\mathbf{P}(i) \geq \rho_B(i)} (\rho_B\mathbf{P}(i) - \rho_B(i))$$

$$= \sum_{i;\pi_B\mathbf{P}(i) \geq \pi_B(i)} (\pi_B\mathbf{P}(i) - \pi_B(i)).$$

Since $\pi_B(i) = 0$ on $\overline{B}$, and $\pi_B(i) = \pi(i)$ on B,

$$\pi_B\mathbf{P}(i) = \sum_{j \in B} \pi_B(j)p_{ji} \leq \sum_{j \in E} \pi(j)p_{ji} = \pi(i)$$

and therefore, if $i \in B$,

$$\pi_B\mathbf{P}(i) \leq \pi_B(i),$$

and for $i \notin B$, since $\pi_B(i)$ is then null,

$$\pi_B\mathbf{P}(i) \geq 0 = \pi_B(i).$$

Therefore, from

$$\pi(B)d_V(\rho_B\mathbf{P}, \rho_B) = \sum_{i \in \overline{B}}(\pi_B\mathbf{P}(i) - \pi_B(i)).$$

Because $\pi_B(i) = 0$ outside B, the right-hand side reduces to

$$\sum_{i \in \overline{B}}\sum_{j \in B}\pi(j)p_{ji} = Q(B, \overline{B}),$$

and, dividing by $\pi(B)$

$$d_V(\rho_B\mathbf{P}, \rho_B) = \Phi(B).$$

Now, since for all $n \geq 0$, $d_V(\rho_B\mathbf{P}^{n+1}, \rho_B\mathbf{P}^n) \leq d_V(\rho_B\mathbf{P}, \rho_B)$,

$$d_V(\rho_B\mathbf{P}^{n+1}, \rho_B\mathbf{P}^n) \leq \Phi(B).$$

By the triangle inequality applied to the sum $\rho_B\mathbf{P}^n - \rho_B = \sum_{k=0}^{n-1}\rho_B\mathbf{P}^{k+1} - \rho_B\mathbf{P}^k$,

$$d_V(\rho_B\mathbf{P}^n, \rho_B) \leq n\Phi(B). \tag{$\star$}$$

If $\pi(B) \leq \frac{1}{2}$,

$$d_V(\rho_B, \pi) \geq \pi(\overline{B}) - \rho_B(\overline{B}) = \pi(\overline{B}) = 1 - \pi(B) \geq \frac{1}{2}.$$

By the triangle inequality

$$\frac{1}{2} \leq d_V(\rho_B, \pi) \leq d_V(\rho_B, \rho_B\mathbf{P}^n) + d_V(\rho_B\mathbf{P}^n, \pi).$$

Letting $n = t_{mix}$, and using $(\star)$,

$$\frac{1}{2} \leq t_{mix}\Phi(B) + \frac{1}{4}$$

from which the result follows. $\qquad\qquad\qquad\qquad\qquad\qquad\qquad\qquad\qquad\square$

EXAMPLE 9.3.8: TOP TO RANDOM CARD SHUFFLING, TAKE 3. ([12]) Examples
9.2.11 and 9.3.5 continued. We prove that for any $\varepsilon > 0$, there exists a constant
c_0 such that for all $c \geq c_0$, for sufficiently large N,

$$d(N \log N - cN) \geq 1 - \varepsilon.$$

---

[12][Aldous and Diaconis, 1987].

In particular, there exists a constant c such that for sufficiently large N,

$$t_{mix} \geq N \log N - cN.$$

Proof. Let A_j denote the event that the original bottom j cards are in their relative original order. Denote by σ_0 the original configuration of the deck.

Let τ_j be the time it takes for the j-th card from the bottom to reach the top of the deck, and let $\tau_{j,i}$ the time it takes for this card to pass from position i to position $i+1$ (positions are counted from the bottom up). Then

$$\tau_j = \sum_{i=j}^{N-1} \tau_{j,i}.$$

The $\tau_{j,i}$'s are independent geometric random variables with parameter $p := \frac{i}{N}$. In particular, $\mathrm{E}\,[\tau_{j,i}] = \frac{N}{i}$ and $\mathrm{Var}(\tau_{j,i}) \leq \frac{N^2}{i^2}$, and therefore

$$\mathrm{E}\,[\tau_j] = \sum_{i=j}^{N-1} \frac{N}{i} \geq N(\log N - \log j - 1),$$

and

$$\mathrm{Var}(\tau_j) \leq N^2 \sum_{i=j}^{\infty} \frac{1}{i(i+1)} \leq \frac{N^2}{j-1}.$$

From these bounds and Chebyshev's inequality,

$$
\begin{aligned}
\mathrm{P}(\tau_j < N \log N - cN) &\leq \mathrm{P}(\tau_j - \mathrm{E}\,[\tau_j] < -N(c - \log j - 1)) \\
&\leq \mathrm{P}(|\tau_j - \mathrm{E}\,[\tau_j]| > N(c - \log j - 1)) \\
&\leq \frac{\mathrm{Var}(\tau_j)}{N^2(c - \log j - 1)^2} \\
&\leq \frac{\frac{N^2}{j-1}}{N^2(c - \log j - 1)^2} \\
&\leq \frac{1}{j-1} \times \frac{1}{N^2(c - \log j - 1)^2} \\
&\leq \frac{1}{j-1}
\end{aligned}
$$

(provided that $c \geq \log j + 2$ for the last inequality).

If $\tau_j \geq N \log N - cN$, the original j bottom cards are still in their original relative order, and therefore, for $c \geq \log j + 2$,

$$\delta_{\sigma_0} \mathbf{P}^{N \log N - cN}(A_j) \geq \mathrm{P}(\tau_j \geq N \log N - cN) \geq 1 - \frac{1}{j-1}.$$

Now for the stationary distribution π, here the uniform distribution on the set of permutations $\mathcal{S}_N$, $\pi(A_j) = \frac{1}{j!} \leq \frac{1}{j-1}$. Therefore, for $c \geq \log j + 2$,

$$d(N \log N - cN) \geq d_V\left(\delta_{\sigma_0}\mathbf{P}^{N \log N - cN}, \pi\right) \geq \delta_{\sigma_0}\mathbf{P}^{N \log N - cN}(A_j) - \pi(A_j) \geq 1 - \frac{2}{j-1}.$$

With $j = e^{c-2}$, if $N \geq e^{c-2}$,

$$d(N \log N - cN) \geq 1 - \frac{2}{e^{c-2} - 1}.$$

Denoting by $g(c)$ the right-hand side of the above inequality, we have that

$$\liminf_{N \uparrow \infty} d(N \log N - cN) \geq g(c),$$

where $\lim_{c \uparrow \infty} g(c) = 1$. $\qquad\square$

Summarizing in rough terms the results of Examples 9.3.5 and 9.3.8: in order to shuffle a deck of N cards by the top-to-random method, "$N \log N$ shuffles suffice, but no less".

9.4 Exercises

Exercise 9.4.1. $p_{ii}(2)$
Show that for an irreducible recurrent positive reversible HMC, the probability that the return time to a given state is equal to 2 is strictly positive.

Exercise 9.4.2. THE χ^2-DISTANCE IN TERMS OF THE EIGENSTRUCTURE
Show that

$$\chi^2(p_{i\cdot}(n); \pi(\cdot)) = \sum_{j=2}^{r} \lambda_j^{2n} v_j(i)^2,$$

where v_j is the jth right-eigenvector associated with the reversible ergodic pair $(\mathbf{P}, \pi)$, and λ_j is the corresponding eigenvalue.

Exercise 9.4.3. A CHARACTERIZATION OF THE SLE
Let $\{X_n\}_{n \geq 0}$ be a stationary HMC corresponding to $(\mathbf{P}, \pi)$. Show that the SLE λ_2 of $\mathbf{P}$ is equal to the maximum of the correlation coefficient between $f(X_0)$ and $f(X_1)$ among all real-valued functions f such that $E[f(X_0)] = 0$.

Exercise 9.4.4. A VERSION OF THEOREM 9.2.4

([13]) Prove the version of Theorem 9.2.4 where Poincaré's coefficient κ is replaced by

$$\widetilde{\kappa} = \max_e Q(e)^{-1} \sum_{\gamma_{ij},\, e \in \gamma_{ij}} |\gamma_{ij}|\pi(i)\pi(j),$$

where $|\gamma|$ is the length of path γ. In the pure random walk case of Example 9.2.6 compare with the Poincaré type bound of Theorem 9.2.4.

Exercise 9.4.5. RANDOM WALK ON THE STAR GRAPH

Consider the random walk on the connected graph, the "star", with one central vertex connected to n outside vertices. Check that the corresponding transition matrix has eigenvalues $+1$, 0 and -1, where 0 has multiplicity $n - 1$. What is the period? To eliminate periodicity, make it a lazy walk with holding probability $p_{ii} = \beta$. Show that eigenvalues the eigenvalues are now $+1$, β and $2\beta - 1$, where β has multiplicity $n - 1$. For small α, compare the exact SLEM with the bound of Theorem 9.2.4.

Exercise 9.4.6. RANDOM WALK ON A BINARY TREE

Consider a random walk on a graph, where the graph is now a full binary tree of depth L.

(i) Show that the second largest eigenvalue λ_2 satisfies

$$\lambda_2 \le 1 - \frac{1}{9L2^{L-1}}.$$

(ii) Explain why formula (9.31) does not apply directly. Show that

$$\lambda_2 \ge 1 - \left(2(2^L - 1) \left(1 - \frac{1}{2^{L+1} - 2} \right) \right)^{-1},$$

which is equivalent for large L to $1 - 2^{-L-1}$. Hint: Use Rayleigh's characterization with x as follows: $x(i) = 0, 1$, or -1 according to whether i is the root, a vertex on the right of the tree, or one on the left.

Exercise 9.4.7. RANDOM WALK ON THE CUBE

Consider the random walk on the N-dimensional cube. Apply the Poincaré type bound of Theorem 9.2.5 with paths γ_x leading from x to those y obtained by changing the coordinates of x whenever they differ from that of y, inspecting the

---

[13][Sinclair, 1990].

coordinates from left to right. Show that

$$\lambda_2 \leq 1 - \frac{2}{N^2}.$$

(In this example, the exact eigenvalues are available: $1 - \frac{2j}{N}$ with multiplicity $\binom{N}{j}$ ($0 \leq j \leq N$), and therefore

$$\lambda_2 = 1 - \frac{2}{N}.$$

Therefore, the bound is off by a factor N.)

Exercise 9.4.8. $d(n)$ AND $\bar{d}(n)$
Refer to Definition 9.37 and denote by $\mathcal{P}(E)$ the collection of all probability distributions on E. Prove that

$$d(n) = \sup_{\mu \in \mathcal{P}(E)} d_V(\mu^T \mathbf{P}^n, \pi)$$

and

$$\bar{d}(n) = \sup_{\mu,\nu \in \mathcal{P}(E)} d_V(\mu^T \mathbf{P}^n, \nu^T \mathbf{P}^n).$$

Exercise 9.4.9. RANDOM WALK ON THE HYPERCUBE
In Example 9.2.12 prove that $t_{mix}(\varepsilon) \leq N \log N - \log(\varepsilon) N$. Compare with the top to random card shuffle of Example 9.3.5. (Hint: the coupon collector.)

Exercise 9.4.10. RANDOM WALK ON A GROUP
Consider the random walk $\{X_n\}_{n \geq 0}$ on a group G (Example 2.4.20) with increment measure μ and transition matrix $\mathbf{P}$. Let $\{\hat{X}_n\}_{n \geq 0}$ be another random walk on G, this time corresponding to the increment measure $\hat{\mu}$, the symmetric of μ (that is, for all $g \in G$, $\hat{\mu}(g) = \mu(g^{-1})$). Let $\hat{\mathbf{P}}$ be the corresponding transition matrix. Then, for all $n \geq 1$, denoting by π the common stationary distribution of the above two chains (the uniform distribution on G),

$$d_V(\delta_e^T \mathbf{P}^n, \pi) = d_V(\delta_e^T \hat{\mathbf{P}}^n, \pi) \quad (n \geq 1).$$

Exercise 9.4.11. MOVE-TO-FRONT POLICY
A professor has N books on his bookshelf. These books are equally interesting for his research, so that when he decides to take one from his library, it is in fact chosen at random and independently of all previous accesses to the library. The thing is that, being lazy or perhaps too busy, he does not put the book back to

where it was, but instead at the beginning of the collection at the left of the other books. The arrangement on the shelf can be represented by a permutation σ of $\{1, 2, \ldots, N\}$ and the evolution of the arrangement is therefore an HMC on the group of permutations $\mathcal{S}_N$.

(i) Show that the corresponding HMC is irreducible and ergodic, and admits the uniform distribution as stationary distribution.

(ii) Inspired by the top-to-random card shuffle Example 9.3.5 and Exercise 9.4.10, show that $t_{mix}(\varepsilon) \leq N \log N - \log(\varepsilon) N$.

Exercise 9.4.12. MIXING TIME OF THE RANDOM WALK ON THE BINARY TREE
Consider the random walk on the rooted binary tree of depth k whose number of edges is therefore $N = 2^{k+1} - 1$. Show that its mixing time satisfies the lower bound
$$t_{mix} \geq \frac{N-2}{2}.$$
(Hint: consider the set $B \subset E$ consisting of the direct descendant of the root to the right, v_R, and of all the descendants of v_R.)

Exercise 9.4.13. ADDITIVE REVERSIBILIZATION
The *additive reversibilization* of $\mathbf{P}$ is, by definition, the matrix $A = A(\mathbf{P}) := \frac{1}{2}(\mathbf{P} + \tilde{\mathbf{P}})$, that is
$$A := \frac{1}{2}\left(\mathbf{P} + D^{-1}\mathbf{P}^T D\right).$$
Show that this matrix is indeed reversible with respect to π.

Exercise 9.4.14. STRONG STATIONARY TIMES
Prove that the times τ in Examples 9.2.11 and 9.2.12 are strong stationary times.

Exercise 9.4.15. SEPARATION
Let $\mathcal{P}(E)$ be the collection of probability distributions on the countable set E. Show that for all $\alpha, \beta \in M_p(E)$,
$$s(\alpha; \beta) = \inf\{s \geq 0 \; ; \; \alpha = (1-s)\beta + s\gamma, \, \gamma \in \mathcal{P}(E)\},$$
where s denotes the separation.

Exercise 9.4.16. CHARACTERIZATIONS OF STRONG STATIONARY TIMES
Show that condition (α) of Definition 9.2.14 is equivalent to either one of the following two conditions:

(β) For all $i \in E$ and all $n \geq 0$,

$$P(X_n = i \,|\, T = n) = \pi(i) .$$

(γ) For all $i \in E$ and all $n \geq 0$,

$$P(X_n = i \,|\, T \leq n) = \pi(i) .$$

Exercise 9.4.17. MIXING TIME OF THE REVERSED RANDOM WALK ON A GROUP

The situation is that prevailing in Theorem 2.4.21. Let now μ be a not necessarily symmetric probability distribution on G, and define its inverse $\hat{\mu}$ by

$$\hat{\mu}(g) = \mu(g^{-1}) .$$

Define the HMC $\{\hat{X}_n\}_{n \geq 0}$ by

$$\hat{X}_{n+1} = \hat{Z}_{n+1} * \hat{X}_n$$

where $\{\hat{Z}_n\}_{n \geq 1}$ is an IID sequence with values in G and distribution $\hat{\mu}$, independent of the initial state $\hat{X}_0$. It turns out that the forward HMC $\{X_n\}_{n \geq 0}$ and the backward HMC $\{\hat{Z}_n\}_{n \geq 1}$ have the same mixing times: $t_{mix} = \hat{t}_{mix}$.

Exercise 9.4.18. ON THE χ^2-DISTANCE
Show that

$$\chi^2(p_{i\cdot}(n); \pi(\cdot)) = \sum_{j=2}^{r} \lambda_j^{2n} v_j(i)^2,$$

where v_j is the jth right-eigenvector associated with the reversible ergodic pair $(\mathbf{P}, \pi)$, and λ_j is the corresponding eigenvalue.

Exercise 9.4.19. MAXIMUM CORRELATION COUPLING
Show that λ_2 is equal to the maximum of the correlation coefficient between $f(X_0)$ and $f(X_1)$ among all real-valued functions f such that $\mathrm{E}[f(X_0)] = 0$, where $\{X_n\}_{n \geq 0}$ is a stationary HMC corresponding to $(\mathbf{P}, \pi)$, and λ_2 is the SLE of $\mathbf{P}$.

Exercise 9.4.20. A VERSION OF THEOREM 9.2.4
Prove the version of Theorem 9.2.4 where Poincaré's coefficient κ is replaced by

$$K = \max_e Q(e)^{-1} \sum_{\gamma_{ij}, e \in \gamma_{ij}} |\gamma_{ij}| \pi(i) \pi(j) ,$$

where $|\gamma|$ is the length of path γ.

Exercise 9.4.21. SLE OF THE RANDOM WALK ON A TREE, I
Consider a random walk on the graph which is a full binary tree of depth L. Show that the second largest eigenvalue λ_2 satisfies

$$\lambda_2 \leq 1 - \frac{1}{9L2^{L-1}}.$$

Exercise 9.4.22. SLE OF THE RANDOM WALK ON A TREE, II
This is a continuation of the previous exercise. Explain why the formula (9.31) does not apply directly. Show that

$$\lambda_2 \geq 1 - 2(2^L - 1)\left(1 - \frac{1}{2^{L+1} - 2}\right),$$

which is equivalent for large L to $1 - 2^{-L-1}$. Hint: Use Rayleigh's characterization with x as follows: $x(i) = 0, 1$, or -1 according to whether i is the root, a vertex on the right of the tree, or one on the left.

Exercise 9.4.23. ON THE SEPARATION PSEUDO-DISTANCE
Let $M_p(E)$ be a collection of probability distributions on the countable set E. Show that for all $\alpha, \beta \in M_p(E)$,

$$s(\alpha; \beta) = \inf\{s \geq 0 \;;\; \alpha = (1 - s)\beta + s\gamma, \, \gamma \in M_p(E)\},$$

where s denotes the separation pseudo-distance.

Exercise 9.4.24. VISITING TIME TO A RANDOM STATE
Let π be the stationary distribution of an ergodic Markov chain with finite state space, and denote by T_i the return time to state i. Let S_Z be the time necessary to visit for the first time the random state Z chosen according to the distribution π, independently of the chain. Show that $E_i[S_Z]$ is independent of i, and give its expression in terms of the fundamental matrix.

Exercise 9.4.25. A MAXIMUM
Show that $\max\{\frac{v(f,\mathbf{P},\pi)}{\text{var}_\pi(f)} ; f \neq 0\} = \frac{1+\lambda_2}{1-\lambda_2}$.

Exercise 9.4.26. TRANSITION MATRICES
Let $\mathbf{P}$ be a transition matrix on E. Show that $\delta(\mathbf{P}) = 0$ iff the rows of $\mathbf{P}$ are identical, and that $\delta(\mathbf{P}) = 1$ iff there exist two rows of $\mathbf{P}$ that are orthogonal.

Exercise 9.4.27. THE SLEM OF AN ERGODIC TRANSITION MATRIX
Let $\mathbf{P}$ be an ergodic transition matrix on the finite set E. Show that its SLEM is bounded above by $\inf_k \delta(\mathbf{P}^k)^{1/k}$.

Exercise 9.4.28. $\ln(\text{SLEM}(\mathbf{P}))$
Let $\mathbf{P}$ be an ergodic transition matrix on the finite state space E. Suppose, moreover, that $\mathbf{P}$ is diagonalizable (for simplicity, since the result is the same for the general case). Show that

$$\ln(\text{SLEM}(\mathbf{P})) = \lim_{k \to \infty} \frac{1}{k} \ln(\delta(\mathbf{P}^k)).$$

Chapter 10

Markov Fields on Graphs

Markov fields are also called Gibbs fields in honor of the founder of Statistical Mechanics.[1] Although they were historically of special interest to physicists, they have recently found applications in other areas, in particular in image processing.

10.1 The Gibbs–Markov Equivalence

10.1.1 Local Characteristics

Let $G = (V, \mathcal{E})$ be a finite graph (see Definition 2.4.18), and let $v_1 \sim v_2$ denote the fact that $\langle v_1, v_2 \rangle$ is an edge of the graph. Such vertices are called *neighbors* (one of the other). One sometimes refers to vertices of V as *sites*. The *boundary* with respect to $\sim$ of a set $A \subset V$ is the set

$$\partial A := \{v \in V \backslash A \,;\, v \sim w \text{ for some } w \in A\}\,.$$

Let Λ be a finite set, called the phase space. A *random field* on V with phases in Λ is a collection $X = \{X(v)\}_{v \in V}$ of random variables with values in Λ. A random field can be regarded as a random variable taking its values in the *configuration space* $E := \Lambda^V$, where a configuration is a function $x : v \in V \mapsto x(v) \in \Lambda$. For a given configuration x and a given subset $A \subseteq V$, let

$$x(A) := (x(v), v \in A)\,,$$

the restriction of x to A. If $V \backslash A$ denotes the complement of A in V, one writes $x = (x(A), x(V \backslash A))$. In particular, for fixed $v \in V$, $x = (x(v), x(V \backslash v))$, where $V \backslash v$ is a shorter way of writing $V \backslash \{v\}$.

---

[1][Gibbs, 1902].

© Springer Nature Switzerland AG 2020
P. Brémaud, *Markov Chains*, Texts in Applied Mathematics 31,
https://doi.org/10.1007/978-3-030-45982-6_10

Of special interest are the random fields characterized by local interactions. This leads to the notion of a Markov random field. The "locality" is in terms of the neighborhood structure inherited from the graph structure. More precisely, for any $v \in V$, $N_v := \{w \in V; w \sim v\}$ is the neighborhood of v. In the following, $\widetilde{N}_v$ denotes the set $N_v \cup \{v\}$.

Definition 10.1.1 *The random field X is called a* Markov random field (MRF) *with respect to $\sim$ if for all sites $v \in V$, the random elements $X(v)$ and $X(V \backslash \widetilde{N}_v)$ are independent given $X(N_v)$.*

In symbols:

$$P(X(v) = x(v) \mid X(V \backslash v) = x(V \backslash v)) = P(X(v) = x(v) \mid X(N_v) = x(N_v)) \quad (10.1)$$

for all $x \in \Lambda^V$ and all $v \in V$. Property (10.1) is of the Markov type in the sense that the distribution of the phase at a given site is directly influenced only by the phases of the neighboring sites.

Remark 10.1.2 Note that any random field is Markovian with respect to the trivial topology, where the neighborhood of any site v is $V \backslash v$. However, the interesting Markov fields (from the point of view of modeling, simulation and optimization) are those with relatively small neighborhoods.

EXAMPLE 10.1.3: MARKOV CHAIN AS MARKOV FIELD. The Markov property of a stochastic sequence $\{X_n\}_{n \geq 0}$ implies (exercise) that for all $n \geq 1$, X_n is independent of $(X_k, k \notin \{n - 1, n, n + 1\})$ given (X_{n-1}, X_{n+1}). If we call n a vertex, X_n the value of the process at vertex n and the set $\{n - 1, n + 1\}$ the neighborhood of vertex n, the above property can be rephrased as: For all $n \geq 1$, the value at vertex n is independent of the values at vertices $k \notin \{n - 1, n, n + 1\}$ given the values in the neighborhood of vertex n.

Definition 10.1.4 *The* local characteristic *of the MRF at site v is the function $\pi^v : \Lambda^V \to [0, 1]$ defined by*

$$\pi^v(x) := P(X(v) = x(v) \mid X(N_v) = x(N_v)).$$

The family $\{\pi^v\}_{v \in V}$ is called the local specification *of the MRF.*

One sometimes writes

$$\pi^v(x) := \pi(x(v) \mid x(N_v))$$

in order to stress the role of the neighborhoods.

Theorem 10.1.5 *Two positive distributions of a random field with a finite con-figuration space Λ^V that have the same local specification are identical.*

Proof. Enumerate V as $\{1, 2, \ldots, K\}$. Therefore a configuration $x \in \Lambda^V$ is represented as $x = (x_1, \ldots, x_{K-1}, x_K)$ where $x_i \in \Lambda$ $(1 \le i \le K)$. The following identity

$$\pi(z_1, z_2, \ldots, z_k) = \prod_{i=1}^{K} \frac{\pi(z_i \mid z_1, \ldots, z_{i-1}, y_{i+1}, \ldots, y_K)}{\pi(y_i \mid z_1, \ldots, z_{i-1}, y_{i+1}, \ldots, y_K)} \pi(y_1, y_2, \ldots, y_k) \quad (\star)$$

holds for any $z, y \in \Lambda^K$. For the proof, write

$$\pi(z) = \prod_{i=1}^{K} \frac{\pi(z_1, \ldots, z_{i-1}, z_i, y_{i+1}, \ldots, y_K)}{\pi(z_1, \ldots, z_{i-1}, y_i, y_{i+1}, \ldots, y_K)} \pi(y)$$

and use Bayes' rule to obtain for each i $(1 \le i \le K)$:

$$\frac{\pi(z_1, \ldots, z_{i-1}, z_i, y_{i+1}, \ldots, y_K)}{\pi(z_1, \ldots, z_{i-1}, y_i, y_{i+1}, \ldots, y_K)} = \frac{\pi(z_i \mid z_1, \ldots, z_{i-1}, y_{i+1}, \ldots, y_K)}{\pi(y_i \mid z_1, \ldots, z_{i-1}, y_{i+1}, \ldots, y_K)} .$$

Let now π and π' be two positive probability distributions on V with the same local specification. Choose any $y \in \Lambda^V$. Identity $(\star)$ shows that for all $z \in \Lambda^V$,

$$\frac{\pi'(z)}{\pi(z)} = \frac{\pi'(y)}{\pi(y)} .$$

Therefore $\frac{\pi'(z)}{\pi(z)}$ is a constant, necessarily equal to 1 since π and π' are probability distributions. $\qquad\square$

Finiteness of the phase space Λ is so far not essential. An immediate extension not requiring a change in the notation is to a denumerable phase space. Arbitrary phase spaces do not present a difficulty except for notation, since $\pi^v(x) = \pi(x(v) \mid x(\mathcal{N}_v))$, considered as a function of $x(v)$ for a fixed $x(\mathcal{N}_v)$, is now interpreted as a density function with respect to some measure. A simple case that contains all that is needed in applications will be briefly outlined.

Here the phase space is $\Lambda = F \times \mathbb{R}^k$, where F is a denumerable set. Thus $x = (x_1, x_2)$, where $x_1 = (x_1(v), v \in V)$, $x_2 = (x_2(s), v \in V)$, $x_1(v) \in F, x_2(v) \in \mathbb{R}^k$. In particular, $X = (X_1, X_2) = \{(X_1(v), X_2(v))\}_{v \in V}$. The local characteristic π^v is now interpreted according to

$$P(X_1(v) = \alpha, X_2(v) \le \beta \mid X_1(\mathcal{N}_v) = x_1(\mathcal{N}_v), X_2(\mathcal{N}_v) = x_2(\mathcal{N}_v))$$

$$= \int_{-\infty}^{\beta} \pi(\alpha, \gamma \mid x_1(\mathcal{N}_v), x_2(\mathcal{N}_v)) \, \mathrm{d}\gamma \,,$$

where the integral is a multiple integral of order k. Thus $\pi(\alpha, \gamma | x_1(\mathcal{N}_v), x_2(\mathcal{N}_v))$ is a probability density with respect to γ for fixed α. The Markov property now reads

$$P(X_1(v) = \alpha, X_2(v) \le \beta \mid X_1(V \backslash v), X_2(V \backslash v))$$
$$= P(X_1(v) = \alpha, X_2(v) \le \beta \mid X_1(\mathcal{N}_v), X_2(\mathcal{N}_v))$$

for all $\alpha \in F$, $\beta \in \mathbb{R}^k$, $v \in V$.

10.1.2   Gibbs Distributions

Consider the probability distribution

$$\pi_T(x) = \frac{1}{Z_T} e^{-\frac{1}{T} U(x)} \tag{10.2}$$

on the configuration space Λ^V, where $T > 0$ is a "*temperature*", $U(x)$ is the "*energy*" of configuration x and Z_T is the normalizing constant, called the *partition function*. Since $\pi_T(x)$ takes its values in $[0, 1]$, necessarily $-\infty < U(x) \le +\infty$. Note that $U(x) < +\infty$ if and only if $\pi_T(x) > 0$. One of the challenges associated with Gibbs models is obtaining explicit formulas for averages, considering that it is generally hard to compute the partition function. (This is however feasible in exceptional cases; see Exercise 10.4.7).

Such distributions are of interest to physicists when the energy is expressed in terms of a potential function describing the local interactions. The notion of clique then plays a central role.

Definition 10.1.6 *Any singleton $\{v\} \subset V$ is a clique. A subset $C \subseteq V$ with more than one element is called a* clique *(with respect to $\sim$) if and only if any two distinct sites of C are mutual neighbors. A clique C is called* maximal *if for any site $v \notin C$, $C \cup \{v\}$ is not a clique.*

The collection of cliques will be denoted by $\mathcal{C}$.

Definition 10.1.7 *A Gibbs potential on Λ^V relative to $\sim$ is a collection $\{V_C\}_{C \subseteq V}$ of functions $V_C : \Lambda^V \to \mathbb{R} \cup \{+\infty\}$ such that*

(i) $V_C \equiv 0$ if C is not a clique, and

(ii) for all $x, x' \in \Lambda^V$ and all $C \subseteq V$,

$$x(C) = x'(C) \Rightarrow V_C(x) = V_C(x').$$

The energy function U is said to derive from the potential $\{V_C\}_{C \subseteq V}$ if

$$U(x) = \sum_C V_C(x).$$

The function V_C depends only on the phases at the sites inside subset C. One could write more explicitly $V_C(x(C))$ instead of $V_C(x)$, but this notation will not be used.

In this context, the distribution in (10.2) is called a Gibbs distribution (with respect to $\sim$).

EXAMPLE 10.1.8: ISING MODEL, TAKE 1. (^{2}) In statistical physics, the following model is regarded as a qualitatively correct idealization of a piece of ferromagnetic material. Here $V = \mathbb{Z}_m^2 = \{(i, j) \in \mathbb{Z}^2, \quad (1 \leq i, j \leq m)\}$ and $\Lambda = \{+1, -1\}$, where ± 1 is the orientation of the magnetic spin at a given site. The figure below depicts two particular neighborhood systems, their respective cliques, and the boundary of a 2×2 square for both cases. The neighborhood system in the original Ising model is as in column (α) of the figure below, and the Gibbs potential is

$$V_{\{v\}}(x) = -\frac{H}{k}x(v),$$
$$V_{\langle v, w \rangle}(x) = -\frac{J}{k}x(v)x(w),$$

where $\langle v, w \rangle$ is the 2-element clique ($v \sim w$). For physicists, k is the *Boltzmann constant*, H is the *external magnetic field*, and J is the *internal energy* of an elementary magnetic dipole. The energy function corresponding to this potential is therefore

$$U(x) = -\frac{J}{k} \sum_{\langle v, w \rangle} x(v)x(w) - \frac{H}{k} \sum_{v \in V} x(v).$$

The Hammersley–Clifford Theorem

Gibbs distributions with an energy deriving from a Gibbs potential relative to a neighborhood system are distributions of Markov fields relative to the same neighborhood system.

---

^{2}[Ising, 1925].

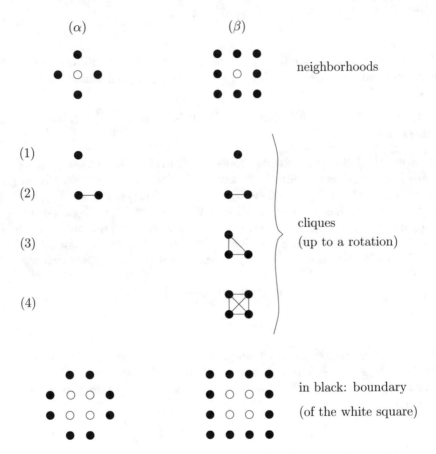

Figure 10.1.1. Two examples of neighborhoods, cliques, and boundaries

Theorem 10.1.9 *If X is a random field with a distribution π of the form $\pi(x) = \frac{1}{Z}e^{-U(x)}$, where the energy function U derives from a Gibbs potential $\{V_C\}_{C \subseteq V}$ relative to $\sim$, then X is a Markov random field with respect to $\sim$. Moreover, its local specification is given by the formula*

$$\pi^v(x) = \frac{e^{-\sum_{C \ni v} V_C(x)}}{\sum_{\lambda \in \Lambda} e^{-\sum_{C \ni v} V_C(\lambda, x(V \setminus v))}}, \tag{10.3}$$

where the notation $\sum_{C \ni v}$ means that the sum extends over the sets C that contain the site v.

Proof. First observe that the right-hand side of (10.3) depends on x only through $x(v)$ and $x(\mathcal{N}_v)$. Indeed, $V_C(x)$ depends only on $(x(w), w \in C)$, and for a clique C, if $w \in C$ and $v \in C$, then either $w = v$ or $w \sim v$. Therefore, if it can be shown that $P(X(v) = x(v) | X(V \setminus v) = x(V \setminus v))$ equals the right-hand side of (10.3), then (Theorem 1.1.30) the Markov property will be proved. By definition of conditional probability,

$$P(X(v) = x(v) \mid X(V \setminus v) = x(V \setminus v)) = \frac{\pi(x)}{\sum_{\lambda \in \Lambda} \pi(\lambda, x(V \setminus v))}. \tag{$\dagger$}$$

But

$$\pi(x) = \frac{1}{Z}e^{-\sum_{C \ni v} V_C(x) - \sum_{C \not\ni v} V_C(x)},$$

and similarly,

$$\pi(\lambda, x(V \setminus v)) = \frac{1}{Z}e^{-\sum_{C \ni v} V_C(\lambda, x(V \setminus v)) - \sum_{C \not\ni v} V_C(\lambda, x(V \setminus v))}.$$

If C is a clique and v is not in C, then $V_C(\lambda, x(V \setminus v)) = V_C(x)$ and is therefore independent of $\lambda \in \Lambda$. Therefore, after factoring out $\exp\left\{-\sum_{C \not\ni v} V_C(x)\right\}$, the right-hand side of ($\dagger$) is found to be equal to the right-hand side of (10.3). □

The local energy at site v of configuration x is

$$U_v(x) = \sum_{C \ni v} V_C(x).$$

With this notation, (10.3) becomes

$$\pi^v(x) = \frac{e^{-U_v(x)}}{\sum_{\lambda \in \Lambda} e^{-U_v(\lambda, x(V \setminus v))}}.$$

EXAMPLE 10.1.10: ISING MODEL, TAKE 2. The local characteristics in the Ising model are

$$\pi_T^v(x) = \frac{\mathrm{e}^{\frac{1}{kT}\left\{J\sum_{w;w\sim v}x(w)+H\right\}x(v)}}{\mathrm{e}^{+\frac{1}{kT}\left\{J\sum_{w;w\sim v}x(w)+H\right\}}+\mathrm{e}^{-\frac{1}{kT}\left\{J\sum_{w;w\sim v}x(w)+H\right\}}}.$$

Theorem 10.1.9 above is the direct part of the Gibbs–Markov equivalence theorem: A Gibbs distribution relative to a neighborhood system is the distribution of a Markov field with respect to the same neighborhood system. The converse part (Hammersley–Clifford theorem) is important from a theoretical point of view, since together with the direct part it concludes that Gibbs distributions and MRFs are essentially the same objects.

Theorem 10.1.11 ([3]) *Let $\pi > 0$ be the distribution of a Markov random field with respect to $\sim$. Then*

$$\pi(x) = \frac{1}{Z}\mathrm{e}^{-U(x)}$$

for some energy function U deriving from a Gibbs potential $\{V_C\}_{C\subseteq V}$ with respect to $\sim$.

Proof. The proof is based on the Möbius formula.

Lemma 10.1.12 *Let Φ and Ψ be two set functions defined on $\mathcal{P}(V)$, the collection of subsets of the finite set V. The two statements below are equivalent:*

$$\Phi(A) = \sum_{B\subseteq A}(-1)^{|A-B|}\Psi(B), \text{ for all } A \subseteq V, \tag{10.4}$$

$$\Psi(A) = \sum_{B\subseteq A}\Phi(B), \text{ for all } A \subseteq V, \tag{10.5}$$

where $|C|$ is the number of elements of the set C.

Proof. We first show that (10.4) implies (10.5). Write the right-hand side of (10.5) using (10.4):

$$\sum_{B\subseteq A}\Phi(B) = \sum_{B\subseteq A}\sum_{D\subseteq B}(-1)^{|B-D|}\Psi(D) = \sum_{D\subseteq A}\left(\sum_{C\subseteq A-D}(-1)^{|C|}\right)\Psi(D).$$

[3][Hammersley and Clifford, 1968].

But if $A - D = \varnothing$,

$$\sum_{C \subseteq A-D} (-1)^{|C|} = (-1)^{|\varnothing|} = (-1)^0 = 1,$$

whereas if $A - D \neq \varnothing$,

$$\sum_{C \subseteq A-D} (-1)^{|C|} = \sum_{k=0}^{|A-D|} (-1)^k \, \text{card} \, \{C; |C| = k, C \subseteq A - D\}$$

$$= \sum_{k=0}^{|A-D|} (-1)^k \binom{|A - D|}{k} = (1 - 1)^{|A-D|} = 0,$$

and therefore

$$\sum_{D \subseteq A} \Psi(D) \sum_{C \subseteq A-D} (-1)^{|C|} = \Psi(A).$$

We now show that (10.5) implies (10.4). Write the right-hand side of (10.4) using (10.5):

$$\sum_{B \subseteq A} (-1)^{|A-B|} \Psi(B) = \sum_{B \subseteq A} (-1)^{|A-B|} \left(\sum_{D \subseteq B} \Phi(D) \right)$$

$$= \sum_{D \subseteq B \subseteq A} (-1)^{|A-B|} \Phi(D) = \sum_{D \subseteq A} \Phi(D) \sum_{C \subseteq A-D} (-1)^{|C|}.$$

By the same argument as above, the last quantity equals $\Phi(A)$. $\qquad \square$

We now prove Theorem 10.1.11. Let 0 be a fixed element of the phase space Λ. Also, let 0 denote the configuration with all phases equal to 0. (The context will prevent confusion between $0 \in \Lambda$ and $0 \in \Lambda^V$.) Let x be a configuration, and let A be a subset of V. Let the symbol x^A represent a configuration of Λ^V coinciding with x on A, and with phase 0 outside A.

Define for $A \subseteq V, x \in \Lambda^V$,

$$V_A(x) := \sum_{B \subseteq A} (-1)^{|A-B|} \log \frac{\pi(0)}{\pi(x^B)}. \tag{10.6}$$

From the Möbius formula,

$$\log \frac{\pi(0)}{\pi(x^A)} = \sum_{B \subseteq A} V_B(x),$$

and therefore, with $A = V$:

$$\pi(x) = \pi(0)e^{-\sum_{A \subseteq V} V_A(x)}.$$

It remains to show (a) that V_A depends only on the phases on A, and (b) that $V_A \equiv 0$ if A is not a clique with respect to $\sim$.

If $x, y \in \Lambda^V$ are such that $x(A) = y(A)$, then for any $B \subseteq A$, $x^B = y^B$, and therefore, by (10.6), $V_A(x) = V_A(y)$. This proves (a).

With t an arbitrary site in A, write (10.6) as follows:

$$
\begin{aligned}
V_A(x) &= \left\{ \sum_{B \subseteq A, B \not\ni t} + \sum_{B \subseteq A, B \ni t} \right\} (-1)^{|A-B|} \log \frac{\pi(0)}{\pi(x^B)} \\
&= \sum_{B \subseteq A \setminus t} (-1)^{|A-B|} \left\{ \log \frac{\pi(0)}{\pi(x^B)} - \log \frac{\pi(0)}{\pi(x^{B \cup t})} \right\}.
\end{aligned}
$$

That is,

$$V_A(x) = \sum_{B \subseteq A \setminus t} (-1)^{|A-B|} \log \frac{\pi(x^{B \cup t})}{\pi(x^B)}. \tag{10.7}$$

Now, if t is not in $B \subseteq A$,

$$\frac{\pi(x^{B \cup t})}{\pi(x^B)} = \frac{\pi^t(x^{B \cup t})}{\pi^t(x^B)},$$

and therefore

$$V_A(x) = \sum_{B \subseteq A \setminus t} (-1)^{|A-B|} \log \frac{\pi^t(x^{B \cup t})}{\pi^t(x^B)},$$

and, by the same calculations that led to (10.7),

$$V_A(x) = -\sum_{B \subseteq A} (-1)^{|A-B|} \log \pi^t(x^B). \tag{10.8}$$

Recall that $t \in A$, and therefore, if A is not a clique, one can find $s \in A$ such that s is not a neighbor of t. Fix such an s, and split the sum in (10.8) as follows:

$$
\begin{aligned}
V_A(x) = &-\sum_{B \subseteq A \setminus \{s,t\}} (-1)^{|A-B|} \log \pi^t(x^B) - \sum_{B \subseteq A \setminus \{s,t\}} (-1)^{|A-(B \cup t)|} \log \pi^t(x^{B \cup t}) \\
&-\sum_{B \subseteq A \setminus \{s,t\}} (-1)^{|A-(B \cup s)|} \log \pi^t(x^{B \cup s}) - \sum_{B \subseteq A \setminus \{s,t\}} (-1)^{|A-(B \cup \{s,t\})|} \log \pi^t(x^{B \cup \{s,t\}}) \\
= &-\sum_{B \subseteq A \setminus \{s,t\}} (-1)^{|A-B|} \log \frac{\pi^t(x^B) \pi^t(x^{B \cup \{s,t\}})}{\pi^t(x^{B \cup s}) \pi^t(x^{B \cup t})}.
\end{aligned}
$$

But since $s \neq t$ and $s \sim t$, we have $\pi^t(x^B) = \pi^t(x^{B \cup s})$ and $\pi^t(x^{B \cup t}) = \pi^t(x^{B \cup \{s,t\}})$, and therefore $V_A(x) = 0$. $\square$

Remark 10.1.13 The energy function U and the partition function are not unique, since adding a constant to the energy function is equivalent to multiplying the normalizing factor by an appropriate constant. Likewise, and more importantly, the Gibbs potential associated with π is not unique.

However, uniqueness can be forced into the result if a certain property is imposed on the potential, namely normalization with respect to a fixed phase value.

Definition 10.1.14 *A Gibbs potential $\{V_C\}_{C \subseteq V}$ is said to be* normalized *with respect to a given phase in Λ, conventionally denoted by 0, if $V_C(x) = 0$ whenever there exists a $t \in C$ such that $x(t) = 0$.*

Theorem 10.1.15 *There exists one and only one potential normalized with respect to a given phase $0 \in \Lambda$ corresponding to a Gibbs distribution.*

Proof. Expression (10.6) gives a normalized potential. In fact, the right-hand side of

$$V_C(x) = \sum_{B \subseteq C \setminus t} (-1)^{|C-B|} \log \frac{\pi\left(x^{B \cup t}\right)}{\pi(x^B)}$$

is independent of t in the clique C, and in particular, choosing any $t \in C$ such that $x(t) = 0$, $x^{B \cup t} = x^B$ for all $B \subseteq C \setminus t$, and therefore $V_C(x) = 0$.

For the proof of uniqueness, suppose that

$$\pi(x) = \frac{1}{Z_1} e^{-U_1(x)} = \frac{1}{Z_2} e^{-U_2(x)}$$

for two energy functions U_1 and U_2 deriving from potentials V_1 and V_2, respectively, both normalized with respect to $0 \in \Lambda$. Since $U_1(0) = \sum_{C \in \mathcal{C}} V_{1,C}(0) = 0$, and similarly $U_2(0) = 0$, it follows that $Z_1 = Z_2 = \pi(0)^{-1}$, and therefore $U_1 \equiv U_2$. Suppose that $V_{1,A} = V_{2,A}$ for all $A \in \mathcal{C}$ such that $|A| \leq k$ (property $\mathcal{P}_k$). It remains to show, in view of a proof by induction, that $\mathcal{P}_k \Rightarrow \mathcal{P}_{k+1}$ and that $\mathcal{P}_1$ is true.

To prove $\mathcal{P}_k \Rightarrow \mathcal{P}_{k+1}$, fix $A \subseteq V$ with $|A| = k+1$. To prove that $V_{1,A} \equiv V_{2,A}$ it suffices to show that $V_{1,A}(x) = V_{2,A}(x)$ for all $x \in \Lambda^V$ such that $x = x^A$. Fix such an x. Then

$$U_1(x) = \sum_C V_{1,C}(x) = \sum_{C \subseteq A} V_{1,C}(x),$$

since x has phase 0 outside A and V_1 is normalized with respect to 0. Also,

$$U_1(x) = \sum_{C \subseteq A} V_{1,C}(x) = V_{1,A}(x) + \sum_{C \subseteq A, |C| \le k} V_{1,C}(x), \qquad (10.9)$$

with a similar equality for $U_2(x)$. Therefore, since $U_1(x) = U_2(x)$, we obtain $V_{1,A}(x) = V_{2,A}(x)$ in view of the induction hypothesis. The root $\mathcal{P}_1$ of the induction hypothesis is true, since when $|A| = 1$, (10.9) becomes $U_1(x) = V_{1,A}(x)$, and similarly, $U_2(x) = V_{2,A}(x)$, so that $V_{1,A}(x) = V_{2,A}(x)$ is a consequence of $U_1(x) = U_2(x)$. $\qquad\square$

Remark 10.1.16 In practice, the potential as well as the topology of V can be obtained directly from the expression of the energy, as the following example shows.

EXAMPLE 10.1.17: MARKOV CHAINS AS MARKOV FIELDS. Let $V = \{0, 1, \ldots N\}$ and $\Lambda = E$, a finite space. A random field X on V with phase space Λ is therefore a vector X with values in E^{N+1}. Suppose that $X_0, \ldots, X_N$ is a homogeneous Markov chain with transition matrix $\mathrm{P} = \{p_{ij}\}_{i,j \in E}$ and initial distribution $\nu = \{\nu_i\}_{i \in E}$. In particular, with $x = (x_0, \ldots, x_N)$,

$$\pi(x) = \nu_{x_0} p_{x_0 x_1} \cdots p_{x_{N-1} x_N},$$

that is,

$$\pi(x) = e^{-U(x)},$$

where

$$U(x) = -\log \nu_{x_0} - \sum_{n=0}^{N-1} (\log p_{x_n x_{n+1}}).$$

Clearly, this energy derives from a Gibbs potential associated with the nearest-neighbor topology for which the cliques are, besides the singletons, the pairs of adjacent sites. The potential functions are:

$$V_{\{0\}}(x) = -\log \nu_{x_0}, \quad V_{\{n,n+1\}}(x) = -\log p_{x_n x_{n+1}}.$$

The local characteristic at site n, $2 \le n \le N - 1$, can be computed from formula (10.3), which gives

$$\pi^n(x) = \frac{\exp(\log p_{x_{n-1} x_n} + \log p_{x_n x_{n+1}})}{\sum_{y \in E} \exp(\log p_{x_{n-1} y} + \log p_{y x_{n+1}})},$$

that is,

$$\pi^n(x) = \frac{p_{x_{n-1} x_n} p_{x_n x_{n+1}}}{p^{(2)}_{x_{n-1} x_{n+1}}},$$

where $p_{ij}^{(2)}$ is the general term of the two-step transition matrix P^2. Similar computations give $\pi^0(x)$ and $\pi^N(x)$. We note that, in view of the neighborhood structure, for $2 \leq n \leq N-1$, X_n is independent of $X_0, \ldots, X_{n-2}, X_{n+2}, \ldots, X_N$ given X_{n-1} and X_{n+1}.

10.2 Specific Models

10.2.1 Random Points

Let $Z := \{Z(v)\}_{v \in V}$ be a random field on V with phase space $\Lambda := \{0,1\}$. Here $Z(v) = 1$ will be interpreted as the presence of a "point" at site v.

Recall that $\mathcal{P}(V)$ denotes the collection of subsets of V, and denote by $\mathbf{x}$ such a subset. A random field $Z \in \{0,1\}^V$ with distribution π being given, we associate to it the random element $\mathbf{X} \in \mathcal{P}(V)$, called a point process on V, by

$$\mathbf{X} := \{v \in V \,;\, Z(v) = 1\}.$$

Its distribution is denoted by ℓ. For any $\mathbf{x} \subseteq V$, $\ell(\mathbf{x})$ is the probability that $Z(v) = 1$ for all $v \in \mathbf{x}$ and $Z(v) = 0$ for all $v \notin \mathbf{x}$.

Let $\mathbf{X}$ be a point process on the finite set V with positive probability distribution $\{\ell(\mathbf{x})\}_{\mathbf{x} \in \mathcal{P}(V)}$. Such point process on V can be viewed as a random field on V with phase space $\Lambda \equiv \{0,1\}$ with probability distribution $\{\pi(z)\}_{z \in \Lambda^V}$. We assume that this random field is Markov with respect to the symmetric relation $\sim$, and then say that $\mathbf{X}$ is Markov with respect to $\sim$. We have the following alternative form of the Hammersley–Clifford theorem:

Theorem 10.2.1 ([4]) *The point process $\mathbf{X}$ on the finite set V with positive probability distribution $\{\ell(\mathbf{x})\}_{\mathbf{x} \in \mathcal{P}(V)}$ is Markov with respect to $\sim$ if and only if there exists a function $\varphi : M_p^f(V) \to (0,1]$ such that*

(i) $\varphi(\mathbf{y}) < 1$ if and only if $\mathbf{y}$ is a clique for $\sim$, and

(ii) for all $\mathbf{x} \in \mathcal{P}(V)$,

$$\ell(\mathbf{x}) = \prod_{\mathbf{y} \subseteq \mathbf{x}} \varphi(\mathbf{y}).$$

Proof. Necessity: The distribution π may be expressed in terms of a potential $\{V_C\}_{C \subseteq V}$ as

$$\pi(z) = \alpha e^{\sum_{C \subseteq V} V_C(z)}. \tag{10.10}$$

---

[4][Ripley and Kelly, 1977].

Take for a potential the (unique) one normalized with respect to phase 0. Identifying a configuration $z \in \{0,1\}^V$ with a subset $\mathbf{x}$ of V, and more generally identifying a subset C of V with a configuration $\mathbf{y} \in \mathcal{P}(V)$, the potential can be represented as a collection of functions $\{V_\mathbf{y}\}_{\mathbf{y} \in \mathcal{P}(V)}$. Note that $V_\mathbf{y}(\mathbf{x}) > 0$ if and only if $\mathbf{y}$ is a clique and $\mathbf{y} \subseteq \mathbf{x}$ (normalized potential), in which case $V_\mathbf{y}(\mathbf{x}) = V_\mathbf{y}(\mathbf{y})$. The result then follows by letting

$$\varphi(\mathbf{y}) := e^{-V_\mathbf{y}(\mathbf{y})} \qquad (\mathbf{y} \neq \varnothing)$$

and

$$\varphi(\varnothing) := \ell(\varnothing) = \alpha.$$

The proof of sufficiency is left for the reader as it follows the same lines as the proof of Theorem 10.1.9.   □

In the case of a positive distribution ℓ of the point process $\mathbf{X}$, let

$$\lambda(u, \mathbf{x}) := \frac{\ell(\mathbf{x} \cup u)}{\ell(\mathbf{x})}$$

if $u \notin \mathbf{x}$, $= 0$ otherwise. For $u \notin \mathbf{x}$,

$$\lambda(u, \mathbf{x}) = \frac{\mathrm{P}(Z(u) = 1, \mathbf{X}\backslash u = \mathbf{x})}{\mathrm{P}(\mathbf{X} = \mathbf{x})} = \frac{\mathrm{P}(Z(u) = 1, \mathbf{X}\backslash u = \mathbf{x})}{\mathrm{P}(\mathbf{X}\backslash u = \mathbf{x})},$$

and therefore

$$\lambda(u, \mathbf{x}) = \mathrm{P}(Z(u) = 1 \mid \mathbf{X}\backslash u = \mathbf{x}),$$

the probability that there is a point at u knowing the point process outside u. This defines the exvisible distribution (on $\{0,1\}$) at point $u \in V$.

Theorem 10.2.2 *Let $g : V \times \mathcal{P}(V) \to \mathbb{R}$ be a non-negative function. Then*

$$\mathrm{E}\left[\sum_{u \in V} g(u, \mathbf{X}\backslash u)\right] = \mathrm{E}\left[\sum_{u \in V} g(u, \mathbf{X})\lambda(u, \mathbf{X})\right].$$

Proof.

$$\mathrm{E}\left[\sum_{u \in V} g(u, \mathbf{X})\lambda(u, \mathbf{X})\right] = \sum_{\mathbf{x} \in \mathcal{P}(V)} \sum_{u \in V} g(u, \mathbf{x})\lambda(u, \mathbf{x})\ell(\mathbf{x})$$

$$= \sum_{\mathbf{x} \in \mathcal{P}(V)} \sum_{u \in V} g(u, \mathbf{x})1_{\{u \notin \mathbf{x}\}}\ell(\mathbf{x} \cup u).$$

With the change of variables $\mathbf{x} \cup u \to \mathbf{y}$, the last quantity is seen to be equal to

$$\sum_{\mathbf{y} \in \mathcal{P}(V)} \sum_{u \in V} g(u, \mathbf{y} \backslash u) \ell(\mathbf{y}) = \mathrm{E} \left[\sum_{u \in V} g(u, \mathbf{X} \backslash u) \right].$$

$\square$

10.2.2 The Auto-binomial Texture Model

For the purpose of image synthesis, one seeks Gibbs distributions describing pictures featuring various textures, lines separating patches with different textures (boundaries), lines per se (roads, rail tracks), randomly located objects (moon craters), etc. The corresponding model is then checked by simulation: images are drawn from the proposed Gibbs distribution, and some tuning of the parameters is done, until the images subjectively correspond to ("look like") the type of image one expects. Image synthesis is an art based on trial and error, and fortunately guided by some general principles. But these principles are difficult to formalize, and we shall mainly resort to simple examples with a pedagogical value. Note, however, that there is a domain of application where the model need not be very accurate, namely Bayesian estimation. As a matter of fact, the models proposed in this section have been devised in view of applications to the Bayesian restoration of degraded pictures.

We shall begin with an all-purpose texture model[5] that may be used to describe the texture of various materials. The set of sites is $V = \mathbb{Z}_m^2$, and the phase space is $\Lambda = \{0, 1, \ldots, L\}$. In the context of image processing, a site v is a pixel (PICTure ELement), and a phase $\lambda \in \Lambda$ is a shade of grey, or a color. The neighborhood system is

$$\mathcal{N}_v = \left\{ w \in V; w \neq v; \|w - v\|^2 \leq d \right\},$$

where d is a fixed positive integer and where $\|w - v\|$ is the euclidean distance between v and w. In this model the only cliques participating in the energy function are singletons and pairs of mutual neighbors. The set of cliques appearing in the energy function is a disjoint sum of collections of cliques

$$\mathcal{C} = \sum_{j=1}^{m(d)} \mathcal{C}_j,$$

where $\mathcal{C}_1$ is the collection of singletons, and all pairs $\{v, w\}$ in $\mathcal{C}_j$, $2 \leq j \leq m(d)$, have the same distance $\|w - v\|$ and the same direction, as shown in the figure

---

[5][Besag, 1974].

below. The potential is given by

$$V_C(x) = \begin{cases} -\log \binom{L}{x(v)} + \alpha_1 x(v) & \text{if } C = \{v\} \in \mathcal{C}_1, \\ \alpha_j x(v) x(w) & \text{if } C = \{v, w\} \in \mathcal{C}_j, \end{cases}$$

where $\alpha_j \in \mathbb{R}$. For any clique C not of type $\mathcal{C}_j$, $V_C \equiv 0$.

The terminology ("auto-binomial") is motivated by the fact that the local system has the form

$$\pi^v(x) = \binom{L}{x(v)} \tau^{x(v)} (1-\tau)^{L-x(v)}, \tag{10.11}$$

where τ is a parameter depending on $x(\mathcal{N}_v)$ as follows:

$$\tau = \tau(\mathcal{N}_v) = \frac{e^{-\langle \alpha, b \rangle}}{1 + e^{-\langle \alpha, b \rangle}}. \tag{10.12}$$

Here $\langle \alpha, b \rangle$ is the scalar product of

$$\alpha = (\alpha_1, \ldots, \alpha_{m(d)}) \text{ and } b = (b_1, \ldots, b_{m(d)}),$$

where $b_1 = 1$, and for all j, $2 \le j \le m(d)$,

$$b_j = b_j(x(\mathcal{N}_v)) = x(u) + x(w),$$

where $\{v, u\}$ and $\{v, w\}$ are the two pairs in $\mathcal{C}_j$ containing v.

Proof. From the explicit formula (10.3) giving the local characteristic at site v,

$$\pi^v(x) = \frac{\exp\left\{\log \binom{L}{x(v)} - \alpha_1 x(v) - \left[\sum_{j=2}^{m(d)} \alpha_j \sum_{v;\{v,w\}\in\mathcal{C}_j} x(w)\right] x(v)\right\}}{\sum_{\lambda\in\Lambda} \exp\left\{\log \binom{L}{\lambda} - \alpha_1 \lambda - \left[\sum_{j=2}^{m(d)} \alpha_j \sum_{t;\{v,w\}\in\mathcal{C}_j} x(w)\right] \lambda\right\}}.$$

The numerator equals

$$\binom{L}{x(v)} e^{-\langle \alpha, b \rangle x(v)},$$

and the denominator is

$$\sum_{\lambda\in\Lambda} \binom{L}{\lambda} e^{(-\alpha,b)\lambda} = \sum_{\ell=0}^{L} \binom{L}{\ell} \left(e^{-\langle\alpha,b\rangle}\right)^\ell = \left(1 + e^{-\langle\alpha,b\rangle}\right)^L.$$

Equality (10.11) then follows. $\square$

Expression (10.11) shows that τ is the average level of grey at site v, given $x(\mathcal{N}_v)$, and expression (10.12) shows that τ is a function of $\langle \alpha, b \rangle$. The parameter α_j controls the bond in the direction and at the distance that characterize $\mathcal{C}_j$.

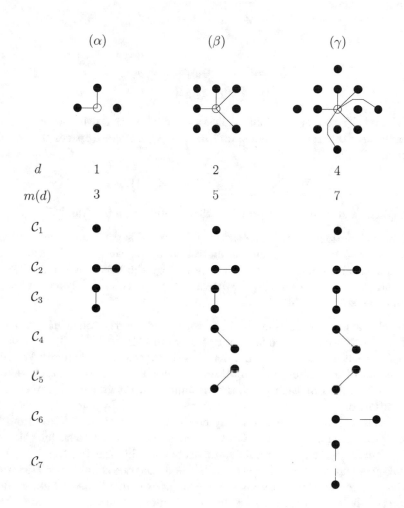

Figure 10.2.1. Neighborhoods and cliques of three auto-binomial models

10.2.3   The Pixel-Edge Model

(^{6}) Let $X = \{X(v)\}_{v \in V}$ be a random field on V with phase space Λ, with the following structure:

$$V = V_1 + V_2 \, , \ \Lambda = \Lambda_1 \cup \Lambda_2,$$

and

$$
\begin{aligned}
X(v) &= Y(v_1) \in \Lambda_1 \text{ if } v = v_1 \in V_1 \\
&= Z(v_2) \in \Lambda_2 \text{ if } v = v_2 \in V_2.
\end{aligned}
$$

Here V_1 and V_2 are two disjoint collections of sites that can be of a different nature, or have different functions, and Λ_1 and Λ_2 are phase spaces that need not be disjoint. Define

$$Y = \{Y(v_1)\}_{v_1 \in V_1}, \ Z = \{Z(v_2)\}_{v_2 \in V_2} \, .$$

The random field X may be viewed as the juxtaposition of Y and Z.

In some situations, Y is the *observed field*, and Z is the *hidden field*. Introduction of a hidden field is in principle motivated by physical considerations. From the computational point of view, it is justified by the fact that the field Y alone usually has a Gibbsian description with large neighborhoods, whereas $X = (Y, Z)$ hopefully has small neighborhoods.

The philosophy supporting the pixel-edge model is the following. A digitized image can be viewed as a realization of a random field on $V^{\mathrm{P}} = \mathbb{Z}_m^2$. A site could be, for instance, a pixel on a digital television screen, and therefore V^{P} will be called the set of *pixel* sites. For an observer, there is, in general, more in an image than just the colors at each pixel. For instance, an image can be perceived as a juxtaposition of zones with various textures separated by lines. However, these lines, or contours, are not seen directly on the pixels, they are inferred from them by some processing in the brain. On the other hand, if one is to sketch the picture observed on the screen, one would most likely start by drawing the lines. In any case, textures and contours are very much linked, and one should seek a description featuring the interaction between them. But as was mentioned, contours do not exist on the digital screen, they are hidden or, more accurately, they are virtual.

EXAMPLE 10.2.3: AN EXAMPLE. In this example, there is a set V^E of *edge* sites, one between each pair of adjacent pixel sites, as indicated in the figure below (a). The possible values of the phase on an edge site are blank or bar: horizontal (resp., vertical) for an edge site between two pixel sites forming a vertical (resp.,

---

^{6}[Geman and Geman, 1984].

horizontal) segment, as shown in the figure below (b). In this figure, not all edge sites between two pixels with a different color have a bar, because a good model should not systematically react to what may be accidents in the global structure.

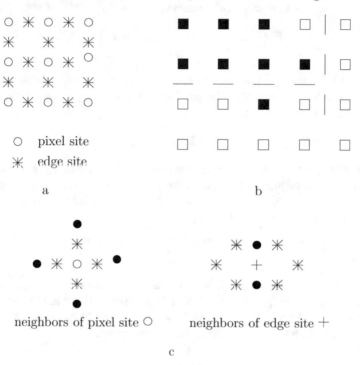

neighbors of pixel site ○   neighbors of edge site +

c

Figure 10.2.2. Example of a neighborhood structure for the pixel-edge model

Let (i, j) denote a pixel site and (α, β) an edge site (these are the coordinates of the sites in two distinct orthogonal frames). The random field on the pixels is denoted by $X^P = \{X_{ij}^P\}_{(i,j)\in V^P}$, and that on the edge sites is $X^E = \{X_{\alpha\beta}^E\}_{(\alpha,\beta)\in V^E}$; X^P is the observed image, and X^E is the hidden, or virtual, line field. The distribution of the field $X = (X^P, X^E)$ is described by an energy function $U(x^P, x^E)$:

$$\pi(x^P, x^E) = \frac{1}{Z} e^{-U(x^P, x^E)},$$

where $x^P = \{x_{ij}^P\}_{(i,j)\in V^P}$, $x^E = \{x_{\alpha\beta}^E\}_{(\alpha,\beta)\in V^E}$. The energy function derives from a potential relative to some neighborhood system, a particular choice of which is pictured in (c) of the figure. The energy function separates into two parts

$$U(x^P, x^E) = U_1(x^P, x^E) + U_2(x^E),$$

where U_1 features only cliques formed by a pair of neighboring pixels and the edge pixel in between, whereas U_2 features only the diamond cliques shown in the figure below. The energy $U_1(x^P, x^E)$ governs the relation between an edge and its two adjacent pixels. For instance, for some real constant $\alpha > 0$ and some function φ,

$$U_1(x^P, x^E) = -\alpha \sum_{\langle 1,2 \rangle} \varphi(x_1^P - x_2^P)\, x_{\langle 1,2 \rangle}^E,$$

where $\langle 1, 2 \rangle$ represents a pair of adjacent pixel sites and $x_{\langle 1,2 \rangle}^E$ is the value of the phase on the edge site between the pixel sites 1 and 2, say 0 for a blank and 1 for a bar. A possible choice of φ is

$$\varphi(x) = 1_{\{x \neq 0\}} - 1_{\{x=0\}}\,.$$

Since the most probable configurations are those with low energy, this model favors bars between two adjacent pixel sites with different colors, as is natural. More sophisticated choices of φ with a similar effect are possible. The organization of the contours is controlled by the energy

$$U_2(x^E) = \beta \sum_D w_D(x^E),$$

where $\beta > 0$ and the sum extends to all diamond cliques, and w_D is a function depending only on the phases of the four edge sites of the diamond clique D. Up to rotations of $\frac{\pi}{2}$, there are six possible values for the four-vector of phases on a given diamond clique D, as shown in the figure below.

   no-line   ending   turn   continuation   3-countries   4-countries

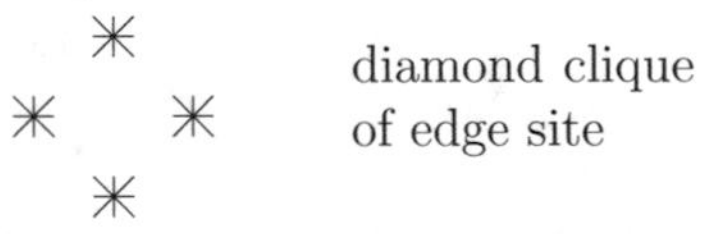

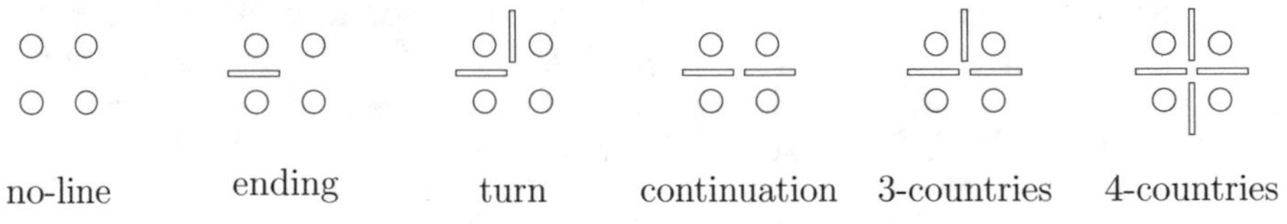

Figure 10.2.3. Six configurations of the diamond clique

If the modeler believes that for the type of images he is interested in the likelihood of the configurations shown in the figure below decreases from left to right, then the values of $w_D(x^E)$ attributed to these configurations will increase from left to right. This is generally the case because four-country border points are rare, broken lines also, and the same is true to a lesser extent for three-country border points. Also, when the picture is not a clutter of lines, the no-line configuration is the most likely.

Remark 10.2.4 The above example admits many variations. However, too many sophisticated features could ruin the model. The purpose of the model is not so much to do image synthesis as to have a reasonable a priori model for the image F in view of Bayesian restoration of this image from a noisy version of it, as will now be explained.

10.2.4 Markov Fields in Image Processing

A number of estimation problems arising in various areas of statistics and engineering, and particularly in image processing, are solved by a method of statistical analysis known as the *maximum a posteriori* (MAP) *likelihood* method. The theory will not be presented, but the examples below will show its substance. These examples compute the a posteriori probability, or conditional probability, of a given MRF X with respect to another MRF Y, the observed field, say,

$$\pi(x \mid y) = \mathrm{P}(X = x \mid Y = y),$$

and the MAP method estimates the nonobservable field X given the observed value y of Y, by $\hat{x} = \hat{x}(y)$, the value of x that maximizes $\pi(x \mid y)$:

$$\hat{x}(y) = \arg\max_x \pi(x \mid y).$$

Usually, this maximization problem is doomed by combinatorial explosion and by the complexity of standard methods of optimization. However, if

$$\pi(x \mid y) \propto e^{-U(x \mid y)}$$

(the proportionality factor depends only on y and is therefore irrelevant to maximization with respect to x) with an energy $U(x \mid y)$ that as a function of x corresponds to a topology N with small neighborhoods, then one can use a simulated annealing algorithm or a related algorithm (see Section 12.2).

EXAMPLE 10.2.5: RANDOM FLIPS OR MULTIPLICATIVE NOISE. Let X, Z be random fields on $V = \mathbb{Z}_m^2$ with phase space $\Lambda = \{-1, +1\}$, and define the field $Y = XZ$ by $y(v) = x(v)z(v), v \in V$. The field Z will be interpreted as multiplicative noise, and one can call it a random flip field, because what it does to X is to flip the phase at site v if $z(v) = -1$.

The computation below uses the fact that if $a, b, c \in \Lambda^V$, where $\Lambda = \{-1, +1\}$, then $ab = c \Leftrightarrow b = ac$:

$$
\begin{aligned}
P(Y = y)P(X = x \mid Y = y) &= P(X = x, Y = y) = P(X = x, ZX = y) \\
&= P(X = x, Zx = y) = P(X = x, Z = yx).
\end{aligned}
$$

In particular, if the noise field Z is independent of the original field X, then

$$
\pi(x \mid y) \propto P(X = x)P(Z = yx).
$$

The random field X has the distribution

$$
P(X = x) \propto e^{-U(x)}.
$$

Suppose that $(Z(v), v \in V)$ is a family of IID random variables, with $P(Z(v) = -1) = p$, $P(Z(v) = +1) = q = 1 - p$. Therefore (Exercise 10.4.4),

$$
P(Z = z) = \prod_{v \in V} P(Z(v) = z(v)) \propto e^{\gamma \sum_{v \in V} z(v)},
$$

where $\gamma = \frac{1}{2} \log \left(\frac{1-p}{p} \right)$. Finally,

$$
\pi(x \mid y) \propto e^{-U(x) + \gamma \sum_{v \in V} y(v) x(v)}.
$$

Note that if $p > \frac{1}{2}$, then $\gamma > 0$.

EXAMPLE 10.2.6: IMAGE RESTORATION. This example refers to the model of Subsection 10.2.3. Recall that we have a random field $X = (X^{\mathrm{P}}, X^E)$ corresponding to some energy function $U(x^{\mathrm{P}}, x^E)$, which need not be made precise here (see however Example 10.2.3). The image X^{P} is degraded into a noisy image Y, and it is this corrupted image that is observed. Degradation combines two effects: blurring, and a possibly nonlinear interaction of the blurred image and the noise. Specifically,

$$
Y_{ij} = \varphi(H(X^{\mathrm{P}})_{ij}, N_{ij}),
$$

where (i, j) is a pixel site and φ, H, and N are defined as follows. First $N = \{N_{ij}\}_{(i,j) \in P_m}$ is, for instance, a family of independent centered Gaussian random variables with common variance σ^2, and is independent of (X^{P}, X^E). As for $H(X^{\mathrm{P}})$, it is the random field obtained by blurring X^{P}, that is,

$$
H(X^{\mathrm{P}})_{ij} = \sum_{k, \ell} H_{k\ell} X^{\mathrm{P}}_{i-k, j-\ell},
$$

where $H = \{H_{k\ell}\}_{-N \leq k, \ell \leq N}$ is the *blurring matrix*. Here, $X^P_{i-k, j-\ell} = 0$ if $(i-k, j-\ell) \notin V^P$. A typical blurring matrix is

$$H = \begin{pmatrix} 1/80 & 1/80 & 1/80 \\ 1/80 & 9/10 & 1/80 \\ 1/80 & 1/80 & 1/80 \end{pmatrix},$$

for which $N = 1$. In this case

$$H(X^P)_{ij} = \frac{9}{10} X^P_{ij} + \frac{1}{80} \left(\sum_\ell X^P_{k,\ell} \right),$$

where the sum extends to the pixel sites adjacent to (i, j). As for φ, it is a function such that for fixed a, the function $b \to \varphi(a, b)$ is invertible. The inverse of this function, for fixed a, is then denoted by $b \to \psi(a, b)$. A typical example for ψ is the additive noise model

$$Y_{ij} = H(X^P)_{ij} + N_{ij}. \tag{10.13}$$

To estimate X given Y, one is led to compute the a posteriori probability of image x given that the noisy image is y:

$$\pi(x^P, x^E \mid y) = P(X^P = x^P, X^E = x^E \mid Y = y).$$

Writing $\pi(y) = P(Y = y)$, we have

$$\begin{aligned} \pi(x^P, x^E \mid y) &= \pi(y) P(X^P = x^P, X^E = x^E, Y = y) \\ &= \pi(y) P(X^P = x^P, X^E = x^E, \varphi(H(X^P), N) = y) \\ &= \pi(y) P(X^P = x^P, X^E = x^E, N = \psi(H(x^P), y)) \\ &= \pi(y) P(X^P = x^P, X^E = x^E) P(N = \psi(H(x^P), y)). \end{aligned}$$

The reader will have noticed the abuse of notation by which the continuous character of N was ignored: The second to fourth terms are actually probability densities, and similarly for

$$P(N = \psi(H(x^P), y)) \propto e^{-\frac{1}{2\sigma^2} \|\psi(H(x^P), y)\|^2}.$$

Using the expression of the distribution of the pixel+line image in terms of the energy function, one finds

$$\pi(x^P, x^E \mid y) \propto e^{-U(x^P, x^E) - \frac{1}{2\sigma^2} \|\psi(H(x^P), y)\|^2}. \tag{10.14}$$

Therefore the *a posteriori* distribution of (X^P, X^E) given $Y = g$ is a Gibbs distribution corresponding to the energy function

$$U(x^P, x^E) = U(x^P, x^E) + \frac{1}{2\sigma^2} \|\psi(H(x^P), y)\|^2. \tag{10.15}$$

For instance, if the noise is additive, as in (10.13), then

$$U(x^P, x^E) = U(x^P, x^E) + \frac{1}{2\sigma^2}\|y - H(x^P)\|^2. \tag{10.16}$$

EXAMPLE 10.2.7: BERNOULLI–GAUSSIAN MODEL. Let $\{Y_n\}_{1\leq n\leq N}$ be a real-valued stochastic process of the form

$$Y_n = \sum_{k=1}^{N} X_k h_{n-k} + Z_n,$$

where $\{Z_n\}_{1\leq n\leq N}$ is a sequence of independent centered Gaussian random variables of variance σ^2, $\{X_n\}_{1\leq n\leq N}$ is an IID sequence of $\{0,1\}$-valued random variables with $P(X_n = 1) = p$, and $\{h_k\}_{k\in\mathbb{Z}}$ is a deterministic function.

This is a particular case of the one-dimensional version of the model in Example 10.2.6. Here $V = \{1, \ldots, N\}$, a configuration $x \in \{0,1\}^N$ is of the form $x = (x_1, x_2, \ldots, x_N)$, X is the original image, Y is the degraded image, Z is the additive noise, and h corresponds to the blurring matrix. For this particular model, the energy of the IID random field X is of the form $\gamma \sum_{i=1}^{N} x_i$ (Exercise 10.4.4), and the energy of the conditional field $x|y$ is

$$\gamma \sum_{i=1}^{N} x_i + \frac{1}{2\sigma^2} \sum_{i=1}^{N} \left| y_i - \sum_{j;1\leq i-j\leq N} h_j x_{i-j} \right|^2.$$

This model is used in problems of detection of reflectors. More precisely, one says that there is a reflector at position i if $X_i = 1$. The function h is a probe signal (radar, sonar) and $\{h_{k-i}\}_{k\in\mathbb{Z}}$ is the signal reflected by the reflector at position i, if any, so that $Y_n = \sum_{k=1}^{N} X_k h_{n-k}$ is the reflected signal from which the map of reflectors X is to be recovered. The process Z is the usual additive noise of signal processing.

Of course, this model can be considerably enriched by introducing random reflection coefficients or by using a more elaborate a priori model for X, say, a Markov chain model.

Penalty Methods

Consider the simple model where the image X is additively corrupted by white Gaussian noise N of variance σ^2, and let Y be the resulting image. Calling $U(x)$

the energy function of the a priori model, the MAP estimate is

$$\hat{x} \;=\; \arg\min_{x} \{U(x) + \frac{1}{\sigma^2}\|y - x\|^2\}.$$

If we take an a priori model where all images are equiprobable, that is, the corresponding energy is null, then the above minimization is trivial, leading one to accept the noisy image as if it were the original image.

A non-trivial a priori model introduces a penalty term $U(x)$ and forces a balance between our belief in the observed image, corresponding to a small value of $\|y-x\|^2$, and our a priori expectation as to what we should obtain, corresponding to a small value of $U(x)$. The compromise between the credibility of the observed image and the credibility of the estimate with respect to the prior distribution is embodied in the criterion $U(x)+\frac{1}{\sigma^2}\|y-x\|^2$. A non-Bayesian mind will, somehow rightly, argue that one cannot even dream of thinking that a correct a priori model is available, and that Gaussian additive white noise is at best an intellectual construction. All that he will retain from the above is the criterion

$$\lambda U(x) + \|y - x\|^2,$$

with the interpretation that the penalty term $\lambda U(x)$ corrects undesirable features of the observed image y. One of these is the usually chaotic aspect at the fine scale. However, he does not attempt to interpret this as due to white noise. In order to correct this effect he introduces a *smoothing* penalty term $U(x)$, which is small when x is smooth, for instance

$$U(x) \;=\; \sum_{\langle s,t \rangle} (x(s) - x(t))^2,$$

where the summation extends over pairs of adjacent pixels. One disadvantage of this smoothing method is that it will tend to blur the boundary between two highly contrasted regions. One must choose an edge-preserving smoothing penalty function, for instance

$$U(x) \;=\; \sum_{\langle s,t \rangle} \Psi(x(s) - x(t)),$$

where, for instance,

$$\Psi(u) = -\left(1 + \left(\frac{u}{\delta}\right)^2\right)^{-1},$$

with $\delta > 0$. This energy function favors large contrasts and therefore prevents to some extent blurring of the edges.

A more sophisticated penalty function would introduce edges, as in Subsection 10.2.3, with a penalty function of the form

$$U(x^{\mathrm{P}}, x^{E}) = U_1(x^{\mathrm{P}}, x^{E}) + U_2(x^{E}),$$

where the term $U_1(x^{\mathrm{P}}, x^{E})$ creates the edges from the pixels, and $U_2(x^{E})$ organizes the edges.

Note that the estimated image now consists of two parts: $\hat{x}^{\mathrm{P}}$ solves the smoothing problem, whereas $\hat{x}^{E}$ extracts the boundaries. If one is really interested in boundary extraction, a sophisticated line-pixel model is desirable. If one is only interested in cleaning the picture, rough models may suffice.

The import of the Gibbs–Bayes approach with respect to the purely deterministic penalty function approach to image restoration lies in the theoretical possibility of the former to tune the penalty function by means of simulation. Indeed, if one is able to produce a typical image corresponding to the energy-penalty function, one will be able to check with the naked eye whether this penalty function respects the constraints one has in mind, and if necessary to adjust the parameters in it. The simulation issue is treated in Chapter 11. Another theoretical advantage of the Gibbs–Bayes approach is the availability of the simulated annealing algorithm (Section 12.2) to solve the minimization problem arising in MAP likelihood method or in the traditional penalty method.

10.3   Phase Transition in the Ising Model

10.3.1   Experimental Results

The first significant success of the Gibbs–Ising model was a qualitative explanation of the phase transition phenomenon in ferromagnetism.[7]

Consider the slightly generalized Ising model of a piece of ferromagnetic material, with spins distributed according to

$$\pi_T(x) = \frac{1}{Z_T} \mathrm{e}^{\frac{-U(x)}{T}}. \tag{10.17}$$

The finite site space is enumerated as $V = \{1, 2, \ldots, N\}$, and therefore a configuration x is denoted by $(x(1), x(2), \ldots, x(N))$. The energy function is

$$U(x) = U_0(x) - \frac{H}{k} \sum_{i=1}^{N} x(i),$$

---

[7][Peierls, 1936].

where the term $U_0(x)$ is assumed symmetric, that is, for any configuration x,

$$U_0(x) = U_0(-x).$$

The constant H is the external magnetic field. The *magnetic moment* of configuration x is

$$m(x) = \sum_{i=1}^{N} x(i),$$

and the *magnetization* is the average *magnetic moment* per site

$$M(H, T) = \frac{1}{N} \sum_{x \in E} \pi_T(x) m(x).$$

We have that $\frac{\partial M(H,T)}{\partial H} \geq 0$ (Exercise 10.4.12), $M(-H, T) = -M(H, T)$ and $-1 \leq M(H, T) \leq +1$. Therefore, at fixed temperature T, the magnetization $M(H, T)$ is a non-decreasing odd function of H with values in $[-1, +1]$. Also,

$$M(0, T) = 0, \qquad\qquad (\diamond)$$

since for any configuration x, $m(-x) = -m(x)$, and therefore $\pi_T(-x) = \pi_T(x)$ when $H = 0$. Moreover, the magnetization is an analytic function of H.

However, the experimental results seem to contradict the last two assertions. Indeed, if an iron bar is placed in a strong magnetic field H parallel to the axis, it is completely magnetized with magnetization $M(H, T) = +1$, and if the magnetic field is slowly decreased to 0, the magnetization decreases, but tends to a limit $M(0, T) = M_0 > 0$, in disagreement with $(\diamond)$. By symmetry, we therefore have a discontinuity of the magnetization at $H = 0$ (Figure 10.3.1 (a)), in contradiction to the theoretical analyticity of the magnetization as a function of H.

This discontinuity is called a *phase transition* by physicists, by analogy with the discontinuity in density at a liquid-gas phase transition. It occurs at room temperature, and if the temperature is increased, the residual, or *spontaneous*, magnetization M_0 decreases until it reaches the value 0 at a certain temperature T_c, called the *critical temperature*. Then, for $T > T_c$, the discontinuity at 0 disappears, and the magnetization curve is smooth (Figure 10.3.1 (c)). At $T = T_c$, the slope at $H = 0$ is infinite, that is, the *magnetic susceptibility* is infinite (Figure 10.3.1 (b)).

The discrepancy between experience and theory below the critical temperature is due to the fact that the experimental results describe a situation at the *thermodynamical limit* $N = \infty$. For fixed but large N the theoretical magnetization

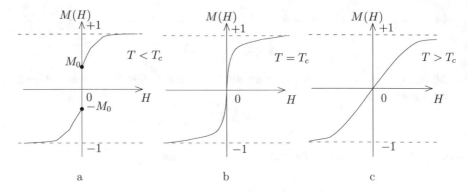

Figure 10.3.1. Magnetization and critical temperature

curve is analytic, but it presents for all practical purposes the same aspect as in Figure 10.3.1(a).

To summarize the experimental results: it seems that below the critical temperature, the spontaneous magnetization has, when no external magnetic field is applied, two "choices." This phenomenon can be explained within the classical Ising model.

The DLR Problem

(^{8}) Consider the Ising model in the absence of an external field ($H = 0$). The energy of a configuration x is of the form

$$U(x) = -J \sum_{\langle v,w \rangle} x(v)x(w),$$

where $\langle v, w \rangle$ represents an unordered pair of neighbors. When the site space V is infinite, the sum in the expression of the energy is not defined for all configurations, and therefore one cannot define the Gibbs distribution π_T on Λ^V by formula (10.17). However, the local specification

$$\pi_T^v(x) = \frac{e^{\beta \sum_{\langle v,w \rangle} x(v)x(w)}}{e^{\beta \sum_{\langle v,w \rangle} x(w)} + e^{-\beta \sum_{\langle v,w \rangle} x(w)}}, \tag{10.18}$$

where β is, up to a factor, the inverse temperature, is well-defined for all configurations and all sites.

---

^{8}[Dobrushin, 1965], [Lanford and Ruelle, 1969].

In the sequel, we shall repeatedly use an abbreviated notation. For instance, if π is the distribution of a random field X under probability P, then $\pi(x(A))$ denotes $P(X(A) = x(A))$, $\pi(x(0) = +1)$ denotes $P(X(0) = +1)$, etc.

A probability distribution π_T on Λ^V is called a *solution of the* DLR *problem* if it admits the local specification (10.18).

When $V = K_N = \mathbb{Z}^2 \cap [-N, +N]^2$, we know that there exists a unique solution, given by (10.17). When $V = \mathbb{Z}^2$, one can prove (this is not done here) the existence of at least one solution of the DLR problem. One way of constructing a solution is to select an arbitrary configuration z, to construct for each integer $N \geq 2$ the unique probability distribution $\pi_T^{(N)}$ on Λ^V such that

$$\pi_T^{(N)}(z(V \backslash K_{N-1})) = 1$$

(the field is frozen at the configuration z outside K_{N-1}) and such that the restriction of $\pi_T^{(N)}$ to K_{N-1} has the required local characteristics (10.18), and then let N tend to infinity. For all configurations x and all *finite* subsets $A \subset V$, the following limit exists:

$$\pi_T(x(A)) = \lim_{N \uparrow \infty} \pi_T^{(N)}(x(A)), \tag{10.19}$$

and moreover, there exists a unique random field X with the local specification (10.18) and such that, for all configurations x and all *finite* subsets $A \subset V$,

$$P(X(A) = x(A)) = \pi_T(x(A)).$$

Note that $\pi_T^{(N)}$ depends on the configuration z only through the restriction of z to the boundary $K_N \backslash K_{N-1}$.

10.3.2 The Peierls Argument

If the DLR problem has more than one solution, one says that a phase transition occurs. The method given by Dobrushin to construct a solution suggests a way of proving phase transition when it occurs. It suffices to select two configurations z_1 and z_2, and to show that for a given finite subset $A \subset S$, the right-hand side of (10.19) is different for $z = z_1$ and $z = z_2$. In fact, for sufficiently small values of the temperature, the phase transition phenomenon is observed. To show this, we apply the above program with z_1 being the configuration with all spins positive and z_2 the all negative configuration, and with $A = \{0\}$, where 0 denotes the central site of $\mathbb{Z}^2$.

Denote then by $\pi_+^{(N)}$ (resp., $\pi_-^{(N)}$) the restriction to K_N of $\pi_T^{(N)}$ when $z = z_1$ (resp., $z = z_2$). We shall prove that if T is large enough, then $\pi_+^{(N)}(x(0) = -1) <$

$\frac{1}{3}$ for all N. By symmetry, $\pi_-^{(N)}(x(0) = +1) < \frac{1}{3}$, and therefore $\pi_-^{(N)}(x(0) = -1) > \frac{2}{3}$. Passing to the limit as $N \uparrow \infty$, we see that $\pi_+(x(0) = -1) < \frac{1}{3}$ and $\pi_-(x(0) = -1) > \frac{2}{3}$, and therefore, the limiting distributions are not identical.

This program for proving the existence of a phase transition is now carried out. We shall prove that

$$\pi_+^{(N)}(x) = \frac{e^{-2\beta n_o(x)}}{Z_+^{(N)}} \quad (x \in \Lambda^{K_N}), \tag{10.20}$$

where $n_o(x)$ is the number of *odd bounds* in configuration x, that is, the number of cliques $\langle v, w \rangle$ such that $x(v) \neq x(w)$, and where $Z_+^{(N)}$ is the normalization factor. In order to prove this, it suffices to observe that

$$-\sum_{\langle v,w \rangle} x(v)x(w) = n_o(x) - n_e(x),$$

where $n_e(x)$ is the number of even bounds, and that $n_e(x) = M - n_o(x)$, where M is the total number of pair cliques. Therefore,

$$U(x) = 2\beta n_o(x) - M,$$

from which (10.20) follows.

Before proceeding to the proof of the announced upper bound for $\pi_+^{(N)}(x(0) = -1)$, a few definitions are needed. In fact, no formal definition will be proposed. Instead, the reader is referred to pictures. Figure 10.3.2 features *circuits C* of various lengths.

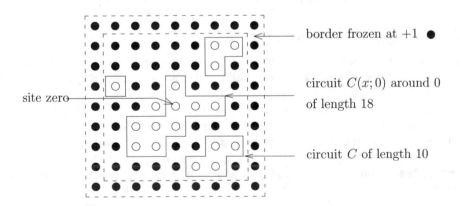

Figure 10.3.2. Circuits in the Ising model

For a given configuration x, $C(x;0)$ denotes the circuit which is the boundary of the largest connected batch of sites with negative phases, containing site 0. It is a *circuit around* 0. If the phase at the central site is positive, then $C(x;0)$ is the empty set.

For a given configuration x, denote by $\widetilde{x}$ the configuration obtained by reversing all the phases inside circuit $C(x;0)$. For a given circuit C around 0,

$$\pi_+^{(N)}(C(x;0) = C) = \frac{\sum_{x\,;\,C(x;0)=C} e^{-2\beta n_o(x)}}{\sum_y e^{-2\beta n_o(y)}}.$$

But

$$\sum_z e^{-2\beta n_o(z)} \geq \sum_{y\,;\,C(y;0)=C} e^{-2\beta n_o(\widetilde{y})}$$

(one can always associate to a configuration y such that $C(y;0) = C$ the configuration $z = \widetilde{y}$, and therefore the sum on the right-hand side is a subsum of the left-hand side). Therefore,

$$\pi_+^{(N)}(C(x;0) = C) \leq \frac{\sum_{x;C(x;0)=C} e^{-2\beta n_o(x)}}{\sum_{x;C(x;0)=C} e^{-2\beta n_o(\widetilde{x})}}.$$

If x is such that $C(x;0) = C$, then $n_0(\widetilde{x}) = n_0(x) - L$, where L is the length of C, and therefore

$$\pi_+^{(N)}(C(x;0) = C) \leq e^{-2\beta L}.$$

In particular,

$$\pi_+^{(N)}(x(0) = -1) \leq \sum r(L)e^{-2\beta L},$$

where the latter summation is over all lengths L of circuits around 0, and $r(L)$ is the number of non-empty circuits around 0 of length L. The possible lengths are $4, 6, \ldots, 2f(N)$, where $f(N) \uparrow \infty$ as $N \uparrow \infty$. In order to bound $r(L)$ from above, observe that a circuit around 0 of length L must have at least one point at a distance smaller than or equal to $\frac{L}{2}$ of the central site 0. There are L^2 ways of selecting such a point, and then at most 4 ways of selecting the segment of C starting from this point, and then at most 3 ways of selecting the next connected segment, and so on, so that

$$r(L) \leq 4L^2 3^L.$$

Therefore,

$$\pi_+^{(N)}(x(0) = -1) \leq \sum_{L=4,6,\ldots} 4L^2(3e^{-2\beta})^L.$$

Now, the series $\sum_{L=4,6,\ldots} L^2 x^L$ has a radius of convergence not less than 1, and therefore, if $3e^{-\beta}$ is small enough, or equivalently if T is large enough, $\pi_+^{(N)}(x(0) = -1) < \frac{1}{3}$ for all N.

10.4   Exercises

Exercise 10.4.1. BOUNDARIES
Define on $V = \mathbb{Z}^2$ the two neighborhood systems of the figure below. Describe the corresponding cliques and give the boundary of a 3×3 square for each case.

Figure 10.4.1. Two neighborhood systems

Exercise 10.4.2. POSITIVE BONDS
Consider the non-oriented graph on $V = \{1, 2, 3, 4, 5, 6, 7\}$ in the figure below. Let the phase space be $\Lambda = \{-1, +1\}$. For a configuration $x \in \Lambda^V$, denote by $n(x)$ the number of positive *bonds*, that is, the number of edges of the graph for which the phases of the adjacent sites coincide. Define a probability distribution π on Λ^V by $\pi(x) = \frac{e^{-n(x)}}{Z}$. Give the value of Z and the local characteristics of this random field.

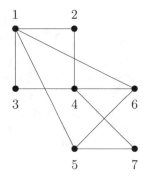

Figure 10.4.2

Exercise 10.4.3. ON THE MARKOV PROPERTY
Let V be a finite set of sites and Λ a finite set of phases. Let $\{X(v)\}_{v \in V}$ be a Markov field with values in Λ^V and admitting a Gibbsian description in terms of the symmetric relation $\sim$, with Gibbs potential $\{V_C\}_{C \subset V}$. Prove that for all

subsets A, B, of S such that

$$A \cap B = \varnothing$$

it holds that for all $x \in \Lambda^V$,

$$P(X(A) = x(A) \mid X(B) = x(B)) = P(X(A) = x(A) \mid X(\partial \bar{B}) = x(\partial \bar{B})).$$

Exercise 10.4.4. IID RANDOM FIELDS
A. Let $(Z(v), v \in V)$ be a family of IID random variables indexed by a finite set V, with $P(Z(v) = -1) = p$, $P(Z(v) = +1) = q = 1 - p$. Show that

$$P(Z = z) = K e^{\gamma \sum_{v \in V} z(s)},$$

for some constants γ and K to be identified.

B. Do the same with $P(Z(v) = 0) = p$, $P(Z(v) = +1) = q = 1 - p$.

Exercise 10.4.5. TWO-STATE HMC AS GIBBS FIELD
Consider an HMC $\{X_n\}_{n \geq 0}$ with state space $E = \{-1, 1\}$ and transition matrix

$$\mathbf{P} = \begin{pmatrix} 1 - \alpha & \alpha \\ \beta & 1 - \beta \end{pmatrix},$$

where $\alpha, \beta \in (0, 1)$, and with the stationary initial distribution

$$(\nu_0, \nu_1) = \frac{1}{\alpha + \beta}(\beta, \alpha).$$

Give a representation of $(X_0, \ldots, X_N)$ as an MRF. What is the normalized potential with respect to phase 1?

Exercise 10.4.6. POISSONIAN VERSION OF BESAG'S MODEL
Consider the model of Example 10.2.2 with the following modifications. Firstly, the phase space is $\Lambda - \mathbb{N}$, and secondly, the potential is now given by

$$V_C(x) := \begin{cases} -\log(g(x(v)) + \alpha_1 x(v) & \text{if } C = \{v\} \in \mathcal{C}_1, \\ \alpha_j x(v) x(w) & \text{if } C = \{v, w\} \in \mathcal{C}_j, \end{cases}$$

where $\alpha_j \in \mathbb{R}$ and $g : \mathbb{N} \to \mathbb{R}$ is strictly positive. As in the auto-binomial model, for any clique C not of the type $\mathcal{C}_j$, $V_C \equiv 0$. For what function g do we have

$$\pi^s(x) = e^{-\rho} \frac{\rho^{x(v)}}{x(v)!},$$

where $\rho = e^{-\langle \alpha, b \rangle}$, and where $\langle \alpha, b \rangle$ is as in Subsection 10.2.2? (This model is the *auto-Poisson model*.)

Exercise 10.4.7. ISING ON THE TORUS
([9]) Consider the classical Ising model of Example 10.1.8, except that the site space $V = \{1, 2, \ldots, N\}$ consists of N points arranged in this order on a circle. The neighbors of site i are $i + 1$ and $i - 1$, with the convention that site $N + 1$ is site 1. The phase space is $\Lambda = \{+1, -1\}$. Compute the partition function. Hint: express the normalizing constant Z_N in terms of the N-th power of the matrix

$$R = \begin{pmatrix} R(+1, +1) & R(+1, -1) \\ R(-1, +1) & R(-1, -1) \end{pmatrix} = \begin{pmatrix} e^{K+h} & e^{-K} \\ e^{-K} & e^{K-h} \end{pmatrix},$$

where $K := \frac{J}{kT}$ and $h := \frac{H}{kT}$.

Exercise 10.4.8. FREEZING
Let V be a finite set of sites and Λ a finite set of phases. Let $\{X(v)\}_{v \in V}$ be a Markov field with values in Λ^V and admitting a Gibbsian description in terms of the neighborhood structure $\sim$, with potential $\{V_C\}_{C \subset V}$. Let $A + B = V$ be a partition of the site. Fix $x(A) = \underline{x}(A)$ and define the distribution π_A on Λ^B by

$$\pi_A(x(B)) = \frac{e^{-U(\underline{x}(A), x(B))}}{\sum_{y(B) \in \Lambda^B} e^{-U(\underline{x}(A), y(B))}},$$

where U is the energy function associated with the potential $\{V_C\}_{C \subset V}$. Show that

$$\pi_A(x(B)) = P(X(B) = x(B) \mid X(A) = \underline{x}(A))$$

and that $\pi_A(x(B))$ is a Gibbs distribution for which you will give the neighborhood system and the corresponding cliques, as well as the local characteristics. (A Markov field with values in Λ^B and with the distribution π_A is called a *version* of $\{X_v\}_{v \in V}$, *frozen* on A at value $\underline{x}(A)$, or *clamped* at $\underline{x}(A)$.)

Exercise 10.4.9. HARD-CORE MODEL
Consider a random field with finite site space V and phase space $\Lambda := \{0, 1\}$ (with the interpretation that if $x(v) = 1$, site v is said to be "occupied", otherwise, it is "vacant") evolving in time. The resulting sequence $\{X_n\}_{n \geq 0}$ is an HMC with state space F, the subset of $E = \{0, 1\}^V$ consisting of the configurations x such that for all $v \in V$, $x(v) = 1$ implies that $x(w) = 0$ for all $w \sim v$. The updating procedure is

---

[9][Baxter, 1965].

the following. If the current configuration is x, choose a site v uniformly at random, and if no neighbor of v is occupied, make v occupied or vacant equiprobably. Show that the HMC so described is irreducible and that its stationary distribution is the uniform distribution on F.

Exercise 10.4.10. NEURAL NETWORK
The graph structure is as in the Ising model, but now the phase space is $\Lambda = \{0, 1\}$. A site v is interpreted as being a *neuron* that is *excited* if $x(v) = 1$ and *inhibited* if $x(v) = 0$. If $w \sim v$, one says that there is a *synapse* from v to w, and such a synapse has a *strength* σ_{vw}. If $\sigma_{vw} > 0$, one says that the synapse is *excitatory*; otherwise it is called *inhibitory*. The energy function is

$$U(x) = \sum_{v \in V} \sum_{w\,;\,w \sim v} \sigma_{wv} x(w) x(v) - \sum_{v \in V} h_v x(v),$$

where h_v is called the *threshold* of neuron v (we shall understand why later).

(a) Describe the corresponding Gibbs potential.

(b) Give the local characteristics.

(c) Describe the Gibbs sampling algorithm.

(d) Show that this procedure can also be described in terms of a random *threshold jitter* Σ with the cumulative distribution function

$$P(\Sigma \leq a) - \frac{e^{-a/T}}{1 + e^{-a/T}},$$

the Gibbs sampler selecting phase 0 if

$$\sum_{w \in N_v} (\sigma_{wv} + \sigma_{vw}) x(w) < h_v + \Sigma,$$

and 1 otherwise. One may interpret h_v as the *nominal threshold* at site v and $h_v + \Sigma$ as the actual (random) threshold. Also the quantity $\sum_{w \in N_v} (\sigma_{wv} + \sigma_{vw}) x(w)$ is the input into neuron v. Thus the excitation of neuron v is obtained by comparing its input to a random threshold (Figure 10.4.3).

Exercise 10.4.11. THERMODYNAMICS, I
Let

$$\pi_T(x) = \frac{1}{Z} e^{\frac{-E(x)}{kT}}$$

be a Gibbs distribution on the finite space $E = \Lambda^V$. Here Z is short for Z_T, and $E(x)$ is the energy of physics, differing from $U(x)$ by the Boltzmann constant k.

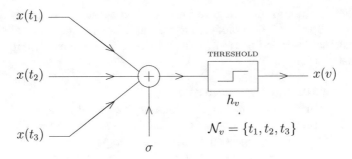

Figure 10.4.3. Jitter sampling of a neural network

For any function $f : E \to \mathbb{R}$, define

$$\langle f \rangle = \sum_{x \in E} \pi(x) f(x).$$

In particular, the *internal energy* is

$$U = \langle E \rangle = \sum_{x \in E} \pi(x) E(x).$$

The *free energy* F is defined by

$$F := -kT \, \log(Z).$$

Show that

$$U = -T^2 \frac{\partial}{\partial T} \left(\frac{F}{T} \right).$$

(This is in agreement with standard thermodynamics.)

Exercise 10.4.12. THERMODYNAMICS, II
(Continuation of Exercise 10.4.11.) For the Ising model, take

$$E(x) = E_0(x) + E_1(x),$$

where $E_0(x)$ is the interaction energy, assumed symmetric, that is, $E_0(-x) = E_0(x)$, and

$$E_1(x) = -Hm(x),$$

where

$$m(x) = \sum_{i=1}^{N} x(i)$$

is the magnetic moment of the configuration $x = (x(1), \ldots, x(N))$ (recall that $S = \{1, 2, \ldots, N\}$), and H is the external magnetic field. The partition function, still denoted by Z, is now a function of T and H. The free energy *per site* is

$$f(H, T) = -kT \frac{1}{N} \log(Z),$$

whereas the *magnetization*

$$M(H, T) = \frac{1}{N} \langle m \rangle$$

is the average magnetic moment per site.

Show that

$$M(H, T) = -\frac{\partial}{\partial H} f(H, T)$$

and

$$\frac{\partial M}{\partial H} = \frac{1}{NkT} \left(\langle m^2 \rangle - \langle m \rangle^2 \right).$$

In particular,

$$\frac{\partial M}{\partial H} \geq 0.$$

Exercise 10.4.13. THERMODYNAMICS, III
(Continuation of Exercise 10.4.12.) Compute $\lim_{N \uparrow \infty} M(H, T)$ for the Ising model on the torus (Exercise 10.4.7). Observe that this limit, as a function of H, is analytic, and null at $H = 0$. In particular, in this model, there is no phase transition.

Chapter 11

Monte Carlo Markov Chains

11.1 General Principles of Simulation

11.1.1 Simulation via the Law of Large Numbers

In view of evaluating the expectation $E[\varphi(Z)]$ of a random vector Z of dimension k with a probability density $f(x)$, where $\varphi : \mathbb{R}^k \to \mathbb{R}$ and $E[|\varphi(Z)|] < \infty$, the formula

$$E[\varphi(Z)] = \int_{\mathbb{R}^k} \varphi(x) f(x)\, dx$$

can be used if one is able to compute the integral analytically, which is seldom the case. A first alternative is numerical integration. A second alternative is to generate a sequence of IID random vectors $\{Z_n\}_{n>1}$ with the same distribution as Z, and to invoke the strong law of large numbers

$$E[\varphi(Z)] = \lim_{n \to \infty} \frac{1}{n} \sum_{i=1}^{n} \varphi(Z_i)$$

to obtain an estimate based on a finite sample $(Z_1, ..., Z_n)$, namely $\frac{1}{n}\sum_{i=1}^{n} \varphi(Z_i)$. The choice of the sample size n can be made via the theory of confidence intervals in order to attain a given accuracy.[1]

This method, known as *Monte Carlo simulation*, must be compared with numerical integration. Monte Carlo simulation is sometimes competitive with numerical integration for large dimension k.

Monte Carlo simulation requires generating IID random samples of a given distribution, that is, producing IID random variables with the said distribution.

---

[1]See, for instance, [Ross, 1987], [Ripley, 1987] or [Fishman, 1996)].

P. Brémaud, *Markov Chains*, Texts in Applied Mathematics 31,
https://doi.org/10.1007/978-3-030-45982-6_11

One of the difficulties is that the probability density $f(x)$ is in important applications known only up to a normalizing factor, that is, $f(x) = K\tilde{f}(x)$, and then, the integral that gives the normalizing factor is difficult or impossible to compute. In physics, this is frequently the case, because the partition function of a Gibbs distribution is usually unavailable in closed form.

Another difficulty concerns round-off errors, or coding problems as we shall see.

11.1.2   Two Methods for Sampling a PDF

Method of the Inverse

The *inverse distribution method* uses a sequence of IID random variables $\{U_n\}_{n \geq 1}$ uniformly distributed on $[0,1]$, and generates $\{Z_n\}_{n \geq 1}$ by

$$Z_n = F^{-1}(U_n),$$

where F^{-1} is the inverse function of $x \mapsto F(x) = \int_{\infty}^{x} f(y)\, \mathrm{d}y$ (we consider the one-dimensional case, but this is not essential in the discussion). (See Figure 11.1.1.) The scope of this method is limited by the fact that the inverse function is often difficult to obtain. In any case the inverse function method requires full knowledge of the probability density function $f(x)$.

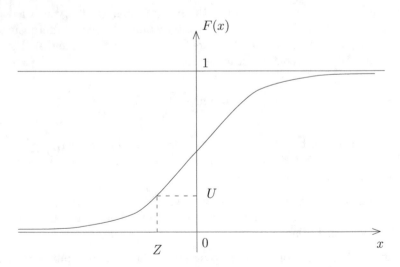

Figure 11.1.1. The inverse method

EXAMPLE 11.1.1: SAMPLING THE EXPONENTIAL DISTRIBUTION. The inverse function of $F : x \mapsto 1 - e^{\theta x}$ is $F^{-1} : y \mapsto -\frac{1}{\theta}\ln(y)$. Therefore, if U is uniformly distributed on $[0,1]$, then $\frac{1}{\theta}\ln(1-U)$ is an exponential random variable with mean θ^{-1}. So is

$$Z = \frac{1}{\theta}\ln(U)\,,$$

since U and $1 - U$ are identically distributed.

The Acceptance-Rejection Method

The *acceptance-rejection* (AR) method works as follows. Suppose one can generate a sequence of IID random vectors $\{Y_n\}_{n\geq 1}$ with the probability distribution $g(x)$ satisfying, for all x,

$$\frac{f(x)}{g(x)} \leq c < \infty\,.$$

Let $\{U_n\}_{n\geq 1}$ be a sequence of IID random variables uniformly distributed on $[0,1]$. If we define τ to be the first index $n \geq 1$ for which

$$U_n \leq \frac{f(Y_n)}{cg(Y_n)}$$

and let

$$Z = Y_\tau\,,$$

then (Exercise 11.5.1) Z admits the probability density function $f(x)$ and

$$E[\tau] = c\,.$$

EXAMPLE 11.1.2: A SMALL EXAMPLE. Figure 11.1.2 summarizes the operations in the AR method in an artificial example (where the AR method is certainly not needed!). Here the target density $f(x)$ is a "triangle" of base $[0,1]$ and summit at $(\frac{1}{2},2)$, and $g(x)$ is the uniform distribution on $[0,1]$. We take the smallest possible c, here $c = 2$. The average number of tests to obtain a sample is therefore $c = 2$.

The Shortcomings

Both the inverse method and the acceptance–rejection method have obvious counterparts when Z is a discrete random variable with values in a finite space $E = \{1, 2, ..., r\}$. We then denote by π the distribution of Z. The inverse method is in

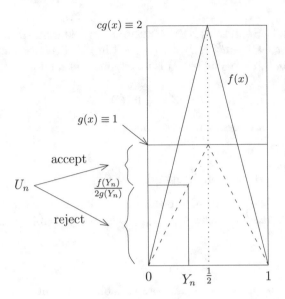

Figure 11.1.2. The acceptance-rejection method

this case always theoretically feasible: It consists in generating a random variable U uniformly distributed on $[0,1]$ and letting $Z = i$ if and only if $\sum_{\ell=1}^{i-1} \pi(\ell) \leq U < \sum_{\ell=1}^{i} \pi(\ell)$. When the size r of the state space E is large, problems arise that are due to the small size of the intervals partitioning $[0,1]$ and to the cost of precision in computing. In random field simulation, another, maybe more important, reason is the necessity to enumerate the configurations, which implies coding and decoding of a mapping from the integers to the configuration space. The decoding part is usually very difficult and a small error may lead to a far-out sample (the configurations corresponding to close integers may be very different, which is a problem in image processing).

The AR method uses a distribution p on E such that $\frac{\pi(i)}{p(i)} \leq c < \infty$ for all $i \in E$. The main problem is that π is often known only up to a normalizing factor. The inverse method suffers from the same ailment.

11.2 Monte Carlo Markov Chain Sampling

11.2.1 The Basic Idea

The methodology of the *Monte Carlo Markov chain* (MCMC) method for sampling a probability distribution π on the finite space E is the following. One constructs an irreducible aperiodic HMC $\{X_n\}_{n\geq 0}$ with state space E admitting π as stationary distribution. Since E is finite, the chain is ergodic and therefore, for any initial distribution μ,

$$\lim_{n\to\infty} P_\mu(X_n = i) = \pi(i) \quad (i \in E)$$

and

$$\lim_{N\to\infty} \frac{1}{N} \sum_{n=1}^{N} \varphi(X_n) = E_\pi[\varphi(X)].$$

When n is "large," we can consider that X_n has a distribution "close" to π. Of course, one would like to know how accurately X_n imitates an E-valued random variable Z with distribution π. For this we shall seek to obtain estimates of the form

$$|\mu \mathbf{P}^n - \pi| \leq A\alpha^n,$$

where $\alpha < 1$. The Perron–Frobenius theorem gives a theoretical answer, in terms of the second-largest eigenvalue modulus (SLEM). To make it applicable, it is necessary to find good upper bounds of the SLEM, since the eigenstructure of a transition matrix is often hard to obtain. Results in this direction were reviewed in Chapter 9. They mainly concern reversible chains, fortunately the most frequent case in MCMC, as we shall now see.

For the time being, the basic problem is that of designing the MCMC algorithm. One looks for an ergodic transition matrix $\mathbf{P}$ on E whose stationary distribution is the *target distribution* π. There are infinitely many such transition matrices, and among them there are infinitely many that correspond to a reversible chain, that is, such that

$$\pi(i)p_{ij} = \pi(j)p_{ji}. \tag{11.1}$$

We seek solutions of the form

$$p_{ij} = q_{ij}\alpha_{ij} \tag{11.2}$$

for $j \neq i$, where $Q = \{q_{ij}\}_{i,j\in E}$ is an arbitrary irreducible transition matrix on E, called the *candidate-generating* matrix: When the present state is i, the next *tentative* state j is chosen with probability q_{ij}. When $j \neq i$, this new state is accepted with probability α_{ij}. Otherwise, the next state is the same state i. Hence, the resulting probability of moving from i to j when $i \neq j$ is given by (11.2). It remains to select the *acceptance probabilities* α_{ij}. A general form of the acceptance probabilities covering most well-known MCMC algorithms is[2]

$$\alpha_{ij} = \frac{s_{ij}}{1 + t_{ij}},$$

where $\Sigma = \{s_{ij}\}_{i,j \in E}$ is a *symmetric* matrix and

$$t_{ij} = \frac{\pi(i)q_{ij}}{\pi(j)q_{ji}}.$$

Of course, Σ must be selected in such a way that the constraint $\alpha_{ij} \in [0,1]$ is respected. One can check that the reversibility condition (11.1) is satisfied and therefore π is the stationary distribution (by Theorem 2.4.12).

EXAMPLE 11.2.1: THE METROPOLIS ALGORITHM. ([3]) In order to satisfy the constraint $\alpha_{ij} \in [0,1]$, one must have

$$s_{ij} \leq 1 + \min(t_{ij}, t_{ji})$$

(recall that Σ is assumed symmetric). Equality corresponds to the Metropolis algorithm

$$\alpha_{ij} = \min\left(1, \frac{\pi(j)q_{ji}}{\pi(i)q_{ij}}\right).$$

A special case is that for which the candidate-generation mechanism is purely random, that is, $q_{ij} = \text{constant}$. Then

$$\alpha_{ij} = \min\left(1, \frac{\pi(j)}{\pi(i)}\right).$$

EXAMPLE 11.2.2: BARKER'S SAMPLER. ([4]) With the choice $s_{ij} = 1$, we have Barker's algorithm, for which

$$\alpha_{ij} = \frac{\pi(j)q_{ji}}{\pi(j)q_{ji} + \pi(i)q_{ij}}.$$

In the special case of a purely random selection of the candidate,

$$\alpha_{ij} = \frac{\pi(j)}{\pi(i) + \pi(j)}.$$

---

[2][Hastings, 1970]; see [Fishman, 1996] for additional bibliography.
[3][Metropolis et al., 1953].
[4][Barker, 1965].

Remark 11.2.3 Hastings algorithms require the knowledge of the target distribution π only up to a normalizing constant, since they depend only on the ratios $\pi(j)/\pi(i)$. The latter statement is true only as long as the candidate-generating matrix Q is known. This is *not* the case in the next example.

11.2.2   Convergence Rates in MCMC

If a Monte Carlo simulation algorithm is stopped at the n-th iteration, the closeness of the sample X_n to the target distribution π is measured by the distance in variation $d_V(\delta_i \mathbf{P}^n, \pi)$, where i is the initial state. Upper bounds on this performance index can in principle be derived with the help of the results of Chapter 9. For instance, assuming that $(\mathbf{P}, \pi)$ is reversible, Theorem 9.1.7 gives

$$4d_V(\delta_i \mathbf{P}^n, \pi)^2 \leq \frac{\rho^{2n}}{\pi(i)}, \tag{11.3}$$

where ρ, the SLEM of $\mathbf{P}$, can be bounded above using any of the techniques (Poincaré, Cheeger, ...) explained in Chapter 9.

EXAMPLE 11.2.4: A SMALL EXAMPLE. ([5]) Let π be the distribution on $E = \{1, \ldots, r\}$ given by

$$\pi(i) = z(a)a^{h(i)},$$

where $a \in (0,1)$, and where

$$h(i+1) - h(i) \geq c \geq 1 \quad (1 \leq i \leq r-1). \tag{11.4}$$

We use the Metropolis algorithm with acceptance probability

$$\alpha_{ij} = \min\left(1, \frac{\pi(j)q_{ij}}{\pi(i)q_{ji}}\right),$$

where the candidate-generating matrix $Q = \{q_{ij}\}_{i,j \in E}$ corresponds to a symmetric random walk on E with holding probability $\frac{1}{2}$ at the extreme states 1 and r. Therefore, the simulation matrix $\mathbf{P}$ is

$$p_{11} = 1 - \frac{a^{h(2)-h(1)}}{2}, \quad p_{12} = \frac{a^{h(2)-h(1)}}{2}, \quad p_{r,r-1} = p_{rr} = \frac{1}{2}$$

and, for $2 \leq i \leq r-1$,

$$p_{i,i-1} = \frac{1}{2}, \, p_{i,i+1} = \frac{a^{h(i+1)-h(i)}}{2}, \, p_{ii} = 1 - p_{i,i-1} - p_{i,i+1}.$$

---

[5][Diaconis and Saloff-Coste, 1996].

We are going to show that

$$\lambda_2 \leq 1 - \frac{(1 - a^{c/2})^2}{2}.$$

According to Theorem 9.2.1, it suffices to show that for all $x \in \mathbb{R}^r$,

$$\mathrm{Var}_\pi(x) \leq A\mathcal{E}_\pi(x, x)$$

for some A such that

$$A \leq \frac{2}{(1 - a^{c/2})^2}. \tag{11.5}$$

For this purpose, one can use an adaptation of the Diaconis–Stroock geometric method. The notations are those of Subsection 9.2.1. Let $\theta \in (0, 1)$. Proceeding as in the proof of Theorem 9.2.4, with $Q(e)^\theta$ replacing $Q(e)^{\frac{1}{2}}$,

$$2\mathrm{Var}_\pi(x) = \sum_{i,j \in E} \left\{ \sum_{e \in \gamma_{ij}} \frac{1}{Q(e)^\theta} Q(e)^\theta (x(e^-) - x(e^+)) \right\}^2 \pi(i)\pi(j)$$

$$\leq \sum_{i,j \in E} \left\{ \sum_{e \in \gamma_{ij}} Q(e)^{2\theta}(x(e^-) - x(e^+))^2 \right\} \left\{ \sum_{e \in \gamma_{ij}} \frac{1}{Q(e)^{2\theta}} \right\} \pi(i)\pi(j),$$

by the Cauchy–Schwarz inequality. Therefore, letting

$$|\gamma_{ij}|_\theta := \sum_{e \in \gamma_{ij}} \frac{1}{Q(e)^{2\theta}},$$

we have

$$2\mathrm{Var}_\pi(x) \leq \sum_{i,j \in E} |\gamma_{ij}|_\theta \sum_{e \in \gamma_{ij}} Q(e)^{2\theta}(x(e^-) - x(e^+))^2 \pi(i)\pi(j)$$

$$= \sum_e (x(e^-) - x(e^+))^2 Q(e)Q(e)^{2\theta - 1} \sum_{\gamma_{ij} \ni e} \pi(i)\pi(j)|\gamma_{ij}|_\theta.$$

With

$$A = \max_e \left\{ Q(e)^{2\theta - 1} \sum_{\gamma_{ij} \ni e} \pi(i)\pi(j)|\gamma_{ij}|_\theta \right\}, \tag{11.6}$$

we finally have

$$\mathrm{Var}_\pi(x) \leq A\mathcal{E}_\pi x, x).$$

It now remains to bound A as in (11.5), making use of the freedom in the choice of the γ_{ij}'s and of θ. First, take $\gamma_{ij} = (i, i+1, \ldots, j-1, j)$ for all $i, j \in E$, $i \geq j$. For the specific algorithm considered,

$$Q(i, i+1) = \pi(i)p_{i,i+1} = z(a)a^{h(i)}\frac{a^{h(i+1)-h(i)}}{2} = z(a)\frac{a^{h(i+1)}}{2} = \frac{\pi(i+1)}{2},$$

and therefore by reversibility,

$$Q(i, i+1) = Q(i+1, i) = \frac{\pi(i+1)}{2}.$$

The dominating term in $|\gamma_{ij}|_\theta$ is $Q(j-1, j)^{-2\theta} = \left(\frac{\pi(j)}{2}\right)^{-2\theta}$. Factor it out to obtain, using (11.4),

$$|\gamma_{ij}|_\theta = \left(\left(\frac{\pi(i+1)}{\pi(j)}\right)^{-2\theta} + \cdots + \left(\frac{\pi(j)}{\pi(j)}\right)^{-2\theta}\right)\left(\frac{\pi(j)}{2}\right)^{-2\theta} \leq \frac{\pi(j)^{-2\theta}}{1 - a^{2c\theta}}.$$

Let $e = (k, k+1)$. We have to bound (see (11.6))

$$\frac{Q(e)^{2\theta} - 1}{1 - a^{2c\theta}} \sum_{0 \leq i \leq k, k+1 \leq j \leq n} \pi(i)\pi(j)^{1-2\theta}.$$

The sum in j is bounded by $\frac{\pi(k+1)^{1-2\theta}}{1-a^{c(1-2\theta)}}$, and the sum in i by 1. Therefore,

$$A \leq \frac{2}{1 - a^{c(1-2\theta)}(1 - a^{2c\theta})},$$

from which (11.5) follows by the choice $\theta = \frac{1}{4}$. The Gershgorin bound (Theorem A.2.4) gives, for the smallest eigenvalue λ_r,

$$\lambda_r \geq -1 + 2\min p_{ii} \geq -1 + 2\left(\frac{1}{2} - \frac{a^c}{2}\right) = a^c,$$

and therefore,

$$\rho = \min(\lambda_2, |\lambda_r|) \leq \min\left(1 - \frac{(1 - a^{c/2})}{2}, a^c\right).$$

As we have already observed, upper bounds are a necessary recourse, since the eigenstructure of a large transition matrix, even a reversible one, is usually out of reach. There is, however, a notable exception concerning the Metropolis algorithm (Exercise 11.5.6).

EXAMPLE 11.2.5: THE DISCRETE EXPONENTIAL DISTRIBUTION. ([6]) Consider the probability distribution π on $E = \{1, \dots, r\}$ given by

$$\pi(j) = \frac{\theta^{j-1}(1-\theta)}{1-\theta^r}$$

where $\theta \in (0,1)$. Generate candidates randomly, that is, $p(j) = \frac{1}{r}$. Here the states are naturally ordered as in Exercise 11.5.6, $w(j) = r\pi(j)$ and

$$\lambda_2 = \sum_{j=1}^{r} \left(\frac{1}{r} - \frac{\pi(j)}{r\pi(1)} \right) = 1 - \frac{1}{r}\frac{1-\theta^r}{1-\theta}.$$

Since in this algorithm $\rho = \lambda_2$ and all the eigenvalues are non-negative, the upper bound (11.3) gives, after some pertinent bounding,

$$4d_V(\delta_r \mathbf{P}^n, \, \pi)^2 \leq \theta^{-r}\left(1 - \frac{2}{r}\right)^{2n}.$$

11.2.3   Variance of Monte Carlo Estimators

So far we have been concerned with assessing the quality of an MCMC in terms of the closeness of its output distribution from the target distribution. We now consider the problem of evaluating expectations with respect to the target distribution by ergodic estimates.

The asymptotic variance

$$v(f, \mathbf{P}, \pi) := \lim_{n\to\infty} \frac{1}{n} \mathrm{var}_\mu\left(\sum_{k=1}^{n} f(X_k)\right)$$

was computed in Theorem 6.3.7:

$$v(f, \mathbf{P}, \pi) = 2\langle f, \mathbf{Z}f\rangle_\pi - \langle f, (I+\Pi)f\rangle_\pi. \tag{11.7}$$

Here $\{X_n\}$ is an ergodic HMC with finite state space E, transition matrix $\mathbf{P}$ and stationary distribution π, and

$$\mathbf{Z} := (I - \mathbf{P} + \Pi)^{-1}, \tag{11.8}$$

---

[6][Diaconis and Saloff-Coste, 1996].

where $\Pi := \mathbf{1} \cdot \pi^T$ is the second fundamental matrix.

In this section, the reversible transition matrices considered are those corresponding to the MCMC simulation algorithms. We are interested in designing the best simulation algorithm in the sense that $v(f, \mathbf{P}, \pi)$ is to be minimized with respect to $\mathbf{P}$, uniformly in f, and of course for a fixed π. The following result answers the question in general terms.

Theorem 11.2.6 ([7]) *Let $\mathbf{P}_1$ and $\mathbf{P}_2$ be reversible ergodic transition matrices on the finite state space E, with the same stationary distribution π. If $\mathbf{P}_1$ has all its off-diagonal terms greater than or equal to the corresponding off-diagonal terms of $\mathbf{P}_2$, then*

$$v(f, \mathbf{P}_1, \pi) \leq v(f, \mathbf{P}_2, \pi)$$

for all $f : E \to \mathbb{R}$.

Proof. Let $k, \ell \in E$ with $k \neq \ell$. From (11.7) we have

$$\frac{\partial}{\partial p_{k\ell}} v(f, \mathbf{P}, \pi) = 2 \left\langle f, \frac{\partial \mathbf{Z}}{\partial p_{k\ell}} f \right\rangle_\pi .$$

From $\mathbf{Z}\mathbf{Z}^{-1} = I$, it follows that $\left(\frac{\partial}{\partial p_{k\ell}} \mathbf{Z} \right) \mathbf{Z}^{-1} + \mathbf{Z} \left(\frac{\partial}{\partial p_{k\ell}} \mathbf{Z}^{-1} \right) = 0$, and therefore

$$\frac{\partial \mathbf{Z}}{\partial p_{k\ell}} = -\mathbf{Z} \frac{\partial \mathbf{Z}^{-1}}{\partial p_{k\ell}} \mathbf{Z} ,$$

so that

$$\frac{\partial}{\partial p_{k\ell}} v(f, \mathbf{P}, \pi) = -2 \left\langle f, \left(\mathbf{Z} \frac{\partial \mathbf{Z}^{-1}}{\partial p_{k\ell}} \mathbf{Z} \right) f \right\rangle_\pi .$$

Since $\mathbf{P}$ is autoadjoint in $\ell^2(\pi)$, so is $\mathbf{Z}$, and therefore

$$\frac{\partial}{\partial p_{k\ell}} v(f, \mathbf{P}, \pi) = -2 \left\langle \mathbf{Z}f, \left(\frac{\partial \mathbf{Z}^{-1}}{\partial p_{k\ell}} \right) \mathbf{Z}f \right\rangle_\pi = -2(\mathbf{Z}f)^T d(\Pi) \frac{\partial \mathbf{Z}^{-1}}{\partial p_{k\ell}} \mathbf{Z}f .$$

Now, from (11.8),

$$\frac{\partial \mathbf{Z}^{-1}}{\partial p_{k\ell}} = -\frac{\partial \mathbf{P}}{\partial p_{k\ell}} ,$$

and therefore

$$\frac{\partial}{\partial p_{k\ell}} v(f, \mathbf{P}, \pi) = 2(\mathbf{Z}f)^T d(\Pi) \frac{\partial \mathbf{P}}{\partial p_{k\ell}} \mathbf{Z}f .$$

---

[7][Peskun, 1973].

Observe that since $\mathbf{P}$ is a stochastic matrix and $(\mathbf{P}, \pi)$ is reversible, the free parameters are the $p_{k\ell}$'s $(k < \ell)$. In view of the reversibility condition, the only non-null
elements of $d(\Pi)\frac{\partial \mathbf{P}}{\partial p_{k\ell}}$ are the (ℓ, ℓ), (ℓ, k), (k, ℓ) and (k, k) elements, respectively
equal to $-\pi(k), +\pi(k), +\pi(k)$ and $-\pi(k)$. Therefore $d(\Pi)\frac{\partial \mathbf{P}}{\partial p_{k\ell}}$ is a negative definite
symmetric matrix and

$$\frac{\partial}{\partial p_{k\ell}} v(f, \mathbf{P}, \pi) \leq 0 \,,$$

from which the conclusion follows. $\square$

EXAMPLE 11.2.7: OPTIMALITY OF THE METROPOLIS SAMPLER. In the Hastings algorithms

$$p_{ij} = q_{ij} \frac{s_{ij}}{1 + t_{ij}} \,,$$

where t_{ij} depends on $Q = \{q_{ij}\}$ and π only. We would like to find the best MCMC
algorithm in the Hastings class where Q is fixed. We have observed that from the
constraints $\leq \alpha_{ij} \in (0, 1)$ and the required symmetry of $\{s_{ij}\}$,

$$s_{ij} \leq 1 + \min(t_{ij}, t_{ji}) \,,$$

with equality for the Metropolis algorithm. It follows from Peskun's result that
the Metropolis algorithm is optimal with respect to asymptotic variance in the
class of Hastings algorithms with fixed candidate-generating matrix Q.

Remark 11.2.8 It is interesting to compare a given MCMC algorithm corresponding to a reversible pair $(\mathbf{P}, \pi)$ to independent sampling for which $\mathbf{P} = \pi$.
From the variance point of view, it follows from (11.7) that an MCMC algorithm
based on $\mathbf{P}$ performs better than independent sampling uniformly in f if and only
if

$$\langle f, \mathbf{Z}f \rangle_{\pi} \leq \langle f, f \rangle_{\pi} \tag{11.9}$$

for all $f : E \to \mathbb{R}$. From (11.7), $\langle f, \mathbf{Z}f \rangle_{\pi} \geq 0$ for all f, and we have already
observed that Z is self-adjoint in $\ell^2(\pi)$. Therefore its eigenvalues are real and
non-negative. Condition (11.9) is equivalent to the fact that these eigenvalues are
less than or equal to 1. Therefore, in view of (11.8), (11.9) is equivalent to $\mathbf{P} - \Pi$
having all its characteristic roots non-positive.

EXAMPLE 11.2.9: BARKER VS. INDEPENDENT SAMPLING. The trace of a matrix is by definition the sum of its diagonal elements. For a stochastic matrix it is therefore the sum of its elements minus the sum of its off-diagonal elements. In particular, tr $(\mathbf{P}) = r - \sum_{i>j}(p_{ij} + p_{ji})$. Since tr $(\Pi) = 1$, we have

$$\text{tr } (\mathbf{P} - \Pi) = r - 1 - \sum_{i>j}(p_{ij} + p_{ji}).$$

One can verify that for Barker's algorithm

$$\min(q_{ij}, q_{ji}) \le p_{ij} + p_{ji} \le \max(q_{ij}, q_{ji})$$

with equality if Q is symmetric. Therefore, in the case where Q is symmetric,

$$\text{tr } (\mathbf{P} - \Pi) = r - 1 + \sum_{i>j} q_{ij} \ge \frac{1}{2}(r - 2).$$

Thus, if $r \ge 2$, the sum of the characteristic roots of $\mathbf{P} - \Pi$ is positive, which implies that at least one characteristic root is positive.

Remark 11.2.10 A consequence of the previous example is that Barker's algorithm is not uniformly better than independent sampling. This does not mean that Barker's algorithm cannot perform better than independent sampling for a specific f. Moreover, and more importantly, the fact that an MCMC algorithm performs not as well as independent sampling is not too alarming, since MCMC algorithms are used when independent sampling cannot be implemented.

We now give a lower bound for the asymptotic variance of any MCMC estimator. Let $(\mathbf{P}, \pi)$ be a reversible pair, where $\mathbf{P}$ is irreducible. Its r (real) eigenvalues are ordered as follows:

$$\lambda_1 = 1 > \lambda_2 \ge \lambda_3 \ge \ldots \ge \lambda_r \ge -1.$$

For a given f, the formula

$$v(f, \mathbf{P}, \pi) = \sum_{j=1}^{r} \frac{1 + \lambda_j}{1 - \lambda_j} |\langle f, v_j \rangle_\pi|^2$$

obtained in Theorem 6.3.7 fully accounts for the interaction between f and $\mathbf{P}$, in terms of the asymptotic variance of the ergodic estimate of $\langle f \rangle_\pi$. Since the function $x \to \frac{1+x}{1-x}$ is increasing in $(0, 1]$, and λ_2 is the second largest eigenvalue of $\mathbf{P}$, the worst (maximal) value of the performance index

$$\gamma(f, \mathbf{P}, \pi) = \frac{v(f, \mathbf{P}, \pi)}{\mathrm{var}_\pi(f)} = \frac{\sum_{j=2}^{r} \frac{1+\lambda_j}{1-\lambda_j} |\langle f, v_j \rangle_\pi|^2}{\sum_{j=2}^{r} |\langle f, v_j \rangle_\pi|^2}$$

is attained for $f = v_2$, and is then equal to

$$\gamma(\mathbf{P}, \pi) = \frac{1 + \lambda_2}{1 - \lambda_2}. \tag{11.10}$$

Let $M(\pi)$ be the collection of irreducible transition matrices $\mathbf{P}$ such that the pair $(\mathbf{P}, \pi)$ is reversible, and denote by $\lambda_2(\mathbf{P})$ the second largest eigenvalue of $\mathbf{P}$. Assume that

$$\pi(1) \le \pi(2) \le \cdots \le \pi(r).$$

In particular, $0 < \pi(1) \le \frac{1}{2}$.

Theorem 11.2.11 ([8]) *Let $\mathbf{P} \in M(\pi)$. Then*

$$\lambda_2(\mathbf{P}) \ge -\frac{\pi(1)}{1 - \pi(1)}, \tag{11.11}$$

and the bound is attained for some $\mathbf{P} \in M(\pi)$. In particular,

$$\inf_{\mathbf{P} \in M(\pi)} \sup_{f \ne 0} \gamma(f, \mathbf{P}, \pi) = 1 - 2\pi(1). \tag{11.12}$$

Proof. Clearly, (11.12) follows from (11.10) and the first part of the theorem. In order to prove (11.11), we use Rayleigh's representation

$$\lambda_2 = \sup \left\{ \frac{\langle \mathbf{P}f, f \rangle_\pi}{\|f\|_\pi^2}; \ \langle f \rangle_\pi = 0, f \ne 0 \right\}$$

and exhibit a vector $f \ne 0$ such that $\langle f \rangle_\pi = 0$ and

$$\frac{\langle \mathbf{P}f, f \rangle_\pi}{\|f\|_\pi^2} = -\frac{\pi(1)}{1 - \pi(1)}.$$

We try

$$v_2 = \delta_1 - \langle \delta_1 \rangle_\pi \mathbf{1} = (1 - \pi(1), -\pi(1), \dots, -\pi(1))^T,$$

---

[8][Frigesi, Hwang and Younès, 1992].

where δ_i has zero entries except for the ith one, which is equal to 1. By the reversibility of $(\mathbf{P}, \pi)$ and the fact that $\mathbf{P}$ is a stochastic matrix, the i-th coordinate of $\mathbf{P}v_2$ is

$$p_{i1}(1 - \pi(1)) - \sum_{j=2}^{r} p_{ij}\pi(1) \;=\; p_{i1}(1 - \pi(1)) - (1 - p_{i1})\pi(1)$$

$$=\; p_{i1} - \pi(1) = \frac{\pi(1)}{\pi(i)}p_{1i} - \pi(1)\,.$$

Therefore,

$$\langle \mathbf{P}v_2, v_2 \rangle_\pi = (p_{11} - \pi(1))(1 - \pi(1))\pi(1) - \sum_{j=2}^{r} \left(\frac{\pi(1)}{\pi(j)}p_{1j} - \pi(1) \right) \pi(1)\pi(j)$$

$$= (p_{11} - \pi(1))\pi(1)\,.$$

Also,

$$\|v_2\|_\pi^2 = \pi(1)(1 - \pi(1))\,.$$

Therefore,

$$\frac{\langle \mathbf{P}v_2, v_2 \rangle_\pi}{\|v\|_\pi^2} = \frac{p_{11} - \pi(1)}{1 - \pi(1)} \geq -\frac{\pi(1)}{1 - \pi(1)}\,.$$

This proves (11.11).

We now prove that the bound is attained. This is done by explicitly constructing $\mathbf{P}$ such that $\lambda_2(\mathbf{P}) = -\frac{\pi(1)}{1-\pi(1)}$. In view of the last display, this implies that

$$p_{11} = 0\,. \tag{11.13}$$

If $\lambda_2(\mathbf{P}) = -\frac{\pi(1)}{1-\pi(1)}$ (and for this (11.13) must hold), we necessarily have

$$\frac{\langle \mathbf{P}v_2, v_2 \rangle_\pi}{\|v_2\|_\pi^2} = \lambda_2(\mathbf{P}),$$

and therefore (Theorem 9.2.1) v_2 is the corresponding eigenvector, as anticipated by the notation. Writing down explicitly

$$\mathbf{P}v_2 = -\frac{\pi(1)}{(1 - \pi(1))}v_2\,,$$

we obtain for $i > 1$,

$$\frac{\pi(1)}{\pi(i)}p_{1i} - \pi(1) = -\frac{\pi(1)}{(1 - \pi(1))}(-\pi(1))\,,$$

that is,

$$p_{1i} = \frac{\pi(i)}{(1 - \pi(1))}.$$

By reversibility, this implies

$$p_{i1} = \frac{\pi(1)}{(1 - \pi(1))}. \tag{11.14}$$

In summary, imposing the second eigenvalue to be equal to $\frac{\pi(1)}{(1-\pi(1))}$ determines the first row and first column of $\mathbf{P}$.

We must now complete the construction by determining the restriction $\mathbf{P}_2$ of $\mathbf{P}$ to $\{2, \ldots, r\}$. In view of (11.14),

$$\mathbf{P}_2 = \left(1 - \frac{\pi(1)}{1 - \pi(1)}\right) \mathbf{P}',$$

where $\mathbf{P}'$ is a stochastic matrix. Define

$$\pi' = \left(\frac{\pi(2)}{1 - \pi(1)}, \ldots, \frac{\pi(r)}{1 - \pi(1)}\right)^T.$$

Since $\mathbf{P}$ is in detailed balance with π, $\mathbf{P}'$ is in detailed balance with π', and therefore π' is a stationary distribution of $\mathbf{P}'$.

$\mathbf{P}_2$ and $\mathbf{P}'$ have the same eigenvectors. In particular, a right-eigenvector of $\mathbf{P}_2$ different from $v_1' = (1, \ldots, 1)^T \in \mathbb{R}^{r-1}$ must be π'-orthogonal to v_1'.

Now, finding a right-eigenvector of $\mathbf{P}_2$ that is π'-orthogonal to v_1' is equivalent to finding a right-eigenvector of $\mathbf{P}$ π-orthogonal to v_1 and v_2. Indeed, if v is a right-eigenvector of $\mathbf{P}$ π-orthogonal to v_1 and v_2, it must be π-orthogonal to $v_2 - v_1 = \delta_1$, that is, its first coordinate is 0, that is, $v = (0, x_2, \ldots, x_r)^T$, and $v' = (x_2, \ldots, x_r)^T$ is π'-orthogonal to v_1', and the latter is easily seen to be a right-eigenvector of $\mathbf{P}_2$ corresponding to the same eigenvalue.

Conversely, if $v' = (x_2, \ldots, x_r)^T$ is a right-eigenvector of $\mathbf{P}_2$ that is π'-orthogonal to v_1', then $v = (0, x_2, \ldots, x_r)^T$ is a right-eigenvector of $\mathbf{P}$ corresponding to the same eigenvalue.

This shows that the second eigenvalue of $\mathbf{P}_2$ is the third eigenvalue of $\mathbf{P}$.

We choose $\mathbf{P}_2$ such that its second eigenvalue is the smallest possible, and therefore the third eigenvalue of $\mathbf{P}$ is the smallest possible, under the constraint that its second eigenvalue is the smallest possible. Since the eigenvalues of $\mathbf{P}$ and $\mathbf{P}_2$ are proportional, it suffices to perform on the pair $(\mathbf{P}', \pi')$ the same treatment as for $(\mathbf{P}, \pi)$. This leads to

$$\lambda_2' = -\frac{\pi'(2)}{1 - \pi'(2)} = -\frac{\pi(2)}{\pi(3) + \cdots + \pi(r)},$$

and the corresponding eigenvector is

$$v_2' = (1 - \pi'(2), -\pi'(2), \ldots, -\pi'(2))^T =$$

$$\left(1 - \frac{\pi(2)}{\pi(2) + \cdots + \pi(r)}, -\frac{\pi(2)}{\pi(2) + \cdots + \pi(r)}, \ldots, -\frac{\pi(2)}{\pi(2) + \cdots + \pi(r)}\right)^T.$$

Therefore

$$\lambda_3 = -\frac{\pi(2)}{\pi(3) + \cdots + \pi(r)}\left(1 - \frac{\pi(1)}{\pi(2) + \cdots + \pi(r)}\right)$$

and

$$v_3 = \left(0, 1 - \frac{\pi(2)}{\pi(2) + \cdots + \pi(r)}, -\frac{\pi(2)}{\pi(2) + \cdots + \pi(r)}, \ldots, -\frac{\pi(2)}{\pi(2) + \cdots + \pi(r)}\right)^T$$

and the first column of $\mathbf{P}_2$ is

$$(0, -\lambda_3, \ldots, -\lambda_3)^T.$$

By reversibility, the first row of $\mathbf{P}_2$ is also determined. At this point it therefore remains to determine $\mathbf{P}_3$, the restriction of $\mathbf{P}$ to $\{3, \ldots, r\}$. The iteration is now clear and the end product is a stochastic matrix with the following properties:

(i) Its eigenvalues are $1 = \lambda_1 > 0 > \lambda_2 \geq \cdots \geq \lambda_r$, such that for $i \in [1, r-1]$, λ_{i+1} is the smallest $\lambda_i(\mathbf{P})$ for all $\mathbf{P} \in M(\pi)$ that already have $1, \ldots, \lambda_i$ for eigenvalues.

(ii) Its diagonal elements are all null except perhaps the last one.

(iii) It has constant entries under the diagonal for each column, namely $\lambda_2, \ldots, \lambda_r$.

(iv) Its right-eigenvectors are $v_1 = \mathbf{1}$ and for $k \geq 1$

$$v_{k+1} =$$

$$\left(0, \ldots, 0, 1 - \frac{\pi(k)}{\pi(k) + \cdots + \pi(r)}, -\frac{\pi(k)}{\pi(k) + \cdots + \pi(r)}, \ldots, -\frac{\pi(k)}{\pi(k) + \cdots + \pi(r)}\right)^T$$

where the first $k - 1$ entries are null. The eigenvalues are, for $k \geq 1$,

$$\lambda_{k+1} = -\frac{\pi(k)}{\pi(k+1) + \cdots + \pi(r)}\Pi_{\ell=1}^{k-1}\left(1 - \frac{\pi(\ell)}{\pi(\ell) + \cdots + \pi(r)}\right).$$

Under condition (i), the above matrix is unique. $\qquad\qquad\square$

11.3   The Gibbs Sampler

11.3.1   Simulation of Random Fields

Consider a random field that changes randomly with time. In other words, we have a stochastic process $\{X_n\}_{n \geq 0}$ where

$$X_n = (X_n(v), v \in V)$$

and $X_n(v) \in \Lambda$. The state at time n of this process is a random field on V with phases in Λ, or equivalently, a random variable with values in the state space $E = \Lambda^V$, which for simplicity we assume finite. The stochastic process $\{X_n\}_{n \geq 0}$ will be called a *dynamical random field*.

The purpose of the current section is to show how a given random field with probability distribution

$$\pi(x) = \frac{1}{Z} e^{-\mathcal{E}(x)} \tag{11.15}$$

can arise as the stationary distribution of a field-valued Markov chain.

The *Gibbs sampler* uses a strictly positive probability distribution $(q_v, v \in V)$ on V, and the transition from $X_n = x$ to $X_{n+1} = y$ is made according to the following rule.

The new state y is obtained from the old state x by changing (or not) the value of the phase at *one site only*. The site v to be changed (or not) at time n is chosen independently of the past with probability q_s. When site v has been selected, the current configuration x is changed into y as follows: $y(V \backslash v) = x(V \backslash v)$, and the new phase $y(v)$ at site v is selected with probability $\pi(y(v) \mid x(V \backslash v))$. Thus, configuration x is changed into $y = (y(s), x(S \backslash s))$ with probability $\pi(y(v) \mid x(V \backslash v))$, according to the local specification at site v. This gives for the non-null entries of the transition matrix

$$P(X_{n+1} = y \mid X_n = x) = q_v \pi(y(v) \mid x(V \backslash v)) 1_{y(V \backslash v) = x(V \backslash v)}. \tag{11.16}$$

The corresponding chain is irreducible and aperiodic. To prove that π is the stationary distribution, we use the detailed balance test. For this, we have to check that for all $x, y \in \Lambda^V$,

$$\pi(x) P(X_{n+1} = y \mid X_n = x) = \pi(y) P(X_{n+1} = x \mid X_n = y),$$

that is, in view of (11.16), for all $v \in V$,

$$\pi(x) q_v \pi(y(v) \mid y(V \backslash v)) = \pi(y) q_v \pi(x(v) \mid x(V \backslash v)).$$

But the last equality is just

$$\pi(x)q_v \frac{\pi(y(v), x(V\setminus v))}{P(X(V\setminus v) = x(V\setminus v))} = \pi(y(v), x(V\setminus v))q_v \frac{\pi(x)}{P(X(V\setminus v) = x(V\setminus v))}.$$

EXAMPLE 11.3.1: SIMULATION OF THE ISING MODEL. The local specification at site v depends only on the local configuration $x(\mathcal{N}_v)$. Note that small neighborhoods speed up computations. Note also that the Gibbs sampler is a natural sampler, in the sense that in a piece of ferromagnetic material, for instance, the spins are randomly changed according to the local specification. When nature decides to update the orientation of a dipole, it does so according to the law of statistical mechanics. It computes the local energy for each of the two possible spins, $E_+ = E(+1, x(\mathcal{N}_v))$ and $E_- = E(-1, x(\mathcal{N}_v))$, and takes the corresponding orientation with a probability proportional to e^{E_+} and e^{E_-}, respectively.

EXAMPLE 11.3.2: PERIODIC GIBBS SAMPLER. In practice, the updated sites are not chosen at random, but instead in a well-determined order $v(1), v(2), \ldots, v(N)$, where $\{v(i)\}_{1 \leq i \leq N}$ is an enumeration of all the sites of V, called a *scanning policy*. The sites are visited in this order periodically. The state of the random field after the n-th sweep is $Z_n = X_{nN}$, where X_k denotes the image before the k-th update time. At time k, site $s(k \bmod N)$ is updated to produce the new image X_{k+1}. If $X_k = x$ and $s(k \bmod N) = s$, then $X_{k+1} = (y(s), x(S\setminus s))$ with probability $\pi(y(s) \mid x(S\setminus s))$. The Gibbs distribution π is stationary for $\{X_k\}_{k \geq 0}$, in the sense that if $P(X_k = \cdot) = \pi$, then $P(X_{k+1} = \cdot) = \pi$. In particular, π is a stationary distribution of the irreducible aperiodic Markov chain $\{Z_n\}_{n \geq 0}$, and $\lim_{n \uparrow \infty} P(Z_n = \cdot) = \pi$.

The transition matrix $\mathbf{P}$ of $\{Z_n\}_{n \geq 0}$ is

$$\mathbf{P} = \prod_{k=1}^{N} \mathbf{P}_{v(k)}, \tag{11.17}$$

where $\mathbf{P}_v = \{p_{xy}^v\}_{x,y \in \Lambda^V}$ and the entry p_{xy}^v of $\mathbf{P}_v$ is non-null if and only if $y(V\setminus v) = x(V\setminus v)$, and then

$$p_{xy}^v = \frac{e^{-\mathcal{E}(y(v), x(V\setminus v))}}{\sum_{\lambda \in \Lambda} e^{-\mathcal{E}(\lambda, x(V\setminus v))}}. \tag{11.18}$$

EXAMPLE 11.3.3: GIBBS SAMPLER FOR RANDOM VECTORS. Clearly, Gibbs
sampling applies to any multivariate probability distribution

$$\pi(x(1), \ldots, x(N))$$

on a set $E = \Lambda^N$, where Λ is countable (but this restriction is not essential, as
we mentioned earlier). This trivial remark is intended to remind us that there are
many applications of Gibbs sampling, and more generally of Monte Carlo Markov
chain simulation, outside physics or image processing, and especially in statistics.

The basic step of the Gibbs sampler for the multivariate distribution π consists in selecting a coordinate number i ($1 \leq i \leq N$) at random, and then choosing the new value $y(i)$ of the corresponding coordinate, given the present values
$x(1), \ldots, x(i-1), x(i+1), \ldots, x(N)$ of the other coordinates, with probability

$$\pi(y(i) \mid x(1), \ldots, x(i-1), x(i+1), \ldots, x(N)).$$

One checks as above that π is the stationary distribution of the corresponding
chain.

11.3.2   Convergence Rate of the Gibbs Sampler

One can, in principle, use the algebraic tools of Chapter 9 to obtain rates of
convergence for the Gibbs sampler.

EXAMPLE 11.3.4: PERIODIC GIBBS SAMPLER. The expression (11.18) will be
used to produce a geometric rate of convergence of the periodic Gibbs sampler,
namely,

$$mu\mathbf{P}^n - \pi| \; \leq \; \frac{1}{2}|\mu - \pi|(1 - e^{-N\Delta})^n \tag{11.19}$$

where

$$\Delta := \sup_{v \in V} \delta_v$$

and

$$\delta_v := \sup\{|\mathcal{E}(x) - \mathcal{E}(y)| \; ; \; x(V \backslash v) = y(V \backslash v)\}.$$

The proof of (11.19) uses Theorem 4.3.15, which gives

$$|\mu\mathbf{P}^n - \pi| \; \leq \; \frac{1}{2}|\mu - \pi|\delta(\mathbf{P})^n.$$

By (4.20), it follows that for any transition matrix $\mathbf{P}$ on a finite state space E,

$$\delta(\mathbf{P}) = 1 - \inf_{i,j \in E} \sum_{k \in E} p_{ik} \wedge p_{jk} \leq 1 - |E| \left(\inf_{i,j \in E} p_{ij} \right). \tag{11.20}$$

If we define $m_v(x) = \inf\{\mathcal{E}(y); y(V \backslash v) = x(V \backslash v)\}$, it follows from (11.18) that

$$p_{xy}^v = \frac{\exp\{-(\mathcal{E}(y(v), x(V \backslash v)) - m_v(x))\}}{\sum_{z(v) \in \Lambda} \exp\{-(\mathcal{E}(z(v), x(V \backslash v)) - m_v(x))\}} \geq \frac{e^{-\delta_s}}{|\Lambda|},$$

and therefore, from (11.17),

$$\min_{x,y \in \Lambda^S} p_{xy} \geq \prod_{k=1}^{N} \frac{e^{-\delta_{s(k)}}}{|\Lambda|} \geq \frac{e^{-N\Delta}}{|\Lambda|^N}.$$

Using (11.20),

$$\delta(\mathbf{P}) \leq 1 - |\Lambda|^N \frac{e^{-N\Delta}}{|\Lambda|^N} = 1 - e^{-N\Delta},$$

and (11.19) follows.

Remark 11.3.5 Convergence in variation of the Gibbs sampler to the target distribution takes place with geometric speed. This is a common feature of all the Monte Carlo Markov chain simulation algorithms, and more generally of all *finite* ergodic HMCs. However, the Gibbs sampler has a special feature. It turns out that the distance in variation to the target distribution *decreases* to 0 (Exercise 11.5.3).

11.4 Exact sampling

11.4.1 The Propp–Wilson algorithm

We attempt to obtain an *exact sample* of a given distribution π on a finite state space E, that is, a random variable Z such that $P(Z = i) = \pi(i)$ for all $i \in E$. The following algorithm[9] is based on a coupling idea. One starts as usual from an *ergodic* transition matrix $\mathbf{P}$ with stationary distribution π, just as in the classical MCMC method.

The algorithm is based on a representation of $\mathbf{P}$ in terms of a recurrence equation, that is, for given a function f and an IID sequence $\{Z_n\}_{n \geq 1}$ independent of the initial state, the chain satisfies the recurrence

$$X_{n+1} = f(X_n, Z_{n+1}). \tag{11.21}$$

---

[9][Propp and Wilson, 1993].

The algorithm constructs a family of HMCs with this transition matrix with the
help of a unique IID sequence of random vectors $\{Y_n\}_{n\in\mathbb{Z}}$, called the *updating
sequence*, where $Y_n = (Z_{n+1}(1), \cdots, Z_{n+1}(r))$ is an r-dimensional random vector,
and where the coordinates $Z_{n+1}(i)$ have a common distribution, that of Z_1. For
each $N \in \mathbb{Z}$ and each $k \in E$, a process $\{X_n^N(k)\}_{n\geq N}$ is defined recursively by:

$$X_N^N(k) = k,$$

and, for $n \geq N$,

$$X_{n+1}^N(k) = f(X_n^N(k), Z_{n+1}(X_n^N(k))).$$

(Thus, if the chain is in state i at time n, it will be at time $n+1$ in state $j =
f(i, Z_{n+1}(i))$.) Each of these processes is therefore an HMC with the transition
matrix $\mathbf{P}$. Note that for all $k, \ell \in E$, and all $M, N \in \mathbb{Z}$, the HMCs $\{X_n^N(k)\}_{n\geq N}$
and $\{X_n^M(\ell)\}_{n\geq M}$ use at any time $n \geq \max(M, N)$ the same updating random
vector Y_{n+1}.

If, in addition to the independence of $\{Y_n\}_{n\in\mathbb{Z}}$, the components $Z_{n+1}(1)$, $Z_{n+1}(2)$,
..., $Z_{n+1}(r)$ are, for each $n \in \mathbb{Z}$, independent, we say that the updating is *componentwise independent*.

Definition 11.4.1 *The random time*

$$\tau^+ = \inf\{n \geq 0; X_n^0(1) = X_n^0(2) = \cdots = X_n^0(r)\}$$

is called the forward coupling time (Figure 11.4.1). The random time

$$\tau^- = \inf\{n \geq 1; X_0^{-n}(1) = X_0^{-n}(2) = \cdots = X_0^{-n}(r)\}$$

is called the backward coupling time (Figure 11.4.1).

   Thus, τ^+ is the first time at which the chains $\{X_n^0(i)\}_{n\geq 0}$, $1 \leq i \leq r$, coalesce.

Lemma 11.4.2 *When the updating is componentwise independent, the forward
coupling time τ^+ is almost surely finite.*

Proof. Consider the (immediate) extension of Theorem 4.2.1 to the case of r
independent HMCs with the same transition matrix. It cannot be applied directly
to our situation, because the chains are not independent. However, the probability
of coalescence in our situation is bounded below by the probability of coalescence
in the completely independent case. To see this, first construct the independent
chains model, using r independent IID componentwise independent updating sequences. The difference with our model is that we use too many updatings. In

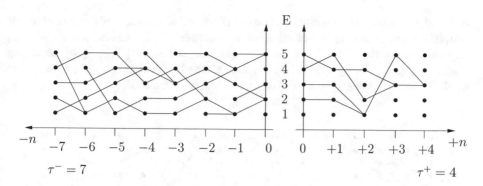

Figure 11.4.1. Backward and forward coupling

order to construct from this a set of r chains as in our model, it suffices to use for two chains the same updatings as soon as they meet. Clearly, the forward coupling time of the so modified model is smaller than or equal to that of the initial completely independent model. □

For a simpler notation, let $\tau^- := \tau$. Let

$$Z = X_0^{-\tau}(i).$$

(This random variable is independent of i. In Figure 1, $Z = 2$.) Then,

Theorem 11.4.3 *With a componentwise independent updating sequence, the backward coupling time τ is almost surely finite. Also, the random variable Z has the distribution π.*

Proof. We shall show at the end of the current proof that for all $k \in \mathbb{N}$, $P(\tau \le k) = P(\tau^+ \le k)$, and therefore the finiteness of τ follows from that of τ^+ proven in the last lemma. Now, since for $n \ge \tau$, $X_0^{-n}(i) = Z$,

$$\begin{aligned}
P(Z = j) &= P(Z = j, \tau > n) + P(Z = j, \tau \le n) \\
&= P(Z = j, \tau > n) + P(X_0^{-n}(i) = j, \tau \le n) \\
&= P(Z = j, \tau > n) - P(X_0^{-n}(i) = j, \tau > n) + P(X_0^{-n}(i) = j) \\
&= P(Z = j, \tau > n) - P(X_0^{-n}(i) = j, \tau > n) + p_{ij}(n) \\
&= A_n - B_n + p_{ij}(n).
\end{aligned}$$

But A_n and B_n are bounded above by $P(\tau > n)$, a quantity that tends to 0 as $n \uparrow \infty$ since τ is almost-surely finite. Therefore

$$P(Z = j) = \lim_{n \uparrow \infty} p_{ij}(n) = \pi(j).$$

It remains to prove the equality of the distributions of the forwards and backwards coupling time. For this, select an arbitrary integer $k \in \mathbb{N}$. Consider an updating sequence constructed from a *bona fide* updating sequence $\{Y_n\}_{n\in\mathbb{Z}}$, by replacing $Y_{-k+1}, Y_{-k+2}, \ldots, Y_0$ by $Y_1, Y_2, \ldots, Y_k$. Call τ' the backwards coupling time in the modified model. Clearly τ an τ' have the same distribution.

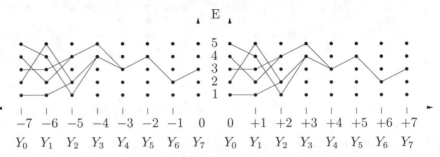

Figure 11.4.2. $\tau^+ \leq k$ implies $\tau' \leq k$

Suppose that $\tau^+ \leq k$. Consider in the modified model the chains starting at time $-k$ from states $1, \ldots, r$. They coalesce at time $-k + \tau^+ \leq 0$ (see Figure 11.4.2), and consequently $\tau' \leq k$. Therefore $\tau^+ \leq k$ implies $\tau' \leq k$, so that

$$\mathrm{P}(\tau^+ \leq k) \leq \mathrm{P}(\tau' \leq k) = \mathrm{P}(\tau \leq k).$$

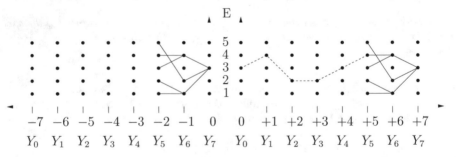

Figure 11.4.3. $\tau' \leq k$ implies $\tau^+ \leq k$

Now, suppose that $\tau' \leq k$. Then, in the modified model, the chains starting at time $k - \tau'$ from states $1, \ldots, r$ must at time $-k + \tau^+ \leq 0$ coalesce at time k. Therefore (Figure 11.4.3), $\tau^+ \leq k$. Therefore $\tau' \leq k$ implies $\tau^+ \leq k$, so that

$$\mathrm{P}(\tau \leq k) = \mathrm{P}(\tau' \leq k) \leq \mathrm{P}(\tau^+ \leq k).$$

$\square$

Remark 11.4.4 Note that the coalesced value at the forward coupling time is not a sample of π (see Exercise 11.5.4).

11.4.2 Sandwiching

The above exact sampling algorithm is often prohibitively time-consuming when the state space is large. However, if the algorithm required the coalescence of *two*, instead of r processes, then it would take less time. The Propp and Wilson algorithm does this in a special, yet not rare, case.

It is now assumed that there exists a partial order relation on E, denoted by $\preceq$, with a minimal and a maximal element (say, respectively, 1 and r), and that we can perform the updating in such a way that for all $i, j \in E$, all $N \in \mathbb{Z}$, all $n \geq N$,

$$i \preceq j \Rightarrow X_n^N(i) \preceq X_n^N(j).$$

However we do not require componentwise independent updating (but the updating vectors sequence remains IID). The corresponding sampling procedure is called the *monotone Propp–Wilson algorithm*.

Define the backwards *monotone* coupling time

$$\tau_m = \inf\{n \geq 1; X_0^{-n}(1) = X_0^{-n}(r)\}.$$

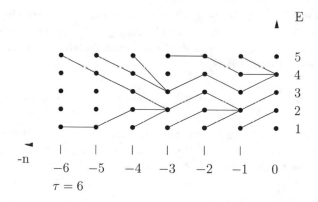

Figure 11.4.4. The Monotone Propp–Wilson algorithm

Theorem 11.4.5 *The monotone backwards coupling time τ_m is almost surely finite. Also, the random variables $X_0^{-\tau_m}(1) = X_0^{-\tau_m}(r)$ has the distribution π.*

Proof. We can use most of the proof of Theorem 11.4.3. We need only to prove independently that τ^+ is finite. It is so because τ^+ is dominated by the first time $n \geq 0$ such that $X_n^0(r) = 1$, and the latter is finite in view of the recurrence assumption.   □

Monotone coupling will occur with representations of the form (11.21) such that for all z,

$$i \preceq j \Rightarrow f(i, z) \preceq f(j, z),$$

and if for all $n \in \mathbb{Z}$, all $i \in \{1, \ldots, r\}$,

$$Z_{n+1}(i) = Z_{n+1}.$$

EXAMPLE 11.4.6: A DAM MODEL. We consider the following model of a dam reservoir. The corresponding HMC, with values in $E = \{0, 2, \ldots, r\}$, satisfies the recurrence equation

$$X_{n+1} = \min\{r, \max(0, X_n + Z_{n+1})\},$$

where, as usual, $\{Z_n\}_{n \geq 1}$ is IID. In this specific model, X_n is the content at time n of a dam reservoir with maximum capacity r, and $Z_{n+1} = A_{n+1} - c$, where A_{n+1} is the input into the reservoir during the time period from n to $n+1$, and c is the maximum release during the same period. The updating rule is then monotone.

Instead of trying the times $-1, -2$, etc., we use successive starting times of the form $\alpha^r T_0$. Let k be the first k for which $\alpha^k T_0 \geq \tau_-$. The number of simulation steps used is $2\left(T_0 + \alpha T_0 + \cdots + \alpha^k T_0\right)$ (the factor 2 accounts for the fact that we ale running two chains), that is,

$$2T_0 \left(\frac{\alpha^{k+1} - 1}{\alpha - 1}\right) < 2T_0 \left(\frac{\alpha^2}{\alpha - 1}\right) \alpha^{k-1} \leq 2\tau_- \frac{\alpha^2}{\alpha - 1}$$

steps, where we have assumed that $T_0 \leq \tau_-$. In the best case, supposing we are informed of the exact value of τ_- by some oracle, the number of steps is $2\tau_-$. The ratio of the worst to best cases is $\frac{\alpha^2}{\alpha-1}$ which is minimized for $\alpha = 2$. This is why one usually suggests to start the successive attempts of backward coalescence at times of the form $-2^k T_0$ $(k \geq 0)$.

We shall now relate the average backward recurrence time to the mixing time of the chain.

Let $(E, \preceq)$ be a partially ordered set. A subset A of E is called a chain if $(A, \preceq)$ is totally ordered. Denote by $\ell := \ell(E)$ the length of the longest chain. For instance, if $E = \{0,1\}^N$, and if $\preceq$ is defined by

$$(x_1, \ldots, x_N) \preceq (y_1, \ldots, y_N) \iff x_i \leq y_i \,(1 \leq i \leq N),$$

$\ell = N$ (start with the maximal element $(1, \ldots, 1)$ and successively change the 1's into 0's until you reach the minimal state $(0, \ldots, 0)$).

Theorem 11.4.7
$$\frac{P(\tau_+ > k)}{\ell} \leq \bar{d}(k) \leq P(\tau_+ > k).$$

Proof. Let $h(x)$ denote the length of the longest chain whose maximal element is x. In the example, it is the Hamming weight of x, that is, the number of 1's in it. If $X_0^k(1) \neq X_0^k(r)$, then $h(X_0^k(1)) + 1 \leq h(X_0^k(r))$, and if $X_0^k(1) = X_0^k(r)$, then $h(X_0^k(1)) \leq h(X_0^k(r))$. Therefore

$$1_{\{X_0^k(1) \neq X_0^k(r)\}} \leq h(X_0^k(r)) - h(X_0^k(1)).$$

In particular,

$$P(\tau_+ > k) = P(X_0^k(1) \neq X_0^k(r)) \leq E\left[h(X_0^k(r)) - h(X_0^k(1))\right].$$

Denoting by ρ_i^k the distribution $\delta_i^T \mathbf{P}^k$ of the chain with initial state i at time k,

$$E\left[h(X_0^k(r)) - h(X_0^k(1))\right] = E_{\rho_r^k}[h(X_0)] - E_{\rho_1^k}[h(X_0)]$$
$$\leq d_V(\rho_r^k, \rho_1^k)\,(\max h(x) - \min h(x)) < \ell\bar{d}(k).$$

This proves the first inequality. For the second, observe that the event that two chains starting in arbitrary initial distributions μ and ν will disagree at time k implies that $\tau_+ > k$. Therefore $d_V(\mu\mathbf{P}^k, \nu\mathbf{P}^k) \leq P(\tau_+ > k)$ and the last inequality follows since $\bar{d}(k) := \sup_{\mu,\nu} d_V(\mu\mathbf{P}^k, \nu\mathbf{P}^k)$. $\quad\square$

The next result states that the function $k \to P(\tau_+ > k)$ is submultiplicative.

Theorem 11.4.8 *Let k_1 and k_2 be integer-valued random variables. Then*

$$P(\tau_+ > k_1 + k_2) \leq P(\tau_+ > k_1)P(\tau_+ > k_2).$$

Proof. Exercise 11.5.7. $\quad\square$

Lemma 11.4.9
$$kP(\tau_+ > k) \leq E[\tau_+] \leq \frac{k}{P(\tau_+ \leq k)}.$$

Proof. The first inequality is just Markov's inequality. By the telescope formula,

$$\mathrm{E}\left[\tau_+\right] = \mathrm{P}(\tau_+ \geq 0) + \mathrm{P}(\tau_+ \geq 1) + \cdots$$

$$= \sum_{i=0}^{\infty} \left(\mathrm{P}(\tau_+ > ik + 1) + \cdots + \mathrm{P}(\tau_+ > (i+1)k) \right)$$

$$\leq 1 + \sum_{i=0}^{\infty} k \mathrm{P}(\tau_+ > ik).$$

By submultiplicativity of $k \to \mathrm{P}(\tau_+ > k)$, $\mathrm{P}(\tau_+ > ik) \leq \mathrm{P}(\tau_+ > k)^i$. Therefore,

$$\mathrm{E}\left[\tau_+\right] \leq k \sum_{i=0}^{\infty} \mathrm{P}(\tau_+ > k)^i = k \frac{1}{1 - \mathrm{P}(\tau_+ > k)} = k \frac{1}{\mathrm{P}(\tau_+ \leq k)}.$$

$\square$

Define the mixing time of the chain T_{mix} to be the first time k such that $\overline{d}(k) \leq \frac{1}{e}$. Recall that $k \to \overline{d}(k)$ is submultiplicative and therefore, after $k = T_{mix}(1 + \log \ell)$ steps,

$$\overline{d}(k) \leq \overline{d}(T_{mix})^{1+\log \ell} = \left(\frac{1}{e}\right)^{1+\log \ell} \frac{1}{e \times e^{\log \ell}} = \frac{1}{e\ell}.$$

By Theorem 11.4.7,

$$\mathrm{P}(\tau_+ > k) \leq \overline{d}(k) \times \ell \leq \frac{1}{e}.$$

Therefore, in view of the Lemma 11.4.9

$$\mathrm{E}\left[\tau_+\right] \leq \frac{k}{\mathrm{P}(\tau_+ \leq k)} \leq \frac{k}{1 - 1/e} \leq 2k = 2T_{mix}(1 + \log \ell).$$

Suppose we perform m independent experiments, called the *reference experiments*, resulting in the IID forward coalescence time sequence $T_1, \ldots, T_m$. We now would like to have an idea of the time of forward coalescence τ_+ of the experiment (independent of the reference experiments) we are about to perform. By the submultiplicativity property of Theorem 11.4.8,

$$\mathrm{P}(\tau_+ > T_1 + \ldots + T_m) \leq \mathrm{P}(\tau_+ > T_1) \times \cdots \times \mathrm{P}(\tau_+ > T_m) \leq \mathrm{P}(\tau_+ > T_1)^m.$$

By symmetry, since τ_+ and T_1 are independent and identically distributed, $\mathrm{P}(\tau_+ > T_1) = \frac{1}{2}$, and therefore

$$\mathrm{P}(\tau_+ > T_1 + \ldots + T_m) \leq \frac{1}{2^m}.$$

Recalling that the forward and backward coalescence times have the same distribution, we also have that

$$P(\tau_- > T_1 + \ldots + T_m) \le \frac{1}{2^m}.$$

11.5 Exercises

Exercise 11.5.1. THE ACCEPTANCE-REJECTION METHOD
Prove the statements concerning the acceptation-rejection sampling method of Subsection 11.1.2 that were given without proof.

Exercise 11.5.2. METROPOLIS AND BARKER ALGORITHMS: IRREDUCIBILITY
Show that for both the Metropolis and Barker samplers, if Q is irreducible and U is not a constant, then $\mathbf{P}(T)$ is irreducible and aperiodic for all $T > 0$.

Exercise 11.5.3. MONOTONICITY OF THE GIBBS SAMPLER
Let μ be an arbitrary probability measure on Λ^S and let ν be the probability measure obtained by applying the Gibbs sampler at an arbitrary site $s \in S$. Show that $d_V(\nu, \pi) \le d_V(\mu, \pi)$.

In fact it is quite unlikely that a good simulation algorithm will choose a short-sighted strategy in which the distance in variation is decreased at every step. It is an experimental fact that the Gibbs sampler is not the best Monte Carlo Markov chain simulation algorithm.

Exercise 11.5.4. A COUNTEREXAMPLE
Let Z^+ be the common value of the coalesced chains at the forwards coupling time τ^+ for the usual two-state ergodic HMC. Is the distribution of Z^+ the stationary distribution?

Exercise 11.5.5. MONOTONE PROPP–WILSON ALGORITHM
Consider the classical Ising model. Define on the state space $E = \{-1, +1\}^V$ the partial order relation $\preceq$ defined as follows: $x = (x(v), v \in V) \preceq y = (y(v), v \in V)$ if and only if for all $v \in V$, $x(v) \le y(v)$. Give conditions guaranteeing that the monotone Propp–Wilson algorithm can be applied.

Exercise 11.5.6. EIGENSTRUCTURE OF THE METROPOLIS MCMC
([10]) Let π and p be two strictly positive probability distributions on $E = \{1, 2, \ldots, r\}$, and let $w(i) := \frac{\pi(i)}{p(i)}$. Consider the Metropolis algorithm corresponding to the

---

[10][Liu, 1995].

candidate-generating matrix Q given by $q_{ij} = p_j$ for all $i, j \in E$ with acceptance probability

$$\min\left(1, \frac{w(j)}{w(i)}\right) \quad (i \neq j).$$

Assume that the states of E are ordered in such a way that

$$w(1) \geq w(2) \geq \cdots \geq w(r).$$

Verify that the eigenvalues λ_k $(1 \leq k \leq r)$ and the corresponding right-eigenvectors v_k $(1 \leq k \leq r)$ of the transition matrix of the above Metropolis algorithm $\mathbf{P}$ are $\lambda_1 = 1$, $v_1 = \mathbf{1}$, and for $k \geq 1$,

$$\lambda_{k+1} = \sum_{j \geq k} \pi(j) \left(\frac{1}{w(j)} - \frac{1}{w(k)}\right),$$

$$v_{k+1} = \left(0, \ldots, 0, \sum_{\ell=k+1}^{r} \pi(\ell), -\pi(k), \ldots, -\pi(k)\right)^T,$$

where the first $k - 1$ entries of v_{k+1} are null.

Exercise 11.5.7. PROOF OF THEOREM 11.4.8
Prove Theorem 11.4.8.

Chapter 12

Non-homogeneous Markov Chains

12.1 Weak and Strong Ergodicity

For non-homogeneous Markov chains (NHMC), the Markov property is retained but the transition probabilities may depend on time. This section gives conditions guaranteeing the existence of a limit in variation of such chains, with their application to simulated annealing in view. When the state space E is finite and the chain is ergodic, Dobrushin's ergodic coefficient (Subsection 4.3.4) is the basic tool to obtain a necessary and sufficient condition of weak ergodicity (yet to be defined) of non-homogeneous Markov chains.

12.1.1 Ergodicity of Non-Homogeneous Markov Chains

Let $\{X_n\}_{n\geq 0}$ be a stochastic process with values in a countable set E and such that

$$P(X_{n+1} = j | X_n = i, X_0 = i_0, \ldots, X_{n-1} = i_{n-1}) = P(X_{n+1} = j | X_n = i)$$

for all $n \geq 0$ and all states $i, j, i_0, i_1, \ldots, i_{n-1}$. In other words, $\{X_n\}_{n\geq 0}$ is a Markov chain. In this section, we consider *non-homogeneous* Markov chains, that is, Markov chains for which the quantity

$$p_{n,i,j} := P(X_{n+1} = j | X_n = i)$$

may depend on n. The matrix

$$\mathbf{P}(n) = \{p_{n,i,j}\}_{i,j\in E}$$

is called the *transition matrix at time n*. We introduce the notation

$$\mathbf{P}(m, k) = \mathbf{P}(m)\mathbf{P}(m + 1)\cdots \mathbf{P}(k - 1) \quad (k > m \geq 0).$$

© Springer Nature Switzerland AG 2020
P. Brémaud, *Markov Chains*, Texts in Applied Mathematics 31,
https://doi.org/10.1007/978-3-030-45982-6_12

From Bayes' sequential rule, we have that if the distribution of X_m is μ_m, the distribution of X_k is $\mu_m^T \mathbf{P}(m, k)$.

Definition 12.1.1 *The above non-homogeneous Markov chain is called* weakly ergodic *if for all $m \geq 0$,*

$$\limsup_{k \uparrow \infty} \, \sup_{\mu, \nu} d_V \left(\mu^T \mathbf{P}(m, k), \nu^T \mathbf{P}(m, k) \right) = 0 \,,$$

where the supremum is taken over all the probability distributions μ, ν on E.

Definition 12.1.2 *The chain is called* strongly ergodic *if there exists a probability distribution π on E such that for all $m \geq 0$,*

$$\limsup_{k \uparrow \infty} \, \sup_{\mu} d_V \left(\mu^T \mathbf{P}(m, k), \pi \right) = 0 \,, \tag{12.1}$$

where the supremum is taken over all the probability distributions μ on E.

One also says that the family of transition matrices $\{\mathbf{P}(n)\}_{n \geq 0}$ (rather than the chain) is weakly ergodic (resp., strongly ergodic).

Strong ergodicity implies weak ergodicity, since

$$d_V \left(\mu^T \mathbf{P}(m, k), \nu^T \mathbf{P}(m, k) \right) \leq d_V \left(\mu^T \mathbf{P}(m, k), \pi \right) + d_V \left(\nu^T \mathbf{P}(m, k), \pi \right) \,.$$

However, there are weakly ergodic chains that are not strongly ergodic, as the following example shows.

EXAMPLE 12.1.3: WEAKLY BUT NOT STRONGLY ERGODIC. Here the state space has two elements, $\mathbf{P}(0) = I$ (the identity) and for $n \geq 1$,

$$\mathbf{P}(2n) = \begin{pmatrix} 1/2n & 1 - 1/2n \\ 1/2n & 1 - 1/2n \end{pmatrix}, \quad \mathbf{P}(2n+1) = \begin{pmatrix} 1 - 1/(2n+1) & 1/(2n+1) \\ 1 - 1/(2n+1) & 1/(2n+1) \end{pmatrix}.$$

Elementary computations show that for any probability distribution μ on $E = \{1, 2\}$,

$$\mu^T \mathbf{P}(m, 2k+1) = \left(1 - \frac{1}{2k+1}, \frac{1}{2k+1} \right), \quad \mu^T \mathbf{P}(m, 2k) = \left(\frac{1}{2k}, 1 - \frac{1}{2k} \right),$$

and therefore, for all $k \geq m$,

$$\mu^T \mathbf{P}(m, k) - \nu^T \mathbf{P}(m, k) = 0.$$

Thus, the chain is weakly ergodic. But it cannot be strongly ergodic, since $\mu^T \mathbf{P}(m, k)$ has, as $k \to \infty$, two limit vectors: $(1, 0)$ and $(0, 1)$.

Remark 12.1.4 If the chain is homogeneous and ergodic, then it is strongly ergodic in the sense of Definition 12.1.2, by the theorem of convergence to steady state for ergodic HMCs.

12.1.2   The Block Criterion of Weak Ergodicity

As one might guess, weak ergodicity is in general not easy to check directly from the definition. Fortunately, there is a somewhat useable criterion in terms of Dobrushin's ergodic coefficient.

Theorem 12.1.5 *The chain is weakly ergodic if and only if for all $m \geq 0$,*

$$\lim_{k \uparrow \infty} \delta(\mathbf{P}(m,k)) = 0. \tag{12.2}$$

Proof. By Theorem 4.3.15 and the observation $d_V(\mu, \nu) \leq 1$,

$$d_V(\mu^T \mathbf{P}(m,k), \nu^T \mathbf{P}(m,k)) \leq d_V(\mu, \nu)\delta(\mathbf{P}(m,k)) \leq \delta(\mathbf{P}(m,k)).$$

Therefore (12.2) implies weak ergodicity. Now we have

$$\begin{aligned} \delta(\mathbf{P}(m,k)) &= \frac{1}{2} \sup_{i,j \in E} \sum_{\ell \in E} |p_{i\ell}(m,k) - p_{j\ell}(m,k)| \\ &\leq \frac{1}{2} \sup_{\mu, \nu} |\mu^T \mathbf{P}(m,k) - \nu^T \mathbf{P}(m,k)|, \end{aligned}$$

from which it follows that weak ergodicity implies (12.2). □

By Dobrushin's inequality, $\delta(\mathbf{P}(m,k)) \leq \prod_{r=m}^{k-1} \delta(\mathbf{P}(r))$, and therefore nullity of the infinite product $\prod_{r \geq 1} \delta(\mathbf{P}(r))$ is enough to guarantee weak ergodicity. However, in a number of applications, weak ergodicity occurs without the above infinite product diverging to zero. It turns out that the consideration of blocks gives a useful necessary and sufficient condition, called the *block criterion of weak ergodicity*.

Theorem 12.1.6 ([1]) *The chain is weakly ergodic if and only if there exists a strictly increasing sequence of integers $\{n_s\}_{s \geq 0}$ such that*

$$\sum_{s=0}^{\infty} (1 - \delta(\mathbf{P}(n_s, n_{s+1}))) = \infty. \tag{12.3}$$

Proof. Since $0 \leq \delta(\mathbf{P}(n_s, n_{s+1})) \leq 1$, (12.3) is equivalent to nullity of the infinite product $\prod_{s \geq 0} \delta(\mathbf{P}(n_s, n_{s+1}))$ (Theorem A.1.11). But denoting by i the first integer

---

[1][Hajnal, 1958].

s such that $n_s \geq m$, and by j the last integer s such that $n_s \leq k - 1$, Dobrushin's inequality gives

$$
\begin{aligned}
\delta(\mathbf{P}(m,k)) &\leq \delta\mathbf{P}(m,n_i)) \left\{ \prod_{s=i}^{j-1} \delta(\mathbf{P}(n_s, n_{s+1})) \right\} \delta(\mathbf{P}(n_j, k)) \\
&\leq \prod_{s=i}^{j-1} \delta(\mathbf{P}(n_s, n_{s+1})).
\end{aligned}
$$

Since $j \to \infty$ as $k \to \infty$, we see that (12.3) implies weak ergodicity, by Theorem 12.1.5.

Conversely, if we suppose weak ergodicity, then by Theorem 12.1.5 we can inductively construct for any $\gamma \in (0,1)$ a strictly increasing sequence of integers $\{n_s\}_{s \geq 0}$ by

$$
n_0 = 1, n_{s+1} = \inf\{k > n_s;\ \delta(\mathbf{P}(n_s, k)) \leq 1 - \gamma\}.
$$

For such sequences, the product $\prod_{s \geq 0} \delta(\mathbf{P}(n_s, n_{s+1}))$ is null, and this is equivalent to (12.3). $\square$

12.1.3   A Sufficient Condition for Strong Ergodicity

Knowing that a given non-homogeneous Markov chain is weakly ergodic, it remains to decide whether it is strongly ergodic or not. No useful criterion of strong ergodicity is available, and we have to resort to sufficient conditions.

Recall the definition of the norm of a matrix (not necessarily a square matrix):

$$
|A| := \max_i \sum_j |a_{ij}|.
$$

Theorem 12.1.7 *If the chain is weakly ergodic and if there exists for all $n \geq 0$ a probability distribution $\pi(n)$ on E such that*

$$
\pi^T(n) = \pi^T(n)\mathbf{P}(n)
$$

and

$$
\sum_{n=0}^{\infty} |\pi(n+1) - \pi(n)| < \infty, \tag{12.4}
$$

then the chain is strongly ergodic.

Proof. Condition (12.4) implies the existence of a probability distribution π such that

$$\lim_{n\uparrow\infty} |\pi(n) - \pi| = 0. \qquad (12.5)$$

Define Π_n (resp., Π) to be the matrix with all rows equal to $\pi(n)$ (resp., π). In particular, $|\Pi_n - \Pi| = |\pi(n) - \pi|$ and similarly $|\Pi_{n+1} - \Pi_n| = |\pi(n+1) - \pi(n)|$.

Also, for any probability distribution μ on E, $\mu^T \Pi = \pi$, and therefore (12.1) is equivalent to

$$\lim_{k\uparrow\infty} \sup_{\mu} \left|\mu^T \left(\mathbf{P}(m, k) - \Pi\right)\right| = 0,$$

which is in turn implied by

$$\lim_{k\uparrow\infty} |\mathbf{P}(m, k) - \Pi| = 0. \qquad (12.6)$$

For the proof of (12.6), write

$$\mathbf{P}(m, k) - \Pi = \mathbf{P}(m, \ell)\mathbf{P}(\ell, k) - \Pi_{\ell+1}\mathbf{P}(\ell, k)$$
$$+ \Pi_{\ell+1}\mathbf{P}(\ell, k) - \Pi_k + \Pi_k - \Pi.$$

The triangle inequality for matrix norms gives

$$|\mathbf{P}(m, k) - \Pi| \leq |\mathbf{P}(m, \ell)\mathbf{P}(\ell, k) - \Pi_\ell\mathbf{P}(\ell, k)|$$
$$+|\Pi_\ell\mathbf{P}(\ell, k) - \Pi_{k-1}| + |\Pi_{k-1} - \Pi| = A + B + C. \qquad (12.7)$$

An upper bound for A is given by Dobrushin's inequality (Theorem 4.3.14):

$$A \leq |\mathbf{P}(m, \ell) - \Pi_\ell|\delta(\mathbf{P}(\ell, k)) \leq 2\delta(\mathbf{P}(\ell, k)),$$

where the last inequality follows from the definition of the matrix norm.

In order to bound B from above, we first observe that $\Pi_\ell\mathbf{P}(\ell) = \Pi_\ell$ and therefore

$$\Pi_\ell\mathbf{P}(\ell, k) = (\Pi_\ell - \Pi_{\ell+1})\mathbf{P}(\ell+1, k) + \Pi_{\ell+1}\mathbf{P}(\ell+1, k).$$

By iteration, we obtain

$$\Pi_\ell\mathbf{P}(\ell, k) = \sum_{j=\ell+1}^{k-1} (\Pi_{j-1} - \Pi_j)\mathbf{P}(j, k) + \Pi_{k-1},$$

and therefore

$$B \leq \sum_{j=\ell+1}^{k-1} |\Pi_{j-1} - \Pi_j|\delta(\mathbf{P}(j, k)) \leq \sum_{j=\ell+1}^{k-1} |\pi(j-1) - \pi(j)|,$$

where we have used the triangle inequality, Theorem 4.3.15, (12.4), and the observation $|\Pi_{j-1} - \Pi_j| = |\pi(j-1) - \pi(j)|$. As for C, we have, in view of the last observation,

$$C = |\pi(k-1) - \pi|.$$

The rest of the proof is now clear: For a given $\epsilon > 0$, fix ℓ such that $B \leq \frac{\epsilon}{3}$ for all $k \geq \ell$ (use (12.4)), and take k large enough so that $A \leq \frac{\epsilon}{3}$ (use Dobrushin's inequality, Theorem 4.3.14) and $C \leq \frac{\epsilon}{3}$ (use (12.5)). $\qquad\square$

Remark 12.1.8 It is *not* required that $\mathbf{P}(n)$ be an ergodic stochastic matrix, or that $\pi(n)$ be a unique stationary probability of $\mathbf{P}(n)$.

Bounded Variation Extensions

The question is: How useful is Theorem 12.1.7? It seems that in order to satisfy (12.4), one has to obtain a closed-form expression for $\pi(n)$, or at least sufficient information about $\pi(n)$. How much information? It turns out that very little is needed in practice. More precisely, a qualitative property of $\{\pi(n)\}_{n\geq 0}$ in terms of bounded variation extensions (to be defined) is sufficient to guarantee (12.4), and therefore strong ergodicity, if the chain is weakly ergodic.

For the results below we recall a definition:

A function $f : (0,1] \to \mathbb{R}$ is said to be of *bounded variation* (BV) if

$$\sup\left\{ \sum_{i=1}^{\infty} |f(x_i) - f(x_{i-1})|; 0 < x_i < \cdots < x_1 = 1 \text{ and } \lim_{i\to\infty} x_i = 0 \right\} < \infty.$$

A vector function $\mu : (0,1] \to \mathbb{R}^E$ (or equivalently a vector $\mu(c) = \{\mu_i(c)\}_{i\in E}$) is said to be of bounded variation if

$$\sup\left\{ \sum_{i=1}^{\infty} |\mu(x_i) - \mu(x_{i-1})|; 0 < x_i < \cdots < x_i = 1 \text{ and } \lim_{i\to\infty} x_i = 0 \right\} < \infty.$$

Definition 12.1.9 *The vector function* $\bar{\pi} : (0,1] \to \mathbb{R}^E$ *is called a* bounded variation extension *of* $\{\pi(n)\}_{n>0}$ *if there exists a sequence* $\{c_n\}_{n\geq 0}$ *in* $(0,1]$ *decreasing to 0 and such that* $\bar{\pi}(c_n) = \pi(n)$ *for all* $n \geq 0$.

Theorem 12.1.10 $(^2)$ *Suppose that* $\{\mathbf{P}(n)\}_{n\geq 0}$ *is weakly ergodic and that for all* $n \geq 0$, *there exists a probability vector* $\pi(n)$ *such that* $\pi(n)\mathbf{P}(n) = \pi(n)$. *If there exists a bounded variation extension* $\bar{\pi}(c)$ *of* $\{\pi(n)\}_{n\geq 0}$, *the chain is strongly ergodic.*

---

[2][Anily and Federgruen, 1987].

Proof. We have

$$\sum_{n\geq 0} |\pi(n+1) - \pi(n)| = \sum_{n\geq 0} |\bar{\pi}(c_{n+1}) - \bar{\pi}(c_n)| < \infty,$$

since $\bar{\pi}(c)$ is a bounded variation extension of $\{\pi(n)\}_{n\geq 0}$ and of bounded variation, and the conclusion follows by Theorem 12.1.7. $\qquad\square$

Let $\bar{\mathbf{P}}(c)$ be a bounded variation extension of $\{\mathbf{P}(n)\}_{n\geq 0}$, that is, there exists a sequence $\{c_n\}_{n\geq 0}$ in $(0, 1]$, decreasing to 0 as n goes to infinity and such that for all $n \geq 0$,

$$\bar{\mathbf{P}}(c_n) = \mathbf{P}(n).$$

Suppose that for each $c \in (0, 1]$, there exists a probability vector $\bar{\pi}(c)$ such that

$$\bar{\pi}(c)\bar{\mathbf{P}}(c) = \bar{\pi}(c).$$

Is it enough for $\bar{\pi}(c)$ to be of bounded variation that $\bar{\mathbf{P}}(c)$ be of bounded variation, that is, that

$$\sup\left\{\sum_{i=1}^{\infty} |\bar{\mathbf{P}}(x_{i+1}) - \bar{\mathbf{P}}(x_i)|; 0 < x_i < \cdots < x_1 = 1 \text{ and } \lim_{i\in\infty} x_i = 0\right\} < \infty?$$

A simple counterexample shows that this is not the case.

EXAMPLE 12.1.11: COUNTEREXAMPLE. Let

$$P(n) = \begin{pmatrix} 1 - e^{-n} & e^{-n} \\ e^{-n}\sin^2\left(\frac{n\pi}{2}\right) & 1 - e^{-n}\sin^2\left(\frac{n\pi}{2}\right) \end{pmatrix},$$

$$\bar{\mathbf{P}}(c) = \begin{pmatrix} 1 - e^{-1/c} & e^{-1/c} \\ e^{-1/c}\sin^2\left(\frac{\pi}{2c}\right) & 1 - e^{-1/c}\sin^2\left(\frac{\pi}{2c}\right) \end{pmatrix}.$$

Clearly, $\bar{\mathbf{P}}(c)$ is a bounded variation extension of $\{\mathbf{P}(n)\}_{n\geq 0}$. If the corresponding stationary probability $\bar{\pi}(c)$ were of bounded variation, then as shown in the proof of Theorem 12.1.10, $\sum_{n\geq 0}|\pi(n+1) - \pi(n)|$ would be finite. Computations give for the second coordinate of $\pi(n)$

$$\pi(n)_2 = \left(1 + \sin^2\left(\frac{n\pi}{2}\right)\right)^{-1},$$

a quantity that oscillates between 1 and $\frac{1}{2}$. Therefore, $\sum_{n\geq 0}|\pi(n+1) - \pi(n)| = \infty$.

In order to give conditions on $\bar{\mathbf{P}}(c)$ ensuring that $\bar{\pi}(c)$ is of bounded variation, we shall first give a precise description of $\bar{\pi}(c)$ in terms of the entries of $\bar{\mathbf{P}}(c)$. This can be done in the case where E is finite, identified with $\{1, \ldots, N\}$ for simplicity. Indeed $\bar{\pi}(c)$ is a solution of the balance equations

$$\bar{\pi}(c)_i = \sum_{j=1}^{N} \bar{\mathbf{P}}(c)_{ji} \bar{\pi}(c)_j \quad (1 \le i \le N-1),$$

together with the normalizing equation

$$\sum_{i=1}^{N} \bar{\pi}(c)_i = 1.$$

That is, in matrix form,

$$\bar{\pi}(c) \begin{pmatrix} 1 - \bar{\mathbf{P}}(c) & \cdots & -\bar{\mathbf{P}}(c)_{N-1,1} & -\bar{\mathbf{P}}(c)_{N,1} \\ -\bar{\mathbf{P}}(c) & \cdots & -\bar{\mathbf{P}}(c)_{N-1,2} & -\bar{\mathbf{P}}(c)_{N,2} \\ \vdots & & \vdots & \vdots \\ -\bar{\mathbf{P}}(c)_{1,N-1} & \cdots & 1 - \bar{\mathbf{P}}(c)_{N-1,N-1} & -\bar{\mathbf{P}}(c)_1 \\ 1 & & 1 & 1 \end{pmatrix} = \begin{pmatrix} 0 \\ 0 \\ \vdots \\ 0 \\ 1 \end{pmatrix}.$$

By Cramer's rule,

$$\bar{\pi}(c)_i = \frac{A_i(c)}{B_i(c)},$$

where $A_i(c)$ and $B_i(c)$ are finite sums and differences of finite products of the entries of $\bar{\mathbf{P}}(c)$.

EXAMPLE 12.1.12: RATIONAL POLYNOMIAL–EXPONENTIAL TRANSITION MA-TRICES. This case covers most applications. The elements of $\bar{\mathbf{P}}(c)$ are ratios of functions of the type

$$\sum_{j=1}^{n} Q_j \left(\frac{1}{c}\right) e^{\lambda_j/c}, \tag{12.8}$$

where the Q_j are polynomial functions and the λ_j are real numbers. Then so are the elements of $\bar{\pi}(c)$, as well as their derivatives with respect to c. But ratios of terms of type (12.8) have for sufficiently small $c > 0$ a constant sign. Therefore a given element $\bar{\pi}(c)_i = \psi(c)$ is such that

(α) for some $c^* > 0$, $\psi : (0, c*] \to \mathbb{R}$ is monotone and bounded;

(β) $\psi : (0, 1] \to \mathbb{R}$ is continuously differentiable.

Properties (α) and (β) are sufficient to guarantee that $\psi : (0,1] \to \mathbb{R}$ is of bounded variation.

___

We have spent some time explaining how the sufficient condition of strong ergodicity (12.4) can be checked. One may wonder whether this is really worthwhile, and whether a weaker and easier to verify condition is available. A natural candidate for a sufficient condition of strong ergodicity, *given weak ergodicity*, is

$$\lim_{n\uparrow\infty} |\pi(n) - \pi| = 0, \tag{12.9}$$

for some probability π.

___

EXAMPLE 12.1.13: COUNTEREXAMPLE. Define for all $n \geq 1$,

$$\mathbf{P}(2n-1) = \begin{pmatrix} 0 & 1 \\ 1 & 0 \end{pmatrix}, \ \mathbf{P}(2n) = \begin{pmatrix} 0 & 1 \\ 1 - \frac{1}{2n} & \frac{1}{2n} \end{pmatrix}.$$

The sequence $\{\mathbf{P}(n)\}_{n\geq 1}$ is weakly ergodic (Exercise 12.3.3). The corresponding stationary distributions are

$$\pi(2n-1) = \left(\frac{1}{2}, \frac{1}{2} \right) \ \pi(2n+1) = \left(\frac{2n-1}{4n-1}, \frac{2n}{4n-1} \right),$$

and therefore (12.9) is satisfied with $\pi = (\frac{1}{2}, \frac{1}{2})$. On the other hand, if we define for all $k \geq 1$

$$R(k) = \mathbf{P}(2k)\mathbf{P}(2k+1) = \begin{pmatrix} 1 - \frac{1}{2k} & \frac{1}{2k} \\ 0 & 1 \end{pmatrix}$$

and

$$S(k) = \mathbf{P}(2k-1)\mathbf{P}(2k) = \begin{pmatrix} 1 & 0 \\ \frac{1}{2k} & 1 - \frac{1}{2k} \end{pmatrix},$$

then the sequences $\{R(k)\}_{k\geq 1}$ and $\{S(k)\}_{k\geq 1}$ are weakly ergodic (exercise), and their stationary distributions are constant, equal to $(1,0)$ and $(0,1)$, respectively. Therefore, by Theorem 12.1.7, they are strongly ergodic, and in particular,

$$\lim_{k\uparrow} \mathbf{P}(1)\mathbf{P}(2) \cdots \mathbf{P}(2k-1)\mathbf{P}(2k) = \begin{pmatrix} 0 & 1 \\ 0 & 1 \end{pmatrix}$$

and

$$\lim_{k\uparrow} \mathbf{P}(1) \left(\mathbf{P}(2)\mathbf{P}(3) \cdots \mathbf{P}(2k)\mathbf{P}(2k+1) \right) = \begin{pmatrix} 1 & 0 \\ 1 & 0 \end{pmatrix}.$$

Therefore, the sequence $\{\mathbf{P}(n)\}_{n\geq 1}$ is not strongly ergodic.

___

Exercise 12.3.5 shows what can be added to (12.9) to obtain a sufficient condition of strong ergodicity.

We shall quote without proof the following natural result.

Theorem 12.1.14 ([3]) *Let $\{\mathbf{P}(n)\}_{n\geq 1}$ be a sequence of transition matrices each having at least one stationary distribution, and such that*

$$\lim_{n\uparrow\infty} |\mathbf{P}(n) - \mathbf{P}| = 0 \qquad (12.10)$$

for some ergodic transition matrix $\mathbf{P}$. This sequence is strongly ergodic.

Remark 12.1.15 Note the requirement that $\mathbf{P}$ be ergodic. This condition will be found too strong in the study of the convergence of simulated annealing algorithms, where typically the limit transition matrix is reducible (see the next section).

12.2   Simulated Annealing

12.2.1   Stochastic Descent and Cooling

Let E be a finite set and $U : E \to \mathbf{R}$ a function, called the *cost function*, to be minimized. More precisely, one is looking for any element $i_0 \in E$ such that

$$U(i_0) \leq U(i) \text{ for all } i \in E.$$

Such an element is called a *global minimum* of the cost function U. The so-called *descent algorithms* define for each state $i \in E$ a subset $N(i)$ of E, called the *neighborhood* of i, and proceed iteratively as follows: Suppose that at a given stage solution i is examined. At the next stage one examines a solution $j \in N(i)$ chosen according to a rule specific to each algorithm, and compares $U(i)$ to $U(j)$. If $U(i) \leq U(j)$, the procedure stops and i is the retained solution. Otherwise a new solution $k \in N(j)$ is examined and compared to j, and so on. The algorithm eventually comes to a stop and produces a solution, since E is finite. However, this solution is usually not optimal, due to the possible existence of local minima that are not global minima, where i is called a *local minimum* if

$$U(i) \leq U(j) \text{ for all } j \in N(i).$$

In most interesting situations, local optima exist and are sometimes abundant, and the algorithms often become trapped at one of these local minima.

---

[3]Theorem V.4.5 of [Isaacson and Madsen, 1976].

EXAMPLE 12.2.1: THE TRAVELING SALESMAN. A salesman must find the short-est route visiting each of K cities exactly once. Here E is the set of the $K!$ admissible routes, i is one such route, and $U(i)$ is the length of route i. One popular choice for the neighborhood $N(i)$ of route i is all the routes j obtained from i by a 2-change, as explained in Figure 13.2.1.

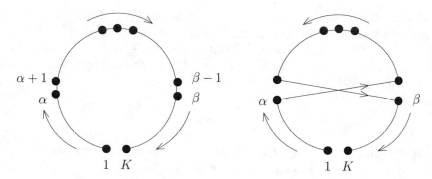

Figure 12.2.1. A 2-change of the traveling salesman's route

If the cities are numbered $1, 2, \ldots, K$ a route i can be identified with a per-mutation σ of $\{1, 2, \ldots, K\}$, where $\sigma(\alpha)$ is the order of the visit to city α in the route corresponding to σ. By renaming the cities, there is no loss of generality in supposing that σ is the identity. The 2-change involving cities α and β consists in cutting the segments $(\alpha, \alpha+1)$ and $(\beta-1, \beta)$, and in replacing them by the new segments $(\alpha, \beta-1)$ and $(\beta-1, \alpha)$ (Figure 12.2.1). In this construction $|\beta-\alpha| \geq 3$ and $K \geq 4$, and one can count exactly $K(K-1)+1$ neighbors of a given route. The sizes of the neighborhoods are therefore reasonable in comparison to the size of the total search space. Note also that the computation of $U(j)$ from $U(i)$ when $j \in N(i)$ involves only four intercity distances.

Typically, and this is indeed the case in the traveling salesman problem, the neighborhood structure is communicating.

Definition 12.2.2 *Let E be a finite set and let $\{N(i), i \in E\}$ be a collection of subsets of E satisfying the condition*

$$i \notin N(i).$$

Such a collection is called a neighborhood structure. *If for all pairs of states $i, j \in E$ there exists a path from i to j, that is, a sequence of states $i_1, \ldots, i_m \in E$*

such that $i_1 \in N(i), i_2 \in N(i_1), \ldots, j \in N(i_m)$, the neighborhood structure is said to be communicating.

The basic idea of stochastic combinatorial optimization is to leave a possibility to escape from a local minimum trap.

A canonical form of the stochastic descent algorithm is as follows: Let $Q = \{q_{ij}\}$ be an irreducible transition matrix on E. Also, for each parameter value T, and all states $i, j \in E$, let $\alpha_{ij}(T)$ be a probability. Calling X_n the current solution at stage n, the process $\{X_n\}_{n \geq 0}$ is a homogeneous Markov chain with state space E and transition matrix $\mathbf{P}(T)$ of general off-diagonal term

$$p_{ij}(T) = q_{ij}\alpha_{ij}(T) . \tag{12.11}$$

We assume the chain irreducible. It is positive recurrent, since the state space is finite. Therefore, it has a unique stationary distribution $\pi(T)$.

One possible choice of the candidate-generating matrix Q consists in first choosing a communicating neighborhood structure such as in Definition 12.2.2, and taking $q_{ij} > 0$ only if $i = j$ or $j \in N(i)$. The matrix Q is then irreducible. Conversely, one can associate with an irreducible transition matrix Q a communicating neighborhood structure defined by $N(i) := \{j; j \neq i, q_{ij} > 0\}$.

The reader may already be aware of the connection with the framework of Monte Carlo Markov chain simulation (Chapter 11).

EXAMPLE 12.2.3: METROPOLIS SAMPLER, TAKE 1. Suppose that the current solution at stage n is i. At stage $n+1$, a tentative solution j is selected according to the probability q_{ij}. This solution is accepted with probability

$$\alpha_{ij}(T) = \min\{1, e^{\frac{(U(i)-U(j))}{T}}\} = e^{-\frac{(U(j)-U(i))^+}{T}} ,$$

where T is a positive constant; otherwise, j is rejected. Therefore, if $U(i) \geq U(j)$, then j is accepted, whereas if $U(i) < U(j)$, a chance is left to the solution j, although it is worse than i. In this particular algorithm as well as in others, the chance left to a candidate j that is worse than i diminishes as its deviation from i, measured by $U(j) - U(i)$, increases.

Suppose that the matrix Q is symmetric. With this special structure, the stationary distribution $\pi(T)$ does not depend on Q and is given by

$$\pi_i(T) = \frac{e^{-U(i)/T}}{\sum_{k \in E} e^{-U(k)/T}}.$$ (12.12)

EXAMPLE 12.2.4: BARKER'S SAMPLER. This simulation algorithm differs from the Metropolis sampler by the choice of the acceptance probability, here

$$a_{ij}(T) = \frac{1}{1 + e^{-(U(i)-U(j))/T}}.$$

The usual reversibility argument shows that the target distribution of Barker's algorithm is the same as that of the Metropolis algorithm.

For both the Metropolis and Barker samplers, if Q is irreducible and U is not a constant, then $\mathbf{P}(T)$ is irreducible and aperiodic for all $T > 0$ (Exercise).

EXAMPLE 12.2.5: GIBBS SAMPLER AS A STOCHASTIC DESCENT ALGORITHM. First, we shall see how the Gibbs sampler connects to the general descent algorithm corresponding to the transition probabilities (12.11). This is only a matter of notation. For simplicity, we assume that the phase space is $\Lambda = \{0, 1\}$.

In this context, a "solution" i is a value $x \in \Lambda^V$ of the random field. The neighborhood $N(x)$ of the solution $x \in \Lambda^V$ (not to be confused with a neighborhood $\mathcal{N}_v$ of some site $v \in V$) is the set of configurations $y \in \Lambda^V$ such that y and x agree on all sites except one:

$$y = (\bar{x}(v), x(V \setminus v)),$$

where $\bar{x}(v) = 1 - x(v)$. This configuration y will be denoted by $\bar{x}^v$. Therefore,

$$N(x) = \{\bar{x}^v; v \in V\}.$$

The Gibbs sampler does this: After examining the image $x \in \Lambda^V$ at time n, it selects $\bar{x}^v \in N(x)$ with probability q_v (with the notations of (12.11), $i = x, j = \bar{x}^v$, and $q_{ij} = q_v$). The image $\bar{x}^v$ is accepted with probability $\pi(\bar{x}(v) \mid x(\mathcal{N}_v))$ (with the notations of (12.11), $a_{ij}(T) = \pi(\bar{x}(v) \mid x(\mathcal{N}_v))$).

EXAMPLE 12.2.6: BARKER'S ALGORITHM FOR NEURAL NETWORKS. The model and the notation are those of Exercise 10.4.10:

$$\pi_T(\bar{x}(v) \mid x(\mathcal{N}_v)) = \frac{e^{-\frac{1}{T}(\sum_{t \in \mathcal{N}_v}(w_{tv}+w_{vt})x(t)-h_v)(1-x(v))}}{1 + e^{-\frac{1}{T}(\sum_{t \in \mathcal{N}_v}(w_{tv}+w_{vt})x(t)-h_v)}}.$$ (12.13)

In the notations of (12.11), with $i = x, j = \bar{x}^v$, the right-hand side of (12.13) is $a_{ij}(T) = a_{x,\bar{x}^v}(T)$. In this case, the simulation algorithm is a special case of Barker's algorithm. Indeed,

$$\mathcal{E}(x) - \mathcal{E}(\bar{x}^v) = \left\{ \sum_{t \in \mathcal{N}_v} (w_{tv} + w_{vt}) x(t) - h_v \right\} (x(v) - \bar{x}(v)).$$

Therefore, if $x(v) = 1$, then $\mathcal{E}(x) - \mathcal{E}(\bar{x}^v) = \sum_{t \in \mathcal{N}_v} ((w_{tv} + w_{vt}) x(t) - h_v)$, and if $x(v) = 0$, then $\mathcal{E}(x) - \mathcal{E}(x^v) = -\sum_{t \in \mathcal{N}_v} ((w_{tv} + w_{vt}) x(t) - h_v)$. Therefore, the quantity on the left-hand side of (12.13) is equal to

$$\frac{1}{1 + e^{-\frac{1}{T}(\mathcal{E}(x) - \mathcal{E}(\bar{x}^v))}}$$

if $x(v) = 1$, and to

$$\frac{e^{\frac{1}{T}(\mathcal{E}(x) - \mathcal{E}(\bar{x}^v))}}{1 + e^{\frac{1}{T}(\mathcal{E}(x) - \mathcal{E}(\bar{x}^v))}} = \frac{1}{1 + e^{-\frac{1}{T}(\mathcal{E}(x) - \mathcal{E}(\bar{x}^v))}}$$

if $x(v) = 0$. In either case, the probability of acceptance of $\bar{x}^v$ is equal to

$$\frac{1}{1 + e^{-\frac{1}{T}(\mathcal{E}(x) - \mathcal{E}(\bar{x}^v))}} \, .$$

EXAMPLE 12.2.7: COMPARISON OF THE METROPOLIS AND BARKER SAMPLERS. Theorem 9.2.2 will be applied to the comparison of the Metropolis and Barker reversible Markov chains (see Examples 12.2.3 and 12.2.4) with the same stationary distribution

$$\pi(i) = \widetilde{\pi}(i) = \frac{e^{-U(i)}}{Z} \, ,$$

where $U : E \to \mathbb{R}$ and $Z := \sum_i e^{-U(i)}$ is the normalizing constant. For Barker's algorithm, with $z := \frac{e^{-U(j)}}{e^{-U(i)}}$,

$$p_{ij} = \frac{1}{1 + z} \, ,$$

whereas for the Metropolis algorithm,

$$\widetilde{p}_{ij} = 1 \wedge z \, .$$

Straightforward computations show that

$$\frac{1}{2} \leq \frac{\mathcal{E}_{\mathbf{P}}(x, x)}{\mathcal{E}_{\widetilde{\mathbf{P}}}(x, x)} \leq 1 \, ,$$

and therefore $\beta_2 \leq \tilde{\beta}_2 \leq 2\beta_2$.

Consider the chain with transition probabilities (12.11), and suppose that the corresponding HMC is irreducible, so that the stationary distribution $\pi(T)$ exists and is unique. Suppose, moreover, that the chain is aperiodic. Therefore it converges to steady state, that is, for all $i \in E$ and all initial distributions μ,

$$\lim_{n \to \infty} P_\mu(X_n = i) = \pi_i(T).$$

We see, in particular, that if the stationary distribution $\pi(T)$ puts most of its mass on the states minimizing the cost function $U(i)$, then if one stops the algorithm at a sufficiently large stage n, the current solution X_n will be a global minimum with high probability.

EXAMPLE 12.2.8: METROPOLIS SAMPLER, TAKE 2. Example 12.2.3 continued. The stationary probability is given by (12.12). Define the set of *global minima*

$$H = \{i \in E; U(i) \leq U(j) \text{ for all } j \in E\}. \tag{12.14}$$

Then clearly, $\pi_i(T)$ is maximal on $i \in H$. But there is more:

$$\lim_{T \downarrow 0} \pi_i(T) = \begin{cases} \frac{1}{|H|} & \text{if } i \in H, \\ 0 & \text{if } i \notin H. \end{cases} \tag{12.15}$$

To see this, call $m = \min_{i \in E} U(i)$, and write the right-hand side of (12.15), after division of its numerator and denominator by $e^{-\frac{m}{T}}$, as

$$\frac{e^{-\frac{(U(i)-m)}{T}}}{|H| + \sum_{k \notin H} e^{-\frac{(U(k)-m)}{T}}}.$$

The result follows, since as $T \downarrow 0$, $e^{-\frac{U(k)-m}{T}}$ tends to 0 if $U(k) > m$, and to 1 if $U(k) = m$.

This observation suggests the following heuristic procedure.[4] Start the algorithm with the value $T = a_0$ of the parameter, and wait a sufficiently long time for the chain to get close to its stationary regime. Then set $T = a_1 < a_0$ and again wait for the steady state. Then set $T = a_2 < a_1$, etc. At the kth change of the parameter T, the chain will be close to the stationary regime $\pi(a_k)$, and therefore if $\lim_{k \uparrow \infty} a_k = 0$, one expects that for large n, X_n will be with very high probability in H, the set of global minima.

---

[4][Kirkpatrick, Gelatt and Vecchi, 1982]; see [Aarts and Cors, 1989] for additional details.

However, for this to happen, the times in between the parameter changes must be sufficiently long for the chain to come close to the stationary distribution corresponding to the current value of the parameter. What is "sufficiently long"?

Simulated annealing algorithms all have a *cooling schedule*, that is, a sequence $\{T_n\}_{n\geq 0}$ of positive numbers decreasing to 0 controlling the transition rates of $\{X_n\}_{n\geq 0}$: At time $n, P(X_{n+1} = j \mid X_n = i) = p_{ij}(T_n)$. The question becomes, *How slowly* must $\{T_n\}_{n\geq 0}$ converge to zero so that

$$\lim_{n\uparrow\infty} P(X_n = i) = \begin{cases} 0 \text{ if } i \notin H, \\ \frac{1}{|H| \text{ if } i \in H}. \end{cases} \tag{12.16}$$

Since $\{X_n\}_{n\geq 0}$ is now a non-homogeneous Markov chain, the results concerning convergence in distribution of such chains of Section 12.1 will be especially useful. They will be applied in Subsection 12.2.2 to obtain cooling schedules with the desired convergence property in various situations.

The next example features an avatar of the simulated annealing algorithm involving on-line estimation of the cost function.

EXAMPLE 12.2.9: THE STOCHASTIC RULER ALGORITHM. In a number of situations the cost function U is not directly computable, but of the form

$$U(i) = \mathrm{E}[H(i)],$$

where $H(i)$ is a non-negative random variable. Moreover, it is assumed that one can generate at will, for any $i \in E$, a random variable $H(i)$, but that one is not able to compute its expectation. Therefore, the simulated annealing procedure as described before cannot be applied as it is. In one way or another, the expectation will have to be estimated from independent samples. The stochastic ruler algorithm[5] does this in an indirect manner.

It works under the assumption that the domains of the random variables $H(i)$ are all contained in a fixed interval, that is, for all $i \in E$,

$$-\infty < a \leq H(i) \leq b < \infty,$$

and it uses at each step a random variable U, called the *ruler*, which is uniform on $[a, b]$:

$$P(U \leq x) = \frac{x - a}{b - a},$$

---

[5][Yan and Mukai, 1992].

for all $x \in [a, b]$. The stochastic ruler algorithm is based on the observation that the simulated annealing algorithm (12.11) with the acceptance probability

$$\alpha_{ij}(T) = p(j)^{1/T} = e^{-\frac{1}{T}\{-\log p(j)\}}, \tag{12.17}$$

where

$$p(j) := \mathrm{P}(H(j) \le U),$$

leads, with a good cooling schedule, to the minimum of $g(i) = -\log p(i)$. But if $H(j)$ and U are independent, then

$$p(j) = E\left[\frac{1}{b-a}\int_a^b 1_{\{u \ge H(i)\}}\mathrm{d}u\right] = \frac{1}{b-a}\mathrm{E}[b - H(i)] = \frac{b - U(i)}{b - a}.$$

Therefore, the simulated annealing with acceptance probability (12.17) will provide, for a good cooling schedule, a minimum of the original cost function U.

The problem now is: how to implement the acceptance rule based on (12.17), since $p(j)$ is not computable? Of course, one also has to give a cooling schedule that makes simulated annealing work. For the time being, suppose that we have a cooling schedule $\{T_k\}$ of the form $T_k = \frac{1}{M_k}$ for some integer M_k. The problem is therefore to realize for any integer M the acceptance probability $p(j)^M$ without computing $p(j)$. A solution is as follows. Generate $2M$ independent random variables $V_1, \dots, V_M, H_1(j), \dots, H_M(j)$, where the V_k are uniform on $[a, b]$ and the $H_k(j)$ have the same distribution as $H(j)$. Then

$$\mathrm{P}(H_1(j) \le V_1, \dots, H_M(j) \le V_M) = \mathrm{P}(H(j) \le V)^M = p(j)^M.$$

The acceptance rule for j is therefore to accept if $H_k(j) \le V_k$ for $k = 1, \dots, M$.

In theory, this procedure works, but it requires more and more time as M increases to ∞. As the algorithm makes progress, more and more care must be taken for the estimation of $U(i)$.

12.2.2   Convergence of Simulated Annealing

The general theory will follow the next introductory example.

EXAMPLE 12.2.10: ANNEALED GIBBS SAMPLER. (^{6}) We use periodic scanning of Example 11.3.2, only at the n-th sweep, we introduce a temperature T_n. Thus $\{Z_n\}_{n\geq 0}$ is a non-homogeneous Markov chain, the transition matrix at time n being

$$\mathbf{P}(n) = \prod_{k=1}^{N} \mathbf{P}_{v(k)}^{T_n},$$

where the (x, y)-entry of $\mathbf{P}_v^T$ is

$$\frac{\exp\left\{-\frac{1}{T}\mathcal{E}(y(v), x(V\setminus v))\right\}}{\sum_{\lambda\in\Lambda}\exp\left\{-\frac{1}{T}\mathcal{E}(\lambda, x(V\setminus v))\right\}}$$

if $y = (y(v), x(V\setminus v))$. The bound of Example 11.3.4 gives

$$\delta(\mathbf{P}(n)) \leq 1 - e^{-\frac{N\Delta}{T_n}}.$$

In particular, by the block criterion of weak ergodicity,

$$\sum_{n=1}^{\infty} e^{-\frac{N\Delta}{T_n}} = \infty \tag{12.18}$$

is a sufficient condition of weak ergodicity.

Now, $\mathbf{P}(n)$ admits the stationary distribution

$$\pi_{T_n}(x) = \frac{e^{-\frac{1}{T_n}\mathcal{E}(x)}}{Z_{T_n}}.$$

Also, for all $x \in \Lambda^S$, $\lim_{T\downarrow 0} \pi_T(x) = \frac{1}{|H|}$ if $x \in H$ and is 0 otherwise, where $H = \{x \in \Lambda^V; \mathcal{E}(x) = \min\}$. Moreover, it can be shown that for $x \in H$, the quantity $\pi_T(x)$ increases as $T \downarrow 0$, whereas for $x \notin H$, it eventually decreases, and this guarantees that

$$\sum_{n=1}^{\infty} |\pi_{T(n+1)} - \pi_{T_n}| < \infty.$$

---

^{6}[Geman and Geman, 1984].

Therefore, by Theorem 12.1.7, if $T_n \downarrow 0$ in such a way that (12.18) is respected, then the non-homogeneous Markov chain $\{Z_n\}_{n \geq 0}$ is strongly ergodic, with the limit distribution uniform on H.

The general results of Section 12.1 will be applied to the simulated annealing algorithm corresponding to the transition matrix $\mathbf{P}(T)$ given by (12.11).

([7]) The transition matrix $\mathbf{P}(T)$ is assumed *uniformly irreducible* for sufficiently small $T \in (0,1]$. This means that for all ordered pairs of states (i,j), there is a $\mathbf{P}(T)$-path from i to j which is independent of $T \in (0,c]$ for some $c > 0$. This is always satisfied in practice. For instance, for the Metropolis or Barker samplers, it suffices that Q is irreducible and that U is not a constant (see Exercise 12.2.1).

Define

$$d = \inf\{q_{ij}; \quad j \neq i, p_{ij}(T) > 0\}, \tag{12.19}$$

a positive quantity, since the state space is finite.

The crucial assumption is the following: There exists a $T^* \in (0,1]$ such that on $(0, T^*]$,

$$\alpha_{ij}(T) \downarrow 0 \text{ as } T \downarrow 0 \text{ if } U(j) > U(i), \tag{12.20}$$

$$\alpha_{ij}(T) \uparrow 1 \text{ as } T \downarrow 0 \text{ if } U(j) < U(i), \tag{12.21}$$

and

$$\lim_{T \downarrow 0} \alpha_{ij}(T) > 0 \text{ exists if } U(i) = U(j). \tag{12.22}$$

Assumptions (12.20) and (12.21) imply, in particular, that in the vicinity of 0, the functions $\alpha_{ij}(T)$ are monotonic if $U(i) \neq U(j)$. They force the algorithm to be less permissive as T approaches 0, rejecting more often the nonlocally optimal solutions. Define for each $T \in (0,1]$

$$\underline{\alpha}(T) = \inf_{i \in E, j \in N(i)} \alpha_{ij}(T). \tag{12.23}$$

Assumptions (12.20)–(12.22) imply that in the vicinity of 0,

$$\inf_{i \in E, j \neq i} \alpha_{ij}(T) = \inf_{\substack{i \in E, j \in N(i) \\ U(j) > U(i)}} \alpha_{ij}(T),$$

and therefore, in the vicinity of 0, $\underline{\alpha}(T)$ is *decreasing* to zero.

We are now ready for the main result.

---

[7]We follow [Anily and Federgruen, 1987].

Theorem 12.2.11 *Let* $\{\mathbf{P}(T)\}_{T \in (0,1]}$ *satisfy the above assumptions. Let* $\{T_n\}_{n \geq 0}$ *be a sequence of numbers in* $(0, 1]$ *decreasing to zero as* $n \to \infty$. *Then if*

$$\sum_{k=0}^{\infty} \left(\underline{\alpha}(T_{kN})\right)^N = \infty, \tag{12.24}$$

$\{\mathbf{P}(T_n)\}_{n \geq 0}$ *is weakly ergodic.*

Proof. Define, with a slight and innocuous notational ambiguity, $\mathbf{P}(n) := \mathbf{P}(T_n)$. The uniform irreducibility assumption guarantees the existence, for all ordered pair of states i, j), of a path $i_0 = i, i_1, ..., i_N = j$ such that

$$p_{i_j, i_{j+1}}(kN + j, kN + j + 1) = p_{i_j, i_{j+1}}(T_{kN+j}) > 0.$$

But $p_{kl}(T) > 0$ implies $p_{kl}(T) \geq d\underline{\alpha}(T)$, and therefore

$$p_{i_j, i_{j+1}}(kN + j, kN + j + 1) \geq d\underline{\alpha}(T_{kN+j}).$$

Since $\underline{\alpha}(T)$ is, in the vicinity of 0, monotone decreasing, then for sufficiently large k

$$p_{i_j, i_{j+1}}(kN + j, kN + j + 1) \geq d\underline{\alpha}(T_{(k+1)N}),$$

and therefore

$$p_{ij}(kN, (k + 1)N) \geq d^N \left(\underline{\alpha}(T_{(k+1)N})\right)^N.$$

Therefore, in view of (7.3) of Chapter 6,

$$1 - \delta(\mathbf{P}(kN, (k + 1)N)) \geq d^N \left(\underline{\alpha}(T_{(k+1)N})\right)^N.$$

Therefore, (12.24) implies

$$\sum_{k=1}^{\infty} \left(1 - \delta(\mathbf{P}(kN, (k + 1)N))\right) = \infty,$$

and the conclusion follows from the block criterion.   □

EXAMPLE 12.2.12: METROPOLIS SAMPLER, TAKE 3. Examples 12.2.3 and 12.2.8 continued. Recall the acceptance probabilities of the Metropolis sampler:

$$\alpha_{ij}(T) = \begin{cases} e^{\{U(j)-U(i)\}/T} & \text{if } U(j) > U(i), \\ 1 & \text{if } U(j) \leq U(i). \end{cases}$$

We see that conditions (12.20), (12.21), and (12.22) are satisfied. We have

$$\underline{\alpha}(T) = \inf_{\substack{j \in N(i) \\ U(i) < U(j)}} e^{-\{U(j)-U(i)\}/T},$$

and therefore

$$\underline{\alpha}(T) \geq e^{-\frac{\Delta}{T}}, \tag{12.25}$$

where

$$\Delta = \sup\{U(j) - U(i); j \in N(i)\}. \tag{12.26}$$

It follows that

$$\sum_{k=0}^{\infty} \{\underline{\alpha}(T_{kN})\}^N \geq \sum_{k=0}^{\infty} e^{-\frac{N\Delta}{T_{kN}}}.$$

For a cooling schedule $\{T_k\}_{k \geq 0}$ satisfying

$$T_k \geq \frac{N\Delta}{\log(k)}, \tag{12.27}$$

we see that

$$\sum_{k=1}^{\infty} \{\underline{\alpha}(T_{kN})\}^N \geq \sum_{k=1}^{\infty} \frac{1}{kN} = \infty,$$

and therefore, $\{\mathbf{P}(T_n)\}_{n \geq 1}$ is weakly ergodic.

Therefore, by the block criterion, $\{\mathbf{P}(T_n)\}$ is strongly ergodic, and the limiting probability vector puts all its mass uniformly on the set H of global minima. Therefore a cooling schedule verifying (12.27) guarantees convergence in distribution to the set of global minima.

————

Fast Cooling

We shall now see the effects of fast cooling. Denote by $\mathbf{P}(\lim)$ the transition matrix corresponding to the limit case $T \downarrow 0$. In particular, $p_{ij}(\lim) = 0$ if $U(i) < U(j)$. Call R_1 the recurrent communication class of some global minimum, and R_2 the recurrent communication class of some strictly local minimum. Note that R_1 only contains global minima, and in particular, R_1 and R_2 are disjoint. Define

$$\bar{\alpha}(2, T) = \sup_{i \in R_2, j \in N(i)} \alpha_{ij}(T). \tag{12.28}$$

Since for $j \in R_2$,

$$\sum_{\ell \in R_2} p_{j,\ell}(T_k) = 1 - \sum_{\substack{\ell \notin R_2 \\ j \in N(i)}} q_{j\ell}\alpha_{j\ell}(T_k) \geq 1 - \bar{\alpha}(2, T_k),$$

the probability of staying in R_2 forever is bounded from below by $\prod_{k=1}^{\infty}(1 - \bar{\alpha}(2, T_k))$. This infinite product is strictly positive if $\sum_{k=1}^{\infty} \bar{\alpha}(2, T_k) < \infty$.

Therefore, if the chain has at least one strictly local minimum, then under the condition

$$\sum_{k=1}^{\infty} \bar{\alpha}(2, T_k) < \infty, \tag{12.29}$$

the probability that it stays eternally in R_2 is strictly positive. In particular, since no globally optimal solution is in R_2, with positive probability the algorithm will never visit a globally optimal state.

EXAMPLE 12.2.13: METROPOLIS SAMPLER, TAKE 4. Examples 12.2.3, 12.2.8 and 12.2.12 continued. Define

$$\delta_2 = \inf\{U(j) - U(i); i \in R_2, j \notin R_2, j \in N(i)\}$$

and suppose that

$$\delta_2 > 0.$$

Since $\bar{\alpha}(2, T) \le e^{-\frac{\delta_2}{T}}$, we have

$$\sum_{k=0}^{\infty} \bar{\alpha}(2, T_k) \le \sum_{k=1}^{\infty} e^{-\delta_2/T_k}.$$

Therefore, with a cooling schedule such that

$$T_k \le \frac{\delta_2 - \alpha}{\log k}$$

for some $\alpha > 0$ such that $\delta_2 - \alpha > 0$, we have

$$\sum_{k=1}^{\infty} \bar{\alpha}(2, T_k) \le \sum_{k=1}^{\infty} e^{-(\log k)(1+\epsilon)},$$

where $1 + \epsilon = \frac{1}{1 - \frac{\alpha}{\delta_2}}$ (and therefore $\epsilon > 0$). Thus

$$\sum_{k=1}^{\infty} \bar{\alpha}(2, T_k) \le \sum_{k=1}^{\infty} \frac{1}{k^{1+\epsilon}} < \infty,$$

which implies that the cooling schedule does not yield convergence in distribution toward the set of global minima.

For the simulated annealing algorithm based on the Metropolis sampler, there exists a necessary and sufficient condition of convergence.[8] It says that there exists

---

[8][Hajek, 1988].

a constant γ such that a necessary and sufficient condition for the convergence of the Metropolis simulated annealing algorithm, whatever the initial state, is

$$\sum_{k=1}^{\infty} e^{-\frac{\gamma}{T_k}} = \infty.$$

In particular, a logarithmic cooling schedule $T_k = \frac{a}{\ln(k+1)}$ yields convergence if and only if $a \geq \gamma$.

The results of convergence given in the present section are of theoretical and qualitative interest only. Practical algorithms use faster than logarithmic schedules on a finite horizon. The theory and the performance evaluation of these algorithms is more difficult and outside the scope of this book.

12.3 Exercises

Exercise 12.3.1. PÓLYA'S URN, TAKE 1

An urn initially contains a white and b black balls. At each time $n \geq 1$, a ball is drawn at random, and is replaced in the urn together with s additional balls of the same color as the one of the ball just drawn. The state of the urn at time n is the number of white balls drawn before ($\leq$) time n. In particular, $X_0 = 0$ (nothing is done at time 0) and $0 \leq X_n \leq n$. Show that $\{X_n\}_{n \geq 0}$ is a non-homogeneous Markov chain and give its transition matrices.

Exercise 12.3.2. EXAMPLE OF WEAK ERGODICITY

Let

$$\mathbf{P}(2n-1) := \begin{pmatrix} \frac{1}{2} & \frac{1}{2} \\ 1 & 0 \end{pmatrix}, \ \mathbf{P}(2n) = \begin{pmatrix} 0 & 1 \\ 1 & 0 \end{pmatrix} \quad (n \geq 0).$$

Show that the non-homogeneous Markov chain with transition matrices $\{\mathbf{P}(n)\}_{n \geq 0}$ is *not* strongly ergodic (Hint: take $\mu = (0, 1)$ as the initial distribution and compute the distributions of the chain at even and odd times). Show that the chain is weakly ergodic.

Exercise 12.3.3. EXAMPLE OF STRONG ERGODICITY

Let

$$\mathbf{P}(2n) := \begin{pmatrix} \frac{1}{2} & 0 & \frac{1}{2} \\ \frac{1}{2} & \frac{1}{2} & 0 \\ 0 & \frac{1}{2} & 0 \\ 0 & \frac{1}{2} & \frac{1}{2} \end{pmatrix}, \ \mathbf{P}(2n+1) = \begin{pmatrix} \frac{1}{3} & \frac{1}{3} & \frac{1}{3} \\ \frac{1}{6} & \frac{1}{3} & \frac{1}{2} \\ \frac{1}{2} & \frac{1}{3} & \frac{1}{6} \end{pmatrix} \quad (n \geq 0).$$

Show directly that the non-homogeneous Markov chain with transition matrices $\{\mathbf{P}(n)\}_{n\geq 0}$ is strongly ergodic.

Exercise 12.3.4. ANOTHER EXAMPLE OF STRONG ERGODICITY
Show that the sequence

$$\mathbf{P}(n) := \begin{pmatrix} \frac{1}{3}+\frac{1}{n} & \frac{2}{3}-\frac{1}{n} \\ \frac{1}{2} & \frac{1}{2} \end{pmatrix} \quad (n \geq 0)$$

is strongly ergodic.

Exercise 12.3.5. A SUFFICIENT CONDITION FOR STRONG ERGODICITY
([9]) Let $\{\mathbf{P}(n)\}_{n\geq 0}$ be a sequence of transition matrices such that for some $D < \infty$ and for all $k \geq 1$,

$$\sum_{j=0}^{k-1} \delta\left(\mathbf{P}(j,k)\right) \leq D.$$

Show that this sequence is weakly ergodic.

Show that it is strongly ergodic if in addition there exists a probability vector π such that (12.9) holds, where $\pi(n)$ is a stationary distribution, assumed to exist but not necessarily unique, of $\mathbf{P}(n)$. (Hint: Mimic the proof of Theorem 12.1.7, only adapting the argument for the term B thereof.) Show that

$$\mathbf{P}(n) = \begin{pmatrix} \frac{1}{n} & 1-\frac{1}{n} \\ 1 & 0 \end{pmatrix}$$

satisfies the conditions of Theorem 12.1.5, but not those of the result just proved.

Exercise 12.3.6. PÓLYA'S URN, TAKE 2
Discuss the ergodicity (weak and strong) of the non-homogeneous Markov chain of Exercise 12.3.1.

Exercise 12.3.7. THE STOCHASTIC RULER
Discuss the convergence of the stochastic ruler algorithm of Example 12.2.9.

---

[9][Isaacson and Madsen, 1976].

Chapter 13

Continuous-Time Markov Chains

13.1 Poisson Processes

13.1.1 Point Processes

This subsection introduces *random point processes* of which the simplest example is the *homogeneous Poisson process*, a central object in the study of continuous-time HMCs.

A random point process is, roughly speaking, a countable random set of points of the real line. In most applications to engineering and operations research, a *point* of a point process is the time of occurrence of some event, and this is why points are also called *events*. For instance, the arrival times of customers at the desk of a post office or of jobs at the central processing unit of a computer are point process events. In biology, an event can be the time of birth of an organism. In physiology, the firing time of a neuron is also an event.

Definition 13.1.1 *A* random point process *on the positive half-line is a sequence* $\{T_n\}_{n \geq 0}$ *of non-negative random variables such that, almost surely,*

(i) $T_0 \equiv 0$,

(ii) $0 < T_1 < T_2 < \cdots$, *and*

(iii) $\lim_{n \uparrow \infty} T_n = +\infty$.

Remark 13.1.2 The usual definition of a random point process is less restrictive. In particular, condition (ii) is relaxed in the more general definition, where multiple points (simultaneous arrivals to a ticket booth, for instance) are allowed. When condition (ii) holds, one speaks of a *simple point process*. Also, condition (iii) is not required in the more general definition which allows with positive probability

© Springer Nature Switzerland AG 2020
P. Brémaud, *Markov Chains*, Texts in Applied Mathematics 31,
https://doi.org/10.1007/978-3-030-45982-6_13

an *explosion*, that is, an accumulation of events in finite time. However, conditions (ii) and (iii) fit the special case of homogeneous Poisson processes, the center of interest in this section.

The sequence $\{S_n\}_{n \geq 1}$ defined by

$$S_n = T_n - T_{n-1}$$

is called the *inter-event* sequence or, in the appropriate context, the *inter-arrival* sequence. For any interval $(a, b]$ in $\mathbb{R}_+$,

$$N((a, b]) := \sum_{n \geq 1} 1_{(a,b]}(T_n)$$

is an integer-valued random variable counting the events occurring in the time interval $(a, b]$. For typographical simplicity, it will be occasionally denoted by $N(a, b]$, omitting the external parentheses. For $t \geq 0$, let

$$N(t) := N(0, t].$$

In particular, $N(0) = 0$ and $N(a, b] = N(b) - N(a)$. Since the interval $(0, t]$ is closed on the right, the trajectories (or sample paths) $t \mapsto N(t, \omega)$ are right-continuous. They are non-decreasing, have limits on the left at every time t and jump one unit upwards at each event of the point process. The family of random variables $N := \{N(t)\}_{t \geq 0}$ is called the *counting process* of the point process $\{T_n\}_{n \geq 1}$. Since the sequence of events can be recovered from N, the latter also receives the appellation "point process."

13.1.2   The Counting Process of an HPP

There exist several equivalent definitions of a Poisson process. The one adopted here is the most practical.

Definition 13.1.3 *A point process N on the positive half-line is called a* homogeneous Poisson process (HPP) *with intensity $\lambda > 0$ if*

(α) *for all times t_i ($1 \leq i \leq k$) such that $0 \leq t_1 \leq t_2 \leq \cdots \leq t_k$, the random variables $N(t_i, t_{i+1}]$ ($1 \leq i \leq k$) are independent, and*

(β) *for any interval $(a, b] \subset \mathbb{R}_+$, $N(a, b]$ is a Poisson random variable with mean $\lambda(b - a)$.*

In particular,

$$P(N(a, b] = k) = e^{-\lambda(b-a)} \frac{[\lambda(b-a)]^k}{k!} \quad (k \geq 0)$$

and

$$E[N(a, b]] = \lambda(b - a).$$

In this sense, λ is the average density of points.

Condition (α) is the property of *independence of increments* of Poisson processes. It implies in particular that for any interval $(a, b]$, the random variable $N(a, b]$ is independent of $(N(s), s \in (0, a])$. For this reason, Poisson processes are sometimes called *memoryless*. A more precise statement is "the *increments* of homogeneous Poisson processes have no memory of the past".

Remark 13.1.4 The definition adopted for random point processes does not allow multiple points or explosions. But suppose it did. It turns out that requirements (α) and (β) in Definition 13.1.3 suffice to prevent such occurrences. The proof is as follows. Since $E[N(a)] = \lambda a < \infty$, $N(a) < \infty$ almost surely. Since this is true for all $a \geq 0$, $\lim_{n \uparrow \infty} T_n = \infty$ almost surely. Simplicity will follow from $P(D(a)) = 0$ for all $a \geq 0$, where

$$D(a) := \{\text{there exists multiple points in } (0, a]\}.$$

We prove this for $D = D(1)$ (without loss of generality). The event

$$D_n := \left\{ N\left(\frac{i}{2^n}, \frac{i+1}{2^n}\right] \geq 2 \text{ for some } i \ (1 \leq i \leq 2^n - 1) \right\}$$

decreases to D as n tends to infinity and therefore, by the monotone sequential continuity of probability,

$$P(D) = \lim_{n \uparrow \infty} P(D_n) = 1 - \lim_{n \uparrow \infty} P(\overline{D}_n).$$

But

$$P(\overline{D}_n) = P\left(\bigcap_{i=0}^{2^n-1} \left\{ N\left(\frac{i}{2^n}, \frac{i+1}{2^n}\right] \leq 1 \right\} \right) = \prod_{i=0}^{2^n-1} P\left(N\left(\frac{i}{2^n}, \frac{i+1}{2^n}\right] \leq 1 \right)$$

$$= \prod_{i=0}^{2^n-1} e^{-\lambda 2^{-n}}(1 + \lambda 2^{-n}) = e^{-\lambda}(1 + \lambda 2^{-n})^{2^n}.$$

The limit of the latter quantity is 1 as $n \uparrow \infty$, and therefore, $P(D) = 0$.

Theorem 13.1.5 *The inter-event sequence* $\{S_n\}_{n\geq 1}$ *of an* HPP *on the positive half-line with intensity* $\lambda > 0$ *is* IID *with a common exponential distribution of parameter* λ.

The cumulative distribution function of an arbitrary inter-event time is therefore

$$P(S_n \leq t) = 1 - e^{-\lambda t}.$$

Recall that

$$E[S_n] = \lambda^{-1},$$

that is, the average number of events per unit of time equals the inverse average inter-event time.

Proof. Suppose we can show that for any $n \geq 1$, the random vector $T := (T_1, \ldots, T_n)$ admits the probability density function

$$f_T(t_1, \ldots, t_n) = \lambda^n e^{-\lambda t_n} 1_C(t_1, \ldots, t_n), \tag{13.1}$$

where $C := \{(t_1, \ldots, t_n); 0 < t_1 < \cdots < t_n\}$. Since

$$S_1 = T_1, \quad S_2 = T_2 - T_1, \ldots, \quad S_n = T_n - T_{n-1},$$

the formula of smooth change of variables gives for the PDF of $S = (S_1, \ldots, S_n)$

$$f_S(s_1, \ldots, s_n) = f_T(s_1, s_1 + s_2, \ldots, s_1 + \cdots + s_n) = \prod_{i=1}^{n} \{\lambda e^{-\lambda s_i} 1_{\{s_i > 0\}}\}.$$

It remains to prove (13.1). The PDF of T at $t = (t_1, \ldots, t_n)$ is obtained as the limit as $h_1, \ldots, h_n \in \mathbb{R}_+$ tend to 0 of the quantity

$$\frac{P(\cap_{i=1}^{n}\{T_i \in (t_i, t_i + h_i]\})}{\prod_{i=1}^{n} h_i}, \tag{13.2}$$

where it suffices to consider those $(t_1, \ldots, t_n)$ inside C since the points $T_1, \ldots, T_n$ are strictly ordered in increasing order. For sufficiently small $h_1, \ldots, h_n$, the event $\cap_{i=1}^{n}\{T_i \in (t_i, t_i + h_i]\}$ is the intersection of the events $\{N(0, t_1] = 0\}$, $\cap_{i=1}^{n-1}\{N(t_i, t_i + h_i] = 1, N(t_i + h_i, t_{i+1}] = 0\}$ and $\{N(t_n, t_n + h_n] \geq 1\}$, and therefore the numerator of (13.2) equals

$$P(N(0, t_1] = 0) \left(\prod_{i=1}^{n-1} P(N(t_i, t_i + h_i] = 1, N(t_i + h_i, t_{i+1}] = 0) \right) \times \cdots$$

$$\cdots \times P(N(t_n, t_n + h_n] \geq 1)$$

$$= e^{-\lambda t_1} \prod_{i=1}^{n-1} \left(e^{-\lambda h_i} \lambda h_i e^{-\lambda(t_{i+1} - t_i - h_i)} \right) (1 - e^{-\lambda h_n}) = \lambda^{n-1} e^{-\lambda t_n} h_1 \cdots h_{n-1}(1 - e^{-\lambda h_n}).$$

Dividing by $h_1 \cdots h_n$ and taking the limit as $h_1, \ldots, h_n$ tend to 0, we obtain $\lambda^n e^{-\lambda t_n}$. $\square$

13.1.3   Competing Poisson Processes

Let $\{T_n^1\}_{n \geq 1}$ and $\{T_n^2\}_{n \geq 1}$ be two *independent* HPPs on $\mathbb{R}_+$ with respective intensities $\lambda_1 > 0$ and $\lambda_2 > 0$. Their *superposition* is defined to be the sequence $\{T_n\}_{n \geq 1}$ formed by merging the two sequences $\{T_n^1\}_{n \geq 1}$ and $\{T_n^2\}_{n \geq 1}$ (see Figure 13.1.1). We shall prove that

(i) the point processes $\{T_n^1\}_{n \geq 1}$ and $\{T_n^2\}_{n \geq 1}$ have no points in common, and

(ii) the point process $\{T_n\}_{n \geq 1}$ is an HPP with intensity $\lambda = \lambda_1 + \lambda_2$.

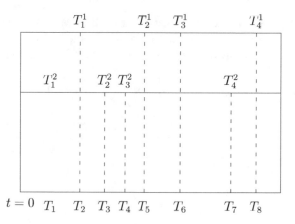

Figure 13.1.1. Superposition, or sum, of two point processes

Indeed, defining N by

$$N(a, b] = N_1(a, b] + N_2(a, b],$$

we see that condition (α) of Definition 13.1.3 is satisfied, in view of the independence of N_1 and N_2. Also, $N(a, b]$ being the sum of two independent Poisson random variables of mean $\lambda_1(b - a)$ and $\lambda_2(b - a)$ is a Poisson variable of mean $\lambda(b - a)$ where $\lambda = \lambda_1 + \lambda_2$, and therefore, condition (β) of Definition 13.1.3 is satisfied. This proves (ii). By Remark 13.1.4, N is simple and therefore (i) is true.

The above result can be extended to several – possibly infinitely many – homogeneous Poisson processes as follows:

Theorem 13.1.6 *Let* $\{N_i\}_{i\geq 1}$ *be a family of independent* HPPs *with respective positive intensities* $\{\lambda_i\}_{i\geq 1}$. *Then,*

  (i) two distinct HPPs *of this family have no points in common, and*

  (ii) if $\lambda := \sum_{i=1}^{\infty} \lambda_i < \infty$, *then* $N(t) := \sum_{i=1}^{\infty} N_i(t)$ $(t \geq 0)$ *defines the counting process of an* HPP *with intensity* λ.

Proof. Assertion (ii) has already been proven. Observe that for all $t \geq 0$, $N(t)$ is almost surely finite since

$$E[N(t)] = \sum_{i=1}^{\infty} E[N_i(t)] = \left(\sum_{i=1}^{\infty} \lambda_i \right) t < \infty.$$

In particular, $N(a, b]$ is almost surely finite for all $(a, b] \subset \mathbb{R}_+$. The proof of lack of memory of N is the same as in the case of two superposed Poisson processes. Finally, $N(a, b]$ is a Poisson random variable of mean $\lambda(b - a)$ since

$$
\begin{aligned}
P(N(a, b] = k) &= \lim_{n\uparrow\infty} P\left(\sum_{i=1}^{n} N_i(a, b] = k \right) \\
&= \lim_{n\uparrow\infty} e^{-(\sum_{i=1}^{n} \lambda_i (b-a))} \frac{[\sum_{i=1}^{n} \lambda_i (b - a)]^k}{k!} \\
&= e^{-\lambda(b-a)} \frac{[\lambda(b - a)]^k}{k!}.
\end{aligned}
$$

$\square$

    The next result is called the *competition theorem* because it features HPPs competing for the production of the first event.

Theorem 13.1.7 *In the situation of Theorem 13.1.6, where* $\lambda := \sum_{i=1}^{\infty} \lambda_i < \infty$, *denote by* Z *the first event time of* $N = \sum_{i=1}^{\infty} N_i$ *and by* J *the index of the* HPP *responsible for it* (Z *is the first event of* N_J). *Then*

$$P(J = i, Z \geq a) = P(J = i)P(Z \geq a) = \frac{\lambda_i}{\lambda} e^{-\lambda a}. \tag{13.3}$$

In particular, J and Z are independent, $P(J = i) = \frac{\lambda_i}{\lambda}$ *and Z is exponential with mean* λ^{-1}.

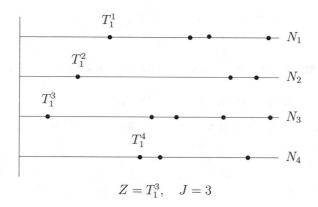

$$Z = T_1^3, \quad J = 3$$

Figure 13.1.2. Competition among four point processes

Proof.

A. We first prove the result for a finite number of Poisson processes. We have to show that if $X_1, \ldots, X_K$ are K independent exponential variables with means $\lambda_1^{-1}, \ldots, \lambda_K^{-1}$ and if J_K is defined by $X_{J_K} = Z_K := Z_K := \inf(X_1, \ldots, X_K)$, then

$$P(J_K = i, Z_K \geq a) = \frac{\lambda_i}{\lambda_1 + \cdots + \lambda_K} \exp\{-(\lambda_1 + \cdots + \lambda_K)a\}. \qquad (\star)$$

First observe that

$$P(Z_K \geq a) = P\left(\cap_{j=1}^K \{X_j \geq a\}\right) = \prod_{j=1}^K P(X_j \geq a) - \prod_{j=1}^K e^{-\lambda_j a} = e^{-(\lambda_1 + \cdots + \lambda_K)a}.$$

Letting $U := \inf(X_2, \ldots, X_K)$, we have

$$
\begin{aligned}
P(J_K = 1, Z_K \geq a) &= P(a \leq X_1 < U) \\
&- \int_a^\infty P(U > x)\lambda_1 e^{-\lambda_1 x}\mathrm{d}x = \int_a^\infty e^{-(\lambda_2 + \cdots + \lambda_K)x}\lambda_1 e^{-\lambda_1 x}\mathrm{d}x \\
&= \frac{\lambda_1}{\lambda_1 + \cdots + \lambda_K} e^{-(\lambda_1 + \cdots + \lambda_K)a}.
\end{aligned}
$$

This gives $(\star)$. Letting $a - 0$ yields $P(J_K = 1) = \frac{\lambda_1}{\lambda_1 + \cdots + \lambda_K}$. This, together with $(\star)$ and the expression for $P(Z_K \geq a)$ gives (13.3), for $i = 1$, without loss of generality.

B. Suppose the result true for a finite number of HPPs. Since the event $\{J_K = 1, Z_K \geq a\}$ decreases to $\{J = 1, Z \geq a\}$ as $K \uparrow \infty$, we have

$$P(J = 1, Z \geq a) = \lim_{K\uparrow\infty} P(J_K = 1, Z_K \geq a),$$

from which (13.3) follows, using the result of part A of the proof.   □

13.2   Distribution of a Continuous-Time HMC

13.2.1   The Transition Semi-Group

The traditional approach to continuous-time Markov chains is based on the *transition semi-group*, and the principal mathematical object is then the *infinitesimal generator*. The transition semi-group is the continuous-time analogue of the iterates of the transition matrix in discrete time. The infinitesimal generator, however, has no analogue because it is essentially a continuous-time notion involving derivatives.

   The transition semi-group approach is mainly analytical, and we shall propose in Chapter 14 a sample path approach that describes continuous-time Markov chains in terms of a stochastic differential equation, the analogue of the recurrence equation associated with a discrete-time Markov chain (Theorem 2.2.1). These two approaches are complementary. We begin with the semi-group approach because it makes the continuous-time theory look more like a natural extension of the discrete-time theory.

   We now proceed with the basic definitions concerning continuous-time HMCs. Let E be a countable set, called the state space, and let $\{X(t)\}_{t\geq 0}$ be an E-valued stochastic process, that is, a family of random variables $X(t)$ indexed by $\mathbb{R}_+$ and taking their values in E. The probability distribution of $\{X(t)\}_{t\geq 0}$ consists of the data $P(X(t_1) = i_1, \ldots, X(t_k) = i_k)$ for all non-negative $t_1, \ldots, t_k$ and all states $i_1, \ldots, i_k$.

Definition 13.2.1 *The E-valued stochastic process $\{X(t)\}_{t\geq 0}$ is called a* continuous-time Markov chain *if for all $i, j, i_1, \ldots, i_k \in E$, all $t, s \geq 0$ and all $s_1, \ldots, s_k \geq 0$ such that $s_\ell \leq s$ $(1 \leq \ell \leq k)$,*

$$P(X(t + s) = j \mid X(t) = i, X(s_1) = i_1, \ldots, X(s_k) = i_k)$$
$$= P(X(t + s) = j \mid X(s) = i),$$

whenever both sides are well defined. If the right-hand side is independent of s, this continuous-time Markov chain is called homogeneous.

   In the homogeneous case, let

$$\mathbf{P}(t) := \{p_{ij}(t)\}_{i,j\in E},$$

where

$$p_{ij}(t) := P(X(t + s) = j \mid X(s) = i).$$

The family $\{\mathbf{P}(t)\}_{t \geq 0}$ is the *transition semi-group* of the continuous-time HMC.

The *Chapman–Kolmogorov equation*

$$p_{ij}(t + s) = \sum_{k \in E} p_{ik}(t) p_{kj}(s),$$

that is, in compact form,

$$\mathbf{P}(t + s) = \mathbf{P}(t)\mathbf{P}(s),$$

is obtained in the same way as for discrete-time Markov chains. Also, clearly,

$$\mathbf{P}(0) = I,$$

where I is the identity matrix.

The distribution at time t of $X(t)$, that is, the vector $\mu(t) := \{\mu_i(t)\}_{i \in E}$, where $\mu_i(t) = P(X(t) = i)$, is obtained from the initial distribution by the formula

$$\mu(t)^T = \mu(0)^T \mathbf{P}(t). \tag{13.4}$$

Also, for all $t_1, \ldots, t_k$ such that $0 \leq t_1 \leq t_2 \leq \cdots \leq t_k$, and for all states $i_0, i_1, \ldots, i_k,$

$$P\left(\bigcap_{j=1}^{k} \{X(t_j) = i_j\}\right) = \sum_{i_0 \in E} P(X(0) - i_0) \prod_{j=1}^{k} p_{i_{j-1} i_j}(t_j - t_{j-1}). \tag{13.5}$$

Formulas (13.4) and (13.5) are proven in the same manner as the corresponding results in discrete time. Formula (13.5) shows in particular that the probability distribution of a continuous-time HMC is entirely determined by its initial distribution and its transition semi-group.

Notation: Recall that $P_i(\cdot)$ is an abbreviated notation for $P(\cdot \mid X(0) = i)$, where i is a state. If μ is a probability distribution on E,

$$P_\mu(A) := \sum_{i \in E} \mu(i) P(A \mid X(0) = i)$$

is the probability of A when the distribution of the initial state is μ.

EXAMPLE 13.2.2: POISSON IS MARKOVIAN. Let N be an HPP on the positive half-line with intensity $\lambda > 0$. The counting process $\{N(t)\}_{t \geq 0}$ is a continuous-time HMC. Indeed, with $C := \{N(s_1) = i_1, \ldots, N(s_k) = i_k\}$,

$$
\begin{aligned}
P(N(t+s) = j \mid N(s) = i, C) &= \frac{P(N(t+s) = j, N(s) = i, C)}{P(N(s) = i, C)} \\
&= \frac{P(N(s, s+t] = j - i, N(s) = i, C)}{P(N(s) = i, C)}.
\end{aligned}
$$

But $N(s, s+t]$ is independent of $N(s)$ and of $N(s_\ell)$ when $s_\ell \leq s$, and therefore,

$$
P(N(s, s+t] = j - i, N(s) = i, C) = P(N(s, s+t] = j - i)P(N(s) = i, C),
$$

so that

$$
P(N(t+s) = j \mid N(s) = i, C) = P(N(s, s+t] = j - i).
$$

Similarly,

$$
P(N(t+s) = j \mid N(s) = i) = P(N(s, s+t] = j - i).
$$

Therefore, $\{N(t)\}_{t \geq 0}$ is a continuous-time Markov chain. It is homogeneous, since $P(N(s, s+t] = j - i)$ does not depend on s. The transition semi-group $\{\mathbf{P}(t)\}_{t \geq 0}$ is given by

$$
p_{ij}(t) = P(N(0, t] = j - i) = e^{-\lambda t} \frac{(\lambda t)^{j-i}}{(j-i)!},
$$

for $j \geq i$. Otherwise, if $j > i$, then $p_{ij}(t) = 0$.

EXAMPLE 13.2.3: FLIP-FLOP. Let N be the Poisson process of intensity λ considered in the previous example, and define the *flip-flop process* $\{X(t)\}_{t \geq 0}$ with state space $E = \{+1, -1\}$ by

$$
X(t) = X(0) \times (-1)^{N(t)},
$$

where $X(0)$ is an E-valued random variable independent of the counting process N. (The flip-flop process switches between -1 and $+1$ at each event of N.) The value $X(t+s)$ depends on $N(s, s+t]$ and $X(s)$. Also, $N(s, s+t]$ is independent of $X(0), N(s_1), \ldots, N(s_k)$ when $s_\ell \leq s$ $(1 \leq \ell \leq k)$, and the latter random variables determine $X(s_1), \ldots, X(s_k)$. Therefore, $X(t+s)$ is independent of $X(s_1), \ldots, X(s_k)$ given $X(s)$, that is, $\{X(t)\}_{t \geq 0}$ is a Markov chain. Moreover,

$$
\begin{aligned}
P(X(t+s) = 1 \mid X(s) = -1) &= P(N(s, s+t] = \text{ odd }) \\
&= \sum_{k=0}^{\infty} e^{-\lambda t} \frac{(\lambda t)^{2k+1}}{(2k+1)!} = \frac{1}{2}(1 - e^{-2\lambda t}),
\end{aligned}
$$

that is, $p_{-1,+1}(t) = \frac{1}{2}(1 - e^{-2\lambda t})$. Similar computations give for the transition semi-group

$$\mathbf{P}(t) = \frac{1}{2} \begin{pmatrix} 1 + e^{-2\lambda t} & 1 - e^{-2\lambda t} \\ 1 - e^{-2\lambda t} & 1 + e^{-2\lambda t} \end{pmatrix}.$$

Definition 13.2.4 *Let $\{\hat{X}_n\}_{n \geq 0}$ be a discrete-time HMC with countable state space E and transition matrix $\mathbf{K} = \{k_{ij}\}_{i,j \in E}$ and let $\{T_n\}_{n \geq 1}$ be an HPP on $\mathbb{R}_+$ with intensity $\lambda > 0$ and associated counting process N. Suppose that $\{\hat{X}_n\}_{n \geq 0}$ and N are independent. The process $\{X(t)\}_{t \geq 0}$ with values in E, defined by*

$$X(t) := \hat{X}_{N(t)}$$

(Figure 13.2.1), is called a uniform Markov chain. *The Poisson process N is called the* clock *and the chain $\{\hat{X}_n\}_{n \geq 0}$ is called the* subordinated chain.

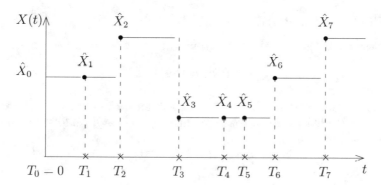

Figure 13.2.1. Uniform Markov chain

Remark 13.2.5 Observe that $X(T_n) = \hat{X}_n$ for all $n \geq 0$. Observe also that the discontinuity times of the uniform chain are all events of N but that not all events of N are discontinuity times, since it may well occur that $\hat{X}_{n-1} = \hat{X}_n$ (a transition of type $i \to i$ of the subordinated chain).

The process $\{X(t)\}_{t \geq 0}$ is a continuous-time HMC (Exercise 13.6.8) with transition semi-group

$$\mathbf{P}(t) = \sum_{n=0}^{\infty} e^{-\lambda t} \frac{(\lambda t)^n}{n!} \mathbf{K}^n, \tag{13.6}$$

that is,

$$p_{ij}(t) = \sum_{n=0}^{\infty} e^{-\lambda t} \frac{(\lambda t)^n}{n!} k_{ij}(n). \tag{13.7}$$

Indeed,

$$P_i(X(t) = j) = P_i(\hat{X}_{N(t)} = j) \quad = \quad \sum_{n=0}^{\infty} P_i(\hat{X}_n = j, N(t) = n)$$

$$= \quad \sum_{n=0}^{\infty} P_i(\hat{X}_n = j) P_i(N(t) = n).$$

13.2.2   The Infinitesimal Generator

Let $\{\mathbf{P}(t)\}_{t \geq 0}$ be a *transition semi-group* on E, that is, for each $t, s \geq 0$,

(a) $\mathbf{P}(t)$ is a stochastic matrix,

(b) $\mathbf{P}(0) = I$, and

(c) $\mathbf{P}(t + s) = \mathbf{P}(t)\mathbf{P}(s)$.

Suppose moreover that the semi-group is *continuous at the origin*, that is,

$$\lim_{h \downarrow 0} \mathbf{P}(h) = \mathbf{P}(0) = I \,,$$

where the convergence therein is pointwise and for each entry. In Exercise 13.6.4, the reader is invited to prove that continuity at the origin implies continuity at any time $t \geq 0$, that is,

$$\lim_{h \to 0} p_{ij}(t + h) = p_{ij}(t) \,,$$

for all states i, j.

The result to follow is purely analytical: it does not require $\{\mathbf{P}(t)\}_{t \geq 0}$ to be the transition semi-group of some continuous-time HMC.

Theorem 13.2.6 *Let $\{\mathbf{P}(t)\}_{t \geq 0}$ be a continuous transition semi-group on the countable state space E. For any state i, there exists*

$$q_i := \lim_{h \downarrow 0} \frac{1 - p_{ii}(h)}{h} \in [0, \infty] \,, \tag{13.8}$$

and for any pair i, j of different states, there exists

$$q_{ij} := \lim_{h \downarrow 0} \frac{p_{ij}(h)}{h} \in [0, \infty) \,. \tag{13.9}$$

Proof. For $t \geq 0$ and $n \geq 1$, $\mathbf{P}(t) = [\mathbf{P}\left(\frac{t}{n}\right)]^n$ and therefore $p_{ii}(t) \geq [p_{ii}\left(\frac{t}{n}\right)]^n$ for all $i \in E$. Since $\lim_{h \downarrow 0} p_{ii}(h) = 1$, there exists an $\epsilon > 0$ such that $p_{ii}(h) > 0$ for all $h \in [0, \epsilon]$. For n sufficiently large, $\frac{t}{n} \in [0, \epsilon]$. Therefore $p_{ii}(t) > 0$ for all $t \geq 0$, and one can define the function

$$f_i(t) := -\log p_{ii}(t),$$

which is real-valued, non-negative and such that $\lim_{h \downarrow 0} f_i(h) = 0$. Also, from $\mathbf{P}(t)\mathbf{P}(s) = \mathbf{P}(t+s)$, we see that $p_{ii}(t+s) \geq p_{ii}(t)p_{ii}(s)$ and therefore f_i is subadditive, that is,

$$f_i(t+s) \leq f_i(t) + f_i(s)$$

for all $s, t \in \mathbb{R}_+$. Define the (possibly infinite) non-negative real number

$$q_i = \sup_{t > 0} \frac{f_i(t)}{t}.$$

By the *subadditive function theorem* (Theorem A.1.13)

$$\lim_{h \downarrow 0} \frac{f_i(h)}{h} = q_i,$$

and therefore

$$\lim_{h \downarrow 0} \frac{1 - p_{ii}(h)}{h} = \lim_{h \downarrow 0} \frac{1 - e^{-f_i(h)}}{f_i(h)} \frac{f_i(h)}{h} = q_i,$$

which proves the first equality in (13.8). For future reference, we shall observe that

$$\frac{1 - p_{ii}(h)}{h} = \frac{1 - e^{-f_i(h)}}{f_i(h)} \frac{f_i(h)}{h} \leq \frac{f_i(h)}{h},$$

and therefore for all $h > 0$,

$$\frac{1 - p_{ii}(h)}{h} \leq q_i. \tag{13.10}$$

It now remains to prove (13.9).

Take two different states i and j. Since $p_{ii}(t)$ and $p_{jj}(t)$ tend to 1 as $t > 0$ tends to 0, for any $c \in (\frac{1}{2}, 1)$, there exists a $\delta > 0$ such that for $t \in [0, \delta]$, $p_{ii}(t) > c$ and $p_{jj}(t) > c$. Let $n > 0$ be an integer and $h > 0$ be such that $0 \leq nh \leq \delta$. Denote by $\{X_n\}_{n \geq 0}$ the discrete-time HMC defined by $X_n := X(nh)$, with transition matrix $\mathbf{P}(h)$. One way to go from $X_0 = i$ to $X_n = j$ is to go from $X_0 = i$ to $X_r = i$ ($0 \leq r = \leq n - 1$) without visiting state j meanwhile, then to go from $X_r = i$ to $X_{r+1} = j$, and then from $X_{r+1} = j$ to $X_n = j$. The paths corresponding to different values of r are different but do not exhaust the possibilities of going from $X_0 = i$ to $X_n = j$. Therefore,

$$p_{ij}(nh) \geq \sum_{r=0}^{n-1} P(\cap_{\ell=1}^{r-1}\{X_\ell \neq j\}, X_r = i \mid X_0 = i) p_{ij}(h) P(X_n = j \mid X_{r+1} = j).$$

The parameters δ, n, and h are such that $P(X_n = j \mid X_{r+1} = j) \geq c$. Also

$$P(\cap_{\ell=1}^{r-1}\{X_\ell \neq j\}, X_r = i \mid X_0 = i)$$
$$= P(X_r = i \mid X_0 = i) - \sum_{k<r} P(\cap_{\ell=1}^{k-1}\{X_\ell \neq j\}, X_k = j \mid X_0 = i) P(X_r = i \mid X_k = j)$$
$$\geq c - (1 - c) \sum_{k<r} P(\cap_{\ell=1}^{k-1}\{X_\ell \neq j\}, X_k = j \mid X_0 = i) \geq c - (1 - c) = 2c - 1,$$

where we have observed that for $i \neq j$,

$$P(X_r = i \mid X_k = j) + P(X_r = j \mid X_k = j) \leq 1,$$

and therefore

$$P(X_r = i \mid X_k = j) \leq 1 - P(X_r = j \mid X_k = j) \leq 1 - c.$$

Therefore,

$$p_{ij}(nh) \geq c(2c - 1) n p_{ij}(h).$$

Let now $t < \delta$ and $h < \delta$, and take for n the integer part of t/h. Writing the last inequality as

$$\frac{p_{ij}(h)}{h} \leq \frac{1}{c(2c-1)} \frac{p_{ij}(nh)}{nh},$$

and since $\lim_{h\downarrow 0} nh = t$, we see that $\lim_{h\downarrow 0} \frac{p_{ij}(nh)}{nh} = \frac{p_{ij}(t)}{t}$, so that

$$\limsup_{h\downarrow 0} \frac{p_{ij}(h)}{h} \leq \frac{1}{c(2c-1)} \frac{p_{ij}(t)}{t} < \infty,$$

which in turn gives

$$\limsup_{h\downarrow 0} \frac{p_{ij}(h)}{h} \leq \frac{1}{c(2c-1)} \liminf_{t\downarrow 0} \frac{p_{ij}(t)}{t} < \infty.$$

Since c can be chosen arbitrarily close to 1, we have

$$\limsup_{h\downarrow 0} \frac{p_{ij}(h)}{h} \leq \liminf_{t\downarrow 0} \frac{p_{ij}(t)}{t} < \infty,$$

and this implies the existence of $\lim_{h\downarrow 0} \frac{p_{ij}(h)}{h}$ and the finiteness of this limit. $\qquad \square$

For each state i, let

$$q_{ii} := -q_i \,.$$

Definition 13.2.7 *The quantities* q_{ij} *($i, j \in E$) are called the* local characteristics *of the semi-group, or of the corresponding continuous-time* HMC. *The matrix*

$$\mathbf{A} := \{q_{ij}\}_{i,j \in E}$$

is called the infinitesimal generator *of the semi-group, or of the continuous-time* HMC.

In compact notation,

$$\mathbf{A} = \lim_{h\downarrow 0} \frac{\mathbf{P}(h) - \mathbf{P}(0)}{h} \,,$$

whose meaning is given by (13.8) and (13.9). Thus, in this sense, the infinitesimal generator $\mathbf{A}$ is the derivative at 0 of the matrix function $t \mapsto \mathbf{P}(t)$.

EXAMPLE 13.2.8: THE INFINITESIMAL GENERATOR OF THE UNIFORM HMC. From the expression (13.6), or (13.7), of the transition semi-group of the uniform HMC of Definition 13.2.4, we easily obtain its infinitesimal generator (Exercise 13.6.9)

$$\mathbf{A} = \lambda(\mathbf{K} - I) \,, \tag{13.11}$$

that is,

$$q_i = \lambda(1 - k_{ii}) \,, \tag{13.12}$$

and for $i \neq j$,

$$q_{ij} = \lambda k_{ij} \,. \tag{13.13}$$

Definition 13.2.9 *If for all states* i,

$$q_i < \infty \,, \tag{13.14}$$

the semi-group $\{\mathbf{P}(t)\}$ *is called* stable. *If for all states* i,

$$q_i = \sum_{\substack{j \in E \\ j \neq i}} q_{ij} \,, \tag{13.15}$$

it is called conservative.

The reason for the last appellation comes from the conservation equality

$$\sum_{j \in E} p_{ij}(h) = 1$$

or, equivalently,

$$\frac{1 - p_{ii}(h)}{h} = \sum_{\substack{j \in E \\ j \neq i}} \frac{p_{ij}(h)}{h},$$

which yields

$$q_i = \lim_{h \downarrow 0} \sum_{\substack{j \in E \\ j \neq i}} \frac{p_{ij}(h)}{h}.$$

Therefore, *if* the interchange of summation and limit is allowed, we obtain (13.15).

Interchange of sums and limits is always possible if E is finite. In this case (13.15) holds and, consequently, (13.14) holds because q_{ij} is finite for all pairs of different sites i, j. Also, the semi-group of a uniform HMC is stable and conservative, even when E is infinite (use (13.12) and (13.13)).

As a matter of fact, a very general class of Markov chains, namely, the *regular jump Markov chains*, are stable and conservative.

Definition 13.2.10 *A stochastic process $\{X(t)\}_{t \geq 0}$ taking its values in the (not necessarily countable) state space E is called a* jump process *if for almost all $\omega \in \Omega$ and all $t \geq 0$, there exists $\epsilon(t, \omega) > 0$ such that*

$$X(t + s, \omega) = X(t, \omega) \text{ for all } s \in [t, t + \epsilon(t, \omega)).$$

It is called a regular *jump process if in addition, for almost all $\omega \in \Omega$, the set $A(\omega)$ of discontinuities of the function $t \to X(t, \omega)$ is σ-discrete, that is, for all $c \geq 0$,*

$$|A(\omega) \cap [0, c]| < \infty,$$

where $|B|$ denotes the cardinality of the set B. A regular jump homogeneous Markov chain *is by definition a continuous-time HMC that is also a regular jump process.*

Note that for a jump process (not necessarily regular), there exists a sequence of times $\{\tau_n\}_{n \geq 0}$ where

$$\tau_0 = 0 < \tau_1 < \tau_2 < \tau_3 < \cdots$$

and a sequence $\{X_n\}_{n\geq 0} \in E$ such that

$$X(t) = X_n \text{ if } \tau_n \leq t < \tau_{n+1}.$$

This describes $\{X(t)\}_{t\geq 0}$ on the interval $[0, \tau_\infty)$, where

$$\tau_\infty := \lim \uparrow \tau_n$$

is the *explosion time*. If moreover the process is regular, then $\tau_\infty = \infty$, and $\{X(t)\}_{t\geq 0}$ is right-continuous.

Theorem 13.2.11 *A regular jump* HMC *is stable and conservative.*

The proof is postponed to Chapter 14.

When the chain is stable and conservative,

$$P(X(t+h) = i \mid X(t) = i) = 1 - q_i h + o(h), \tag{13.16}$$

and if $i \neq j$,

$$P(X(t+h) = j \mid X(t) = i) = q_{ij} h + o(h). \tag{13.17}$$

(Recall that the $o(h)$ symbol represents a function defined in a neighborhood of 0 for which $\lim_{h\to 0} \frac{|o(h)|}{h} = 0$.)

Definition 13.2.12 *A continuous-time birth-and-death process is a regular jump* HMC *taking its values in $\mathbb{N}$ and with an infinitesimal generator of the form*

$$q_{i,i+1} = \lambda_i, \quad q_{i,i-1} = \mu_i 1_{\{i\geq 1\}},$$

and $q_{ij} = 0$ if $j \notin \{i-1, i, i+1\}$.

The parameters λ_i and μ_i are the *birth and death parameters* respectively. In view of (13.17),

$$P(X(t+h) = i+1 \mid X(t) = i) = \lambda_i h + o(h) \tag{13.18}$$

and

$$P(X(t+h) = i-1 \mid X(t) = i) = \mu_i 1_{\{i\geq 1\}} h + o(h). \tag{13.19}$$

By conservation,

$$P(X(t+h) = i \mid X(t) = i) = 1 - (\lambda_i + \mu_i 1_{\{i\geq 1\}})h + o(h).$$

Birth-and-death processes are important models in biology, and also in operations research, in particular in queueing theory where they appear as $M/M/1/\infty$, $M/M/K/0$ queues, among many other models of waiting lines (see Chapter 14).

Remark 13.2.13 Observe that we have included in the definition the regularity of birth-and-death processes. However, in modeling, it is the birth-and-death parameters that are given. The question is: Given the birth-and-death parameters, does there exist a *regular* jump HMC satisfying (13.18) and (13.19)? This problem is not trivial when the birth-and-death parameters are not uniformly bounded. This issue will be examined in Subsection 13.3.3.

13.2.3   Kolmogorov's Differential Systems

In view of the semi-group properties, for all $t \geq 0$ and all $h \geq 0$

$$\frac{\mathbf{P}(t+h) - \mathbf{P}(t)}{h} = \mathbf{P}(t)\frac{\mathbf{P}(h) - I}{h} = \frac{\mathbf{P}(h) - I}{h}\mathbf{P}(t). \tag{13.20}$$

Therefore, if the passage to the limit in (13.20) is allowed, which is the case when the state space E is finite, we obtain the differential system

$$\frac{\mathrm{d}}{\mathrm{d}t}\mathbf{P}(t) = \mathbf{P}(t)\mathbf{A} = \mathbf{A}\mathbf{P}(t), \tag{13.21}$$

where $\mathbf{A}$ is the infinitesimal generator. The equation

$$\frac{\mathrm{d}}{\mathrm{d}t}\mathbf{P}(t) = \mathbf{A}\mathbf{P}(t) \tag{13.22}$$

can be written explicitly. For all $i, j \in E$,

$$\frac{\mathrm{d}}{\mathrm{d}t}p_{ij}(t) = -q_i p_{ij}(t) + \sum_{\substack{k \in E \\ k \neq i}} q_{ik} p_{kj}(t).$$

System (13.22) is *Kolmogorov's backward differential system*. The *forward differential system* is

$$\frac{\mathrm{d}}{\mathrm{d}t}\mathbf{P}(t) = \mathbf{P}(t)\mathbf{A}, \tag{13.23}$$

that is, for all $i, j \in E$,

$$\frac{\mathrm{d}}{\mathrm{d}t}p_{ij}(t) = -p_{ij}(t)q_j + \sum_{\substack{k \in E \\ k \neq j}} p_{ik}(t)q_{kj}.$$

When the state space is finite, a solution of (13.22) or (13.23) with the initial condition $\mathbf{P}(0) = I$ is

$$\mathbf{P}(t) = \mathrm{e}^{t\mathbf{A}},$$

where the exponential of a finite-dimensional matrix $\mathbf{C}$ is defined by

$$e^{\mathbf{C}} := \sum_{n=0}^{\infty} \frac{\mathbf{C}^n}{n!}$$

(Theorem A.2.2). This is the *unique* solution with the given initial data.[1]

When the state space is not finite, difficulties may arise in the passage to the limit $h \downarrow 0$ in (13.20) because of the possibly infinite sums involved. However, for the backward system, there is a positive result:

Theorem 13.2.14 *If the continuous semi-group $\{\mathbf{P}(t)\}_{t \geq 0}$ is stable and conservative, Kolmogorov's backward differential system (13.22) is satisfied.*

Proof. Write the equality involving the extreme terms of (13.20):

$$\frac{p_{ij}(t+h) - p_{ij}(t)}{h} = \frac{p_{ii}(h) - 1}{h} p_{ij}(t) + \sum_{\substack{k \in E \\ k \neq i}} \frac{p_{ik}(h)}{h} p_{kj}(t). \qquad (13.24)$$

It will be shown that the limit as $h \downarrow 0$ of the sum appearing in the above equality exists and is equal to

$$\sum_{\substack{k \in E \\ k \neq i}} q_{ik} p_{kj}(t).$$

This means that $p_{ij}(t)$ has a right-hand derivative equal to

$$-q_i p_{ij}(t) + \sum_{\substack{k \in E \\ k \neq i}} q_{ik} p_{kj}(t).$$

Since for all $k, j \in E$, $p_{kj}(t)$ is a continuous function, and since $\sum_{\substack{k \in E \\ k \neq i}} q_{ik} < \infty$, the right-hand derivative is a continuous function, by the dominated convergence theorem for series (see the Appendix). It is therefore also the left-hand derivative, since a continuous function with a continuous right-hand derivative is differentiable.

To prove that the limit as $h \downarrow 0$ of the sum in (13.24) exists, start from the inequality

$$\sum_{\substack{k \in E \\ k \neq i}} \frac{p_{ik}(h)}{h} p_{kj}(t) \geq \sum_{\substack{k=1 \\ k \neq i}}^{N} \frac{p_{ik}(h)}{h} p_{kj}(t),$$

---

[1]See [Hirsch and Smale, 1974] for more on linear systems of differential equations.

where N is an arbitrary integer (E is identified with the set of integers). Therefore, the lim inf as $h \downarrow 0$ of the left-hand side is larger than or equal to the limit of the right-hand side. Letting then N go to ∞ after the passage to the limit $h \downarrow 0$ yields

$$\liminf_{h \downarrow 0} \sum_{\substack{k \in E \\ k \neq i}} \frac{p_{ik}(h)}{h} p_{kj}(t) \geq \sum_{\substack{k \in E \\ k \neq i}} q_{ik} p_{kj}(t). \tag{13.25}$$

Next, observe that for $N > i$,

$$\sum_{\substack{k \in E \\ k \neq i}} \frac{p_{ik}(h)}{h} p_{kj}(t) \leq \sum_{\substack{k=0 \\ k \neq i}}^{N} \frac{p_{ik}(h)}{h} p_{kj}(t) + \sum_{k > N} \frac{p_{ik}(h)}{h}$$

$$= \sum_{\substack{k=0 \\ k \neq i}}^{N} \frac{p_{ik}(h)}{h} p_{kj}(t) + \frac{1 - p_{ii}(h)}{h} - \sum_{\substack{k=0 \\ k \neq i}}^{N} \frac{p_{ik}(h)}{h},$$

and therefore,

$$\limsup_{h \downarrow 0} \sum_{\substack{k \in E \\ k \neq i}} \frac{p_{ik}(h)}{h} p_{kj}(t) \leq \sum_{\substack{k=0 \\ k \neq i}}^{N} q_{ik} p_{kj}(t) + q_i - \sum_{\substack{k=0 \\ k \neq i}}^{N} q_{ik}.$$

By letting N go to ∞ and in view of the stability and conservation hypothesis,

$$\limsup_{h \downarrow 0} \sum_{\substack{k \in E \\ k \neq i}} \frac{p_{ik}(h)}{h} p_{kj}(t) \leq \sum_{\substack{k \in E \\ k \neq i}} q_{ik} p_{kj}(t). \tag{13.26}$$

Equality (13.22) then follows from (13.25) and (13.26).   □

For the forward system, in the absence of regularity assumptions on the trajectories of the Markov chain, the result is considerably less general.

Theorem 13.2.15 *Under the assumptions of Theorem 13.2.14 and if moreover*

$$\sum_{k \in E} p_{ik}(t) q_k < \infty \quad (i \in E, \, t \geq 0), \tag{13.27}$$

then Kolmogorov's forward differential system (13.23) is satisfied.

Proof. Using the estimate (13.10), we see that

$$\frac{p_{ki}(h)}{h} \leq \frac{1 - p_{kk}(h)}{h} \leq q_k.$$

The first equality in (13.20) reads

$$\frac{p_{ij}(t+h) - p_{ij}(t)}{h} = \sum_{\substack{k \in E \\ k \neq i}} p_{ik}(t) \frac{p_{kj}(h) - \delta_{kj}}{h}.$$

Each term of the series on the right-hand side is bounded by the corresponding term of the convergent series $\sum_{k \in E} p_{ik}(t) q_k$. Passage to the limit $h \downarrow 0$ is therefore allowed by Lebesgue's dominated convergence theorem, and this leads to the announced result, again after observing that the continuous right-hand derivative of a continuous function is also the left-hand derivative. $\square$

The distribution at time t of the chain is the column vector $\mu(t) := \{\mu_i(t)\}_{i \in E} = \{P(X(t) = i)\}_{i \in E}$. It satisfies, for all $t, s \geq 0$,

$$\mu^T(t+s) = \mu^T(t)\mathbf{P}(s), \tag{13.28}$$

that is,

$$\mu_i(t+s) = \sum_{j \in E} \mu_j(t) p_{ji}(s) \quad (i, j \in E).$$

Theorem 13.2.16 *Under the assumptions of Theorem 13.2.14 and if moreover*

$$\sum_{i \in E} q_i \mu_i(t) < \infty \quad (i \in E, \, t \geq 0), \tag{13.29}$$

then Kolmogorov's global differential system

$$\frac{\mathrm{d}}{\mathrm{d}t} \mu^T(t) = \mu^T(t)\mathbf{A} \tag{13.30}$$

is satisfied, that is,

$$\frac{\mathrm{d}}{\mathrm{d}t} \mu_i(t) = -\mu_i(t) q_i + \sum_{\substack{j \in E \\ j \neq i}} \mu_j(t) q_{ji} \quad (i, j \in E).$$

Proof. From (13.28), for all $i \in E$, all $t \geq 0$, and all $h > 0$,

$$\mu_i(t+h) = \mu_i(t) p_{ii}(h) + \sum_{\substack{j \in E \\ j \neq i}} \mu_j(t) p_{ji}(h),$$

that is, after rearrangement,

$$\frac{\mu_i(t+h) - \mu_i(t)}{h} = -\mu_i(t) \frac{1 - p_{ii}(h)}{h} + \sum_{\substack{j \in E \\ j \neq i}} \mu_j(t) \frac{p_{ji}(h)}{h}.$$

The rest of the proof is then the same as that of Theorem 13.2.15. $\square$

Remark 13.2.17 Condition (13.27) is trivially satisfied when the state space is finite or when

$$\sup_{i \in E} q_i < \infty .$$

If this last condition is satisfied, the chain has the same distribution as a uniform chain, as we shall prove later, in Corollary 13.3.8.

Definition 13.2.18 *A stationary distribution of the semi-group $\{\mathbf{P}(t)\}_{t \geq 0}$ is any probability distribution π on E such that for all $t \geq 0$,*

$$\pi^T \mathbf{P}(t) = \pi^T . \tag{13.31}$$

If $X(0)$ is distributed according to π, then so is $X(t)$ for all $t > 0$, because from (13.28),

$$\mu^T(t) = \pi^T \mathbf{P}(t) = \pi^T.$$

The chain is then said to be *in a stationary regime* or *in equilibrium* because for all times $t_1, \ldots, t_k, t$ such that $t \geq 0$ and $0 \leq t_1 \leq t_2 \leq \cdots \leq t_k$, and all states $i_1, \ldots, i_k \in E$,

$$\mathrm{P}(X(t_1 + t) = i_1, \ldots, X(t_k + t) = i_k) = \mathrm{P}(X(t_1) = i_1, \ldots, X(t_k) = i_k) .$$

This is proved by computing the left-hand side using Bayes' sequential rule,

$$\pi(i_1) p_{i_1 i_2}(t_2 - t_1) \cdots p_{i_{k-1} i_k}(t_k - t_{k-1}) ,$$

and by observing that this quantity does not depend upon t.

In general, as for discrete-time HMCs, a stationary distribution need not exist, and if it exists, it need not be unique.

Continuous time brings something new into the picture, namely a local characterization of the stationary distributions, which is often more practical than the global characterization (13.31).

If π is a stationary distribution of a stable and conservative continuous-time HMC and if $\sum_{i \in E} \pi(i) q_i < \infty$, then, according to Theorem 13.2.16, π satisfies the global balance equation

$$\pi^T \mathbf{A} = 0 , \tag{13.32}$$

that is, in expanded form,

$$\pi(i) q_i = \sum_{\substack{j \in E \\ j \neq i}} \pi(j) q_{ji}.$$

The above result is not stated as a theorem because it is too weak. First we need to assume stability and conservation, then we must introduce an assumption on the stationary distribution itself. Also, we lack the principal result, namely the converse: If a probability distribution π satisfies (13.32), then it is a stationary distribution. However, there are interesting cases where a satisfying result can be obtained at low cost.

Remark 13.2.19 If the state space is finite, we know that the system is stable and conservative, and that both the forward and backward Kolmogorov systems are satisfied, as well as (13.30). Also, if there exists a stationary distribution π, (13.29) is satisfied, and therefore, necessarily $\pi^T \mathbf{A} = 0$.

Conversely, suppose that $\pi^T \mathbf{A} = 0$ for some probability distribution π on E. Then $\mu(t) - \pi$ is a solution of the system (13.30). But (13.30) is a finite linear system of differential equations and it has a unique solution $\mu(t)$ such that $\mu(0) = \pi$. Therefore, if $X(0)$ is distributed according to π, then so is $X(t)$ for all $t \geq 0$. Thus π is a stationary distribution.

In summary: For a continuous-time HMC on the finite state space E with infinitesimal generator $\mathbf{A}$, the condition $\pi^T \mathbf{A} = 0$ is necessary and sufficient for the probability distribution π on E to be a stationary distribution.

EXAMPLE 13.2.20: STATIONARY UNIFORM HMC. For the uniform Markov chain, $q_i = \lambda(1 - k_{ii})$ and therefore $\sup q_i < \infty$. Therefore, the conditions of Theorems 13.2.14, 13.2.15 and 13.2.16 are satisfied. Therefore, a stationary distribution π, *if it exists*, is a solution of $\pi^T \mathbf{A} = 0$. In view of (13.11), $\pi^T \mathbf{A} = 0$ is equivalent to

$$\pi^T = \pi^T \mathbf{K}. \tag{13.33}$$

(This shows, in particular, that π is a stationary distribution of the subordinated chain.)

Conversely, suppose that $\pi^T \mathbf{A} = 0$, and therefore (13.33) holds. From (13.6),

$$\pi^T \mathbf{P}(t) = \sum_{n=0}^{\infty} e^{-\lambda t} \frac{(\lambda t)^n}{n!} \pi^T \mathbf{K}^n = \left(\sum_{n=0}^{\infty} e^{-\lambda t} \frac{(\lambda t)^n}{n!} \right) \pi^T = \pi^T.$$

In other words, π is a stationary distribution of the uniform HMC $\{X(t)\}_{t\geq0}$. Therefore, in this case also, $\pi^T \mathbf{A} = 0$ is a necessary and sufficient condition for the probability distribution π to be a stationary distribution.

Regular Jumps

In the practice of operations research and of engineering, as well as in the biological and social sciences, the Markov chains that are likely to be encountered are regular jump processes. For such chains, the situation is just as simple as for finite state space or uniform Markov chains.

Theorem 13.2.21 *Let* $\{X(t)\}_{t \geq 0}$ *be a regular jump* HMC *with countable state space* E *and transition semi-group* $\{\mathbf{P}(t)\}_{t \geq 0}$. *Then,*

(i) *the semi-group* $\{\mathbf{P}(t)\}$ *is continuous, stable and conservative,*

(ii) *the backward and forward differential systems of Kolmogorov are satisfied, as well as the global balance differential systems, and*

(iii) *a necessary and sufficient condition for a probability distribution* π *on* E *to be a stationary distribution of the chain is* $\pi^T \mathbf{A} = 0$, *where* $\mathbf{A}$ *is the infinitesimal generator.*

The proof is technical and is omitted.[2] In Section 13.5, we shall obtain a very close result, which is sufficient for practical purposes.

EXAMPLE 13.2.22: GLOBAL BALANCE FOR BIRTH-AND-DEATH PROCESSES. Recall Definition 13.2.12, where a birth-and-death process with parameters λ_n and μ_n is defined to be a regular jump HMC taking its values in $\mathbb{N}$ and with an infinitesimal generator of the form

$$\mathbf{A} = \begin{pmatrix} -\lambda_0 & \lambda_0 & 0 & 0 & \cdots \\ \mu_1 & -(\lambda_1 + \mu_1) & \lambda_1 & 0 & 0 & \cdots \\ 0 & \mu_2 & -(\lambda_2 + \mu_2) & \lambda_2 & 0 & \cdots \\ 0 & 0 & \mu_3 & -(\lambda_3 + \mu_3) & \lambda_3 & \cdots \\ \vdots & \vdots & \vdots & \vdots & \vdots \end{pmatrix}.$$

The differential system satisfied by $p_n(t) = \mathrm{P}(X(t) = n)$ is

$$\dot{p}_0(t) = -\lambda_0 p_0(t) + \mu_1 p_1(t),$$
$$\dot{p}_n(t) = \lambda_{n-1} p_{n-1}(t) - (\lambda_n + \mu_n) p_n(t) + \mu_{n+1} p_{n+1}(t) \quad \text{(for } n \geq 1\text{)}.$$

A necessary and sufficient condition for a probability π to be a stationary distribution of the birth-and-death process is

$$0 = \lambda_0 \pi(0) - \mu_1 \pi(1),$$
$$0 = \lambda_{n-1} \pi(n-1) - (\lambda_n + \mu_n) \pi(n) + \mu_{n+1} \pi(n+1) \quad \text{(for } n \geq 1\text{)}.$$

---

[2]See for instance Chapter 1 of [Anderson, 1991].

For fixed $\pi(0)$, there exists one and only one solution for this system, namely,

$$\pi(n) = \pi(0)\frac{\lambda_0\lambda_1\cdots\lambda_{n-1}}{\mu_1\mu_2\cdots\mu_n} \tag{13.34}$$

for $n \geq 1$. If we require π to be a probability distribution, we must have

$$\pi(0)\left(1 + \sum_{n=1}^{\infty}\frac{\lambda_0\lambda_1\cdots\lambda_{n-1}}{\mu_1\mu_2\cdots\mu_n}\right) = 1, \tag{13.35}$$

and this is possible if and only if

$$1 + \sum_{n=1}^{\infty}\frac{\lambda_0\lambda_1\cdots\lambda_{n-1}}{\mu_1\mu_2\cdots\mu_n} < \infty, \tag{13.36}$$

in which case the unique stationary probability is given by (13.34) and (13.35).

13.3 The Regenerative Structure

13.3.1 The Strong Markov Property

A regular jump HMC has the strong Markov property. For a precise statement of this result, we introduce stopping times in continuous time, the definition of which is analogous to the discrete-time definition.

Definition 13.3.1 *Let $\{X(t)\}_{t\geq 0}$ be a stochastic process with values in E. A random variable τ taking its values in $\overline{\mathbb{R}}_+$ is called a stopping time with respect to $\{X(t)\}_{t\geq 0}$ if for all $t \geq 0$, the event $\{\tau \leq t\}$ is expressible in terms of $X(s)$ $(0 \leq s \leq t)$. This is denoted by*

$$\{\tau \leq t\} \in X_0^t.$$

EXAMPLE 13.3.2: ESCAPE AND RETURN TIMES. Let $\{X(t)\}_{t\geq 0}$ be a regular jump HMC with countable state space E. The *escape time* from state i is

$$E_i := \inf\{t \geq 0; X(t) \neq i\},$$

with $E_i = \infty$ if $X(t) = i$ for all $t \geq 0$, and the *return* time to i is

$$R_i := \inf\{t > 0; t > E_i \text{ and } X(t) = i\},$$

with $R_i = \infty$ if $E_i = \infty$ or $X(t) \neq i$ if for all $t \geq E_i$. Then E_i and R_i are stopping times with respect to $\{X(t)\}_{t\geq 0}$ (exercise).

_____

Let the *process after τ* be defined as in discrete time:

$$\{X(t+\tau)\}_{t\geq 0}\,,$$

with the convention $X(\infty) = \Delta$, where Δ is an arbitrary element not in E. The *process before τ* is

$$\{X(t \wedge \tau)\}_{t\geq 0}\,.$$

Theorem 13.3.3 *Let $\{X(t)\}_{t\geq 0}$ be a regular jump* HMC *with state space E and transition semi-group $\{\mathbf{P}(t)\}_{t\geq 0}$, and let τ be a stopping time with respect to $\{X(t)\}_{t\geq 0}$. Then, given that $X(\tau) = k$, an arbitrary state,*

(α) *the chain after τ and the chain before τ are independent, and*

(β) *the chain after τ is a regular jump* HMC *with transition semi-group $\{\mathbf{P}(t)\}_{t\geq 0}$.*

Proof. It suffices to show that for all states k, all positive times $t_1, \ldots, t_n$, $s_1, \ldots, s_p$, all real numbers $u_1, \ldots, u_n, v_1, \ldots, v_p$, and all initial distributions μ

$$\mathrm{E}_\mu\left[\exp\left\{i\sum_{\ell=1}^n u_\ell X(\tau+t_\ell) + i\sum_{m=1}^p v_m X(\tau_\wedge s_m)\right\} 1_{\{X(\tau)=k\}}\right]$$

$$= \mathrm{E}_k\left[\exp\left\{i\sum_{\ell=1}^n u_\ell X(t_\ell)\right\}\right] \mathrm{E}_\mu\left[\exp\left\{i\sum_{m=1}^p v_m X(\tau_\wedge s_m)\right\} 1_{\{X(\tau)=k\}}\right]. \quad (13.37)$$

Indeed, if (13.37) is satisfied, we obtain by fixing $v_1 = \cdots = v_p = 0$.

$$\mathrm{E}_\mu\left[\exp\left\{i\sum_{\ell=1}^n u_\ell X(\tau+t_\ell)\right\} 1_{\{X(\tau)=k\}}\right]$$

$$= \mathrm{E}_k\left[\exp\left\{i\sum_{\ell=1}^n u_\ell X(t_\ell)\right\}\right] \mathrm{P}_\mu(X(\tau) = k)\,,$$

that is,

$$\mathrm{E}_\mu\left[\exp\left\{i\sum_{\ell=1}^n u_\ell X(\tau+t_\ell)\right\} \mid X(\tau) = k\right]$$

$$= \mathrm{E}_k\left[\exp\left\{i\sum_{\ell=1}^n u_\ell X(t_\ell)\right\}\right] = \mathrm{E}_\mu\left[\exp\left\{i\sum_{\ell=1}^n u_\ell X(t_\ell)\right\} \mid X(0) = k\right],$$

and this shows that given $X(\tau) = k$, $\{X(\tau + t)\}_{t \geq 0}$ had the same distribution as $\{X(t)\}_{t \geq 0}$ given $X(0) = k$. We therefore will have proven (β). For (α), it suffices to rewrite (13.37) as follows, using the previous equality:

$$E_\mu \left[\exp \left\{ i \sum_{\ell=i}^{n} u_\ell X(\tau + t_\ell) + i \sum_{m=1}^{n} v_m X(\tau_\wedge s_m) \right\} \mid X(\tau) = k \right]$$

$$= E_\mu \left[\exp \left\{ i \sum_{\ell=1}^{n} u_\ell X(\tau + t_\ell) \right\} \mid X(\tau) = k \right] \times \cdots$$

$$\cdots E_\mu \left[\exp \left\{ i \sum_{n=1}^{n} v_m X(\tau_\wedge s_m) \right\} \mid X(\tau) = k \right].$$

It remains to prove (13.37). For the sake of simplicity, we do the case where $n = m = 1$, and let $u_1 = u, t_1 = t, v_1 = v, s_1 = s$.

We first treat the case where τ takes a countable number of values a_j $(j \geq 1)$. Then

$$E_\mu[\exp\{iuX(\tau + t) + ivX(\tau_\wedge s)\}1_{\{X(\tau)=k\}}]$$

$$= \sum_{j=1}^{\infty} E_\mu[\exp\{iuX(a_j + t) + ivX(a_j \wedge s)\}1_{\{X(a_j)=k\}}1_{\{\tau=a_j\}}].$$

For all $j \geq 1$,

$$E_\mu[\exp\{iuX(a_j + t) + ivX(a_j \wedge s)\}1_{\{X(a_j)=k\}}1_{\{\tau=a_j\}}]$$
$$- E_\mu[\exp\{iuX(a_j + t)\}1_{\{X(a_j)=k\}} \exp\{ivX(a_j \wedge s)\}1_{\{\tau=a_j\}}]$$
$$= E_\mu[\exp\{iuX(a_j + t)\} \mid X(a_j) = k]E_\mu[\exp\{ivX(a_j \wedge s)\}1_{\{\tau=a_j\}}1_{\{X(a_j)=k\}}],$$

where for the last equality, we have used the fact that $1_{\{\tau=a_j\}}$ is a function $X(u), 0 \leq u \leq a_j$, and the Markov property at time a_j.

Therefore, for all $j \geq 1$,

$$E_\mu[\exp\{iuX(a_j + t) + ivX(a_j \wedge s)\}1_{\{X(a_j)=k\}}1_{\{\tau=a_j\}}]$$
$$- E_k[\exp\{iuX(t)\}]E_\mu[\exp\{ivX(a_j \wedge s)\}1_{\{X(a_j)=k\}}1_{\{\tau=a_j\}}].$$

Summing with respect to j, we obtain the equality corresponding to (13.37). To pass from the case where the stopping time τ takes a countable number of values to the general case, define for each $n \geq 1$ the following approximation $\tau(n)$ to the arbitrary stopping time τ (with respect to $\{X(t)\}_{t \geq 0}$):

$$\tau(n, \omega) = \begin{cases} 0 & \text{if } \tau(\omega) = 0, \\ \frac{\ell+1}{2^n} & \text{if } \frac{\ell}{2^n} < \tau(\omega) \leq \frac{\ell+1}{2^n}, \\ +\infty & \text{if } \tau(\omega) = \infty. \end{cases}$$

Then $\tau(n)$ is a stopping time (exercise) with a countable number of values, and therefore, (13.37) is satisfied for $\tau(n)$. Now, $\lim_{n\uparrow\infty} \downarrow \tau(n,\omega) = \tau(\omega)$, and therefore

$$\lim_{n\uparrow\infty} X(\tau(n)\wedge a) = X(\tau\wedge a),\ \lim_{n\uparrow\infty} X(\tau(n)+b) = X(\tau+b),\ \lim_{n\uparrow\infty} 1_{\{X(\tau(n))=k\}} = 1_{\{X(\tau)=k\}}$$

(a regular jump process is right-continuous). Therefore, letting n go to ∞ in (13.37) with τ replaced by $\tau(n)$, we obtain the result for τ itself, by dominated convergence. $\square$

13.3.2   Embedded Chain and Transition Times

Let $\{\tau_n\}_{n\geq 0}$ be the non-decreasing sequence of transition times of the regular jump process $\{X(t)\}_{t\geq 0}$, where $\tau_0 = 0$ and $\tau_n = \infty$ if there are strictly fewer than n transitions in $(0,\infty)$.

For each $n \geq 0$, τ_n is a stopping time with respect to $\{X(t)\}_{t\geq 0}$ (exercise).

The sequence $\{X_n\}_{n\geq 0}$ with values in $E_\Delta = E \cup \{\Delta\}$, where Δ is an arbitrary element not in E, is defined by

$$X_n = X(\tau_n),\tag{13.38}$$

with the convention $X(\infty) = \Delta$, and it is called the *embedded sequence* of the jump process. If $\{X(t)\}_{t\geq 0}$ is a regular jump HMC, it follows from the strong Markov property that given $X(\tau_n) = k \in E$, $\{X(\tau_n + t)\}_{t\geq 0}$ is independent of $\{X(\tau_n \wedge t)\}_{t\geq 0}$, and therefore, given $X_n = k$, the variables $(X_{n+1}, X_{n+2}, \ldots)$ are independent of $(X_0, \ldots, X_n)$, that is, $\{X_n\}_{n\geq 0}$ is a Markov chain. It is clearly homogeneous because the distribution of $\{X(\tau_n + t)\}_{t\geq 0}$ given $X(\tau_n) = k$ is independent of n, being identical with the distribution of $\{X(t)\}_{t\geq 0}$ given $X(0) = k$.

Theorem 13.3.4 *Let $\{X(t)\}_{t\geq 0}$ be a regular jump HMC, with infinitesimal generator $\mathbf{A}$, transition times sequence $\{\tau_n\}_{n\geq 0}$, and embedded process $\{X_n\}_{n\geq 0}$. Then*

(α) *$\{X_n\}_{n\geq 0}$ is a discrete-time HMC with state space $E_\Delta = E \cup \{\Delta\}$ with transition matrix given by $p_{\Delta\Delta} = 1$, $p_{i\Delta} = 1$ if $i \in E$ and $q_i = 0$; by $p_{i\Delta} = 0$ if $i \in E$ and $q_i > 0$; and if $q_i > 0$ and $j \neq i$, by $p_{ij} = \frac{q_{ij}}{q_i}$, and*

(β) *given $\{X_n\}_{n\geq 0}$, the sequence $\{\tau_{n+1} - \tau_n\}_{n\geq 0}$ is independent, and for all $n \geq 0$ and all $a \in \mathbb{R}_+$. Also,*

$$P(\tau_{n+1} - \tau_n \leq a \mid \{X_k\}_{k\geq 0}) = 1 - e^{-q_{X_n}a}.\tag{13.39}$$

Proof. The complete proof will be given in Section 13.5. We shall for the time being prove the following partial result:

(α') $\{X_n\}_{n \geq 0}$ is a discrete-time HMC on $E_\Delta = E \cup \{\Delta\}$,

(β') given $\{X_n\}_{n \geq 0}$, the sequence $\{\tau_{n+1} - \tau_n\}_{n \geq 0}$ is independent, and

(β'') there exists for each $i \in E$ a real number $\lambda(i) \geq 0$ such that for all $n \geq 0$ and all $a \in \mathbb{R}_+$,

$$P(\tau_{n+1} - \tau_n \leq a \mid \{X_k\}_{k \geq 0}) = 1 - e^{-\lambda(X_n)a}.$$

We have already proved (α'). Call $p_{ij} = P_i(X(\tau_1) = j)$ the transition probability of $\{X_n\}_{n \geq 1}$ from i to j. To prove (β') and (β''), it suffices to show that

$$P_i(X_1 = i, \ldots, X_n = i_n, \tau_1 - \tau_0 > a_1, \ldots, \tau_n - \tau_{n-1} > a_n)$$
$$= e^{-\lambda(i)a_1} p_{ii_1} e^{-\lambda(i_1)a_2} p_{i_1 i_2} \cdots e^{-\lambda(i_{n-1})a_n} p_{i_{n-1} i_n}$$

for all $i, i_1, \ldots, i_n \in E$, $a_1, \ldots, a_n \in \mathbb{R}_+$, and for some function $\lambda : E \to \mathbb{R}_+$. By the strong Markov property, it suffices, in fact, to show that for all $i, j \in E$, $a \in \mathbb{R}_+$, there exists $\lambda(i) \geq 0$ such that

$$P_i(X_1 = j, \tau_1 - \tau_0 > a) = P_i(X_1 = j)e^{-\lambda(i)a}. \tag{13.40}$$

Let $g(t) := P_i(\tau_1 > t)$. For $t, s \geq 0$, using the obvious set identities,

$$
\begin{aligned}
g(t+s) &= P_i(\tau_1 > t + s) \\
&= P_i(\tau_1 > t + s, \tau_1 > t, X(t) = i) \\
&= P_i(X(t+u) - i \, (0 \leq u \leq t), \tau_1 > t, X(t) = i).
\end{aligned}
$$

The last expression is, in view of the Markov property at time t and using the fact that $\{\tau_1 > t\}$ is expressible in terms of $(X(v), 0 \leq v \leq t)$,

$$
\begin{aligned}
P_i(X(t+u) &= i(u \in [0, s]) \mid X(t) = i)P_i(\tau_1 > t, X(t) = i) \\
&= P_i(X(u) = i \,(0 \leq u \leq t) \mid X(0) = i)P_i(\tau_1 > t) \\
&= P_i(\tau_1 > s)P_i(\tau_1 > t),
\end{aligned}
$$

where the last two equalities again follow from the obvious set identities. Therefore, for all $s, t \geq 0$,

$$g(t+s) = g(t)g(s).$$

Also, $t \mapsto g(t)$ is non-increasing and $\lim_{t \downarrow 0} g(t) = 1$ (the semi group is continuous). It follows from the subadditive function theorem (Theorem A.1.13) that there exists $\lambda(i) \in [0, \infty)$ such that $g(t) = e^{-\lambda(i)t}$, that is, $P_i(\tau_1 > t) = e^{-\lambda(i)t}$, for all $t \geq 0$.

Now, by the Markov property and the appropriate set identities,

$$
\begin{aligned}
P_i(X_1 = j, \tau_1 > t) &= P_i(X(\tau_1) = j, \tau_1 > t, X(t) = i) \\
&= P_i(\text{first jump of } \{X(t+s)\}_{s \geq 0} \text{ is } j, \tau_1 > t, X(t) = i) \\
&= P_i(\text{first jump of } \{X(t+s)\}_{s \geq 0} \text{ is } j \mid X(t) = i)P_i(\tau_1 > t, X(t) = i) \\
&= P_i(\text{ first jump of } \{X(s)\}_{s \geq 0} \text{ is } j \mid X(0) = i)P_i(\tau_1 > t) \\
&= P_i(X(\tau_1) = j)P_i(\tau_1 > t),
\end{aligned}
$$

and this is (13.40). □

Definition 13.3.5 *A state $i \in E$ such that $q_i = 0$ is called* permanent. *Otherwise, it is called* essential.

In view of (13.39), if $X(\tau_n) = i$, a permanent state, then $\tau_{n+1} - \tau_n = \infty$; that is, there is no more transition at finite distance, hence the terminology.

EXAMPLE 13.3.6: REGENERATIVE STRUCTURE OF A UNIFORM HMC. For the uniform HMC, the embedded process $\{X_n\}_{n \geq 0}$ is an HMC with state space $E_\Delta = E \cup \{\Delta\}$, and if $i \in E$ is not permanent (that is, in this case if $k_{ii} < 1$), then for $j \neq i$,

$$
p_{ij} = \frac{k_{ij}}{1 - k_{ii}}.
$$

Indeed, $\{X_n\}$ is obtained from $\{\hat{X}_n\}$ by considering only the "real" transitions (see Exercise 2.6.30).

An immediate consequence of Theorem 13.3.4 is:

Theorem 13.3.7 *Two regular jump HMCs with the same infinitesimal generator and the same initial distribution are probabilistically equivalent.*

Another way to state this is as follows: Two regular jump HMCs with the same infinitesimal generator have the same transition semi-group.

Corollary 13.3.8 *A regular jump HMC with infinitesimal generator $\mathbf{A}$ such that $\sup_{i \in E} q_i \leq \infty$ has the same transition semi-group as a uniform chain.*

Proof. Select any real number $\lambda > \sup_{i \in E} q_i$, and define the transition matrix $\mathbf{K}$ by (13.12) and (13.13). Then, the uniform chain corresponding to $(\lambda, \mathbf{K})$ has the infinitesimal generator $\mathbf{A}$. $\qquad\square$

Definition 13.3.9 *A continuous time* HMC *with an infinitesimal generator such that*

$$\sup_{i \in E} q_i < \infty$$

is uniformizable.

Any pair $(\lambda, \mathbf{K})$ as in the proof of Corollary 13.3.8 gives rise to a uniform version of the chain. The *minimal uniform version* is that with $\lambda = \sup_{i \in E} q_i$.

13.3.3 Explosions

Definition 13.3.10 *Let* $\mathbf{A} = \{q_{ij}\}_{i,j \in E}$ *be a matrix with entries satisfying for all* $i, j \in E$,

$$q_i \in [0, \infty), \quad q_{ij} \in [0, \infty), \quad \sum_{\substack{k \in E \\ k \neq i}} q_{ik} = q_i \,,$$

where $q_i = -q_{ii}$. *This matrix is called a stable and conservative generator on* E; *and it is called an* essential generator *if moreover* $q_i > 0$ *for all* $i \in E$.

Note that at this point no reference is made to a continuous-time HMC.

Theorem 13.3.4 suggests a way of constructing a regular jump HMC $\{X(t)\}_{t \geq 0}$ with values in a countable state space E and admitting a *given* generator $\mathbf{A} = \{q_{ij}\}$ that is stable and conservative (we shall also suppose for simplicity that it is *essential*). For this, we construct a sequence $\tau_0 = 0, X_0, \tau_1 - \tau_0, X_1, \tau_2 - \tau_1, X_2, \ldots$ according to

$$\begin{cases} P(\tau_{n+1} - \tau_n \leq x \mid X_0, \ldots, X_n, \tau_0, \ldots, \tau_n) = 1 - e^{-q_{X_n} x} \\ P(X_{n+1} = j \mid X_0, \ldots, X_n, \tau_0, \ldots, \tau_{n+1}) = q_{X_n j}/q_{X_n}, \end{cases}$$

the initial state X_0 being chosen at random, with arbitrary distribution. The value of $X(t)$ for $\tau_n \leq t < \tau_{n+1}$ is then X_n.

The problem with the above regenerative construction is that

$$\tau_\infty := \lim_{n \uparrow \infty} \uparrow \tau_n$$

may not be almost surely finite. We say that $\mathbf{A}$ is *non-explosive generator* if

$$P_\mu(\tau_\infty = \infty) = 1,$$

for any initial distribution μ and all regular jump HMCs associated to a semi-group with generator $\mathbf{A}$. It will be proven in Chapter 14 that in the non-explosive case, the above construction indeed produces a regular jump HMC.

The following result is a criterion of non-explosion.

Theorem 13.3.11 *Let $\mathbf{A}$ be the stable and conservative infinitesimal generator of a continuous-time HMC $\{X(t)_{t\geq0}$ on E. It is non-explosive if and only if for any real $\lambda > 0$, the system of equations*

$$(\lambda + q_i)x_i = \sum_{\substack{j\in E \\ j\neq i}} q_{ij}x_j \quad (i \in E) \tag{13.41}$$

admits no non-negative bounded solution other than the trivial one.

Proof. The quantity

$$g_i(\lambda) := E_i\left[\exp\left\{-\lambda\sum_{k=1}^\infty S_k\right\}\right]$$

is uniformly bounded in $\lambda > 0$ and $i \in E$, and if $P_i(\tau_\infty = \infty) < 1$, it is strictly positive. Also, $x_i = g_i(\lambda)$ $(i \in E)$ is a solution of (13.41), as follows from the calculations below:

$$\begin{aligned}
g_i(\lambda) &= E_i\left[\exp\{-\lambda S_1\}\exp\left\{-\lambda\sum_{k=2}^\infty S_k\right\}\right] \\
&= \left(\int_0^\infty e^{-\lambda t}q_i e^{-q_i t}\mathrm{d}t\right)E_i\left[\exp\left\{-\lambda\sum_{k=2}^\infty S_k\right\}\right] \\
&= \frac{q_i}{\lambda + q_i}E_i\left[\exp\left\{-\lambda\sum_{k=2}^\infty S_k\right\}\right],
\end{aligned}$$

and by first-step analysis,

$$E_i\left[\exp\left\{-\lambda\sum_{k=2}^\infty S_k\right\}\right] = \sum_{\substack{j\in E \\ j\neq i}} E_j\left[\exp\left\{-\lambda\sum_{k=2}^\infty S_k\right\}\right]\frac{q_{ij}}{q_i} = \sum_{\substack{j\in E \\ j\neq i}} g_j(\lambda)\frac{q_{ij}}{q_i}.$$

Therefore, if **A** is explosive there exists a non-trivial bounded solution of (13.41).

To prove the converse, let $\{g_i(\lambda)\}_{i \in E}$ be a bounded solution of (13.41) for a fixed real $\lambda > 0$. We have

$$g_i(\lambda) = \mathrm{E}[\exp\{-\lambda S_1\}g_{X_1}(\lambda) \mid X_0 = i], \qquad (13.42)$$

since the right-hand side is equal to that of (13.41), by first-step analysis. We prove by induction that

$$g_i(\lambda) = \mathrm{E}\left[\exp\left\{-\lambda \sum_{k=1}^{n} S_k\right\} g_{X_n}(\lambda) \mid X_0 = i\right]. \qquad (13.43)$$

For this, rewrite (13.42) as

$$g_i(\lambda) = \mathrm{E}[\exp\{-\lambda S_{n+1}\}g_{X_{n+1}}(\lambda) \mid X_n = i],$$

that is,

$$g_{X_n}(\lambda) = \mathrm{E}[\exp\{-\lambda S_{n+1}\}g_{X_{n+1}}(\lambda) \mid X_n].$$

If this expression of $g_{X_n}(\lambda)$ is used in (13.43), then it follows that

$$g_i(\lambda) = \mathrm{E}\left[\exp\left\{-\lambda \sum_{k=1}^{n+1} S_k\right\} g_{X_{n+1}}(\lambda) \mid X_0 = i\right]. \qquad (13.44)$$

Therefore, (13.43) implies (13.44) (the forward step in the induction argument). Since (13.43) is true for $n = 1$ (Eqn. (13.42)), it is true for all $n \geq 1$, and therefore, since $K := |g_i(\lambda)| < \infty$,

$$|g_i(\lambda)| \leq K\mathrm{E}_i\left[\exp\left\{-\lambda \sum_{k=1}^{\infty} S_k\right\}\right].$$

Therefore, if $\{g_i(\lambda)\}_{i \in E}$ is not trivial, it must hold that $\mathrm{P}_i(\sum_{k=1}^{\infty} S_k < \infty) > 0$, or equivalently, $\mathrm{P}_i(\tau_\infty = \infty) < 1$ for some $i \in E$. $\qquad \square$

Exercise 13.6.12 gives sufficient conditions of non-explosion that are satisfied in many cases. We shall see in the next chapter that for continuous-time HMCs arising in a queueing context, non-explosion is easy to prove. This seems to diminish the impact of the previous criterion. However, there exists a very important situation where it is irreplaceable. The following result is *Reuter's criterion*.

Theorem 13.3.12 *Let* **A** *be generator on* $E = \mathbb{N}$ *defined by* $q_{n,n+1} = \lambda_n$ *and* $q_{n,n-1} = \mu_n 1_{n \geq 1}$, *where the birth parameters* λ_n *are strictly positive. A necessary and sufficient condition of non-explosion of this generator is*

$$\sum_{n=1}^{\infty} \left[\frac{1}{\lambda_n} + \frac{\mu_n}{\lambda_n \lambda_{n-1}} + \cdots + \frac{\mu_n \cdots \mu_1}{\lambda_n \cdots \lambda_1 \lambda_0}\right] = \infty. \qquad (13.45)$$

Proof. We start with preliminary remarks concerning the system of equations (13.41) in the particular case of a birth-and-death generator

$$\begin{cases} \lambda x_0 = -\lambda_0 x_0 + \lambda_0 x_1, \\ \lambda x_k = \mu_k x_{k-1} - (\lambda_k + \mu_k) x_k + \lambda_k x_{k+1} \quad (k \geq 1). \end{cases} \tag{13.46}$$

For any fixed x_0, this system admits a unique solution, that is identically null if and only if $x_0 = 0$. If $x_0 \neq 0$, the solution is such that x_k/x_0 does not depend on x_0, and therefore only the case where $x_0 = 1$ needs to be treated.

Writing $y_k = x_{k+1} - x_k$, we obtain from (13.46)

$$y_k = \frac{\lambda}{\lambda_k} x_k + \frac{\mu_k}{\lambda_k} \frac{\lambda}{\lambda_{k-1}} x_{k-1} + \cdots + \frac{\mu_k \cdots \mu_2}{\lambda_k \cdots \lambda_2} \frac{\lambda}{\lambda_1} x_1 + \frac{\mu_k \cdots \mu_1}{\lambda_k \cdots \lambda_1} y_0 \tag{13.47}$$

and $y_0 = \frac{\lambda}{\lambda_0}$. From this we deduce that if $\lambda > 0$, then $y_k > 0$ and therefore $\{x_k\}_{k \geq 0}$ is a strictly increasing sequence.

Therefore, with $y_0 = \frac{\lambda}{\lambda_0}$ in (13.47),

$$y_k \geq \lambda \left[\frac{1}{\lambda_k} + \frac{\mu_k}{\lambda_k \lambda_{k-1}} + \cdots + \frac{\mu_k \cdots \mu_1}{\lambda_k \cdots \lambda_1 \lambda_0} \right].$$

Thus, a necessary condition for $\{x_k\}_{k \geq 0}$ to be bounded is that the left-hand side of (13.45) be finite. This proves the sufficiency of (13.45) for non-explosion.

We now turn to the proof of necessity. For $i \leq k$, bounding in (13.47) x_i by x_k yields the majoration

$$y_k \leq \left[\frac{\lambda}{\lambda_k} + \cdots + \frac{\mu_k \cdots \mu_1 \lambda}{\lambda_k \cdots \lambda_1 \lambda_0} \right] x_k,$$

and therefore, since $y_k = x_{k+1} - x_k$,

$$\begin{aligned} x_{k+1} &\leq \left[1 + \frac{\lambda}{\lambda_k} + \cdots + \frac{\mu_k \cdots \mu_1 \lambda}{\lambda_k \cdots \lambda_1 \lambda_0} \right] x_k \\ &\leq x_k \exp \left\{ \lambda \left[\frac{1}{\lambda_k} + \cdots + \frac{\mu_k \cdots \mu_1}{\lambda_k \cdots \lambda_0} \right] \right\}. \end{aligned}$$

Since $x_0 = 1$, this leads to

$$x_n \leq \exp \left\{ \lambda \sum_{k=1}^{n} \left[\frac{1}{\lambda_k} + \cdots + \frac{\mu_k \cdots \mu_1}{\lambda_k \cdots \lambda_0} \right] \right\}.$$

Therefore, a sufficient condition for the solution $\{x_n\}_{n \geq 0}$ to be bounded is that the left-hand side of (13.45) be finite. $\square$

EXAMPLE 13.3.13: PURE BIRTH PROCESS. A *pure birth generator* $\mathbf{A}$ is a birth-and-death generator with all $\mu_n = 0$. The necessary and sufficient condition of non-explosion (13.45) reads in this case

$$\sum_{n=0}^{\infty} \frac{1}{\lambda_n} = \infty. \qquad (13.48)$$

EXAMPLE 13.3.14: ABSORBING BIRTH-AND-DEATH PROCESS   Let $\mathbf{A}$ be a birth-and-death generator with parameters λ_n and μ_n, where $\lambda_n > 0$ and $\mu_n > 0$ for all $n \geq 1$, but

$$\lambda_0 = 0$$

(a condition that makes 0 an absorbing state). Starting from the initial state $i \geq 1$, we seek to obtain the probability of absorption u_i in state 0. Clearly, this probability is the same as the probability of absorption by 0 of the embedded chain, with transition matrix

$$\mathbf{P} = \begin{pmatrix} 1 & 0 & 0 & 0 & \cdots \\ \frac{\mu_1}{\lambda_1 + \mu_1} & 0 & \frac{\lambda_1}{\lambda_1 + \mu_1} & 0 & \cdots \\ 0 & \frac{\mu_2}{\lambda_2 + \mu_2} & 0 & \frac{\lambda_2}{\lambda_2 + \mu_2} & \cdots \\ \vdots & \vdots & \vdots & \vdots & \end{pmatrix}$$

By first-step analysis,

$$u_i = \frac{\mu_i}{\lambda_i + \mu_i} u_{i-1} + \frac{\lambda_i}{\lambda_i + \mu_i} u_{i+1} \quad (i \geq 1)$$

or equivalently

$$u_{i+1} - u_i = \frac{\mu_i}{\lambda_i}(u_i - u_{i-1}) \quad (i \geq 1),$$

with the boundary condition $u_0 = 1$. It follows from this that

$$u_n - u_1 = -(1 - u_1) \sum_{i=1}^{n-1} \gamma_i \quad (n \geq 1), \qquad (13.49)$$

where

$$\gamma_0 = 1, \quad \gamma_i = \frac{\mu_1 \mu_2 \cdots \mu_i}{\lambda_1 \lambda_2 \cdots \lambda_i} \quad (i \geq 1).$$

Since u_n is bounded by 1 for all $i \geq 0$ (it is a probability), we see from (13.49) that if

$$\sum_{i=1}^{\infty} \gamma_i = \infty, \tag{13.50}$$

then necessarily $u_1 = 1$, and consequently $u_n = u_1 = 1$ ($n \geq 0$). This means that under condition (13.50), the process is ultimately absorbed by state 0, whatever the initial state. Suppose now that $u_1 \in (0,1)$ and consequently

$$\sum_{i=1}^{\infty} \gamma_i < \infty.$$

By (13.49), u_n is decreasing to some $u \geq 0$ as $n \to \infty$. Necessarily, $u = 0$, since otherwise $u_n \geq u > 0$ for all n, and a Borel–Cantelli argument then shows that $u_n = 1$ for all n, a contradiction with $u_1 \in (0,1)$.

Equation (13.49) gives at the limit $n \to \infty$,

$$u_1 = \frac{\sum_{i=1}^{\infty} \gamma_i}{1 + \sum_{i=1}^{\infty} \gamma_i},$$

and therefore, from (13.49) again,

$$u_n = \frac{\sum_{i=n}^{\infty} \gamma_i}{1 + \sum_{i=1}^{\infty} \gamma_i}. \tag{13.51}$$

We now compute m_i, the mean time to extinction (absorption by 0) when starting from state $i \geq 0$, supposing that absorption takes place almost surely, that is, condition (13.50) is satisfied. When in state $i \geq 1$, the process remains there for an exponential time with mean $\frac{1}{\lambda_i + \mu_i}$, and then jumps to $i+1$ (resp., $i-1$) with the probability $\frac{\lambda_i}{\lambda_i + \mu_i}$ (resp., $\frac{\mu_i}{\lambda_i + \mu_i}$). First-step analysis therefore gives

$$m_i = \frac{1}{\lambda_i + \mu_i} + \frac{\mu_i}{\lambda + \mu_i} m_{i-1} + \frac{\lambda_i}{\lambda_i + \mu_i} m_{i+1} \quad (i \geq 1)$$

with the boundary condition $m_0 = 0$. This can be written

$$m_i - m_{i+1} = \frac{1}{\lambda_i} + \frac{\mu_i}{\lambda_i} (m_{i-1} - m_i) \quad (i \geq 1).$$

Straightforward manipulations yield

$$\frac{1}{\gamma_n} (m_n - m_{n+1}) = \sum_{i=1}^{n} \frac{1}{\lambda_i \gamma_i} - m_1 \quad (n \geq 1). \tag{13.52}$$

Observe that $m_n \leq m_{n+1}$ for all n. If

$$\sum_{i=1}^{\infty} \frac{1}{\lambda_i \gamma_i} = \infty \,,$$

then necessarily $m_1 = \infty$. Consequently, $m_n = \infty$ $(n \geq 1)$. Suppose now that

$$\sum_{i=1}^{\infty} \frac{1}{\lambda_i \gamma_i} < \infty \,.$$

Then necessarily $m_1 < \infty$. Consequently, $m_n < \infty$ for all $n \geq 1$. From (13.52), letting $n \uparrow \infty$ therein,

$$m_1 = \sum_{r=1}^{\infty} \frac{1}{\lambda_i \gamma_i} - \lim_{n \to \infty} \frac{1}{\gamma_n} (m_n - m_{n-1}).$$

It can be shown that

$$\lim_{n \to \infty} \frac{1}{\gamma_n} (m_n - m_{n-1}) = 0 \,.$$

This gives

$$m_1 = \sum_{i=0}^{\infty} \frac{1}{\lambda_i \gamma_i} \,,$$

and going back to (13.52), for $n \geq 1$,

$$m_n = \sum_{i=1}^{\infty} \frac{1}{\lambda_i \gamma_i} + \sum_{k=1}^{n-1} \gamma_k \sum_{j=k+1}^{\infty} \frac{1}{\lambda_j \gamma_j} \,. \tag{13.53}$$

EXAMPLE 13.3.15: LINEAR BIRTH-AND-DEATH PROCESS. The above results will be applied to the linear birth-and-death process for which

$$\lambda_n = n\lambda, \quad \mu_n = n\mu \,,$$

and $\lambda_0 = 0$, where $\lambda > 0$ and $\mu > 0$. In this case $\gamma_i = \left(\frac{\mu}{\lambda}\right)^i$, and therefore $\sum_{i=1}^{\infty} \gamma_i < \infty$, if and only if $\mu < \lambda$. Therefore, $\mu \geq \lambda$ implies eventual extinction, as expected. For $\mu < \lambda$, the probability of extinction when starting from state n is u_n, given by formula (13.51). In this case, for $n \geq 0$,

$$u_n = \left(\frac{\mu}{\lambda}\right)^n \,.$$

In the case where $\mu \geq \lambda$, which implies eventual extinction from any initial state n, formula (13.53) gives the mean time to extinction. We consider the case with a single ancestor

$$m_1 = \sum_{i=1}^{\infty} \frac{1}{\lambda_i \gamma_i} = \frac{1}{\lambda} \sum_{i=1}^{\infty} \frac{1}{i} \left(\frac{\lambda}{\mu} \right)^i,$$

and therefore, $m_1 = \infty$ if $\lambda = \mu$, and if $\mu > \lambda$,

$$m_1 = \frac{1}{\lambda} \log \left(\frac{\mu}{\mu - \lambda} \right),$$

where we have used the expansion

$$\sum_{i=1}^{\infty} \frac{x^i}{i} = \log \left(\frac{1}{1-x} \right) \quad (x \in (0,1)).$$

13.4   Recurrence and Long-Run Behavior

13.4.1   Stationary Distribution Criterion of Ergodicity

Irreducibility, recurrence, transience and positive recurrence will now be defined for a regular jump HMC.

Definition 13.4.1 *A regular jump* HMC *is called* irreducible *if and only if the embedded discrete-time* HMC *is irreducible.*

Definition 13.4.2 *A state i is called* recurrent *if and only if it is recurrent for the embedded chain. Otherwise, it is called* transient.

A recurrent state $i \in E$ is called t-positive recurrent if and only if $\mathrm{E}_i[R_i] < \infty$, where R_i is the return time to state i. Otherwise, it is called t-null recurrent.

Remark 13.4.3 *We shall soon see that t-positive recurrence and n-positive recurrence (the latter is positive recurrence of the embedded chain) are not equivalent concepts. Also, observe that recurrence of a given state implies that this state is essential. Finally, in the same vein, note that irreducibility implies that all states are essential.*

Definition 13.4.4 *A t-invariant measure is a non-trivial vector $\nu = \{\nu(i)\}_{i \in E}$ such that for all $t \geq 0$,*

$$\nu^T \mathbf{P}(t) = \nu^T. \tag{13.54}$$

Of course, an n-invariant measure is, by definition, an invariant measure for the embedded chain.

Theorem 13.4.5 *Let the regular jump* HMC $\{X(t)\}_{t\geq 0}$ *with infinitesimal generator* $\mathbf{A}$ *be irreducible and recurrent. Then, there exists a unique (up to a multiplicative factor) t-invariant measure such that* $\nu(i) > 0$ $(i \in E)$. *Moreover,* ν *is obtained in one of the following ways. (i):*

$$\nu(i) = E_0 \left[\int_0^{R_0} 1_{\{X(s)=i\}} ds \right], \qquad (13.55)$$

where 0 is an arbitrary state and R_0 *is the return time to state 0, or (ii):*

$$\nu(i) = \frac{\mu(i)}{q_i} = \frac{E_0 \left[\sum_{n=1}^{T_0} 1_{[X_n=i]} \right]}{q_i}, \qquad (13.56)$$

where μ *is the canonical invariant measure of the embedded chain relative to state 0, and* T_0 *is the return time to 0 of the embedded chain, or (iii):*

$$\nu^T \mathbf{A} - 0. \qquad (13.57)$$

Proof.

(α) We first show that (13.55) defines an invariant measure, that is, for all $j \in E$ and all $t \geq 0$,

$$\nu(j) - \sum_{k \in E} \nu(k) p_{kj}(l).$$

The right-hand side of the above equality is equal to

$$
\begin{aligned}
A &= \sum_{k \in E} E_0 \left[\int_0^\infty 1_{[X(s)=k]} 1_{\{s \leq R_0\}} ds \right] p_{kj}(t) \\
&= \int_0^\infty \sum_{k \in E} P_0(X(t+s) = j \mid X(s) = k) P_0(X(s) = k, s \leq R_0) ds.
\end{aligned}
$$

From the Markov property, since $\{s \leq R_0\}$ is an event expressible in terms of $(X(u), 0 \leq u \leq s)$,

$$P_0(X(t+s) = j \mid X(s) = k) = P_0(X(t+s) = j \mid X(s) = k, \ s \leq R_0).$$

Therefore,

$$A = \int_0^\infty \sum_{k \in E} \mathrm{P}_0(X(t+s) = j \mid X(s) = k, s \leq R_0) \mathrm{P}_0(X(s) = k, s \leq R_0) \, \mathrm{d}s$$

$$= \int_0^\infty \mathrm{P}_0(X(t+s) = j, s \leq R_0) \, \mathrm{d}s = \int_0^\infty \mathrm{E}_0 \left[1_{\{X(t+s)=j\}} 1_{\{s \leq R_0\}} \right] \, \mathrm{d}s$$

$$= \mathrm{E}_0 \left[\int_0^{R_0} 1_{\{X(t+s)=j\}} \, \mathrm{d}s \right] = \mathrm{E}_0 \left[\int_t^{t+R_0} 1_{\{X(u)=j\}} \, \mathrm{d}u \right]$$

$$= \mathrm{E}_0 \left[1_{\{t \leq R_0\}} \int_t^{R_0} 1_{\{X(u)=j\}} \, \mathrm{d}u \right] - \mathrm{E}_0 \left[1_{\{t > R_0\}} \int_{R_0}^t 1_{\{X(u)=j\}} \, \mathrm{d}u \right] + \cdots$$

$$\cdots \mathrm{E}_0 \left[\int_{R_0}^{R_0+t} 1_{\{X(u)=j\}} \, \mathrm{d}u \right] .$$

From the strong Markov property applied to $\tau = R_0$,

$$\mathrm{E}_0 \left[\int_{R_0}^{R_0+t} 1_{\{X(u)=j\}} \mathrm{d}u \right] = \mathrm{E}_0 \left[\int_0^t 1_{\{X(u)=j\}} \mathrm{d}u \right] .$$

Therefore,

$$A = \mathrm{E}_0 \left[1_{\{t \leq R_0\}} \int_t^{R_0} \cdots \right] - \mathrm{E}_0 \left[1_{\{t > R_0\}} \int_{R_0}^t \cdots \right] + \mathrm{E}_0 \left[\int_0^t \cdots \right]$$

$$= \mathrm{E}_0 \left[\int_0^{R_0} 1_{\{X(u)=j\}} \mathrm{d}u \right] = \nu(j).$$

(β) We now show uniqueness. For this consider the *skeleton chain* $\{X(n)\}_{n \geq 0}$, which is irreducible, since the continuous-time HMC is irreducible (Exercise 13.6.17). It is also recurrent. Indeed for a fixed state i, consider the sequence $Z_1, Z_2, \ldots$ of successive sojourn times in state i of the state process. This sequence is infinite because the embedded chain is recurrent, IID, exponential with mean $\frac{1}{q_i}$. Therefore, the event $\{Z_n > 1\}$ occurs infinitely often, and this implies that $\{X(n) = i\}$ also occurs infinitely often. Therefore, the skeleton being irreducible and recurrent has one and only one invariant measure. Since an invariant measure of the continuous-time chain is an invariant measure of the skeleton, uniqueness follows.

(γ) We now show that if ν is the (essentially unique) invariant measure of the continuous-time HMC, then μ defined by (13.56) is the (essentially unique) invariant measure of the embedded chain. For this, call T_0 the return time to 0 of the embedded chain. Then

$$
\nu(i) = E_0\left[\int_0^{R_0} 1_{\{X(s)=i\}}ds\right] = E_0\left[\sum_{n=0}^{T_0-1} S_{n+1} 1_{\{X_n=i\}}\right]
$$

$$
= E_0\left[\sum_{n=0}^{\infty} S_{n+1} 1_{\{X_n=i\}} 1_{\{n<T_0\}}\right] = \sum_{n=0}^{\infty} E_0[S_{n+1} 1_{\{X_n=i\}} 1_{\{n<T_0\}}]
$$

$$
= \sum_{n=0}^{\infty} E_0[S_{n+1} \mid X_n = i, n < T_0] E_0[1_{\{X_n=i\}} 1_{\{n<T_0\}}]
$$

$$
= \sum_{n=0}^{\infty} E_0[S_{n+1} \mid X_n = i] E_0[1_{\{X_n=i\}} 1_{\{n<T_0\}}],
$$

where the last equality follows from the strong Markov property at time τ_n. But $E_0[S_{n+1} \mid X_n = i] = \frac{1}{q_i}$, since conditionally on $X_n = i$, S_{n+1} is exponential of mean $1/q_i$. Therefore,

$$
\nu(i) = \frac{1}{q_i} \sum_{n=0}^{\infty} E_0[1_{\{X_n=i\}} 1_{\{n<T_0\}}] = \frac{1}{q_i} E_0\left[\sum_{n=0}^{T_0-1} 1_{\{X_n=i\}}\right] = \frac{\mu(i)}{q_i}.
$$

(δ) We show that ν given by (13.56), where μ is an invariant measure of the embedded chain, also satisfies (13.57). The transition matrix $\mathbf{P}$ of the embedded chain is

$$
p_{ii} = 0, \quad p_{ij} = \frac{q_{ij}}{q_i} \text{ if } i \neq j,
$$

and the balance equation $\mu^T = \mu^T \mathbf{P}$ reads

$$
\frac{\mu(i)}{q_i} q_i = \sum_{\substack{j \in E \\ j \neq i}} \frac{\mu(j)}{q_j} q_{ji},
$$

which is just (13.57). $\qquad\square$

Theorem 13.4.6 *An irreducible recurrent regular jump* HMC *with invariant measure ν is t-positive recurrent if and only if*

$$
\sum_{i \in E} \nu(i) < \infty.
$$

In this case the stationary probability π is related to the mean return times by

$$
\pi_i E_i[R_i] = 1.
$$

Proof. Observing that

$$q \sum_{i \in E} \nu(i) = \mathrm{E}_0[R_0] \, ,$$

the proof is the same as for the corresponding result in discrete time.   □

Remark 13.4.7 We can give consistency to Remark 13.4.3 by using the relationship between the t-invariant measure ν and the n-invariant measure μ of the embedded chain in (13.56), which shows that there exist versions of the invariant measure μ of the embedded chain and ν of the continuous-time chain related by

$$\mu(i) = q_i \nu(i) \, .$$

All four possibilities concerning the convergence or divergence of the series $\sum_{i \in E} \mu(i)$ and $\sum_{i \in E} \nu(i)$ are open. The mean sojourn times q_i^{-1} make the difference, as is natural since they embody the deformation of the time scale when passing from the discrete time of the embedded chain to the continuous time of the regular jump HMC.

Definition 13.4.8 *An irreducible regular jump* HMC *is called* ergodic *if it is t-positive recurrent.*

Remark 13.4.9 The notion of periodicity is irrelevant in continuous time. Note also that the embedded chain of an ergodic regular jump HMC may well be non-ergodic, for two reasons. The first is the possibility that the embedded chain is null recurrent, as explained in Remark 13.4.7, while the second is the possibility that the embedded chain is periodic.

Theorem 13.4.10 *The irreducible regular jump* HMC *with infinitesimal generator* $\mathbf{A}$ *is ergodic if and only if there exists a probability π on E such that*

$$\pi^T \mathbf{A} = 0. \tag{13.58}$$

Proof. In view of Theorem 13.4.5, only sufficiency has to be proved. Therefore, suppose that (13.58) holds for a probability distribution π. We shall prove that $\pi^T \mathbf{P}(t) = \pi^T$ for all $t \geq 0$. For this, define

$$p_{ij}^{(n)}(t) = \mathrm{P}_i(X(t) = j, t < \tau_n),$$

where the τ_n are the transition times. A trajectory starting from state i contributes to $p_{ij}^{(n)}(t)$ either if it has no jump before t and $i = j$, or if it has a last jump (say from k to j) at a time $s \leq t$, and therefore at most $n-1$ jumps before s. Therefore (Exercise 13.6.18),

$$p_{ij}^{(n)}(t) = \delta_{ij} \exp(-q_i t) + \int_0^t \sum_{k \in E} p_{ik}^{(n-1)}(s) q_{kj} \exp(-q_j(t-s)) ds. \qquad (13.59)$$

Thus

$$\sum_{i \in E} \pi(i) p_{ij}^{(n)}(t) = \pi(j) \exp(-q_j t) + \int_0^t \exp(-q_j(t-s)) \sum_{k \in E} q_{kj} \sum_{i \in E} \pi(i) p_{ik}^{(n-1)}(s) ds.$$

Now,

$$\sum_{i \in E} \pi(i) p_{ik}^{(1)}(s) = \pi(k) \exp(-q_k s) \leq \pi(k)$$

that is, $\pi^T \mathbf{P}^{(1)}(s) \leq \pi$, with the obvious notation. By induction, $\pi^T \mathbf{P}^{(n)}(s) \leq \pi^T$. Indeed, if the latter is true, then

$$\sum_{i \in E} \pi(i) p_{ij}^{(n+1)}(s) \leq \pi(j) \exp(-q_j t) + \int_0^t \exp(-q_j(t-s)) \sum_{k \in E} q_{kj} \pi(k) ds$$

$$= \pi(j) \exp(-q_j t) + \pi(j) q_j \int_0^t \exp(-q_j s) ds = \pi(j).$$

Since the process is non-explosive,

$$\lim_{n \uparrow \infty} p_{ij}^{(n)}(t) = p_{ij}(t) \text{ and } \sum_{j \in E} p_{ij}(t) = 1.$$

Therefore, by dominated convergence,

$$\sum_{i \in E} \pi(i) p_{ij}(t) \leq \pi(j).$$

Summation of both sides of the inequality shows that equality must hold. Therefore, π is a stationary distribution.

If the chain is transient, $\lim_{t \uparrow \infty} 1_{X(t)=j} = 0$, and therefore, by dominated convergence,

$$\lim_{t \to \infty} p_{ij}(t) = 0.$$

In particular, by dominated convergence, $(\pi^T \mathbf{P}(t))_j$ would tend to 0 as $t \to \infty$, a contradiction with $\pi^T \mathbf{P}(t) = \pi^T$. The chain is therefore recurrent. It is t-positive recurrent by Theorem 13.4.5. $\qquad \square$

Time Reversal

The concept of reversibility in continuous time is basically the same as in discrete time. Let $\{X(t)\}_{t\geq 0}$ be a regular jump HMC with countable state space E, transition semi-group $\{\mathbf{P}(t)\}_{t\geq 0}$, and infinitesimal generator $\mathbf{A}$. Assume, moreover, that this chain is irreducible and ergodic, and in equilibrium, with the stationary distribution π. For arbitrary but fixed $T > 0$, consider the process $\{\tilde{X}(t)\}_{t\in[0,T]}$ obtained by time reversal of $\{X(t)\}_{t\in[0,T]}$, that is,

$$\tilde{X}(t) = X(T - t).$$

Then $\{\tilde{X}(t)\}$ is an HMC with transition semi-group $\{\tilde{\mathbf{P}}(t)\}_{t\geq 0}$ given by

$$\pi(j)\tilde{p}_{ij}(t) = \pi(j)p_{ji}(t) \tag{13.60}$$

(the proof is analogous to that of the corresponding result in discrete time), and therefore with infinitesimal generator $\tilde{\mathbf{A}}$, where

$$\pi(j)\tilde{q}_{ij} = \pi(j)\tilde{q}_{ji}, \tag{13.61}$$

as differentiation in (13.60) immediately shows.

Remark 13.4.11 In order not to depend on the choice of T, we may extend the definition of $X(t)$ to negative times by letting $\{X(-t)\}_{t\geq 0}$ be an HMC with transition semi-group $\{\tilde{\mathbf{P}}(t)\}_{t\geq 0}$ and independent of $\{X(t)\}_{t\geq 0}$ given $X(0)$. In this manner, we obtain, as in discrete time, a stationary HMC $\{X(t)\}_{t\in\mathbb{R}}$ with transition semi-group $\{\mathbf{P}(t)\}_{t\geq 0}$ and infinitesimal generator $\mathbf{A}$. (It will be assumed that the process has been modified so as to guarantee that the trajectories are right-continuous. In general, we shall take whatever convention is suitable as to right- or left-continuity, since this does not change the semi-group or the infinitesimal generator.) The reversed process $\{X(-t)\}_{t\in\mathbb{R}}$ is therefore a regular jump HMC with transition semi-group $\{\tilde{\mathbf{P}}(t)\}_{t\in\mathbb{R}}$ and infinitesimal generator $\mathbf{A}$ given by (13.60) and (13.61), respectively.

Theorem 13.4.12 *Let* $\{X(t)\}_{t\in\mathbb{R}}$ *be a stationary regular jump* HMC *on the countable state space* E, *and assume that it is irreducible and ergodic with stationary distribution* π. *If its infinitesimal generator* $\mathbf{A}$ *satisfies the* detailed balance equations

$$\pi(i)q_{ij} = \pi(j)q_{ji}, \tag{13.62}$$

then the reversed process $\{X(-t)\}_{t\in\mathbb{R}}$ *properly modified to be right-continuous is distributionwise equivalent to the direct process* $\{X(t)\}_{t\in\mathbb{R}}$.

Proof. Comparing (13.61) and (13.62), we see that $\tilde{\mathbf{A}} \equiv \mathbf{A}$, and therefore, the direct and the reversed processes have the same infinitesimal generator. $\qquad\square$

We shall not give examples of reversible chains, since we shall see a number of them when queueing networks are treated in Chapter 14.

For future reference, the reversal test analogous to the one in discrete time will be given because it will be needed in Chapter 14, again in connection with queueing networks.

Theorem 13.4.13 *Let $\mathbf{A}$ be a stable and conservative generator on the countable state space E, and let π be a strictly positive probability distribution on E. Define $\tilde{\mathbf{A}}$ by (13.61). If for all $i \in E$*

$$\sum_{j \in E, j \neq i} \tilde{q}_{ij} = q_i \,, \tag{13.63}$$

then $\pi^T \mathbf{A} = \pi$.

The proof is the same as that of the corresponding result in discrete time.

If $\{X(t)\}_{t>0}$ is a regular jump HMC on E that is irreducible and stationary, with stationary distribution π and infinitesimal generator $\mathbf{A}$ satisfying the detailed balance equation (13.63), then it is called *reversible*. Sometimes, the pair $(\mathbf{A}, \pi)$ is called reversible. The idea of reversibility finds important applications in queueing theory (see Chapter 14).

13.4.2   Ergodicity

Theorem 13.4.14 *Let $\{X(t)\}_{t \geq 0}$ be an ergodic regular jump HMC with state space E and transition semi-group $\{\mathbf{P}(t)\}_{t \geq 0}$. Then,*

$$\lim_{t \to \infty} p_{ij}(t) = \pi(j) \quad (i, j \in E) \,,$$

where π is the (unique) stationary distribution.

Proof. Observe that the *skeleton* $\{X(n)\}_{n \geq 0}$ is an irreducible HMC (see the argument given in the proof of Theorem 13.4.5). It is positive recurrent, since it has π for a stationary distribution. Even though the embedded chain might be periodic, the skeleton $\{X(n)\}$ is not, because the sojourn times of the continuous-time chain in a given state i are IID exponentials, and this eliminates periodic behavior for the skeleton (Exercise 13.6.17).

It is clear that two independent continuous-time HMCs with the same transition semi-group $\{\mathbf{P}(t)\}_{t\geq 0}$, but possibly different initial distributions, will meet at a finite integer random time, since their skeletons do. From this observation the result follows by the same coupling argument as in the discrete time case. $\square$

Theorem 13.4.15 *Let* $\{X(t)\}$ *be ergodic and let* π *be its stationary distribution. Then for any initial distributions* μ *and all* $f : E \to \mathbb{R}$ *such that* $\sum_{i\in E} |f(i)|\pi(i) < \infty$,

$$\lim_{t\uparrow\infty} \frac{1}{t} \int_0^t f(X(s))\mathrm{d}s = \sum_{i\in E} f(i)\pi(i), \quad \mathrm{P}_\mu \ a.s.$$

Proof. The proof is analogous to that of the discrete-time case and is left to the reader. $\square$

Convergence Rates

Any method giving an estimate of the geometric rate of convergence for a discrete time chain with finite state space can in principle be used to obtain an analogous estimate for a continuous time chain with finite state space. The trick is uniformization. Indeed, suppose $\{\mathbf{P}(t)\}_{t\geq 0}$ is the transition semi-group of an irreducible ergodic continuous-time HMC with infinitesimal generator $\mathbf{A}$ and stationary distribution π. If the state space is finite, the chain is uniformizable, that is, its infinitesimal generator can be written

$$\mathbf{A} = \lambda(\mathbf{K} - I),$$

where $\lambda > 0$ and $\mathbf{K}$ is an irreducible transition matrix. Moreover, the semi-group takes the form

$$\mathbf{P}(t) = \sum_{n=0}^\infty \mathrm{e}^{-\lambda t}\frac{(\lambda t)^n}{n!}\mathbf{K}^n.$$

Also, $\mathbf{K}$ has π as stationary distribution. Suppose that for an initial distribution μ it holds that for some $\alpha \in (0,1)$ and some constant $C > 0$,

$$d_V(\mu^T\mathbf{K}^n, \pi) \leq C\alpha^n.$$

Since

$$\mu^T\mathbf{P}(t) - \pi^T = \sum_{n=0}^\infty \mathrm{e}^{-\lambda t}\frac{(\lambda t)^n}{n!}(\mu^T\mathbf{K}^n - \pi^T),$$

it follows from the result of Exercise 4.4.1 that

$$d_V(\mu^T \mathbf{P}(t), \pi) = \sum_{n=0}^{\infty} e^{-\lambda t} \frac{(\lambda t)^n}{n!} d_V(\mu^T \mathbf{K}^n, \pi),$$

and therefore

$$d_V(\mu^T \mathbf{P}(t), \pi) \leq \sum_{n=0}^{\infty} e^{-\lambda t} \frac{(\lambda t)^n}{n!} C\alpha^n.$$

Finally,

$$d_V(\mu^T \mathbf{P}(t), \pi) \leq Ce^{-\lambda(1-\alpha)t}.$$

13.4.3   Absorption

We consider *finite state space continuous-time* HMCs, which can always be assumed to have the uniform structure, that is, with a generator of the form

$$\mathbf{A} = \lambda(\mathbf{K} - I), \tag{13.64}$$

where λ is the intensity of the clock and $\mathbf{K}$ is the transition matrix of the subordinated chain.

Absorption is discussed in terms of the subordinated chain, since it has the same structure as the continuous-time chain in terms of recurrent, transient, and absorbing states. In particular, the subordinated chain and the continuous-time chain have the same absorption probabilities. The difference arises when time intervenes. However, the adaptation of the results to the continuous-time chain is straightforward.

For instance, call $\mathbf{\Gamma} = \{\gamma_{ij}\}_{i,j \in E}$ the *potential matrix* of the continuous-time chain, where

$$\gamma_{ij} = \mathrm{E}_i \left[\int_0^\infty 1_{\{X(s)=j\}} \mathrm{d}s \right].$$

Since

$$\int_0^\infty 1_{\{X(s)=j\}} \mathrm{d}s = \sum_{n=0}^{\infty} (T_{n+1} - T_n) 1_{\{\hat{X}_n = j\}},$$

we have, using the independence of the clock and of the subordinated chain, and $\mathrm{E}[T_{n+1} - T_n] = \frac{1}{\lambda}$,

$$\gamma_{ij} = \frac{1}{\lambda} g_{ij},$$

where $\mathbf{G}$ is the potential matrix of the subordinated chain.

Suppose for simplicity that the state space E consists of one absorbing state, 1, and of the collection $\{2,\ldots,r\}$ of transient states. Let us decompose the infinitesimal generator according to the partition $\{1\} + \{2,\ldots,r\}$:

$$\mathbf{A} = \begin{pmatrix} 0 & 0 \\ a & \mathbf{T} \end{pmatrix}.$$

From the analogous decomposition of $\mathbf{K}$,

$$\mathbf{K} = \begin{pmatrix} 1 & 0 \\ b & \mathbf{Q} \end{pmatrix},$$

and (13.64), it follows that

$$b = \frac{1}{\lambda}a \ , \ \mathbf{Q} = I + \frac{\mathbf{T}}{\lambda}.$$

This identification permits us to adapt the results obtained for the discrete time case to continuous time. For instance, if we call S and Σ the restrictions of $\mathbf{G}$ and $\mathbf{\Gamma}$, respectively, to $\{2,\ldots,r\}$, then

$$\Sigma = \frac{1}{\lambda}S = \frac{1}{\lambda}(I - \mathbf{Q})^{-1} = \frac{1}{\lambda}\left(-\frac{\mathbf{T}}{\lambda}\right)^{-1},$$

that is,

$$\Sigma = -\mathbf{T}^{-1}.$$

Calling τ the time to absorption in $\{2,\ldots,r\}$ and supposing that $X(0)$ is distributed on the transient set according to ν, we have (Exercise 13.6.19)

$$\mathrm{E}_\nu[e^{iu\tau}] = 1 - iu\nu(\mathbf{T} + iuI)^{-1},$$

and consequently,

$$\mathrm{P}_\nu(\tau \le t) = 1 - |\nu e^{t\mathbf{T}}|.$$

13.5   Continuous-Time HMCS from HPPS

13.5.1   The Strong Markov Property of HPPS

In order to develop the theory of this section independently of the previous chapter, a direct proof of the strong Markov property for HPPs will be given. The framework of this technical result is the following: There is an arbitrarily indexed family $\{N_\ell\}_{\ell \in L}$ of independent HPPs with respective intensities $\{\lambda_\ell\}_{\ell \in L}$, and an arbitrary collection of random variables $\mathcal{Y} = \{Y_m\}_{m \in M}$ independent of $\{N_\ell\}_{\ell \in L}$.

For a random variable τ with values in $[0, +\infty]$ and an index $\ell \in L$, one defines N_ℓ^τ to be the restriction of N_ℓ to $\mathbb{R}_+ \cap [0, \tau]$:

$$N_\ell^\tau(a, b] := N_\ell^\tau((a, b] \cap [0, \tau]) \quad ([a, b] \subset \mathbb{R}_+).$$

The point process N_ℓ^τ is "N_ℓ before τ." One also defines the point process $S_\tau N_\ell$, or "N_ℓ after τ," by

$$S_\tau N_\ell(a, b] := N_\ell(\tau + a, \tau + b] \quad ([a, b] \subset \mathbb{R}_+).$$

Definition 13.5.1 *The random variable τ with values in $\mathbb{R}_+ \cup \{+\infty\}$ is called a stopping time with respect to $\{N_\ell\}_{\ell \in L}$ and $\mathcal{Y}$ if for all $t \in \mathbb{R}_+$, the event $\{\tau \leq t\}$ is expressible in terms of $\mathcal{Y}$ and $\{N_\ell^t\}_{\ell \in L}$.*

Theorem 13.5.2 *Let $\{N_\ell\}_{\ell \in L}$ and $\mathcal{Y}$ be as above, and let τ be a stopping time with respect to $\{N_\ell\}_{\ell \in L}$ and $\mathcal{Y}$. Then, given $\{\tau < \infty\}$,*

(α) $\{N_\ell^\tau\}_{\ell \in L}$ and $\mathcal{Y}$ are independent of $\{S_\tau N_\ell\}_{\ell \in L}$, and

(β) $\{S_\tau N_\ell\}_{\ell \in L}$ is a family of independent HPPs with respective intensities $\{\lambda_\ell\}_{\ell \in L}$.

Proof. It suffices to show that for all positive integers J, all $u_1, \ldots, u_J, v_1, \ldots, v_J$, $w \in \mathbb{R}$, all bounded intervals $[a_j, b_j]$ and $[c_j, d_j]$ ($1 \leq j \leq J$) and all real-valued random variables Z that are functionals of $\mathcal{Y}$ and all indices $\ell_j \in L$ ($1 \leq j \leq J$),

$$\mathrm{E}\left[\exp\left\{i\left(\sum_{j=1}^{J} u_j(S_\tau N_{\ell_j})(a_j, b_j] + \sum_{j=1}^{J} v_j N_{\ell_j}^\tau(c_j, d_j] + wZ\right)\right\} 1_{\{\tau < \infty\}}\right]$$

$$= \mathrm{E}\left[\exp\left\{i\sum_{j=1}^{J} u_j N_{\ell_j}(a_j, b_j]\right\}\right] \mathrm{E}\left[\exp\left\{i\sum_{j=1}^{J} v_j N_{\ell_j}^\tau(c_j, d_j] + wZ\right\} 1_{\{\tau < \infty\}}\right].$$

$$\tag{13.65}$$

Indeed, with $v_1 = \cdots = v_J = w = 0$ in the above identity,

$$\mathrm{E}\left[\exp\left\{i\sum_{j=1}^{J} u_j(S_\tau N_{\ell_j})(a_j, b_j]\right\} 1_{\{\tau < \infty\}}\right]$$

$$= \mathrm{E}\left[\exp\left\{i\sum_{j=1}^{J} u_j N_{\ell_j}(a_j, b_j]\right\}\right] \mathrm{P}(\tau < \infty), \tag{13.66}$$

and this implies that given $\{\tau < \infty\})$, $\{S_\tau N_\ell\}_{\ell \in L}$ and $\{N_\ell\}_{\ell \in L}$ have the same distribution. Thus (β) is proven. Next, using (13.66), identity (13.65) becomes

$$
E\left[\exp\left\{i \sum_{j=1}^{J} u_j (S_\tau N_{\ell_j})(a_j, b_j] + i \sum_{j=1}^{J} v_j N_{\ell_j}^\tau (c_j, d_j] + iwZ\right\} 1_{\{\tau < \infty\}}\right] P(\tau < \infty)
$$

$$
= E\left[\exp\left\{i \sum_{j=1}^{J} u_j (S_\tau N_{\ell_j})(a_j, b_j]\right\} 1_{\{\tau < \infty\}}\right] \times \cdots
$$

$$
\cdots E\left[\exp\left\{i \sum_{j=1} v_j N_{\ell_j}^\tau (c_j, d_j] + iwZ\right\} 1_{\{\tau < \infty\}}\right].
$$

This proves statement (α), by Theorem 1.2.16.

We now proceed to the proof of (13.65), actually a simplified version that contains all the ingredients of the complete proof and saves us from notational hell. Calling N one of the N_ℓ's, we prove that

$$
\begin{aligned}
&E[\exp\{iuS_\tau N(a, b] + ivN^\tau(c, d] + iwZ\}1_{\{\tau < \infty\}}] \\
&= E[\exp\{iuN(a, b]\}E[\exp\{ivN^\tau(c, d] + iwZ\}1_{\{\tau < \infty\}}].
\end{aligned} \tag{13.67}
$$

The left-hand side is

$$
E[\exp\{iuN(\tau + a, \tau + b] + ivN((0, \tau] \cap (c, d]) + iwZ\}1_{\{\tau < \infty\}}],
$$

that is, in the special case where τ takes a countable number of values $t_k \in \mathbb{R}_+$ $(k \geq 1)$ and maybe also the value $+\infty$,

$$
\sum_{k=1}^{\infty} E[\exp\{iuN(t_k + a, t_k + b] + ivN((0, t_k] \cap (c, d]) + iwZ\}1_{\{\tau = t_k\}}].
$$

But $1_{\{\tau = t_k\}}$ is a functional of $\mathcal{Y}$ and $\{N_\ell^{t_k}\}_{\ell \in L}$ and therefore the above expression becomes, in view of the independence and homogeneity properties of HPPs,

$$
\sum_{k=1}^{\infty} E[\exp\{iuN(t_k + a, t_k + b]\}]E[\exp\{ivN((0, t_k] \cap (c, d]) + iwZ\}1_{\{\tau = t_k\}}]
$$

$$
= E[\exp\{iuN(a, b]\}]\left(\sum_{k=1}^{\infty} E[\exp\{ivN((0, t_k] \cap (c, d]) + iwZ\}1_{\{\tau = t_k\}}]\right),
$$

and this is the right-hand side of (13.67), as announced. The passage to an arbitrary stopping time is done in exactly the same manner as in the proof of the strong Markov property of regular jump HMCs (Theorem 13.3.3). □

13.5.2   From Generator to Markov Chain

In the following definition of a generator, it is not required that such generator be the *infinitesimal* generator of some continuous-time HMC.

Definition 13.5.3 *Let E be a countable space, and let $\mathbf{A} = \{q_{ij}\}_{i,j\in E}$ be a matrix of real numbers indexed by E, such that for all $i, j \in E$ with $i \neq j$,*

$$q_i \in [0, \infty], \quad q_{ij} \in [0, \infty), \tag{13.68}$$

and

$$q_i < \infty, \quad q_i = \sum_{\substack{k \in E \\ k \neq i}} q_{ik}, \tag{13.69}$$

where $q_i = -q_{ii}$.

We now consider the problem of *realization,* that is, of associating with a generator $\mathbf{A}$ a continuous-time HMC admitting $\mathbf{A}$ for infinitesimal generator. When $\mathbf{A}$ is the infinitesimal generator of some continuous transition semi-group on E, only (13.68) is guaranteed, whereas if it is the transition semi-group of a *regular* jump HMC, the stability and conservation properties (13.69) are also satisfied. We have not yet proved the last result, and we shall therefore be forbidden to use it until we have obtained it without getting into logical circles.

Actually, we are going to *construct* a continuous-time HMC $\{X(t)\}_{t\geq 0}$ admitting $\mathbf{A}$ as an infinitesimal generator. For this, we use a family $\{N_{i,j}\}_{\substack{i,j\in E \\ i\neq j}}$ of independent HPPs with respective intensities $\{q_{ij}\}_{\substack{i,j\in E \\ i\neq j}}$, and an initial state $X(0)$, independent of the above family of HPPs, and taking its values in E. The process is constructed as a jump process:

$$X(t) = X_n \quad (t \in [\tau_n, \tau_{n+1})),$$

where $\{\tau_n, X_n\}_{n\geq 0}$ is defined recursively as follows (see Figure 13.5.1). Let Δ be any dummy element not in E.

First let $\tau_0 \equiv 0$ and $X_0 := X(0)$. If $\tau_n < \infty$ and $X_n = X(\tau_n) = i \in E$, then $\tau_{n+1} - \tau_n$ is defined to be the first event of the competing HPPs $\{S_{\tau_n} N_{ij}\}_{j\in E, j\neq i}$. The random variable $\tau_{n+1} - \tau_n = \infty$ if and only if $q_i = 0$ (no events), and in this case the construction ends by letting $X_{n+m} := \Delta$ and $\tau_{n+m} := \infty$ for all $m \geq 1$. Otherwise, if $\tau_{n+1} - \tau_n < \infty$, X_{n+1} is the index $k \neq i$ such that $S_{\tau_n} N_{ik}$ is the first among the competing HPPs to produce an event.

The process is defined by this procedure up to the explosion time $\tau_\infty := \lim_{n\uparrow\infty} \tau_n$. If $\tau_\infty = \infty$ P-a.s, it is a regular jump process on E.

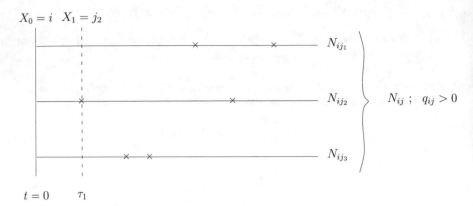

Figure 13.5.1. First transition, obtained by competition

Theorem 13.5.4 *If $\tau_\infty = \infty$ P-a.s, the stochastic process constructed above is a regular jump* HMC *with infinitesimal generator* $\mathbf{A}$.

Proof. The τ_n's form a sequence of stopping times with respect to $\{N_{ij}\}_{\substack{i,j\in E \\ j\neq i}}$ and $X(0)$. In view of the strong Markov property for HPPs,

$$P(X_{n+1} = j, \tau_{n+1} - \tau_n > a \mid X_0, \ldots, X_n, \tau_1, \ldots, \tau_n)$$
$$= P(X_{n+1} = j, \tau_{n+1} - \tau_n > a \mid X_n),$$

and by the competition theorem,

$$P(X_{n+1} = j, \tau_{n+1} - \tau_n > a \mid X_n = i) = e^{-q_i a}\frac{q_{ij}}{q_i}$$

(where $\frac{q_{ij}}{q_i} = 0$ if $q_i = 0$).

By construction, for a given time t, the process after time t depends only upon $X(t)$ and the HPPs $\{N_{ij}\}_{\substack{i,j\in E \\ i\neq j}}$ after t, and is therefore independent of the process before t, given $X(t)$, since the process before t depends only upon $X(0)$ and the HPPs $\{N_{ij}\}_{\substack{i,j\in E \\ i\neq j}}$ before t. This proves the Markov property. Also, for all $t, s \geq 0$,

$$X(t + s) = \psi(X(t), \{S_t N_{k\ell}\}_{\substack{k,\ell\in E \\ k\neq \ell}}, s),$$

for a functional ψ that is given explicitly by the construction. Therefore, on $\{X(t) = i\}$,

$$X(t + s) = \psi(i, \{S_t N_{ij}\}, s)$$

has the same distribution as $\psi(i, \{N_{k,\ell}\}_{\substack{k,\ell\in E \\ k\neq\ell}}, s)$. This proves homogeneity.

We check that $\{X(t)\}_{t\geq 0}$ admits $\mathbf{A}$ as an infinitesimal generator. For instance, we check that $\lim_{t\downarrow 0} \frac{1}{t} P_i(X(t) = j) = q_{ij}$. For this, observe that if $X(t) \neq X(0)$, necessarily $\tau_1 < t$, and write

$$P_i(X(t) = j) = P_i(\tau_2 \leq t, X(t) = j) + P_i(\tau_2 > t, X(t) = j)$$
$$= P_i(\tau_2 \leq t, X(t) = j) + P_i(\tau_2 > t, X_1 = j, \tau_1 < t)$$
$$= P_i(\tau_2 \leq t, X(t) = j) + P_i(X_1 = j, \tau_1 < t) - P_i(\tau_2 \leq t, X_1 = j, \tau_1 \leq t).$$

Now,

$$P_i(X_1 = j, \tau_1 < t) = (1 - e^{-q_i t})\frac{q_{ij}}{q_i},$$

and therefore

$$\lim_{t\downarrow 0} \frac{1}{t} P(X_1 = j, \tau_1 < t) = q_{ij}.$$

It therefore remains to show that $P_i(\tau_2 \leq t, X(t) = j)$ and $P_i(\tau_2 \leq t, X_1 = j, \tau_1 \leq t)$ are $o(t)$. Both terms are bounded by $P_i(\tau_2 < t)$, and

$$P_i(\tau_2 \leq t) \leq P_i(\tau_1 \leq t, \tau_2 - \tau_1 \leq t) = \sum_{\substack{k\in E \\ k\neq j}} P_i(\tau_1 \leq t, X_1 = k, \tau_2 - \tau_1 \leq t)$$

$$= \sum_{\substack{k\in E \\ k\neq j}} (1 - e^{-q_i t}) p_{ik}(1 - e^{-q_k t})$$

$$= (1 - e^{-q_i t}) \sum_{\substack{k\in E \\ k\neq i}} p_{ik}(1 - e^{-q_k t}).$$

The conclusion follows because $(1 - e^{-q_i t})$ is $O(t)$, or identically zero if $q_i = 0$, and because $\lim_{t\downarrow 0} \sum_{k\neq i} p_{ik}(1 - e^{-q_k t}) = 0$, by dominated convergence. $\qquad\square$

EXAMPLE 13.5.5: EXISTENCE OF BIRTH-AND-DEATH PROCESSES. Recall that a birth-and-death process is by definition a regular jump HMC on $E = \mathbb{N}$ with an infinitesimal generator of the form

$$\mathbf{A} = \begin{pmatrix} -\lambda_0 & \lambda_0 & 0 & \cdots \\ \mu_1 & -(\lambda_1 + \mu_1) & \lambda_1 & 0 & \cdots \\ 0 & \mu_2 & -(\lambda_2 + \mu_2) & \lambda_2 & \cdots \\ \vdots & \vdots & \vdots & \vdots \end{pmatrix}.$$

The above construction shows the existence of such a process under Reuter's necessary and sufficient condition of non-explosion.

_____

The construction of the state process on $[0, \tau_\infty)$ is summarized by the following equations: for all $i \in E$,

$$Z_i(t) = Z_i(0) + \sum_{j \in E; j \neq i} \int_{(0,t]} Z_j(s-) N_{ji}(ds) - \sum_{j \in E; j \neq i} \int_{(0,t]} Z_i(s-) N_{ij}(ds). \quad (13.70)$$

Equations (13.70) constitute a system of stochastic differential equations driven by the Poisson processes N_{ij} $(i, j \in E, i \neq j)$ for the processes $\{Z_i(t)\}_{t \geq 0}$ $(i \in E)$.

It is sometimes convenient to state Theorem 13.5.4 as follows:

Theorem 13.5.6 *Let $\{X(t)\}_{t \geq 0}$ be an E-valued regular jump process satisfying (13.70) where $Z_i(t) = 1_{\{X(t)=i\}}$ and where $N_{i,j}$ $(i, j \in E, i \neq j)$ is a family of independent* HPP*s with respective intensities q_{ij} $(i, j \in E, i \neq j)$, and independent of the initial state $X(0)$. Then $\{X(t)\}_{t \geq 0}$ is a regular jump* HMC *with infinitesimal generator* $\mathbf{A}$.

Note that (13.70) is equivalent to the requirement that

$$f(X(t)) - f(X(0)) = \sum_{\substack{i,j \in E \\ i \neq j}} \{f(j) - f(i)\} \int_{(0,t]} 1_{\{X(s-)=i\}} dN_{ij}(s) \quad (13.71)$$

for all non-negative functions $f : E \to \mathbb{R}$.

Paying Two Debts

We are now able to pay two debts, by finishing the proof of Theorem 13.3.3 and by proving Theorem 13.2.11.

First debt: the proof of Theorem 13.2.11. Recall that the starting point was a regular jump HMC $\{X(t)\}_{t \geq 0}$ with infinitesimal generator $\mathbf{A}$. We sought to prove (α) and (β) of Theorem 13.3.3, but were only able to prove something close, namely (α) and (β) where q_i was replaced by $\lambda(i) \in \mathbb{R}_+$ (only known to exist but not yet identified with q_i) and where $\frac{q_{ij}}{q_i}$ was replaced by $P_i(X(\tau_1) = j)$ (not yet identified with $\frac{q_{ij}}{q_i}$). We now proceed with the required identifications.

Let the generator $\mathbf{A}'$ on E be defined as follows:

$$q_i' := \lambda(i), \quad q_{ij}' = \lambda(i) P_i(X(\tau_1) = j).$$

Construct $\{X'(t)\}_{t\geq 0}$, the regular jump HMC associated with $\mathbf{A}'$ via the construction of Section 13.5.2. Then $\{X'(t)\}_{t\geq 0}$ and $\{X(t)\}_{t\geq 0}$ are probabilistically equivalent, by construction. In particular, $\mathbf{A}' = \mathbf{A}$, which is the identification we wanted.

Second debt: Theorem 13.2.11. We claimed that the infinitesimal generator of a regular jump HMC is necessarily stable and conservative. Indeed, a regular jump HMC is strongly Markovian, and therefore has the regenerative structure of the weak version of Theorem 13.3.3, where $\lambda(i)$ replaced q_i and $\lambda(i)P_i(X(\tau_1) = j)$ replaced q_{ij} (we cannot use Theorem 13.3.3 directly because there is an implicit claim that $q_i < \infty$ and $q_i = \sum_{\substack{j\in E \\ j\neq i}} q_{ij}$ in it; the weak version does not have this a priori). But the identification $\mathbf{A}' = \mathbf{A}$ we made above is $q_i = \lambda(i)$, $q_{ij} = \lambda(i)P_i(X(\tau_1) = j)$, and therefore $q_i < \infty, \sum_{\substack{j\in E \\ j\neq i}} q_{ij} = q_i$.

13.5.3 Poisson Calculus and Continuous-time HMCs

Let $\{T_n\}_{n\geq 1}$ be a simple and non-explosive point process on $(0, \infty]$ and let $\{N(t)\}_{t\geq 0}$ be the associated counting function. Let $(a, b] \subset \mathbb{R}_+$ be an arbitrary finite interval and let $\{Z(t)\}_{t\geq 0}$ be an arbitrary complex-valued stochastic process. The following equality is a *definition* of its left-hand side:

$$\int_{(a,b]} Z(t)\mathrm{d}N(t) := \sum_{n\geq 1} Z(T_n)1_{(a,b]}(T_n). \tag{13.72}$$

This quantity is called the *Stieltjes–Lebesgue integral*, or the *counting integral*, of Z with respect to N on $(a, b]$. Clearly, this definition could have been given with any bounded set $C \subset \mathbb{R}_+$ replacing $(a, b]$.

For bounded intervals, this definition does not present difficulties because the sum involved has a finite number of terms. For unbounded intervals, one may have to restrict the integrands in order to have a meaningful sum. For instance,

$$\int_{(0,\infty)} Z(t)\mathrm{d}N(t) = \sum_{n\geq 1} Z(T_n)1_{(0,\infty)}(T_n)$$

is always defined if Z is non-negative, even though it may then be infinite.

Note that if $Z(t)$ is defined on $(a, b]$ by

$$Z(t) = c_j \in \mathbb{C} \text{ on } (t_j, t_{j+1}] \quad (1 \leq j \leq K),$$

where $a = t_0 < t_1 < \cdots < t_{K+1} = b$, then

$$\int_{(a,b]} Z(t)\mathrm{d}N(t) = \sum_{j=0}^{K} c_j \{N(t_{j+1}) - N(t_j)\}.$$

Although there is not much in the definition of a stochastic integral with respect to a point process, the stochastic calculus that such integrals generate is rather powerful. It is based on the result below.

Let $\{N_i\}_{i \in I}$ be an arbitrarily indexed family of homogeneous Poisson processes on $\mathbb{R}_+$ of respective intensities $\{\lambda_i\}_{i \in I}$. Let $\mathcal{Y}$ be an arbitrarily indexed family of random variables independent of $\{N_i\}_{i \in I}$.

Theorem 13.5.7 *Fix $j \in I$. Let $\{Z(t)\}_{t \geq 0}$ be a complex stochastic process with left-continuous trajectories and such that for all $t \geq 0$ the random variable $Z(t)$ is a functional of $\mathcal{Y}$, of $\{N_k\}_{k \in I, k \neq j}$ and of the restriction of N_j to $(0, t]$. Suppose at least one of the following two conditions is satisfied:*

(i) $\{Z(t)\}_{t \geq 0}$ is non-negative, or

(ii) $\{Z(t)\}_{t \geq 0}$ is complex-valued, and $\mathrm{E}\left[\int_0^\infty |Z(t)| N_j(\mathrm{d}t)\right]$ or $\mathrm{E}\left[\int_0^\infty |Z(t)| \lambda_j \mathrm{d}t\right]$ is finite.

Then,

$$\mathrm{E}\left[\int_0^\infty Z(s) \mathrm{d}N_j(s)\right] = \mathrm{E}\left[\int_0^\infty Z(s) \lambda_j \mathrm{d}s\right]. \tag{13.73}$$

Formula (13.73) is called the *smoothing formula*.

Proof.   A. We first treat the case where $\{Z(t)\}_{t \geq 0}$ is real, non-negative and bounded, and prove that

$$\mathrm{E}\left[\int_0^T Z(s) \mathrm{d}N_j(s)\right] = \mathrm{E}\left[\int_0^T Z(s) \lambda_j \mathrm{d}s\right], \tag{13.74}$$

where $T < \infty$. The hypothesis that it has left-continuous trajectories means that for all $\omega \in \Omega$, the function $t \mapsto Z(t, \omega)$ is left-continuous. In particular, if one lets for all $n \geq 1$, all $\omega \in \Omega$ and all $t \geq 0$,

$$Z_n(t, \omega) := \sum_{k=0}^{2^n - 1} Z\left(\frac{kT}{2^n}, \omega\right) 1_{\left(\frac{kT}{2^n}, \frac{(k+1)T}{2^n}\right]}(t),$$

then for all $\omega \in \Omega$ and all $t \in [0, T]$,

$$\lim_{n \uparrow \infty} Z_n(t, \omega) = Z(t, \omega).$$

We first check that (13.74) is true when $\{Z(t)\}_{t\geq 0}$ is replaced by its approximation $\{Z_n(t)\}_{t\geq 0}$. Indeed, the left-hand side of this equality is in this particular case

$$\mathrm{E}\left[\int_0^T Z_n(t)\mathrm{d}N(t)\right] = \mathrm{E}\left[\sum_{k=0}^{2^n-1} Z\left(\frac{kT}{2^n}\right)\left\{N_j\left(\frac{(k+1)T}{2^n}\right) - N_j\left(\frac{kT}{2^n}\right)\right\}\right]$$

$$= \sum_{k=0}^{2^n-1} \mathrm{E}\left[Z\left(\frac{kT}{2^n}\right)\left\{N_j\left(\frac{(k+1)T}{2^n}\right) - N_j\left(\frac{kT}{2^n}\right)\right\}\right].$$

But $Z\left(\frac{kT}{2^n}\right)$ is a functional of $\mathcal{Y}$, of $\{N_k\}_{k\in I, k\neq j}$ and of the restriction of N_j to $\left(0, \frac{kT}{2^n}\right]$, and is therefore independent of $N_j\left(\frac{(k+1)T}{2^n}\right) - N_j\left(\frac{kT}{2^n}\right)$, so that the last term of the above chain of equalities is equal to

$$\sum_{k=0}^{2^n-1} \mathrm{E}\left[Z\left(\frac{kT}{2^n}\right)\right]\mathrm{E}\left[N_j\left(\frac{(k+1)T}{2^n}\right) - N_j\left(\frac{kT}{2^n}\right)\right] = \sum_{k=0}^{2^n-1} \mathrm{E}\left[Z\left(\frac{kT}{2^n}\right)\right]\lambda_j\frac{T}{2^n}$$

$$= \mathrm{E}\left[\sum_{k=0}^{2^n-1} Z\left(\frac{kT}{2^n}\right)\lambda_j\frac{T}{2^n}\right]$$

$$= \mathrm{E}\left[\int_0^T Z_n(s)\lambda_j\mathrm{d}s\right].$$

Therefore,

$$\mathrm{E}\left[\int_0^T Z_n(s)\mathrm{d}N(s)\right] = \mathrm{E}\left[\int_0^T Z_n(s)\lambda_j\mathrm{d}s\right]. \tag{13.75}$$

Calling K the upper bound of $Z(t,\omega)$, we see that $\int_0^t Z_n(t)\lambda_j\mathrm{d}t \leq K\lambda_j T$ and $\int_0^T Z_n(t)\mathrm{d}N_j(t) \leq KN_j(T)$. Also, $\mathrm{E}[N_j(T)] = \lambda_j T < \infty$. We may therefore apply the dominated convergence theorem to both sides of (13.75) to obtain (13.74).

B. For the non-negative *unbounded* case, consider the process $\{g_K(Z(t))\}_{t\geq 0}$, where $K < \infty$ and

$$g_K(x) = \begin{cases} 1 & \text{if } x \leq K, \\ 0 & \text{if } x \geq K+1, \\ -x + K + 1 & \text{if } K \leq x \leq K+1. \end{cases}$$

This process satisfies the conditions for (13.74) and therefore

$$\mathrm{E}\left[\int_0^T g_K(Z(s))\mathrm{d}N_j(s)\right] = \mathrm{E}\left[\int_0^T g_K(Z(s))\lambda_j\mathrm{d}s\right],$$

which yields (13.74) as $K \to \infty$, by monotone convergence. It then suffices to let $T \uparrow \infty$ and to invoke the monotone convergence theorem again to obtain (13.73) in the non-negative unbounded case. Note that the quantities in (13.73) can now very well be infinite.

C. The real-valued case follows easily, by first considering separately the positive and negative parts of the integrand. The complex case is a direct consequence of the real case when one considers separately the real and imaginary parts. $\square$

Watanabe's Characterization of Poisson Processes

Another application of the smoothing formula that is typical of the Poisson calculus is to the *martingale characterization of Poisson processes*.

Theorem 13.5.8 ([3]) *Let $\{N_i\}_{i \in I}$ be an arbitrarily indexed family of simple non-explosive point processes on $\mathbb{R}_+$, and let $\mathcal{Y}$ be a family of random variables independent of $\{N_i\}_{i \in I}$. Let $\{\lambda_i\}_{i \in I}$ be a family of positive real numbers. Suppose that*

$$\mathrm{E}\left[\int_0^T Z(t)\mathrm{d}N_j(t)\right] = \mathrm{E}\left[\int_0^T Z(t)\lambda_j\,\mathrm{d}t\right] \tag{13.76}$$

for all $j \in I$, all $T > 0$ and all non-negative real-valued stochastic processes $\{Z(t)\}_{t \geq 0}$ with left-continuous trajectories, and such that for all $t \geq 0$, the random variable $Z(t)$ is a functional of $\mathcal{Y}$, $\{N_i\}_{i \in I \setminus \{j\}}$ and the restriction of N_j to $(0, t]$.

Then $\{N_i\}_{i \in I}$ is a family of independent homogeneous Poisson processes with respective intensities $\{\lambda_i\}_{i \in I}$, and $\mathcal{Y}$ and $\{N_i\}_{i \in I}$ are independent.

Proof. Take $j = 1$ without loss of generality. We show that it is enough to prove that

$$\mathrm{E}\left[1_A e^{iuN_1(a,b]}\right] = \mathrm{P}(A)\exp\{(e^{iu} - 1)\lambda_1(b - a)\}, \tag{13.77}$$

where $(a, b] \subset \mathbb{R}_+$, $u \in \mathbb{R}$ and A is any event expressible in terms of $\mathcal{Y}$, $\{N_i\}_{i \in I - \{1\}}$ and the restriction of N_1 to $(0, a]$. Indeed, with $A = \Omega$ in (13.77),

$$\mathrm{E}[e^{iuN_1(a,b]}] = \exp\{(e^{iu} - 1)\lambda_1(b - a)\}\,,$$

which is the characteristic function of a Poisson random variable of mean $\lambda_1(b-a)$. Equality (13.77) reads

$$\mathrm{E}[1_A e^{iuN_1(a,b]}] = \mathrm{P}(A)\mathrm{E}[e^{iuN_1(a,b]}]\,.$$

---

[3][Watanabe, 1964].

By Theorem 1.2.16, it follows that $N_1(a, b]$ and A are independent. But A is an arbitrary event expressible in terms of $\mathcal{Y}$, $\{N_i\}_{i \in I-\{1\}}$ and the restriction of N_1 to $(0, a]$. Therefore, $N_1(a, b]$ is independent of $\mathcal{Y}$, $\{N_i\}_{i \in I-\{1\}}$, and the restriction of N_1 to $(0, a]$. Since $(a, b]$ is arbitrary, it follows that N_1 has independent increments and that it is independent of $\mathcal{Y}$ and $\{N_i\}_{i \in I-\{1\}}$.

We now proceed with the proof of (13.77). For this, consider the process

$$X(t) := 1_A e^{iuN_1(a,t]} \quad (t \geq a),$$

and observe that it is piecewise constant with discontinuity times located at the events of N_1, which we denote by $\{T_n^1\}_{n \geq 1}$. Therefore, for $t \geq a$,

$$X(t) = X(a) + \sum_{n>1} \{X(T_n^1) - X(T_{n-}^1)\} 1_{(a,t]}(T_n^1).$$

But $X(a) = 1_A$, and for any $T_n^1 \geq a$,

$$X(T_n^1) = e^{iuN_1(a,T_n^1]} = e^{iu(N_1(a,T_n^1)+1)} = e^{iuN_1(a,T_n^1)} e^{iu} = X(T_{n-}^1) e^{iu}.$$

Therefore,

$$X(t) = 1_A + \sum_{n \geq 1} X(T_{n-}^1)(e^{iu} - 1) 1_{(a,t]}(T_n^1)$$

or, in other notation,

$$X(t) = 1_A + \int_{(a,t]} X(s-)(e^{iu} - 1) \, dN_1(s). \tag{13.78}$$

Since $Z(t) = X(t-)(e^{iu} - 1)$ defines a bounded left-continuous complex-valued stochastic process $\{Z(t)\}_{t \geq 0}$ such that for all $t \geq 0$, the random variable $Z(t)$ is expressible in terms of $\mathcal{Y}$, of $\{N_i\}_{i \in I \setminus \{1\}}$ and of the restriction of N_1 to $(0, t]$, equality (13.76) holds true by hypothesis. Therefore,

$$\mathrm{E}\left[\int_{(a,t]} X(s-)(e^{iu} - 1) \, dN_1(s)\right] = \mathrm{E}\left[\int_a^t X(s-)(e^{iu} - 1)\lambda_1 \, ds\right].$$

Now, for each $\omega \in \Omega$, $X(s-, \omega) = X(s, \omega)$ except on a countable set, and therefore one may replace in the right-hand side of the above equality $X(s-)$ by $X(s)$. Taking expectations in (13.78) therefore yields

$$\mathrm{E}[X(t)] = \mathrm{P}(A) + \int_a^t \mathrm{E}[X(s)](e^{iu} - 1)\lambda_1 \, ds.$$

This being true for all $t \geq a$, (13.77) follows. $\qquad \square$

We now give two applications of the Poisson calculus.

Kolmogorov's Forward System via Poisson Calculus

Our first application of the smoothing formula is to the proof that the forward integral system of Kolmogorov holds true for all regular jump HMCs:

$$\mathbf{P}(t) = I + \int_0^t \mathbf{P}(s)\mathbf{A}\,\mathrm{d}s. \tag{13.79}$$

Proof. We return to the construction in Section 13.5 of a regular jump HMC $\{X(t)\}_{t\geq 0}$ with values in E and with a stable, conservative and non-explosive generator $\mathbf{A}$ by means of a family of independent HPPs $\{N_{ij}\}_{i,j\in E}$ of respective intensities $\{q_{ij}\}_{i,j\in E}$. Let $f : E \to \mathbb{R}$ be a non-negative function. Since the trajectories of $\{f(X(t))\}_{t\geq 0}$ are right-continuous step functions, whose discontinuity times are discontinuity times of $\{X(t)\}_{t\geq 0}$, we may write

$$
\begin{aligned}
f(X(t)) - f(X(0)) &= \sum_{s\in(0,t]} \{f(X(s)) - f(X(s-))\} \\
&= \sum_{\substack{i,j\in E \\ i\neq j}} \int_{(0,t]} \{f(X(s)) - f(X(s-))\}\mathrm{d}N_{ij}(s) \\
&= \sum_{\substack{i,j\in E \\ i\neq j}} \int_{(0,t]} \{f(j) - f(i)\}1_{\{X(s-)=i\}}\mathrm{d}N_{ij}(s).
\end{aligned}
$$

Therefore, reorganizing this equality and taking expectations, we obtain from the smoothing formula

$$
\mathrm{E}_\mu[f(X(t))] + \sum_{\substack{i,j\in E \\ i\neq j}} \mathrm{E}_\mu\left[\int_{(0,t]} f(i)1_{\{X(s-)=i\}}q_{ij}\,\mathrm{d}s\right]
$$
$$
= \mathrm{E}_\mu[f(X(0))] + \sum_{\substack{i,j\in E \\ i\neq j}} \mathrm{E}_\mu\left[\int_{(0,t]} f(j)1_{\{X(s-)=i\}}q_{ij}\,\mathrm{d}s\right],
$$

where μ is the initial distribution of the chain. Since

$$
\mathrm{E}_\mu\left[\int_{(0,t]} 1_{\{X(s-)=k\}}\mathrm{d}s\right] = \int_{(0,t]} \mathrm{E}_\mu[1_{\{X(s)=k\}}]\mathrm{d}s = \int_0^t \mathrm{P}_\mu(X(s)=k)\,\mathrm{d}s,
$$

we have, by the conservation equality $q_i = \sum_{\substack{j\in E \\ j\neq i}} q_{ij}$,

$$
\mathrm{E}_\mu[f(X(t))] + \sum_{i\in E} f(i)q_i \int_0^t \mathrm{P}_\mu(X(s)=i)\,\mathrm{d}s
$$
$$
= \mathrm{E}_\mu[f(X(0))] + \sum_{\substack{i,j\in E \\ i\neq j}} f(j)q_{ij} \int_0^t \mathrm{P}_\mu(X(s)=i)\,\mathrm{d}s. \tag{13.80}
$$

Taking $f(j) = 1_{\{k\}}(j)$ for a fixed state k, and $\mu = \delta_\ell$, the Dirac distribution at ℓ, the above equalities reduce to

$$p_{\ell k}(t) + q_k \int_0^t p_{\ell k}(s)\,\mathrm{d}s = \delta_{\ell k} + \sum_{\substack{i \in E \\ i \neq k}} q_{ik} \int_0^t p_{\ell i}(s)\,\mathrm{d}s\,.$$

Since the left-hand side is finite, so is the right-hand side, and therefore one can reorganize this equality to obtain

$$p_{\ell k}(t) = \delta_{\ell k} + \int_0^t \Big\{ -p_{\ell k}(s)q_k + \sum_{\substack{i \in E \\ i \neq k}} p_{\ell i}(s)q_{ik} \Big\}\,\mathrm{d}s\,. \tag{13.81}$$

This is true for all $\ell, k \in E$ and the corresponding equations are summarized by (13.79).

This is the integral form of the forward Kolmogorov system. Note that (13.81) is of the form

$$f(t) = a + \int_0^t g(s)\,\mathrm{d}s\,,$$

where g is locally integrable. Therefore[4] $f'(t) = y(t)$ *almost everywhere*. Thus Kolmogorov's forward differential system holds *almost everywhere*. Similarly, we find from (13.80) that for any initial distribution $\mu = \{\mu(i)\}_{i \in E}$, the vector $\mu(t) = \{\mu_i(t)\}_{i \in E}$, where $\mu_i(t) = \mathrm{P}(X(t) = i)$, satisfies *almost-everywhere* with respect to the Lebesgue measure

$$\frac{\mathrm{d}}{\mathrm{d}t}\mu^T(t) = \mu^T(t)\mathbf{A}\,.$$

In particular, if a stationary distribution π exists, it satisfies the global balance equation

$$\pi^T \mathbf{A} = 0\,. \tag{13.82}$$

□

Note that in the course of the proof of Kolmogorov's forward system, we have obtained the following result:

Theorem 13.5.9 *Let* $\{X(t)\}_{t \geq 0}$ *be a regular jump* HMC *with state space* E, *infinitesimal generator* $\mathbf{A}$ *and initial distribution* μ. *Let the function* $f : E \to \mathbb{R}_+$ *be such that* $\mathrm{E}_\mu[f(X_0)] < \infty$.

If either one of the following two conditions is satisfied:

---

[4][Rudin, 1966], Theorem 8.17.

(i) $E_\mu \left[\int_0^t q_{X(s)} f(X(s)) ds \right] < \infty$, or

(ii) $E_\mu[f(X(t))] < \infty$,

then the other is satisfied and

$$E_\mu[f(X(t)] = E_\mu[f(X(0))] + E_\mu \left[\int_0^t (\mathbf{A}f)(X(s)) ds \right]$$

for all $t \geq 0$, where

$$(\mathbf{A}f)(j) := \sum_{\substack{i \in E \\ i \neq j}} q_{ij} f(i) - q_j f(j) \, .$$

Aggregation of States

The following result will be obtained as an application of Watanabe's theorem.

Theorem 13.5.10 *Consider a regular jump* HMC $\{X(t)\}_{t \geq 0}$ *with state space* E *and infinitesimal generator* $\mathbf{A}$. *Let* $\tilde{E} = \{\alpha, \beta, \ldots\}$ *be a partition of* E *and define the process* $\{\tilde{X}(t)\}_{t \geq 0}$ *taking its values in* $\tilde{E}$ *by*

$$\tilde{X}(t) = \alpha \iff X(t) \in \alpha \, .$$

Suppose that for all $\alpha \in \tilde{E}$, $i \in \alpha$, $\beta \in \tilde{E}$ *with* $\alpha \neq \beta$,

$$\sum_{j \in \beta} q_{ij} = \tilde{q}_{\alpha\beta} \tag{13.83}$$

(with the meaning that the left-hand side is independent of $i \in \alpha$*). Then* $\{\tilde{X}(t)\}_{t \geq 0}$ *is a regular jump* HMC *with state space* $\tilde{E}$ *and infinitesimal generator* $\tilde{\mathbf{A}}$, *with off-diagonal terms given by (13.83).*

Proof. This statement concerns only the distribution of $\{X(t)\}_{t \geq 0}$ (not its sample paths) and therefore we may suppose that $\{X(t)\}_{t \geq 0}$ is generated as in Section 13.5, using independent HPPs $\{N_{ij}\}_{i,j \in E; i \neq j}$ with respective intensities $\{q_{ij}\}_{i,j \in E; i \neq j}$ and an initial state $X(0)$ independent of the above HPPs. Then, for $f : \tilde{E} \to \mathbb{R}$, $s \in (0, t]$,

$$\begin{aligned}
f(\tilde{X}(t)) &= f(\tilde{X}(s)) + \sum_{\substack{i,j \in E \\ i \neq j}} \int_{(s,t]} \{f(\tilde{X}(u)) - f(\tilde{X}(u-))\} 1_{\{X(u-)=i\}} dN_{ij}(u) \\
&= f(\tilde{X}(s)) + \sum_{\substack{\alpha,\beta \in \tilde{E} \\ \alpha \neq \beta}} \int_{(s,t]} \{f(\beta) - f(\alpha)\} \sum_{i \in \alpha} \left(1_{\{X(u-)=i\}} \left(\sum_{j \in \beta} dN_{ij}(u) \right) \right) \, .
\end{aligned}$$

Define for all $\alpha, \beta \in \tilde{E}, \alpha \neq \beta$, the point process $\tilde{N}_{\alpha\beta}$ by

$$\tilde{N}_{\alpha\beta}(0,t] := \int_{(0,t]} \sum_{i\in\alpha} \left(1_{\{X(s-)=i\}} \left(\sum_{j\in\beta} \mathrm{d}N_{ij}(s)\right)\right) + \int_{(s,t]} 1_{\{\tilde{X}(s-)\neq\alpha\}} \mathrm{d}\hat{N}_{\alpha,\beta}(s),$$

where the (dummy) point processes $\{\hat{N}_{\alpha\beta}\}_{\substack{\alpha,\beta\in\tilde{E}\\\alpha\neq\beta}}$ form an independent family of HPPs with respective intensities $\{\tilde{q}_{\alpha\beta}\}_{\substack{\alpha,\beta\in\tilde{E}\\\alpha\neq\beta}}$, and are independent of $X(s)$ and $\{N_{ij}\}_{\substack{i,j\in E\\i\neq j}}$. Then

$$f(\tilde{X}(t)) = f(\tilde{X}(s)) + \sum_{\substack{\alpha,\beta\in\tilde{E}\\\alpha\neq\beta}} (f(\beta)-f(\alpha)) \int_{(s,t]} 1_{\{\tilde{X}(u-)=\alpha\}} \mathrm{d}\tilde{N}_{\alpha\beta}(u). \quad (13.84)$$

Suppose (the proof comes later) that $\{\tilde{N}_{\alpha,\beta}\}_{\substack{\alpha,\beta\in\tilde{E}\\\alpha\neq\beta}}$ is a family of independent HPPs with respective intensities $\{\tilde{q}_{\alpha\beta}\}_{\substack{\alpha,\beta\in\tilde{E}\\\alpha\neq\beta}}$. It follows from (13.84) that $\{\tilde{X}(t)\}_{t\geq0}$ is a homogeneous Markov chain with transition matrix $\tilde{\mathbf{A}}$.

To prove the assumed result concerning $\{\tilde{N}_{\alpha\beta}\}$, we apply Watanabe's theorem with the family $\mathcal{Y}$ reduced to the single random variable $X(0)$. Take $Z(t)$ to be as specified in Theorem 13.5.7, where the family $\{N_i\}$ is now $\{\tilde{N}_{\alpha\beta}\}$. We obtain

$$\mathrm{E}\left[\int_{(0,T]} Z(t)\mathrm{d}\tilde{N}_{\alpha\beta}(t)\right]$$
$$= \sum_{i\in\alpha}\sum_{j\in\beta} \mathrm{E}\left[\int_{(0,T]} Z(t)1_{\{X(t-)=i\}}\mathrm{d}N_{ij}(t)\right] + \mathrm{E}\left[\int_{(0,T]} Z(t)1_{\{\tilde{X}(t-)\neq\alpha\}}\mathrm{d}\hat{N}_{\alpha\beta}\right],$$

and by the smoothing formula, this quantity equals

$$\sum_{i\in\alpha}\sum_{j\in\beta} \mathrm{E}\left[\int_0^T Z(t)1_{\{X(t-)=i\}}q_{ij}\,\mathrm{d}t\right] + \mathrm{E}\left[\int_0^T Z(t)1_{\{\tilde{X}(t)\neq\alpha\}}\tilde{q}_{\alpha\beta}\,\mathrm{d}t\right]$$
$$= \mathrm{E}\left[\int_0^T Z(t)\left[\left(\sum_{j\in\beta} q_{ij}\right)1_{\{\tilde{X}(t-)=\alpha\}} + \tilde{q}_{\alpha\beta}1_{\{\tilde{X}(t-)\neq\alpha\}}\right]\mathrm{d}t\right]$$
$$= \mathrm{E}\left[\int_0^T Z(t)\tilde{q}_{\alpha\beta}\,\mathrm{d}t\right],$$

where we used (13.83). $\qquad\qquad\Box$

13.6   Exercises

Exercise 13.6.1. FORWARD AND BACKWARD RECURRENCE TIMES OF AN HPP
Let $\{T_n\}_{n\geq 0}$ be an HPP on $\mathbb{R}_+$ with intensity $\lambda > 0$. For fixed $t > 0$, define $T_-(t) := \sup_{n\geq 0}\{T_n; T_n \leq t\}$ and $T_+(t) := \inf_{n\geq 0}\{T_n; T_n > t\}$. Define the *forward recurrence time* $F(t) = T_+(t) - t$ and the *backward recurrence time* $B(t) := t - T_-(t)$. Give the probability distribution of the vector $(B(t), F(t))$, and see what happens when $t \to \infty$. In particular, compare $\lim_{t\to\infty} \mathrm{E}[B(t) + F(t)]$ with $\mathrm{E}[S_n] = \mathrm{E}[T_n - T_{n-1}]$ for fixed $n \geq 1$.

Exercise 13.6.2. APPROXIMATION OF AN HPP
On the interval $[0, T]$ place n points independently and uniformly. Compute for the disjoint intervals $[a_1, b_1], \ldots, [a_k, b_k] \subset [0, T]$ the distribution of $(X_1, \ldots, X_k)$, where X_j is the number of points on $[a_j, b_j]$. What does this distribution become when n and T tend simultaneously to infinity in such a way that $\frac{n}{T}$ remains constant $(= \lambda > 0)$ and the a_i and b_i remain fixed?

Exercise 13.6.3. CHANGE OF INTENSITY OF AN HPP VIA A CHANGE OF PROBABILITY
Let N be an HPP with intensity $\lambda > 0$ defined on the probability space $(\Omega, \mathcal{F}, \mathrm{P})$. Define a set function P' on $\mathcal{F}$ by

$$\mathrm{P}'(A) = \mathrm{E}_\mathrm{P}[L_T 1_A],$$

where E_P refers to expectation with respect to P and

$$L_t := \left(\frac{\lambda'}{\lambda}\right)^{N(t)} \exp\{-(\lambda' - \lambda)t\},$$

where $\lambda' > 0$. Show that P' is a probability, and that under P', N is an HPP with intensity λ'. Show that for all $t \geq 0$,

$$L_t = 1 + \int_{(0,t]} L_{s-} \left(\frac{\lambda'}{\lambda} - 1\right) (\mathrm{d}N(s) - \lambda\mathrm{d}s).$$

Exercise 13.6.4. CONTINUITY OF THE TRANSITION SEMI-GROUP
Show that a transition semi-group that is continuous at the origin is continuous at all times $t \geq 0$. Show that the transition semi-group of a regular jump HMC is continuous.

Exercise 13.6.5. THE SIMPLEST NON-TRIVIAL UNIFORM HMC

Consider the uniform Markov chain with state space $E = \{0, 1\}$; transition matrix $K = \begin{pmatrix} 1 - \alpha & \alpha \\ \beta & 1 - \beta \end{pmatrix}$, where $\alpha, \beta \in (0, 1)$; and intensity $\lambda > 0$ for the underlying HPP. Find the transition semi-group $\{\mathbf{P}(t)\}_{t \geq 0}$. Suppose that $X(0) = 0$. Give the joint probability density of $(\tau_1, \tau_2 - \tau_1, \ldots, \tau_n - \tau_{n-1})$, where $\tau_1, \tau_2, \ldots$ are the successive times when $\{X(t)\}_{t \geq 0}$ switches from one value to the other.

Exercise 13.6.6. A VERY TRIVIAL TRANSITION SEMI-GROUP

Let π be any probability distribution on the countable state space E, and for each $t \in \mathbb{R}_+$, let $X(t)$ be distributed according to π. Also suppose that the family $\{X(t)\}_{t \geq 0}$ is independent. Show that $\{X(t)\}_{t \geq 0}$ is a homogeneous Markov chain. Give its transition semi-group. What about its local characteristics? In particular, is $q_{ij} < \infty$? How does this fit with the main results of Section 13.2?

Exercise 13.6.7. FLIP-FLOP

Show that the flip-flop process is a special case of a uniform HMC. Show that its infinitesimal generator is

$$\mathbf{A} = \begin{pmatrix} -\lambda & \lambda \\ \lambda & -\lambda \end{pmatrix}.$$

Show that the infinitesimal generator of a Poisson counting process of intensity λ is

$$\mathbf{A} = \begin{pmatrix} -\lambda & \lambda & 0 & 0 & \cdots \\ 0 & -\lambda & \lambda & 0 & \cdots \\ 0 & 0 & -\lambda & \lambda & \cdots \\ \vdots & \vdots & \vdots & & \end{pmatrix}.$$

Exercise 13.6.8. THE UNIFORM HMC

Prove that the stochastic process in Definition 13.2.4 is indeed an HMC.

Exercise 13.6.9. PROVE (13.11)

Prove (13.11).

Exercise 13.6.10. PURE BIRTH PROCESSES

Let $\{X(t)\}_{t\geq 0}$ be a *pure birth process*, that is, a regular jump HMC with state space $E = \mathbb{N}$ and infinitesimal generator of the form

$$
\mathbf{A} = \begin{pmatrix}
-\lambda_0 & \lambda_0 & 0 & 0 & \cdots \\
0 & -\lambda_1 & \lambda_1 & 0 & 0 & \cdots \\
0 & 0 & -\lambda_2 & \lambda_2 & 0 & \cdots \\
\vdots & \vdots & \vdots & \vdots & \vdots
\end{pmatrix}.
$$

Suppose that $\lambda_n > 0$ for all $n \geq 0$. Compute $P_0(X(t) = n)$ for $n \geq 0$, $t \geq 0$.

Exercise 13.6.11. MEAN AND VARIANCE OF SOME BIRTH-AND-DEATH PROCESS

Let $\{X(t)\}_{t\geq 0}$ be a birth-and-death process with parameters $\lambda_n = n\lambda + a$ and $\mu_n = n\mu 1_{n\geq 1}$, where $\lambda > 0, \mu > 0, a > 0$. Suppose that $X(0) = i$ is fixed. Show that $M(t) = \mathrm{E}[X(t)]$ satisfies a differential equation and solve this equation. Do the same for the variance $S(t) = \mathrm{E}[X(t)^2] - M(t)^2$.

Exercise 13.6.12. NON-EXPLOSION, TAKE 1

(i) Let $\mathbf{A}$ be an essential generator on E. Define $\mathbf{P} = \{p_{ij}\}_{i,j\in E}$ by $p_{ii} = 0$, $p_{ij} = \frac{q_{ij}}{q_i}$ if $i \neq j$. Show that if the transition matrix $\mathbf{P}$ is irreducible and recurrent, the generator is non-explosive.

(ii) Same question only assuming that the state space is finite ($\mathbf{P}$ is not supposed irreducible).

(iii) Same question without irreducibility and recurrence assumptions, but assuming $\sup_{i\in E} q_i < \infty$.

Exercise 13.6.13. AVERAGE TIME TO REACH A GIVEN LEVEL

Let $\{X(t)\}_{t\geq 0}$ be a birth-and-death process with parameters $\lambda_n > 0$ and $\mu_n > 0$. Let w_n be the average time required to pass from state n to state $n + 1$. Find a recurrence equation for $\{w_n\}_{n\geq 0}$. Deduce from it the average time required to reach state $n \geq 1$ when starting from state 0.

Exercise 13.6.14. A NON-EXPLOSIVE BIRTH-AND-DEATH PROCESS

Show that the birth-and-death generator with birth-and-death parameters $\lambda_0 = \lambda$, $\lambda_n = n\lambda$ ($n \geq 1$) and $\mu_n = n\mu$ ($n \geq 1$), where λ and μ are positive, is non-explosive. When is such a chain positive recurrent?

Exercise 13.6.15. RECURRENCE OF A BIRTH-AND-DEATH PROCESS
Consider an irreducible recurrent birth-and-death process with state space $\mathbb{N}$. Discuss the nature of recurrence (positive or null) of this process and of its embedded chain in terms of the birth-and-death parameters.

Exercise 13.6.16. REVERSIBILITY OF THE BIRTH-AND-DEATH PROCESS
Show that an irreducible ergodic birth-and-death process in steady state is reversible.

Exercise 13.6.17. THE SKELETON
Show that the skeleton of an irreducible regular jump HMC is irreducible. Prove that the skeleton of a regular jump HMC cannot be periodic.

Exercise 13.6.18. PROVE (13.59)
Prove (13.59) formally.

Exercise 13.6.19. PROOFS OF THEOREMS 13.4.14 AND 13.4.15
Give the missing details in the proofs of Theorems 13.4.14 and 13.4.15.

Exercise 13.6.20. ON SIMULATION
The regenerative structure of a regular jump HMC suggests that to simulate such a chain, one needs to generate *two* random variables per transition: One fixes the time of occurrence of the event, and the other one determines what the next state is. Show that for a finite-state regular jump HMC, one really needs to generate one single exponential random variable per transition *in the long run*. Discuss the feasibility when the state space is large in terms of memory requirements.

Exercise 13.6.21. HPPS INSIDE A CONTINUOUS-TIME HMC
Let $\{X(t)\}_{t\geq 0}$ be an irreducible ergodic regular jump HMC, and take for initial distribution $\mu = \pi$, the stationary distribution. Suppose moreover that with each state $i \in E$ is associated a subset $T(i) \subset E$ such that $\sum_{j\in T(i)} q_{ij} = \lambda$. Let N be the counting process recording all transition times τ_n such that $X(\tau_n) \in T(X(\tau_n-))$. Show that N is a Poisson process with intensity λ.

Chapter 14

Markovian Queueing Theory

14.1 Poisson Systems

14.1.1 The Purely Poissonian Description

Most continuous-time Markov chains arising in operations research receive a natural description in terms of a mapping transforming an *input* into an output called the *state process*. In *Poisson systems*, the input consists of HPPs, also called the *driving* HPPs.

The construction of the state process from the basic driving HPPs is similar to the one used in Section 13.5 for jump HMCs, only with more details corresponding to a "natural" description of the system considered.

The rationale behind this construction is that there are sources of events that can be either active or inactive depending on the state of the process. Active sources then compete to produce an event that will cause a transition.

We shall now proceed with the precise definitions.

Let S and E be two countable sets, the set of *event sources* (*sources* for short) and the set of *states*, respectively. With each source-state pair (s, i) is associated a number $c(s, i) \geq 0$ measuring the *intensity of activity* of source s when the state is i. An event source s is said to be *active in state* i if $c(s, i) > 0$. We denote by $A(i)$ the collection of sources s active in state i. With each source-state pair (s, i) is associated a probability distribution $p(s, i, \cdot)$ on E.

The four objects E, S, $p(\cdot, \cdot, \cdot)$ and $c(\cdot, \cdot)$ constitute the *transition mechanism*.

The collection T of triples (s, i, j) such that $c(s, i)p(s, i, j) > 0$ is called the set of *allowed transitions*. If $(s, i, j) \in T$, one says that a transition $i \to j$ can be *triggered* by an event produced by source s.

© Springer Nature Switzerland AG 2020
P. Brémaud, *Markov Chains*, Texts in Applied Mathematics 31,
https://doi.org/10.1007/978-3-030-45982-6_14

With each source $s \in S$ is associated an intensity, or *rate*, $\lambda_s > 0$. Let $\{N_{s,i,j}\}_{(s,i,j)\in T}$ be a family of independent HPPs with respective intensities $\{a_{s,i,j}\}_{(s,i,j)\in T}$ where

$$a_{s,i,j} := \lambda_s c(s, i) p(s, i, j).$$

Let for all pairs of states (i, j)

$$a_{ij} := \sum_{s\in S} \lambda_s c(s, i) p(s, i, j)$$

and

$$a_i := \sum_{j\in E} a_{ij}.$$

We shall suppose that (*stability assumption*)

$$a_i < \infty \quad (i \in E).$$

We first give the purely Poissonian description of the state of the system, which is in general not directly readable from the natural description of the system one wishes to model, but which has the advantage of immediately leading to the infinitesimal generator.

The *state process* $\{X(t)\}_{t\geq 0}$ with values in $E_\Delta = E \cup \{\Delta\}$, where Δ is an element outside E, is constructed as follows.

The initial state $X(0)$ is independent of the HPPs $\{N_{s,i,j}\}_{(s,i,j)\in T}$. Suppose the state at time t is $X(t) = i \in E$. The active sources are then all the sources $s \in A(i)$. The HPPs competing for the determination of the next transition after t are $\{N_{s,i,j}\}_{s\in A(i), p(s,i,j)>0}$. If N_{s_0,i,j_0} is the HPP producing the first event strictly after t, say at time $t + \tilde{\tau}$, then a transition from state i to state j_0 occurs at time $t + \tilde{\tau}$. The stability condition $a_i < \infty$ ensures that $\tilde{\tau} > 0$, since a_i is the sum of the intensities of the $N_{s,i,j}$ in competition when the state is i.

This recursive procedure started from time 0 gives a sequence of transition times $\tilde{\tau}_1, \tilde{\tau}_2, \ldots$. Possibly, there is a random index N such that $\tilde{\tau}_N = \infty$, in which case we let $\tilde{\tau}_k = \infty$ for $k \geq N$. The embedded process is $\tilde{X}_1, \tilde{X}_2, \ldots$, where $\tilde{X}_n = X(\tilde{\tau}_n)$, with the convention $X(\infty) = \Delta$. This procedure defines $X(t)$ for all $t \in [0, \tilde{\tau}_\infty)$, where $\tilde{\tau}_\infty = \lim_{n\uparrow\infty} \tilde{\tau}_n$ is called the *explosion time*. For $t \geq \tilde{\tau}_\infty$, we set $X(t) = \Delta$.

The same arguments as in Section 13.5 show that $\{X(t)\}_{t\geq 0}$ is a strong Markov jump HMC with state space E_Δ. From now on, we shall assume that $\tilde{\tau}_\infty = \infty$, and therefore that $\{X(t)\}_{t\geq 0}$ is a regular jump HMC on E. (In all the queueing networks that we shall consider, this assumption can be checked easily.)

Infinitesimal Generator of a Poisson System

We now focus on the first transition at time $\tilde{\tau}_1$, which is typical of all other transitions. Suppose that $X(0) = i$. The competing processes are the $N_{s,i,j}$ where $s \in A(i)$ and j satisfies $p(s,i,j) > 0$. If one is not interested in the particular source triggering the transition, one can lump all the $N_{s,i,j}$, $s \in A(i)$, together to form a single HPP N_{ij} of intensity a_{ij}. We are then in the same situation as in Section 13.5, with one exception: a_{ii} may be strictly positive. This motivates the notations $\tilde{\tau}_n$ instead of τ_n, $\tilde{X}_n$ instead of X_n, a_{ij} instead of q_{ij}, etc., which are there to warn us that a *pseudo-transition*, that is, a "transition" from i to i may occur. Otherwise, the situation is similar to that of Section 13.5, and we can state the following result, defining $\tilde{X}_0 = X(0)$:

(α) $\{\tilde{X}_n\}_{n\geq 0}$ is a discrete-time HMC with values in $E_\Delta = E \cup \{\Delta\}$ and transition matrix $\{\tilde{p}_{ij}\}_{i,j\in E_\Delta}$ given by $\tilde{p}_{\Delta\Delta} = 1$, $\tilde{p}_{i\Delta} = 1$ if $a_i = 0$, and if $i \in E$, $a_i > 0$, $j \in E$,

$$\tilde{p}_{ij} = \frac{a_{ij}}{a_i}.$$

(β) Given $\{\tilde{X}_n\}_{n\geq 0}$, the sequence $\{\tilde{\tau}_{n+1} - \tilde{\tau}_n\}_{n\geq 0}$ is independent, and for all $n \geq 0$, $i, i_0, i_1, \cdots, i_{n-1} \in E$, $x \in \mathbb{R}_+$,

$$P(\tilde{\tau}_{n+1} - \tilde{\tau}_n \geq x \mid \tilde{X}_n = i, \tilde{X}_{n-1} = i_{n-1}, \cdots, \tilde{X}_0 = i_0) = e^{-a_i x}.$$

So far there is nothing new with respect to the construction of Section 13.5 except that we allow pseudo-transitions which can play a role in some situations, in feedback queues, for instance (see Example 14.1.5 below). Let us denote by $\{\tau_n\}_{n\geq 0}$ the sequence of *true transitions* of the state process (that is, from some state to a different state) and by $\{X_n\}_{n\geq 0}$ the embedded process at the true transition times. If we are interested only in true transitions, it suffices to get rid of all the HPPs $N_{s,i,i}$ because they do not affect the state process. We therefore retrieve the situation of Section 13.5 with

$$q_{ij} = \sum_{s\in S} \lambda_s c(s,i) p(s,i,j), \tag{14.1}$$

where $j \neq i$, and

$$q_i = \sum_{j\in E, j\neq i} \sum_{s\in S} \lambda_s c(s,i) p(s,i,j). \tag{14.2}$$

14.1.2   Markovian Queues as Poisson Systems

One of the simplest examples of a Poisson system is the M/M/1/∞ *queue*, a model of a waiting line that we proceed to describe. In front of a ticket booth with a single attendant, or *server*, customers wait in line. The facility is so large that no bound is imposed on the number of customers waiting for service. In other words, the *waiting room* has *infinite capacity*. Such a system is called a 1/∞ *service system*, where 1 indicates the number of servers and ∞ is the capacity of the waiting room.

Customer arrivals are modeled by a homogeneous Poisson process $\{T_n\}_{n\geq 1}$ of intensity $\lambda > 0$. Customer n arriving at time T_n brings a service request σ_n, which means that the server will need σ_n units of time to process the request of customer n. The sequence $\{\sigma_n\}_{n\geq 1}$ is assumed IID, with exponential distribution of mean μ^{-1}. Also, the arrival sequence $\{T_n\}_{n\geq 1}$ and the service sequence $\{\sigma_n\}_{n\geq 1}$ are assumed to be independent. Such a pattern of arrivals is called an *M/M input*. In this notation[1] M means "Markovian" because the Poisson process is Markovian and exponential distributions are intimately connected with the Markov property.

The server attends one customer at a time and does not remain idle as long as there is at least one customer in the *system* (ticket booth plus waiting room). Once the service of a customer is started it cannot be interrupted before completion.

The above system is called an *M/M/1/∞ queue*. Its description could be complemented by an indication of the *service discipline* used: for instance, FIFO (*first in first out*), where the server, after completion of a service, chooses his next customer at the head of the line. However, we shall see that the service discipline turns out to be irrelevant if one is interested in the congestion process counting the number of customers present in the system (waiting line and ticket booth).

We now describe the *M/M/1/∞*/FIFO queue as a Poisson system.

For this, we take $S = \{\alpha, \delta\}$, where α stands for *arrival* and δ for *departure*. The state space is $E = \mathbb{N} = \{0, 1, 2, \ldots\}$, the state $i \in E$ representing the number of customers present in the system (in the waiting line or being attended by the server). If $i > 0$, $A(i) = \{\alpha, \delta\}$, meaning that when the number of customers in the system is strictly positive, an arrival or a departure could occur. If $i = 0$, $A(0) = \{\alpha\}$ (if the system is empty, one cannot expect a departure, only an arrival).

All the non-null activity intensities are taken equal to 1. Also, $\lambda_\alpha = \lambda$ and $\lambda_\delta = \mu$, because we want the inter-arrival times to be exponential random variables with mean λ^{-1} and the service requests to be exponential with mean μ^{-1}.

---

[1]This notation and similar ones were introduced by [Kendall, 1953].

If the state $i \geq 0$, a transition is triggered by an event of source $s_0 = \alpha \in A(i)$, the next state is $j = i + 1$ (an arrival increases the number of customers by 1); whereas in state $i > 0$, if $s_0 = \delta \in A(i)$, the next state is $j = i - 1$. Thus, $p(\alpha, i, i+1) = 1$ for all $i \geq 0$, and $p(\delta, i, i-1) = 1$ for all $i > 0$, and all other probabilities $p(s, i, j)$ are null.

Figure 14.1.1 depicts a typical evolution of the state process, starting from state $i_0 = 0$.

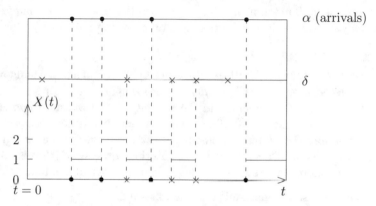

Figure 15.3.2. GSMP construction of M/M/1/∞/FIFO

Formula (14.1) makes the infinitesimal generator readable from the natural description in terms of the rates λ_s, the speeds $c(s, i)$, and the transition probabilities $p(s, i, j)$. For the above M/M/1/∞/FIFO queue, it can be readily checked that

$$q_{i,i+1} = \lambda, \quad q_{i,i-1} = \mu 1_{\{i \geq 0\}}.$$

Remark 14.1.1 Note that even for the simplest example – the M/M/1/∞ queue – the Poisson system description differs from the usual regenerative description of the corresponding Markov chain. The Poisson system description gives all the details about the mechanism of generation of transitions, and its collapse into the usual regenerative description is accompanied by a loss of information concerning the fine details. In fact, one cannot distinguish statistically between the state process of an M/M/1/∞/FIFO queue that has an infinitesimal generator of the form

$$\mathbf{A} = (\lambda + \mu)(\mathbf{K} - I),$$

where

$$\mathbf{K} = \begin{pmatrix} 0 & 1 & 0 & 0 & \cdots \\ \frac{\mu}{\lambda+\mu} & 0 & \frac{\lambda}{\lambda+\mu} & 0 & \cdots \\ 0 & \frac{\mu}{\lambda+\mu} & 0 & \frac{\lambda}{\lambda+\mu} & \cdots \\ \vdots & \vdots & \vdots & \vdots & \end{pmatrix},$$

from the uniform Markov chain constructed from a Poisson process of intensity $\lambda + \mu$ and a random walk with reflecting barrier 0 with the transition matrix $\mathbf{K}$. It is actually not easy to see from the uniform description that one is dealing with an $M/M/1/\infty$ queue. This difficulty increases with the size and the complexity of the system as we shall see in the Section 14.3 on queueing networks.

EXAMPLE 14.1.2: $M/M/K/0$ (THE ERLANG QUEUE). This queueing system has the same description as the $M/M/1/\infty$/FIFO queue except for the following features. There are $K \geq 1$ servers, and the capacity of the waiting room is 0. In particular, an arriving customer finding the K servers busy is not accepted in the system, and therefore the number of customers present in the system at any given time is less than or equal to K. A customer entering the system (that is, arriving and seeing one or more servers idle) will select a free server at random.

One possible transition mechanism is the following:

$$\begin{aligned} S &= \{0, 1, \ldots, K\}, \\ E &= \{\text{all subsets of } S \text{ containing } 0\}, \\ A(i) &= \{\text{all event sources in } i\}, \\ c(s, i) &= 1 \text{ for all } s \in A(i). \end{aligned}$$

The transition probabilities $p(\cdot, \cdot, \cdot)$ will be described in a few lines. The Poisson processes corresponding to event sources $1, 2, \ldots, K$ all have the same intensity $\mu > 0$, whereas the Poisson process of event source 0 has intensity $\lambda > 0$. Source 0 corresponds, as in the previous example, to the arrivals.

For any state $i \in E, i - \{0\}$ is a subset of $\{1, \ldots, K\}$, representing the servers that are busy when the system is in that state. If $X(t) = i$, the number of customers in the system is $|i| - 1$, where $|i|$ is the cardinality of the set i. An event source $s \in [1, K]$ corresponds to server s, and the sequence $S_n^s = T_n^s - T_{n-1}^s (n \geq 1)$ is the sequence of successive service times provided by this server.

In order to obtain the transition probabilities $p(\cdot, \cdot, \cdot)$, two states will be distinguished: $i_0 = \{0\}$ and $i_K = \{0, 1, \ldots, K\}$ corresponding respectively to an empty and a full system.

If $i \neq i_0$ and a transition is triggered on event source $s \in i$ where $s > 0$, this means that server s releases a customer, and the next state is then $i - \{s\}$.

If $i \neq i_K$ and a transition is triggered on event source 0 (which means that a new customer arrives), the next state is $i + \{r\}$, where r is chosen at random in $S - i$, the set of idle servers in state i.

If $i = i_K$ and a transition is triggered on event source 0 (a new customer arrives), the state does not change (the new customer is not accepted, all servers being busy).

The *M/M/K/0 queue* is also called the *Erlang queue* after its inventor, a Danish telephone engineer who used it to quantitatively study a *telephone switch* with K *lines*, or *channels*. Each customer finding a free line is connected and uses it for the time of a conversation. A customer finding all channels busy is rejected, or in the best case routed to another switch. Erlang was able to obtain his famous *blocking formula* giving the probability in stationary regime that a given customer will find all lines busy. This is the first formula of queueing theory (Example 14.2.2 below).

The infinitesimal generator can be computed from formula (14.1) and (exercise) this gives for the non-null terms

$$q_{i,i+\{r\}} = \lambda \frac{1}{K+1-|i|} \qquad \text{if } i \in E, r \in \{1,\dots,K\}, r \notin i,$$
$$q_{i,i-\{r\}} = \mu \qquad \text{if } i \in E, r \in \{1,\dots,K\}, r \in i.$$

Defining $Q(t) = |X(t)| - 1$ (the cardinality of $X(t)$ minus 1, that is, the number of busy servers), we can apply Theorem 13.5.10 to prove that this is a regular jump HMC with state space $\tilde{E} = \{0, 1, \dots, K\}$ and infinitesimal generator $\tilde{A}$ given by

$$\tilde{q}_{n,n+1} = \lambda \text{ if } n \in [0, K-1],$$
$$\tilde{q}_{n,n-1} = n\mu \text{ if } n \in [1, K].$$

A given state $n \in \tilde{E}$ is obtained by grouping the states $i \in E$ such that $|i| - 1 = n$.

In order to show that $\tilde{q}_{n,n-1} = \mu n$ $(1 \le n \le K)$, consider the transitions of $\{X(t)\}$ from i such that $|i| - 1 = n$ to j such that $|j| - 1 = n - 1$. Fixing i such that $|i| - 1 = n$, there are exactly n states j such that $q_{ij} > 0$ and $|j| - 1 = n - 1$, namely all states j of the form $j = i - \{r\}$, where $r \in \{1, \dots, K\}$ and $r \in i$. The corresponding sum $\sum_j q_{ij} = n\mu$ and is independent of i such that $|i| - 1 = n$. Thus condition (13.83) of Theorem 13.5.10 is satisfied for $\alpha = n, \beta = n - 1$. The proof of $\tilde{q}_{n,n+1} = \lambda$ is similar.

EXAMPLE 14.1.3: $M/M/1/\infty/$LIFO PREEMPTIVE RESUME. This queue has a description similar to that of an $M/M/1/\infty/$FIFO, except for the LIFO discipline (*last in first out*) which is moreover of the *preemptive resume* type. More precisely: a customer upon arrival goes right to the ticket booth, and the customer who was receiving service is sent back to the waiting room to stand in front of the line (at least until the time when another rude customer shows up, sending the first rude customer to the front of the queue, and so on). This type of service discipline is called *preemptive*. The phrase preemptive *resume* means that a preempted customer does not have to start from scratch: when the server sees him next time, he will resume work where it was left.

The construction of an $M/M/1/\infty/$LIFO *preemptive resume* as a Poisson system is as follows

$$
\begin{aligned}
S &= \mathbb{N}, \\
E &= \{\text{all finite subsets of } S \text{ containing } 0\}, \\
A(i) &= \{0, i_{n(i)}\}, \text{where } i = \{0, i_1, i_2, \ldots, i_{n(i)}\} \in E, \\
c(s, i) &= 1 \text{ if } s \in A(i), \\
p(0, i, i + \{i_{n(i)} + 1\}) &= 1, \quad p(i_{n(i)}, i, i - \{i_{n(i)}\}) = 1.
\end{aligned}
$$

Source 0 is the arrival source. The last prescription is that when a new customer shows up (that is, a transition is triggered on source 0), the state being i, a new source is added to i to form the next state, the one just above the set i, and this source becomes immediately active. When source $i_{n(i)}$ triggers an event, this means a departure from the queue, and the source $i_{n(i)}$ disappears from state i, and the source $i_{n(i)-1}$ is reactivated. The HPP $\{T_n^0\}_{n \geq 1}$ has intensity $\lambda > 0$, and for all $k \geq 1, \{T_n^k\}_{n \geq 1}$ has intensity $\mu > 0$. When $k \geq 1$, the inter-event times in $\{T_n^k\}_{n \geq 1}$ correspond to service times.

We see that in state i, there are $n(i)$ customers in the system. Thus the congestion process at time t is $|X(t)| - 1$, where $|X(t)|$ is the cardinality of the *set* $X(t)$.

Remark 14.1.4 The preemptive resume discipline is not as unfair as it may appear. First of all, all customers being equally rude, each one endures as much as he hurts the others. A customer with a large service request spends a larger time at the ticket booth and is therefore more exposed than a customer with a modest request, and it is precisely in this sense that the discipline is fair. It makes longer requests which are responsible for congestion wait longer in the system. The precise result is that the expected sojourn time of a customer in the system, given that its service request is x, is equal to $x/(1-\rho)$, where $\rho = \lambda/\mu$ ([2]).

In the queueing literature, especially when applications to communications networks or computer networks are considered, a queueing system is represented by the pictogram of Figure 14.1.2, where the input arrow represents the arrival stream of *jobs* (customers), the output arrow represents the stream of *completed jobs* (served customers), the circle is a *processor* (the service system), and the stack is a *buffer* (a waiting room, where customers wait for a server to be free).

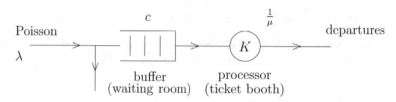

Figure 14.1.2. The holy pictogram of queueing theory

This basic pictogram can be richly adorned. For instance, in Figure 14.1.2 we have a system with K servers, a waiting room of capacity c, λ is the arrival rate of customers, and the derivation shows that the customers finding all K servers busy and a full waiting room are rejected. Also, there is an indication of the average service time, $1/\mu$, and of the fact that the incoming stream is Poisson. In this pictogram, one would take it as implicit that the service times sequence is IID and independent of the arrival process, unless other assumptions are explicitly mentioned. One sometimes also gives the service discipline (LIFO, FIFO, etc.).

EXAMPLE 14.1.5: $\mathrm{M}/\mathrm{M}/1/\infty/$FIFO QUEUE WITH INSTANTANEOUS FEEDBACK. Although one can add and suppress pseudo-transitions at will without altering the state process, pseudo-transitions are not always meaningless. For instance, consider an $\mathrm{M}/\mathrm{M}/1/\infty$ queue with *instantaneous feedback*, where a customer finishing service either leaves the system with probability $1-p$ or is immediately recycled with probability p at the end of the waiting line or at the service booth if there is an empty waiting line, with a new independent exponential service request (see Figure 14.1.3). Then the times of service completion of recycled customers do not correspond to a genuine transition of $\{X(t)\}_{t\geq 0}$, where $X(t)$ is the number of customers present in the system at time t.

---

[2]See [Wolff, 1989].

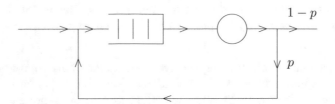

Figure 14.1.3. A queue with instantaneous feedback

14.2   Isolated Markovian Queues

14.2.1   As Birth-and-Death Processes

This section is an introduction to queueing theory, which finds applications in operations research as well as in the performance analysis of communications networks.

For the M/M/1/∞ queue and the M/M/K/0 queue, the congestion processes $\{X(t)\}_{t\geq 0}$ recording the evolution of the number of customers in the system are special cases of birth-and-death processes with an infinitesimal generator of the form

$$
\mathbf{A} = \begin{pmatrix}
-\lambda_0 & \lambda_0 & 0 & 0 & \cdots \\
\mu_1 & -(\lambda_1 + \mu_1) & \lambda_1 & 0 & \cdots \\
0 & \mu_2 & -(\lambda_2 + \mu_2) & \lambda_2 & \cdots \\
\vdots & \vdots & \vdots & \vdots &
\end{pmatrix},
$$

where the state space is $E = \mathbb{N}$, or $E = \{0, 1, \ldots, N\}$ for finite N. In any case, for all the queueing processes we are going to consider, $\lambda_i > 0$ for all $i \in E$ except $i = N$ when $E = \{0, 1, \ldots, N\}$, and $\mu_i > 0$ for all $i \in E$ except $i = 0$. These conditions guarantee irreducibility.

The global balance equations are

$$
\lambda_0 \pi(0) = \mu_1 \pi(1)
$$

and

$$
(\lambda_i + \mu_i)\pi(i) = \lambda_{i-1}\pi(i-1) + \mu_{i+1}\pi(i+1) \quad (i \geq 1),
$$

with the convention $\mu_{N+1} = 0$ if $E = \{0, 1, \ldots, N\}$. In Example 13.2.22, a stationary distribution π was shown to exist if and only if

$$
\sum_{\substack{i \in E \\ i \geq 1}} \prod_{n=1}^{i} \frac{\lambda_{n-1}}{\mu_n} < \infty, \tag{14.3}
$$

in which case, for $i \geq 1$,

$$\pi(i) = \pi(0) \prod_{n=1}^{i} \frac{\lambda_{n-1}}{\mu_n}$$

and

$$\pi(0) = \left(1 + \sum_{\substack{i \in E \\ i \geq 1}} \prod_{n=1}^{i} \frac{\lambda_{n-1}}{\mu_n} \right)^{-1}.$$

The ergodicity condition (14.3) is, of course, automatically satisfied when the state space is finite.

EXAMPLE 14.2.1: M/M/1/∞. Here $E = \mathbb{N}$, $\lambda_i = \lambda > 0$ and $\mu_i = \mu > 0$ for all $i \geq 1$. The ergodicity condition reads $\sum_{i \geq 1} \left(\frac{\lambda}{\mu} \right)^i < \infty$, that is,

$$\rho := \frac{\lambda}{\mu} < 1.$$

This condition says that $\rho = \lambda E[\sigma_1]$, the *traffic intensity*, which is the average rate of work entering the system per time unit should not exceed the maximal speed of service, equal to 1. The solution of the balance equation is

$$\pi(i) = (1 - \rho)\rho^i.$$

EXAMPLE 14.2.2: M/M/K/0 (ERLANG LOSS SYSTEM) Here $E = \{0, \dots, K\}$. The solution of the balance equations is

$$\pi(i) = \frac{\rho^i / i!}{\sum_{n=0}^{K} \rho^n / n!} \quad (1 \leq i \leq K).$$

In particular,

$$\pi(K) = \frac{\rho^K / K!}{\sum_{n=0}^{K} \rho^n / n!}$$

is the *blocking probability*, the probability of finding the K channels busy (a formula due to Erlang).

The distribution π is the Poisson distribution truncated at K since

$$\pi(i) = \mathrm{P}(Z = i \mid Z \leq K),$$

where Z is a Poisson random variable with mean ρ.

EXAMPLE 14.2.3: M/M/∞/∞ OR PURE DELAY. This is not really a queueing system, but a *pure delay system*. The arrival process, the service times sequence and the waiting room are as in the M/M/1/∞ model, but now there is an infinity of servers and therefore no queueing since anyone entering the system finds an idle server. The state space is $E = \mathbb{N}$, and the parameters are $\lambda_i \equiv \lambda > 0$ and $\mu_i = i\mu$. The form of the birth parameter is due to the fact that when $i > 0$ customers are present in the system, they are all being served, and therefore there are i independent exponential random variables (the service times) of mean μ^{-1} being consumed at speed 1. A transition $i \rightarrow i-1$ will take place as soon as one of them is consumed. If $X(t) = i$, the transition $i \rightarrow i-1$ will therefore occur in the time interval $(t, t+h]$ with probability $i\mu h + o(h)$.

The ergodicity condition (14.3) is $\sum_{i=1}^{\infty} \frac{\rho^i}{i!} < \infty$ and is always satisfied. The solution of the balance equations is the Poisson distribution

$$\pi(i) = \mathrm{e}^{-\rho} \frac{\rho^i}{i!}.$$

EXAMPLE 14.2.4: M/M/K/∞. This is almost the M/M/1/∞ queue, except that there are now K servers. Here $\lambda_i \equiv \lambda > 0$ and $\mu_i = \inf(i, K)\mu$. Indeed, if there are $X(t) = i \leq K$ customers, there are i independent exponentials of mean μ^{-1} active in provoking a downward transition. Therefore, the probability that a downward transition occurs in $(t, t+h]$ is $i\mu h + o(h)$. If there are $X(t) = i > K$ customers, only K exponentials are active, since there are only K servers, and therefore a downward transition occurs in the interval $(t, t+h]$ with probability $K\mu h + o(h)$.

The ergodicity condition (14.3) is satisfied only if

$$\rho := \frac{\lambda}{\mu} < K.$$

It says that the average incoming work per unit time should not exceed the maximal service speed K (when all servers are busy).

The stationary distribution is then

$$\pi(i) = \pi(0)\frac{\rho^i}{i!} \quad (1 \le i \le K)$$

and for $i > K$,

$$\pi(i) = \pi(0)\frac{\rho^K}{K!}\left(\frac{\rho}{K}\right)^{i-K},$$

where

$$\pi(0)^{-1} = \sum_{i=0}^{K-1}\frac{\rho^i}{i!} + \frac{\rho^K}{K!}\frac{1}{1-\rho/K}.$$

In this system, the probability of waiting is the probability of entering the system when the K servers are busy, that is, $\pi(\ge K) := \sum_{i\ge K}\pi(i)$. One obtains *Erlang's waiting formula*

$$\pi(\ge K) = \frac{\frac{\rho^K}{K!}\frac{1}{1-\rho/K}}{\sum_{i=0}^{K-1}\frac{\rho^i}{i!} + \frac{\rho^K}{K!}\frac{1}{1-\rho/K}}.$$

EXAMPLE 14.2.5: M/M/1/∞ PROCESSOR SHARING. The arrival process is of the M/M type, as in the M/M/1/∞ queue, and the waiting-room capacity is infinite. Just as a LIFO non-preemptive discipline or a FIFO discipline does not make a difference in the infinitesimal generator, the processor-sharing discipline does not either. In this discipline the server is equally shared among the customers in the system. More precisely, if at time t there are $X(t) = i$ customers in the system, each one is served at speed $1/i$. For a transition $i \to i - 1$, there are i independent exponentials of mean $1/\mu$, consumed at speed $1/i$. Therefore, the probability of such a transition in the time interval $(t, t + h]$ is $i\frac{\mu}{i}h + o(h) = \mu h + o(h)$.

The last example suggests that the statistics of the congestion process $\{X(t)\}_{t\ge 0}$ are independent of the service discipline in an M/M/1/∞ queue, and this is indeed the case if we consider only service disciplines such that the server works at full speed, equal to 1, whenever there is at least one customer in the system.

14.2.2 The M/GI/1/∞/FIFO Queue

Markovian models are easy to handle, but sometimes they are not adequate. Although the Poissonian assumption for the input is often justified, the exponential assumption for service times is usually unrealistic. Note that there are situations where the exponential assumption leads to the correct result even in the case where the service time distribution is arbitrary. The typical case is Erlang's system, in which the substitution of an arbitrary service distribution with the same mean as the original exponential distribution leaves unaltered the stationary distribution of the number of busy lines at an arbitrary time. In particular, Erlang's blocking formula remains valid, and this robustness explains the success of this formula, which was widely applied in the design of early telephone switches. This lucky phenomenon is called *insensitivity*, and is not so rare. For instance, the M/M/∞ queue and the M/M/1/∞/LIFO *preemptive* queue are insensitive. However, since insensitivity is not a general phenomenon, one must work a little more, and we shall consider in this subsection the M/GI/1/∞/FIFO queue, a non-Markovian queue for which the analysis is possible. This queue is exactly like an M/M/1/∞/FIFO queue, only with a general distribution for the IID service times sequence,

$$P(\sigma_1 \leq x) = G(x).$$

The Embedded Congestion Process

The corresponding congestion process $\{X(t)\}_{t\geq 0}$ is no longer Markovian. Fortunately, the process $\{X_n\}_{n\geq 0}$, where

$$X_n = X(\tau_n)$$

and τ_n is the n-th departure time (Figure 14.2.1), $\tau_0 = 0$, is a discrete-time HMC. This is a direct consequence of the extended strong Markov property of HPPs. This discrete-time chain is called the *chain embedded at departure times*. Since the congestion process is taken as right-continuous, we see that X_n is the number of customers that customer n leaves behind him when he has completed service.

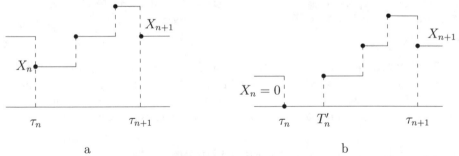

Figure 14.2.1. The chain embedded at departure times

A glance at Figure 14.2.1 reveals that

$$X_{n+1} = (X_n - 1)^+ + Z_{n+1},$$

where Z_{n+1} is the number of customers arriving in the interval $(\alpha_n, \alpha_n + \sigma_{n+1}]$, where $\alpha_n = \tau_n$ if $X_n > 0$ and $\alpha_n = T'_n$ if $X_n = 0$ (Figure 14.2.1).

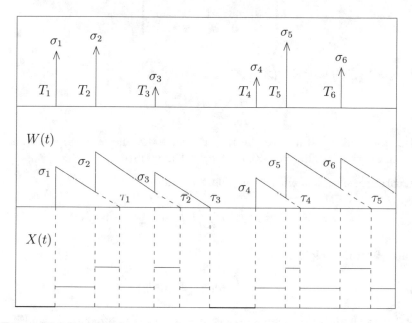

Figure 14.2.2. Arrival process, FIFO workload, and congestion processes

The time α_n is a stopping time with respect to the arrival process $N = N_\alpha$ and the first n service requests $(\sigma_1, \ldots, \sigma_n)$, and the $(n+1)$st service request σ_{n+1} is independent of the arrival process before α_n and of the previous requests $(\sigma_1, \ldots, \sigma_n)$. In particular,

$$Z_{n+1} = N(\alpha_n, \alpha_n + \sigma_{n+1}]$$

is independent of $X_0, \ldots, X_n$, and $\{Z_n\}_{n \geq 1}$ is an IID sequence, with each Z_n distributed as

$$\tilde{Z} = \tilde{N}(0, \tilde{\sigma}],$$

where $\tilde{N}$ is some HPP with intensity λ, and $\tilde{\sigma}$ is a random variable independent of $\tilde{N}$, with cumulative distribution function $G(x)$.

We recognize an avatar of the repair shop example for which we have already proven that $\{X_n\}_{n \geq 0}$ is indeed an HMC with transition matrix

$$\begin{pmatrix} a_0 & a_1 & a_2 & a_3 & \dots \\ a_0 & a_1 & a_2 & a_3 & \dots \\ 0 & a_0 & a_1 & a_2 & \dots \\ 0 & 0 & a_0 & a_1 & \dots \\ \vdots & \vdots & \vdots & \vdots & \end{pmatrix},$$

where $a_i = \mathrm{P}(Z_{n+1} = i) = \mathrm{P}(\tilde{Z} = i)$, that is,

$$a_i = \int_0^\infty \mathrm{e}^{-\lambda t} \frac{(\lambda t)^i}{i!} \mathrm{d}G(t) . \tag{14.4}$$

For this particular instance of the repair shop HMC, the necessary and sufficient condition of irreducibility $a_0 > 0$, $a_0 + a_1 < 1$ is verified.

Recurrence of this chain depends on the value

$$\rho = \mathrm{E}\,[Z_1] = \mathrm{E}[\tilde{Z}] = \sum_{i=1}^\infty i \int_0^\infty \mathrm{e}^{-\lambda t} \frac{(\lambda t)^i}{i!} \mathrm{d}G(t) .$$

A straightforward calculation gives

$$\rho = \lambda \int_0^t t \mathrm{d}G(t) = \lambda \mathrm{E}\,[\sigma] .$$

If $\rho < 1$, the chain is positive recurrent; if $\rho = 1$, it is null recurrent; and if $\rho > 1$, the chain is transient.

Again, this result is natural if we observe that ρ is the average work entering the system per unit of time and remember that the server works at a maximum speed of one unit of work completed per unit of time.

We found earlier that in the positive recurrent case $\rho < 1$, the generating function of the stationary distribution π is given by the formula

$$\sum_{i=0}^\infty \pi(i)z^i = \pi(0)\frac{(z-1)g_Z(z)}{z - g_Z(z)} ,$$

where

$$\pi(0) = 1 - \rho$$

and $g_Z(z)$ is the generating function of Z_n. The latter can be explicitly computed using expression (14.4):

$$g_Z(z) = \int_0^\infty \mathrm{e}^{-\lambda t} \left(\sum_{i=0}^\infty \int_0^\infty \frac{(\lambda t)^i}{i!} z^i \right) \mathrm{d}G(t)$$

that is

$$g_Z(z) = \int_0^\infty e^{-\lambda t(1-z)} \, dG(t) \,.$$

Embedded Sojourn Process

In the $M/GI/1/\infty/$FIFO queueing system, the number of customers $X_n = X(\tau_n)$ left *behind* by the n-th customer when he leaves the system is exactly the number of customers arriving during the time interval $(T_n, \tau_n]$, that is,

$$X_n = N(T_n, T_n + V_n] \,, \tag{$\star$}$$

where V_n is the sojourn time of the n-th customer in the system. Invoking the extended strong Markov property of HPPs and noting that V_n depends only on $(\sigma_1, \ldots, \sigma_n)$ and the past of N at time T_n, it follows from $(\star)$ that at equilibrium (for $\{X_n\}_{n \geq 0}$),

$$\sum_{i=0}^\infty \pi(i) z^i = \mathrm{E}\left[z^{\tilde{N}(V_n)} \right] \,,$$

where $\tilde{N}$ is a Poisson process of intensity λ independent of V_n. Now,

$$
\begin{aligned}
\mathrm{E}\left[z^{\tilde{N}(V_n)} \right] &= \int_0^\infty \mathrm{E}\left[z^{\tilde{N}(v)} \right] dF_{V_n}(v) \\
&= \int_0^\infty e^{\lambda v(z-1)} dF_{V_n}(v) = \Phi_{V_n}(\lambda(z-1)),
\end{aligned}
$$

where $F_{V_n}(v)$ is the CDF of V_n, and Φ_{V_n} is the Laplace transform of V_n. We therefore see that the distribution of V_n is independent of n, and that calling V any random variable with the same distribution as V_n, we have

$$\sum_{i=0}^\infty \pi(i) z^i = \Phi_V(\lambda(z-1)) \,,$$

where

$$\Phi_V(s) = \int e^{\lambda s} dF_V(s) \,.$$

Note that this expresses the fact that at equilibrium, X_n is distributed as

$$X = N(V),$$

where N is an HPP of intensity λ, and V is a random variable independent of N with the distribution of the stationary sojourn time.

14.2.3   The GI/M/1/∞/FIFO Queue

The GI/M/1/∞/FIFO queue is of the same nature as the M/M/1/∞/FIFO queue, except that now the arrival process is not Poissonian, but renewal. The inter-arrival times form an IID sequence with cumulative distribution function $F(x)$ and mean λ^{-1}. The service times are exponential with mean μ^{-1}.

As in the M/GI/1/∞/FIFO queue, the congestion process $\{X(t)\}_{t\geq 0}$ is *not* Markovian, but the system is amenable to Markovian analysis. Indeed, letting

$$X_n := X(T_n-)$$

(the number of customers in the system seen *upon arrival* by the n-th customer), the process $\{X_n\}_{n\geq 1}$ is an HMC. To see this, we can take the GSMP description of the queue analogous to the one given for the M/M/1/∞ model, except that now the input process $\{T_n\}_{n\geq 0}$ corresponding to the source δ is a renewal process. In Figure 14.2.3, not all the crossed events in source δ are used as departure times, but only $X_n + 1$ of them at most. More precisely,

$$X_{n+1} = (X_n + 1 - N_\delta(T_n, T_{n+1}])^+ .$$

Defining

$$Z_{n+1} = N_\delta(T_n, T_{n+1}] ,$$

the sequence $\{Z_n\}_{n\geq 1}$ is IID (use, for instance, the extended strong Markov property for HPPs) and therefore $\{X_n\}_{n\geq 0}$ is an HMC. Moreover, with

$$b_k := \mathrm{P}(Z_{n+1} = k) = \int_0^\infty e^{-\mu t} \frac{(\mu t)^k}{k!}\, dF(t) ,$$

the i-th row of the transition matrix $\mathbf{P}$ is

$$\left(1 - \sum_{k=0}^i b_k,\, b_i,\, b_{i-1},\, \ldots,\, b_0,\, 0,\, 0,\, \ldots\right).$$

Since $b_k > 0$ for all $k \geq 0$, $\mathbf{P}$ is irreducible and aperiodic. We first observe that

$$\sum_{k=0}^\infty k b_k = \rho^{-1} ,$$

where

$$\rho = \frac{\lambda}{\mu}$$

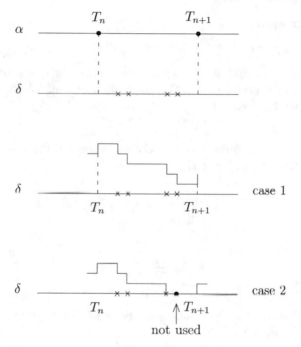

Figure 14.2.3. GSMP construction of GI/M/1/∞/FIFO

is the traffic intensity. Indeed,

$$
\begin{aligned}
\sum_{k=0}^{\infty} k b_k & = & \mathrm{E}[Z_{n+1}] = \mathrm{E}[N_\delta(T_n, T_{n+1}]] \\
& = & \mathrm{E}[\int_0^{\infty} 1_{(T_n, T_{n+1}]}(t) \mathrm{d}N_\delta(t)] = \mathrm{E}[\int_0^{\infty} 1_{(T_n, T_{n+1}]}(t) \mu \, \mathrm{d}t] \\
& = & \mu \mathrm{E}[T_{n+1} - T_n] = \frac{\mu}{\lambda},
\end{aligned}
$$

where we have used the smoothing formula of Poisson calculus. We want to show that the embedded HMC $\{X_n\}_{n\geq 0}$ is

- positive recurrent if $\rho < 1$,

- null recurrent if $\rho = 1$,

- transient if $\rho > 1$.

A. First we show that if $\rho < 1$, the chain is positive recurrent. For this it suffices to prove the existence of a stationary distribution. We make the educated guess that π has the form

$$
\pi(i) = \xi^i (1 - \xi) \tag{14.5}
$$

for some $\xi \in (0, 1)$. In order to check the correctness of this guess, we must find $\xi \in (0, 1)$ such that for all $i \geq 1$,

$$
\sum_{j=i-1}^{\infty} \xi^{j-i+1} b_{j-i+1} = \xi \tag{14.6}
$$

and

$$
\sum_{j=0}^{\infty} (\sum_{k=j+1}^{\infty} b_k) \xi^j = 1, \tag{14.7}
$$

since these equations are the balance equations when π is given by (14.5). Equations (14.6) all reduce to

$$
\xi = g_Z(\xi), \tag{14.8}
$$

where

$$
g_Z(\xi) = \sum_{k=0}^{\infty} b_k \xi^k
$$

is the generating function of Z_1. Since all the b_k's are positive, there is a unique solution ξ_0 of (14.8) in $(0, 1)$ if and only if

$$
g_Z'(1) = \sum_{k=1}^{\infty} k b_k > 1 ,
$$

that is, $\rho < 1$. We now verify (14.7). The left-hand side equals

$$\sum_{k=1}^{\infty}\sum_{j=0}^{k-1} b_k \xi^j = \sum_{k=1}^{\infty} b_k \left(\frac{1-\xi^k}{1-\xi}\right) = \frac{1}{1-\xi}\left(1 - b_0 - \sum_{k=1}^{\infty} b_k \xi^k\right),$$

and in view of (14.8), this equals

$$\frac{1}{1-\xi}(1 - b_0 - (\xi - b_0)) = 1.$$

B. Consider the matrix

$$\tilde{\mathbf{P}} = \begin{pmatrix} b_0 & b_1 & b_2 & \cdots \\ b_0 & b_1 & b_2 & \cdots \\ & h_0 & h_1 & \\ & & b_0 & \cdots \end{pmatrix}.$$

This is the transition matrix of the chain imbedded at the arrival times of an M/GI/1/∞ system with traffic intensity

$$\tilde{\rho} = \frac{\mu}{\lambda} = \frac{1}{\rho}.$$

We know from Theorem 7.3.4, that the irreducible HMC with transition matrix $\mathbf{P}$ is transient if and only if there exists a non-trivial bounded solution $h : E \to \mathbb{R}$ of

$$h(j) = \sum_{k\neq 0} p_{jk} h(k) \quad (j \neq 0), \tag{14.9}$$

where 0 is an arbitrary state. Equation (14.9) is

$$h(j) = \sum_{i=1}^{j+1} b_{j-i+1} h(i) \quad (j \neq 0). \tag{14.10}$$

We can assume without loss of generality that $h(1) \geq 0$ (otherwise replace the function h by $-h$). Defining

$$\tilde{\pi}(0) = h(1)b_0; \quad \tilde{\pi}(1) = h(1)(1-b_0); \quad \tilde{\pi}(j) = h(j) - h(j-1), \quad (j \geq 2),$$

it is easy to verify that

$$\tilde{\pi}\tilde{\mathbf{P}} = \tilde{\pi}. \tag{14.11}$$

We show that $\tilde{\pi} \geq 0$. Indeed, (14.11) is, for $j \geq 0$,

$$\tilde{\pi}(j) = \tilde{\pi}(0)b_j + \sum_{i=1}^{j+1} \tilde{\pi}(j)b_{j-i+1}.$$

Add these equations for $j = 0$ to n and solve to obtain, for $n \geq 0$,

$$\tilde{\pi}(n+1)b_0 = \tilde{\pi}(0)c_n + \sum_{i=1}^{n} \tilde{\pi}(i)c_{n-i+1},$$

where

$$c_n := 1 - b_0 - \ldots - b_n > 0.$$

Therefore, if $\tilde{\pi}(i) \geq 0$ for $i = 0$ to n, then $\tilde{\pi}(n+1) \geq 0$. But $\tilde{\pi}(0) \geq 0$. Therefore, by induction, $\tilde{\pi} \geq 0$.

Equation (14.11) has a non-trivial bounded solution with bounded sum $\sum_{i=0}^{\infty} \tilde{\pi}(i)$ if and only if $\tilde{\rho} < 1$. Equivalently, as we have just shown, (14.10) has a non-trivial bounded solution if and only if (14.11) has a non-trivial bounded solution with bounded sum. Therefore (Theorem 7.3.4) $\mathbf{P}$ is transient if and only if $\tilde{\mathbf{P}}$ is positive recurrent, that is, if and only if $\rho > 1$.

C. It remains to show that $\mathbf{P}$ is recurrent null if $\rho = 1$. From the previous discussion, $\mathbf{P}$ cannot be transient. It suffices therefore to show that $\mathbf{P}$ cannot be recurrent positive. Indeed, if it were, there would be a stationary distribution π for $\mathbf{P}$. Writing

$$\tilde{h}(j) = \pi(0) + \ldots + \pi(j-1), \; j \geq 1, \tag{14.12}$$

and using $\sum_{k=0}^{\infty} b_k = 1$, we can verify that

$$\tilde{h}(j) = \sum_{k \neq 0} \tilde{p}_{jk}\tilde{h}(k), \; j \neq 0. \tag{14.13}$$

Therefore, (14.13) has a non-trivial bounded solution, and by Theorem 3.4 of Chapter 5, this is equivalent to transience of $\tilde{\mathbf{P}}$. But we know that in this case necessarily $\tilde{\rho} > 1$, and therefore $\rho < 1$, a contradiction with our assumption $\rho = 1$. Therefore, for $\rho = 1$, $\mathbf{P}$ is recurrent null.

An Embedded Waiting Time Process

An arriving customer finds $X_n = X(T_n-)$ customers in front of him and therefore, with the FIFO service discipline, his waiting time W_n (before he starts to be served) is the time needed for the server to take care of the X_n customers present at time T_n-. By the extended strong Markov property applied to the HPP N_δ in the GSMP construction (see Figure 14.2.3), the point process N_δ after T_n is an HPP of intensity μ and is independent of X_n. In particular,

$$W_n = \sum_{j=1}^{X_n} Y_j,$$

where the Y_j are IID exponentials of mean μ^{-1} and common characteristic function

$$\mathrm{E}[e^{iuY}] = \frac{\mu}{\mu - iu},$$

and are independent of X_n. Also,

$$\mathrm{P}(X_n = k) = (1 - \xi_0)\xi_0^k.$$

Therefore, in steady state,

$$
\begin{aligned}
\mathrm{E}[e^{iuW_n}] &= \mathrm{E}[\sum_{k=0}^{\infty} e^{iu\sum_{j=1}^{k} Y_j} 1_{\{X_n=k\}}] \\
&= \sum_{k=0}^{\infty} \mathrm{E}[e^{iuY}]^k \mathrm{P}(X_n = k) \\
&= \sum_{k=0}^{\infty} (\frac{\mu}{\mu - iu})^k (1 - \xi_0)\xi_0^k.
\end{aligned}
$$

We find for the characteristic function of the stationary waiting time

$$(1 - \xi_0) + \xi_0 \left(\frac{\mu(1 - \xi_0)}{\mu(1 - \xi_0) + iu} \right).$$

This is the characteristic function of a random variable that is null with probability $1 - \xi_0$ and exponential of mean $[\mu(1 - \xi_0)]^{-1}$ with probability ξ_0.

14.3 Markovian Queueing Networks

14.3.1 The Tandem Network

A seminal result of Queueing theory, *Burke's output theorem*, says in particular that the point process of the departure times of an $M/M/1/\infty$ queue in equilibrium is a homogeneous Poisson process with the same intensity as the arrival process. In fact, the result extends to more general birth and death processes.

Consider a birth-and-death process on $\mathbb{N}$ with birth parameters of the form

$$\lambda_i \equiv \lambda,$$

and suppose that $\mu_i > 0$ for all $i \geq 1$. The corresponding chain is irreducible and we shall assume that it is ergodic. At equilibrium, this chain is reversible since the detailed balance equations $\pi(i + 1)q_{i+1,i} = \pi(i)q_{i,i+1}$, that is, in the present case,

$\pi(i+1)\mu_{i+1} = \pi(i)\lambda$, are satisfied, as can be readily checked using the explicit form of the stationary distribution.

The upward transitions are due to arriving customers and the downward transitions are due to departing customers. Suppose that the queue is in equilibrium, and therefore the reversed process has the same distribution as the direct process. When time is reversed, the point process of departures becomes the point process of arrivals, and therefore, in view of the reversibility property, the reversed process of departures is an HPP. Now, the probabilistic nature of an HPP does not change when time is reversed. The departure process is therefore a Poisson process.

Also, since in direct time, for any time $t \geq 0$, the state $X(t)$ is independent of the future at time t of the arrival process (Markov property of Poisson processes), it follows from reversibility that $X(t)$ is independent of the past at time t of the departure process.

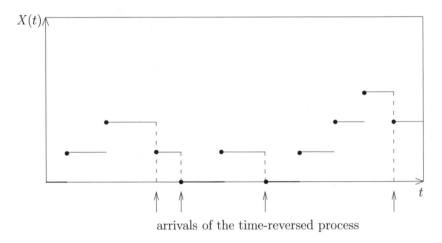

arrivals of the time-reversed process

Figure 14.3.1. Reversibility proof of Burke's theorem

Consider two M/M/1/∞ queues in succession, in the sense that the departure process from the first one is the arrival process into the other. The arrival process, that is, the point process of arrivals into the first queue is an HPP of intensity λ. The service sequences are independent IID exponential sequences with mean $1/\mu_1$ and $1/\mu_2$, respectively. Both service sequences are independent of the arrival process. Let $X_1(t)$ and $X_2(t)$ be the number of customers in the first and the second queueing system, respectively. At time t, $X_2(t)$ depends on the departure process of the first queue before time t and on the second service sequence. Since, by Burke's theorem and the independence property of the service sequence, the latter are independent of $X_1(t)$, it follows that $X_1(t)$ and $X_2(t)$ are independent. Both queues are in isolation M/M/1/∞ with traffic intensities $\rho_1 = \lambda/\mu_1$ and $\rho_2 = \lambda/\mu_2$, and therefore a necessary and sufficient condition of ergodicity of the continuous-time HMC $\{(X_1(t), X_2(t))\}_{t \geq 0}$ is $\lambda < \inf(\mu_1, \mu_2)$, and its stationary distribution has the *product form*

$$\pi(n_1, n_2) = \pi_1(n_1)\pi_2(n_2) = (1 - \rho_1)\rho_1^{n_1}(1 - \rho_2)\rho_1^{n_2}.$$

14.3.2 The Jackson Network

A *Jackson network*[3] is an open network of interconnected queues. More precisely: there are K *stations*, and each station has a 1/∞ service system, that is, a unique server working at unit speed and an infinite waiting room. There are two types of customers queueing at a given station, (1) those which are fed-back, that is, who have received service in another or the same station and are re-routed to the given station for more service, and (2) those who enter the network for the first time (Figure 14.3.2).

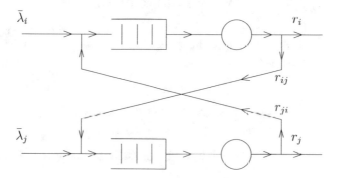

Figure 14.3.2. Jackson network

The *exogenous arrivals* into station i form an HPP, denoted by $\bar{N}_i$, with intensity $\bar{\lambda}_i \in [0, \infty)$. The sequence of service times at station i are exponential random variables of mean $1/\mu_i \in (0, \infty)$.

The service times in the same and in different stations are independent and independent of the exogenous input HPPs $\bar{N}_i$, and the latter HPPs are independent of one another.

The *routing* is of the Bernoulli type. Each customer just completing service in station i tosses a $(K+1)$-faced die with probabilities $r_{i,1}, \ldots, r_{i,K}, r_i$ with the effect

---

[3][Jackson, 1957].

that the customer is sent to station j with probability r_{ij} or leaves the system with probability $r_i = 1 - \sum_{j=1}^{K} r_{ij}$. The matrix

$$\mathbf{R} = \{r_{ij}\}_{1 \leq i,j \leq K}$$

is the *routing matrix*. The successive tosses of the routing dice of all stations are independent, and independent of the exogeneous arrival processes and of all the service times.

This is the original Jackson model, which can be enriched by the introduction of service speeds. If there are n_i customers in station i, the server works at speed $\varphi_i(n_i)$, where $\varphi_i(0) = 0$ and $\varphi_i(n_i) > 0$ for all $n_i \geq 1$.

Remark 14.3.1 *The tandem queue of the previous subsection is clearly a special case of a Jackson network.*

Let $X_i(t)$ be the number of customers in station i at time t, and define

$$X(t) = (X_1(t), \dots, X_k(t)).$$

It can be verified that the process $\{X(t)\}_{t \geq 0}$ is a regular jump HMC with state space $E = \mathbb{N}^K$ and with infinitesimal generator $\mathbf{A} = \{q_{n,n'}\}$, where all the non-null off-diagonal terms are

$$
\begin{aligned}
q_{n,n+e_i} &= \bar{\lambda}_i, \\
q_{n,n-e_i} &= \mu_i \varphi_i(n_i) r_i 1_{\{n_i > 0\}}, \\
q_{n,n-e_i+e_j} &= \mu_i \varphi_i(n_i) r_{ij} 1_{\{n_i > 0\}},
\end{aligned}
$$

where $n = (n_1, \dots, n_K) \in E = \mathbb{N}^K$, and e_i is the ith vector of the canonical basis of $\mathbb{R}^K$.

This generator is independent of the service strategy—LIFO, FIFO, or processor-sharing.

The form of the infinitesimal generator can be obtained from a full description of the network as a Poisson system. However, we shall be content with heuristic arguments. For instance, the expression for $q_{n,n-e_i+e_j}$ follows from the intuitive considerations below. If the state at time t is n, a transfer from station i to station j requires that the exponential random variable with mean $1/\mu_i$ representing the required service of the customer being served at station i at time t be terminated between times t and $t + h$ (probability $\mu_i \varphi_i(n_i) h$ up to the first order in h), and that the corresponding customer be routed to station j (probability r_{ij}).

We shall assume that the chain $\{X(t)\}_{t \geq 0}$ is irreducible. This is the case when

(α) for all $j \in [1, K]$, there exist $i, i_1, \ldots, i_m \in [1, K]$ such that

$$\bar{\lambda}_i r_{i i_1} r_{i_1 i_2} \cdots r_{i_m j} > 0,$$

and

(β) for all $j \in [1, K]$, there exist $j_1, j_2, \ldots, j_\ell, k \in [1, K]$ such that

$$r_{j j_1} r_{j_1 j_2} \cdots r_{j_\ell k} r_k > 0$$

(this is left for the reader to prove).

Recalling that the $\mu_i > 0$ for all stations, condition (α) tells us that any station is *exogenously supplied*, and (β) tells us that any station has an *outlet*.

Consider now the $(K + 1) \times (K + 1)$ matrix

$$\widetilde{\mathbf{R}} = \begin{pmatrix} r_{11} & r_{12} & \cdots & r_{1K} & r_1 \\ r_{21} & r_{22} & \cdots & r_{2K} & r_2 \\ \vdots & \vdots & & \vdots & \vdots \\ r_{K1} & r_{K2} & \cdots & r_{KK} & r_K \\ 0 & 0 & \cdots & 0 & 1 \end{pmatrix}.$$

It can be interpreted as the transition matrix of an HMC on the finite state space $\{1, \ldots, K, K + 1\}$. Condition (β) implies that state $K + 1$ is absorbing and the states $1, \ldots, K$ are transient. It follows that $(1 - \mathbf{R})^{-1}$ exists and therefore the solution of the system of equations

$$\lambda_i = \bar{\lambda}_i + \sum_{j=1}^{K} \lambda_j r_{ji}, \tag{14.14}$$

that is, with obvious notations, $\lambda = \bar{\lambda} + \mathbf{R}\lambda$, has a unique solution

$$\lambda = (1 - \mathbf{R})^{-1} \bar{\lambda} = \left(\sum_{n-0}^{\infty} \mathbf{R}^n \right) \bar{\lambda},$$

where the latter expression shows that this solution is non-negative. Equations (14.14) are called the *traffic equations*, because they give a necessary relation between the average numbers of customers λ_i entering station i in steady state if the network is ergodic. Indeed, λ_i is equal to the exogeneous rate of arrivals $\bar{\lambda}_i$ plus the sum of all average rates of transfer from other stations. From station j, the corresponding rate is $\alpha_j r_{ji}$, where α_j is the average rate of customers finishing service in station j. But at equilibrium $\alpha_j = \lambda_j$, since the average number of customers in station j remains constant, whence the traffic equations (14.14).

Of course, this heuristic argument is not needed, and in particular, one need not attempt to identify λ_i in (14.14) as the average incoming arrival rate in station i. This identification is possible if equilibrium is guaranteed. However, even in the non-ergodic cases, the traffic equations have a solution, and it is unique.

The existence of a probability distribution π on $E = \mathbb{N}^K$ satisfying Kolmogorov's balance equations is a necessary and sufficient condition of ergodicity of the network. We consider the case where the service speeds of all the servers are equal to 1 and refer to Exercise 14.4.10 for the general case.

The global balance equations for the Jackson network are

$$\pi(n)\left\{\sum_{i=1}^{K}(\lambda_i + \mu_i r_i 1_{\{n_i>0\}})\right\} = \sum_{i=1}^{K}\pi(n - e_i)\bar\lambda_i 1_{\{n_i>0\}} + \sum_{i=1}^{K}\pi(n + e_i)\mu_i r_i$$

$$+ \sum_{i=1}^{K}\sum_{j=1}^{K}\pi(n + e_i - e_j)\mu_i r_{ij} 1_{\{n_j>0\}}.$$

It turns out that if the solution of the traffic equation satisfies

$$\rho_i = \frac{\lambda_i}{\mu_i} < 1 \quad (1 \le i \le K), \tag{14.15}$$

then the network is ergodic, and its stationary distribution is given by

$$\pi(n) = \prod_{i=1}^{K}\pi_i(n_i), \tag{14.16}$$

where π_i is the stationary distribution of an $M/M/1/\infty$ queue with traffic intensity ρ_i,

$$\pi_i(n_i) = \rho_i^{n_i}(1 - \rho_i). \tag{14.17}$$

To prove this, we shall apply the reversal test (Theorem 13.4.13). We define the generator $\widetilde{\mathbf{A}}$ on $E = \mathbb{N}^K$ by

$$\pi(n)\widetilde{q}_{n,n'} = \pi(n')q_{n',n}$$

and check that

$$\sum \widetilde{q}_{n,n'} = q_n.$$

Here

$$q_n = \sum_{i=1}^{K}\lambda_i + \mu_i 1_{\{n_i>0\}}.$$

The generator $\widetilde{\mathbf{A}}$ is given by

$$
\begin{aligned}
\pi(n)\widetilde{q}_{n,n+e_i} &= \pi(n+e_i)q_{n+e_i,n} = \pi(n+e_i)\mu_i r_i, \\
\pi(n)\widetilde{q}_{n,n-e_i} &= \pi(n-e_i)q_{n-e_i,n} = \pi(n-e_i)\bar{\lambda}_i 1_{\{n_i>0\}}, \\
\pi(n)\widetilde{q}_{n,n+e_i-e_j} &= \pi(n+e_i-e_j)q_{n+e_i-e_j,n} = \pi(n+e_i-e_j)\mu_i r_{ij} 1_{\{n_j>0\}},
\end{aligned}
$$

and therefore, taking into account the specific form of $\pi(n)$ given by (14.16) and (14.17),

$$
\begin{aligned}
\widetilde{q}_{n,n+e_i} &= \rho_i \mu_i r_i = \lambda_i r_i, \\
\widetilde{q}_{n,n-e_i} &= \frac{\bar{\lambda}_i}{\rho_i} 1_{\{n_i>0\}}, \\
\widetilde{q}_{n,n+e_i-e_j} &= \frac{\rho_i}{\rho_j}\mu_i r_{ij} 1_{\{n_j>0\}} = \frac{1}{\rho_j}\lambda_i r_{ij} 1_{\{n_j>0\}}.
\end{aligned}
$$

We have to verify that

$$
\sum_{i=1}^{K}(\bar{\lambda}_i + \mu_i 1_{\{r_i>0\}}) = \sum_{i=1}^{K}\left(\lambda_i r_i + \frac{\bar{\lambda}_i}{\rho_i}1_{\{n_i>0\}} + \frac{1}{\rho_i}\left(\sum_{i=1}^{K}\lambda_j r_{ji}\right)1_{\{n_i>0\}}\right).
$$

By the traffic equation, $\sum_{i=1}^{K}\lambda_j r_{ji} = \lambda_i - \bar{\lambda}_i$, and therefore the right-hand side of the previous equality is

$$
\sum_{i=1}^{K}\left(\lambda_i r_i + \frac{1}{\rho_i}\lambda_i 1_{\{n_i>0\}}\right) = \sum_{i=1}^{K}(\lambda_i r_i + \mu_i 1_{\{n_i>0\}}),
$$

and it remains to check that

$$
\sum_{i=1}^{K}\bar{\lambda}_i = \sum_{i=1}^{K}\lambda_i r_i.
$$

For this we need only sum the traffic equations

14.3.3 The Gordon–Newell Network

A *closed Jackson network*, also called a *Gordon–Newell network*,[4] is one for which

$$
\bar{\lambda}_i = 0, \; r_i = 0 \quad (1 \le i \le K). \tag{14.18}
$$

---

[4][Gordon and Newell, 1967].

In other words, there is no inlet and no outlet, and therefore the number of customers in the network remains constant. It will be denoted it by N. The state space is

$$E = \left\{ (n_1, \ldots, n_K) \in \mathbb{N}^K, \sum_{i=1}^{K} n_i = N \right\}.$$

The traffic equations are now

$$\lambda = \mathbf{R}\lambda, \tag{14.19}$$

and since $\mathbf{R}$ is a stochastic matrix in this case, which we shall assume irreducible, it has an infinity of solutions, all multiples of the same vector, which is the stationary distribution of $\mathbf{R}$.

It is true that the vector of average traffic through the stations is a solution of (14.19). We do not know which one, in contrast with the open network, where the solution of the traffic equation is unique. Nevertheless, if we take any positive solution, we see by inspection that for all $n \in E$,

$$\pi(n) = \frac{1}{G(N, K)} \prod_{i=1}^{K} \rho_i^{n_i}, \tag{14.20}$$

where $\rho_i = \lambda_i/\mu_i$ is a stationary solution. Here $G(N, K)$ is the *normalizing factor*:

$$G(N, K) = \sum_{\substack{n \in \mathbb{N}^K \\ n_1 + \cdots + n_K = N}} \prod_{i=1}^{K} \rho_i^{n_i}. \tag{14.21}$$

Note that under the irreducibility assumption for the routing matrix $\mathbf{R}$, the chain $\{X(t)\}_{t \geq 0}$ is itself irreducible and therefore, since the state space is finite, it is positive recurrent with a unique stationary distribution. In particular, for closed networks, there is no ergodicity condition.

A practical issue in closed Jackson networks is the computation of the normalizing constant $G(N, K)$. Brute-force summation via (14.21) is virtually infeasible for large populations and/or large networks. Instead, we can use the following algorithm.

Define $G(j, \ell)$ to be the coefficient of z^j in the power series development of

$$g_\ell(z) = \prod_{i=1}^{\ell} \frac{1}{1 - \rho_i z} = \prod_{i=1}^{\ell} \left(\sum_{n_i=0}^{\infty} \rho_i^{n_i} z^{n_i} \right).$$

The normalizing factor is indeed equal to $G(N, K)$. Since

$$g_\ell(z) = g_{\ell-1}(z) + \rho_\ell z g_\ell(z),$$

we find the recurrence equation

$$G(j, \ell) = G(j, \ell - 1) + \rho_\ell G(j - 1, \ell) \qquad (14.22)$$

with the initial conditions

$$G(j, 1) = \rho_1^j \ (j \geq 0),$$

$$G(0, \ell) = 1 \ (\ell \geq 1).$$

It is also of interest to be able to compute the *utilization* of server i, defined by

$$U_i(N, K) = \mathrm{P}(X_i(t) > 0),$$

which gives the average *throughput* from station i to station j,

$$d_{ij}(N, K) = \mu_i U_i(N, K) r_{ij}.$$

This is the average number of customers transferred from station i to station j in one unit of time. Since

$$U_i(N, K) = \sum_{\substack{n_1 + \cdots + n_K = N \\ n_i > 0}} \pi(n) = \frac{1}{G(N, K)} \sum_{\substack{n_1 + \cdots + n_K = N \\ n_i > 0}} \prod_{j=1}^{K} \rho_j^{n_j},$$

we see that $G(N, K)U_i(N, K)$ is the coefficient of z^N in

$$\widetilde{g}_{N,i}(z) = \left(\prod_{\substack{j=1 \\ j \neq i}}^{K} \left(\sum_{n_j = 0}^{\infty} \rho_j^{n_j} z^{n_j} \right) \right) \left(\sum_{n_i = 1}^{\infty} \rho_i^{n_i} z^{n_i} \right).$$

Now,

$$\widetilde{g}_{N,i}(z) = g_N(z)\rho_i z,$$

and therefore $G(N, K)U_i(N, K) = \rho_i G(N - 1, K)$, that is,

$$U_i(N, K) = \rho_i \frac{G(N, K)}{G(N - 1, K)}. \qquad (14.23)$$

Closed Jackson network models arise in the following situation, where an open network is operated with a blocking admission policy: If there are already N customers in the network, the newcomers wait at a gate (in a queue) until one customer is released from the network, at which time one among the blocked customers, if there are any, is admitted. In the network, there are at most N customers. At "saturation," there is always one customer ready to replace a departing customer,

since the gate queue is infinite, by definition of saturation. Therefore, at saturation, or for all practical purposes near saturation, everything looks as if a departing
customer was being immediately recycled.

It is important in practice to be able to compute the average number of customers $d(N)$ passing through the entrance point A per unit of time: This is the
maximum throughput, and therefore it is related to the efficiency of the system,
from the point of view of the operator (who makes money with customers). From
the point of view of the quality of service, an important parameter is $W(N)$, the
average time spent by a customer between A and B. As a matter of fact, N,
$W(N)$, and $d(N)$ are related by

$$d(N)W(N) = N, \tag{14.24}$$

a particular case of Little's formula. One can therefore compute $d(N)$ via formulas such as (14.23) and the algorithm (14.22), and then $W(N)$ via (14.24), and
then choose an operating point N that provides the required balance between the
operator's profit and the customer's comfort.

14.4 Exercises

Exercise 14.4.1. LINEAR BIRTH-AND-DEATH PROCESS

Consider the following model of a population in terms of birth and death. The
lifetimes of the individuals appearing in the population are independent and exponentially distributed with the same mean μ^{-1}, where $\mu \in (0, \infty)$. An individual
introduced into the population (by birth or by immigration) gives birth to children
according to an HPP of intensity $\lambda > 0$ (the reader will interpret this statement)
as long as she is alive. All the birth point processes attached to individuals are
independent, and independent of all the lifetimes. There is also an immigration
process, an HPP of intensity a, $a \in (0, \infty)$, independent of all the rest. Give a
Poisson system description of this population process, and show that the process
counting the number of individuals in the population is a birth-and-death process
with parameters $\lambda_n = n\lambda + a$, $\mu_n = n\mu$.

Exercise 14.4.2. M/M/1/∞/FIFO QUEUE WITH INSTANTANEOUS FEEDBACK

Give a purely Poissonian description of the M/M/1/∞/FIFO queue with instanta-neous feedback and a first-come-first-served discipline. Show that the congestion process counting the number of customers in the system at time t is a regular jump HMC, and give its infinitesimal generator.

Exercise 14.4.3. POISSON SYSTEMS AND DISCRETE-TIME HMCS

Consider the general Poisson system of Section 14.1. Call $\widetilde{\tau}_n$ the n-th transition (maybe a pseudo-transition) time, where $\widetilde{\tau}_0 \equiv 0$. Call $\widetilde{s}_n$ the source responsible for the transition at time $\widetilde{\tau}_n$, with $\widetilde{s}_0 \in S$ arbitrary and independent of the driv-ing HPPs. Show that $\{(\widetilde{s}_n, X(\widetilde{\tau}_n))\}_{n \geq 0}$ forms a discrete-time HMC, and give its transition matrix.

Exercise 14.4.4. M/M/1/∞/LIFO PREEMPTIVE RESUME QUEUE

Consider an M/M/1/∞/LIFO preemptive resume queue. Show that the congestion process (counting the customers in the system: ticket booth plus waiting room) is a regular jump HMC with the same infinitesimal generator as the corresponding process in an M/M/1/∞/FIFO queueing system.

Exercise 14.4.5. M/M/1/∞/FIFO QUEUE WITH INSTANTANEOUS BERNOULLI FEEDBACK

Consider the M/M/1/∞/FIFO queue with instantaneous Bernoulli feedback of Example 14.1.5. Define the process $\{Z(t)\}_{t \geq 0}$ with values in $\{-1, +1\}$ as fol-lows: $Z(0)$ is chosen independently of everything else, and the process $\{Z(t)\}_{t \geq 0}$ switches from one value to the other whenever a customer is fed-back. Show that $Y(t) = (X(t), Z(t))$ defines a regular jump HMC $\{Y(t)\}_{t \geq 0}$, and give its in-finitesimal generator. Give a necessary and sufficient condition of ergodicity and compute the stationary distribution in the ergodic case. Give, at equilibrium, the distribution of the times between transitions of $\{Z(t)\}_{t \geq 0}$. Show that the times of departure of customers definitively leaving the system form a Poisson process. Do the same for the times of end of service.

Exercise 14.4.6. M/GI/1/∞/FIFO QUEUE AT DEPARTURE TIMES

Prove that the congestion process in an M/GI/1/∞/FIFO queue observed at the departure times is a discrete-time HMC.

Exercise 14.4.7. CONSTANT SERVICE TIMES MINIMIZE CONGESTION
Show that for a fixed traffic intensity ρ, constant service times minimize average congestion in the $M/GI/1/\infty$ FIFO queue.

Exercise 14.4.8. WORKLOAD OF $M/M/1/\infty$
Show that the stationary distribution of the workload process of an $M/M/1/\infty$ queue with arrival rate λ and mean service time μ^{-1} such that $\rho = \frac{\lambda}{\mu} < 1$ is

$$F_W(x) = 1 - \rho\left(1 - \exp\{-(\mu - \lambda)x\}\right).$$

Exercise 14.4.9. AVERAGE NUMBER OF TRANSITIONS
Let $\{X(t)\}_{t \geq 0}$ be an irreducible regular jump HMC with state space E and infinitesimal generator $\mathbf{A}$. Suppose that it is ergodic, with stationary probability π. Show that at equilibrium $(\mathrm{P}(X(t) = i) = \pi(i)$ $(t \geq 0, i \in E)$ and that the average number of transitions from i to j in a unit time interval is $\pi(i)q_{ij}$. Use this to interpret the detailed balance equations for a reversible chain.

Exercise 14.4.10. AN EXTENSION OF THE JACKSON NETWORK, TAKE 1
The following modification of the basic Jackson network is considered. For all i $(1 \leq i \leq K)$, the server at station i has a speed of service $\varphi_i(n_i)$ when there are n_i customers present in station i, where $\varphi_i(k) > 0$ for all $k \geq 1$ and $\varphi_i(0) = 0$. The new infinitesimal generator is obtained from the standard one by replacing μ_i by $\mu_i\varphi(n_i)$. Check this and show that if for all i $(1 \leq i \leq K)$,

$$A_i := 1 + \sum_{A_i=1}^{\infty}\left(\frac{\rho_i^{n_i}}{\prod_{k=1}^{n_i}\varphi(k)}\right) < \infty,$$

where $\rho_i = \lambda_i/\mu_i$ and λ_i is the solution of the traffic equation, then the network is ergodic with stationary distribution

$$\pi(n) = \prod_{i=1}^{K}\pi_i(n_i),$$

where

$$\pi_i(n_i) = \frac{1}{A_i}\frac{\rho_i^{n_i}}{\prod_{k=1}^{n_i}\varphi_i(k)}.$$

Exercise 14.4.11. AN EXTENSION OF THE JACKSON NETWORK, TAKE 2
The basic Jackson network of the theory is now modified by assigning K_i servers
with unit speed of service to each station i ($1 \leq i \leq K$). Show that the infinitesimal
generator of this network is the same as that of the network of the previous exercise,
with $\varphi_i(n_i) = \inf(K_i, n_i)$. Give the ergodicity condition corresponding to that of
the previous exercise.

Exercise 14.4.12. THE TANDEM NETWORK
Consider the Jackson network of the theory with $K = 2$, $\bar{\lambda}_2 = 0$, $r_{12} = 1$, $r_2 = 1$
(two queues in series). Show that the point process counting the customers passing
from station 1 to station 2 is a Poisson process, and give its intensity. Show that
for all $t \geq 0$, $X_2(t)$ is independent of $(X_1(s), 0 \leq s \leq t)$.

Exercise 14.4.13. THE GORDON–NEWELL NETWORK WITH A SINGLE CUS-
TOMER
Consider the closed Jackson network of the theory, with $N = 1$ customer. Let
$\{Y(t)\}_{t\geq0}$ be the process giving the position of this customer, that is, $Y(t) = i$
if he/she is in station i at time t. Show that $\{Y(t)\}_{t\geq0}$ is a regular jump HMC,
irreducible, and give its stationary distribution. Observe that $\lambda_i = \alpha_i \mu_i$ ($1 \leq i \leq$
K) is a solution of the traffic equation.

Exercise 14.4.14. THE RING, TAKE 1
Consider the closed Jackson network of the theory, with a pure ring structure, that
is, $r_{i,i+1} = 1$ ($1 \leq i \leq K - 1$) and $r_{K,1} = 1$. Compute the stationary distribution.

Exercise 14.4.15. THE RING, TAKE 2
Consider the closed Jackson network of the figure below where all service times at
different queues have the same (exponential) distribution of mean 3. Compute for
$N = 1, \ldots, 10$ the average time spent by a customer to go from the leftmost point
A to the rightmost point B, and the average number of customers passing by A
per unit time.

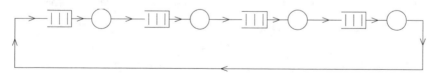

Figure 14.4.1. The ring

Exercise 14.4.16. SUPPRESSED TRANSITIONS

Let $\{X(t)\}_{t \geq 0}$ be an irreducible positive recurrent HMC with infinitesimal generator $\mathbf{A}$ and stationary distribution π. Suppose that $(\mathbf{A}, \pi)$ is reversible. Let S be a subset of the state space E. Define the infinitesimal generator $\tilde{\mathbf{A}}$ by

$$\tilde{q}_{ij} = \begin{cases} \alpha q_{ij} \text{ if } i \in S, j \in E - S \\ q_{ij} \text{ otherwise} \end{cases}$$

when $i \neq j$. The corresponding HMC is irreducible if $\alpha > 0$. If $\alpha = 0$, the state space will be reduced to S, to maintain irreducibility. Show that the continuous time HMC associated to $\tilde{\mathbf{A}}$ admits the stationary distribution $\tilde{\pi}$ given by

$$\tilde{\pi}_i = \begin{cases} \alpha C \times \pi(i) \text{ if } i \in S, \\ C \times \alpha \pi(i) \text{ if } i \in E - S, \end{cases}$$

with the obvious modification when $\alpha = 0$.

Exercise 14.4.17. LOSS NETWORKS (KELLY'S NETWORKS)

([5]) Consider a telecommunications network with K relays. A "route" (through the network) is a succession of relays. An incoming call chooses a "route" r from a set $\mathcal{R}$. The network then reserves the set of relays $r_1, \ldots, r_{k(r)}$ corresponding to this route, and processes the call to destination in an exponential time of mean μ_r^{-1}. The incoming calls with route r form a homogeneous Poisson process of intensity λ_r. It is assumed that all the relays are useful, in that they are part of at least one route in $\mathcal{R}$. (Hint for part (2): Exercise 14.4.16.)

(1) The capacity of the system is for the time being assumed to be infinite, that is, the number $X_r(t)$ of calls on route r at time t can take any integer value. All the usual independence hypotheses are made: the processing times and the Poisson processes are independent. Give the stationary distribution of the continuous time HMC $\{X(t)\}_{t \geq 0}$, where $X(t) = (X_r(t), r \in \mathcal{R})$.

(2) The capacity of the system is now restricted as follows. Consider a given pair (a, b) of relays. It represents a "link" in the network. This link has finite capacity C_{ab}. This means that the total number of calls using this link cannot exceed this capacity, with the consequence that an incoming call requiring a route passing through this link will be lost if the link is saturated when it arrives. The process $\{X(t)\}_{t \geq 0}$ therefore has for state space

$$\tilde{E} = \{n = (n_r; r \in \mathcal{R}); \sum_{r \in \mathcal{R}, (a,b) \in r} n_r \leq C_{ab} \text{ for all links } (a, b)\}.$$

What is the stationary distribution of the chain $\{X(t)\}_{t \geq 0}$?

---

[5][Kelly, 1979].

Appendix A

A.1 Number Theory and Calculus

A.1.1 Greatest Common Divisor

Let $a_1, \ldots, a_k \in \mathbb{N}$ be such that $\max(a_1, \ldots, a_k) > 0$. Their greatest common divisor (gcd) is the largest positive integer dividing all of them. It is denoted by $\gcd(a_1, \ldots, a_k)$. Clearly, removing all zero elements does not change the gcd, so that we may assume without loss of generality that all the a_k's are positive.

Let $\{a_n\}_{n \geq 1}$ be a sequence of positive integers. The sequence $\{d_k\}_{k \geq 1}$ defined by $d_k = \gcd(a_1, \ldots, a_k)$ is bounded below by 1 and is non-increasing, and it therefore has a limit $d \geq 1$, a positive integer called the gcd of the sequence $\{a_n\}_{n \geq 1}$. Since the d_k's are integers, the limit is attained after a finite number of steps, and therefore there exists a positive integer k_0 such that $d = \gcd(a_1, \ldots, a_k)$ for all $k \geq k_0$.

Lemma A.1.1 *Let $S \subset \mathbb{Z}$ contain at least one non-null element and be closed under addition and subtraction. Then S contains a least positive element a, and $S = \{ka \, ; k \in \mathbb{Z}\}$.*

Proof. Let $c \in S$, $c \neq 0$. Then $c - c = 0 \in S$. Also $0 - c = -c \in S$. Therefore, S contains at least one positive element. Denote by a the smallest positive element of S. Since S is closed under addition and subtraction, S contains a, $a + a = 2a, \ldots$ and $0 - a = -a, 0 - 2a = -2a, \ldots$, that is, $\{ka \, ; k \in \mathbb{Z}\} \subset S$.

Let $c \in S$. Then $c = ka + r$, where $k \in \mathbb{Z}$ and $0 \leq r < a$. Since $r = c - ka \in S$, we cannot have $r > 0$, because this would contradict the definition of a as the smallest positive integer in S. Therefore, $r = 0$, that is, $c = ka$. Therefore, $S \subset \{ka \, ; k \in \mathbb{Z}\}$. $\square$

Lemma A.1.2 *Let $a_1, \ldots, a_k$ be positive integers with greatest common divisor d. There exist $n_1, \ldots, n_k \in \mathbb{Z}$ such that $d = \sum_{i=1}^{k} n_i a_i$.*

© Springer Nature Switzerland AG 2020
P. Brémaud, *Markov Chains*, Texts in Applied Mathematics 31,
https://doi.org/10.1007/978-3-030-45982-6

Proof. The set $S = \left\{ \sum_{i=1}^{k} n_i a_i \; ; \; n_1, \ldots, n_k \in \mathbb{Z} \right\}$ is closed under addition and subtraction, and therefore, by Lemma A.1.1, $S = \{ka \; ; \; k \in \mathbb{Z}\}$, where $a = \sum_{i=1}^{k} n_i a_i$ is the smallest positive integer in S.

Since d divides all the a_i's, d divides a, and therefore $0 < d \leq a$. Also, each a_i is in S and is therefore a multiple of a, which implies that $a \leq \gcd(a_1, \ldots, a_k) = d$. Therefore, $d = a$. $\qquad\square$

Theorem A.1.3 *Let d be the gcd of $A = \{a_n \; ; n \geq 1\}$, a set of positive integers that is closed under addition. Then A contains all but a finite number of the positive multiples of d.*

Proof. We may assume without loss of generality that $d = 1$ (otherwise, divide all the a_n by d). For some k, $d = 1 = \gcd(a_1, \ldots, a_k)$, and therefore by Lemma A.1.2,

$$1 = \sum_{i=1}^{k} n_i a_i$$

for some $n_1, \ldots, n_k \in \mathbb{Z}$. Separating the positive from the negative terms in the latter equality, we have $1 = M - P$, where M and P are in A.

Let $n \in \mathbb{N}$, $n \geq P(P-1)$. We have $n = aP + r$, where $r \in [0, P-1]$. Necessarily, $a \geq P - 1$, otherwise, if $a \leq P - 2$, then $n = aP + r < P(P-1)$. Using $1 = M - P$, we have that $n = aP + r(M - P) = (a - r)P + rM$. But $a - r \geq 0$. Therefore, n is in A. We have thus shown that any $n \in \mathbb{N}$ sufficiently large – say $n \geq P(P-1)$ – is in A. $\qquad\square$

A.1.2   Abel's Theorem

Lemma A.1.4 *Let $\{b_n\}_{n \geq 1}$ and $\{a_n\}_{n \geq 1}$ be two sequences of real numbers such that*

$$b_1 \geq b_2 \geq \cdots \geq b_n \geq 0,$$

and such that for some real numbers m and M, and all $n \geq 1$,

$$m \leq a_1 + \cdots + a_n \leq M.$$

Then, for all $n \geq 1$,

$$b_1 m \leq a_1 b_1 + \cdots + a_n b_n \leq b_1 M.$$

The above result is *Abel's lemma*.

Proof. Let $s_n = a_1 + \cdots + a_n$, and use Abel's summation technique to obtain

$$
\begin{aligned}
a_1 b_1 + \cdots + a_n b_n &= b_1 s_1 + b_2 (s_2 - s_1) + \cdots + b_n (s_n - s_{n-1}) \\
&= s_1 [b_1 - b_2] + \cdots + s_{n-1} [b_{n-1} - b_n] + s_n [b_n].
\end{aligned}
$$

The bracketed terms are all non-negative, and therefore replacing each s_i by its lower bound or upper bound yields the result.   □

We recall without proof a standard result of calculus.

Lemma A.1.5 *The sum of a uniformly convergent series of continuous functions is a continuous function.*

The next result is *Abel's theorem*.

Theorem A.1.6 *Let $\{a_n\}_{n \geq 1}$ be a sequence of real numbers such that the radius of convergence of the power series $\sum_{n=0}^{\infty} a_n z^n$ is 1. Suppose that the sum $\sum_{n=0}^{\infty} a_n$ is convergent. Then the power series $\sum_{n=0}^{\infty} a_n x^n$ is uniformly convergent in $[0,1]$ and*

$$
\lim_{x \uparrow 1} \sum_{n=0}^{\infty} a_n x^n = \sum_{n=0}^{\infty} a_n, \tag{A.1}
$$

where $x \uparrow 1$ means that x tends to 1 from below.

Proof. It suffices to prove that $\sum_{n=0}^{\infty} a_n x^n$ is uniformly convergent in $[0,1]$, since (A.1) then follows by Lemma A.1.5. Write $A_n^p = a_n + \cdots + a_p$. By convergence of $\sum_{n=0}^{\infty} a_n$, for all $\epsilon > 0$, there exists an $n_0 \geq 1$ such that $p \geq n \geq n_0$ implies $|A_n^p| \leq \epsilon$, and therefore, since for $x \in [0,1]$, the sequence $\{x^n\}_{n>0}$ is non-increasing, Abel's lemma gives, for all $x \in [0,1]$,

$$
|a_n x^n + \ldots + a_p x^p| \leq \epsilon x^n \leq \epsilon,
$$

from which uniform convergence follows.   □

A.1.3   Cesàro's Lemma

Theorem A.1.7 *Let $\{b_n\}_{n \geq 0}$ be a sequence of real numbers such that*

$$
\lim_{n \uparrow \infty} b_n = 0.
$$

Then

$$
\lim_{n \uparrow \infty} \frac{b_1 + \cdots + b_n}{n} = 0.
$$

Proof. The sequence $\{b_n\}_{n\geq 0}$ is bounded in absolute value, say by K. For fixed arbitrary $\epsilon > 0$, there exists an n_0 such that $n > n_0$ implies $|b_n| \leq \epsilon$, and therefore

$$\left| \frac{b_1 + \cdots + b_n}{n} \right| \leq \left| \frac{b_1 + \cdots + b_{n_0}}{n} \right| + \left| \frac{b_{n_0} + \cdots + b_n}{n} \right|$$

$$\leq \frac{n_0 K}{n} + \frac{n - n_0}{n}\epsilon \leq 2\epsilon,$$

if n is sufficiently large. $\qquad\square$

A.1.4   Lebesgue's Theorems for Series

The results in this subsection concern the validity of exchanging limits and infinite sums.

Theorem A.1.8 *Let $\{a_{nk}\}_{n\geq 1, k\geq 1}$ be an array of real numbers such that for some sequence $\{b_k\}_{k\geq 1}$ of non-negative numbers satisfying*

$$\sum_{k=1}^{\infty} b_k < \infty,$$

it holds that (dominated convergence theorem)

$$|a_{nk}| \leq b_k \quad (n \geq 1, k \geq 1).$$

If moreover

$$\lim_{n\uparrow\infty} a_{nk} = a_k \quad (k \geq 1),$$

then

$$\lim_{n\uparrow\infty} \sum_{k=1}^{\infty} a_{nk} = \sum_{k=1}^{\infty} a_k.$$

Proof. Let $\epsilon > 0$ be fixed. Since $\sum_{k=1}^{\infty} b_k$ is a convergent series, one can find $M = M(\epsilon)$ such that $\sum_{k=M+1}^{\infty} b_k < \frac{\epsilon}{3}$. Since $|a_{nk}| \leq b_k$ and therefore $|a_k| \leq b_k$, we have

$$\sum_{k=M+1}^{\infty} |a_{nk}| + \sum_{k=M+1}^{\infty} |a_k| \leq \frac{2\epsilon}{3}.$$

Now, for sufficiently large n,

$$\sum_{k=1}^{M} |a_{nk} - a_k| \leq \frac{\epsilon}{3}.$$

Therefore, for sufficiently large n,

$$\left| \sum_{k=1}^{\infty} a_{nk} - \sum_{k=1}^{\infty} a_k \right| \leq \sum_{k=1}^{M} |a_{nk} - a_k| + \sum_{k=M+1}^{\infty} |a_{nk}| + \sum_{k=M+1}^{\infty} |a_k| \leq \frac{\epsilon}{3} + \frac{2\epsilon}{3} = \epsilon.$$

$\square$

Theorem A.1.9 *Let* $\{a_{nk}\}_{n \geq 1, k \geq 1}$ *be an array of non-negative real numbers such that for all* $k \geq 1$, *the sequence* $\{a_{nk}\}_{n \geq 1}$ *is non-decreasing and such that*

$$\lim_{n \uparrow \infty} a_{nk} = a_k \leq \infty.$$

Then (monotone convergence theorem)

$$\lim_{n \uparrow \infty} \sum_{k=1}^{\infty} a_{nk} = \sum_{k=1}^{\infty} a_k.$$

Proof. If $\sum_{k=1}^{\infty} a_k < \infty$, the result is a direct application of the dominated convergence theorem.

For the case $\sum_{k=1}^{\infty} a_k = \infty$, let $A > 0$ be fixed, and choose $M = M(A)$ such that $\sum_{k=1}^{M} a_k \geq 2A$. For sufficiently large n, $\sum_{k=1}^{M} (a_k - a_{nk}) \leq A$. Therefore, for sufficiently large n,

$$\sum_{k=1}^{\infty} a_{nk} \geq \sum_{k=1}^{M} a_{nk} + \sum_{k=1}^{M} (a_{nk} - a_k) \geq 2A - A = A.$$

$\square$

Theorem A.1.10 *Let* $\{a_{nk}\}_{n \geq 1, k \geq 1}$ *be an array of non-negative real numbers. Then* (Fatou's lemma)

$$\sum_{k-1}^{\infty} \liminf_{n \uparrow \infty} a_{nk} < \liminf_{n \uparrow \infty} \sum_{k-1}^{\infty} a_{nk}.$$

Proof. By definition of $\liminf$, for fixed k,

$$z_{nk} := \inf(a_{nk}, a_{n+1,k}, \ldots)$$

increases, as $n \uparrow \infty$, to $\liminf_{n \uparrow \infty} a_{nk}$. Therefore, by monotone convergence,

$$\sum_{k=1}^{\infty} \liminf_{n \uparrow \infty} a_{nk} = \lim_{n \uparrow \infty} \uparrow \sum_{k=1}^{\infty} z_{nk}.$$

But since $z_{nk} \le a_{nk}$,

$$\sum_{k=1}^{\infty} z_{nk} \le \sum_{k=1}^{\infty} a_{nk},$$

and therefore

$$\lim_{n\uparrow\infty} \sum_{k=1}^{\infty} z_{nk} \le \liminf_{n\uparrow\infty} \sum_{k=1}^{\infty} a_{nk}.$$

□

A.1.5   Infinite Products

Theorem A.1.11 *Let $\{a_n\}_{n\ge 1}$ be a sequence of numbers of the interval $[0,1)$.*

(a) If $\sum_{n=1}^{\infty} a_n < \infty$, then

$$\lim_{n\uparrow\infty} \prod_{k=1}^{n} (1 - a_k) > 0.$$

(b) If $\sum_{n=1}^{\infty} a_n = \infty$, then

$$\lim_{n\uparrow\infty} \prod_{k=1}^{n} (1 - a_k) = 0.$$

Proof. (a): For any numbers $c_1, \ldots, c_n$ in $[0,1)$, it holds that

$$(1 - c_1)(1 - c_2) \cdots (1 - c_n) \ge 1 - c_1 - c_2 - \cdots - c_n$$

(proof by induction). Since $\sum_{n=1}^{\infty} a_n$ converges, there exists an N such that for all $n \ge N$,

$$a_N + \cdots + a_n < \frac{1}{2}.$$

Therefore, defining $\pi(n) = \prod_{k=1}^{n} (1 - a_k)$, we have that for all $n \ge N$,

$$\frac{\pi(n)}{\pi(N-1)} = (1 - a_N) \cdots (1 - a_n) \ge 1 - (a_N + \cdots + a_n) \ge \frac{1}{2}.$$

Therefore, the sequence $\{\pi(n)\}_{n\ge N}$ is a non-increasing sequence bounded from below by $\frac{1}{2}\pi(N-1) > 0$, so that $\lim_{n\uparrow\infty} \pi(n) > 0$.

(b): Using the inequality $1 - a \le e^{-a}$ when $a \in [0,1)$, we have that $\pi(n) \le e^{-a_1 - a_2 - \cdots - a_n}$, and therefore, if $\sum_{n=1}^{\infty} a_n = \infty$, $\lim_{n\uparrow\infty} \pi(n) = 0$.   □

A.1.6   Tychonov's Theorem

Theorem A.1.12 *Let $\{x_n\}_{n\geq0}$ be a sequence of elements of $[0,1]^{\mathbb{N}}$, that is,*

$$x_n = (x_n(0),\ x_n(1),\ldots),$$

where $x_n(k) \in [0,1]$ for all k, $n \in \mathbb{N}$. There exists a strictly increasing sequence of integers $\{n_l\}_{l\geq0}$ and an element $x \in \{0,1\}^{\mathbb{N}}$ such that

$$\lim_{l\uparrow\infty} x_{n_l}(k) = x(k)$$

for all $k \in \mathbb{N}$.

Proof. Since the sequence $\{x_n(0)\}_{n\geq0}$ is contained in the closed interval $[0,1]$, by the Bolzano–Weierstrass theorem, one can extract a subsequence $\{x_{n_0(l)}(0)\}_{l\geq0}$ such that

$$\lim_{l\uparrow\infty} x_{n_0(l)}(0) = x(0)$$

for some $x(0) \in [0,1]$. In turn, one can extract from $\{x_{n_0(l)}(1)\}_{l\geq0}$ a subsequence $\{x_{n_1(l)}(1)\}_{l\geq0}$ such that

$$\lim_{l\uparrow\infty} x_{n_1(l)}(1) = x(1)$$

for some $x(1) \in [0,1]$. Note that

$$\lim_{l\uparrow\infty} x_{n_1(l)}(0) = x(0)\,.$$

Iterating this process, we obtain for all $j \in \mathbb{N}$ a sequence $\{x_{n_j(l)}\}_{l\geq0}$ that is a subsequence of each sequence $\{x_{n_0(l)}(1)\}_{l\geq0},\ldots,\{x_{n_{j-1}(l)}(1)\}_{l\geq0}$ and such that

$$\lim_{l\uparrow\infty} x_{n_j(l)}(k) = x(k)$$

for all $k \leq j$, where $x(1),\ldots,x(j) \in [0,1]$. The diagonal sequence $n_l = n_l(l)$ then does it. $\square$

A.1.7   Subadditive Functions

Theorem A.1.13 *Let $f : (0,\infty) \to \mathbb{R}$ be a non-negative function such that $\lim_{t\downarrow0} f(t) = 0$, and assume that f is subadditive, that is,*

$$f(t+s) \leq f(t) + f(s)$$

for all $s, t \in (0, \infty)$. Define the (possibly infinite) non-negative real number

$$q = \sup_{t>0} \frac{f(t)}{t}.$$

Then

$$\lim_{h \downarrow 0} \frac{f(h)}{h} = q.$$

Proof. By definition of q, there exists for each $a \in [0, q)$ a real number $t_0 > 0$ such that $\frac{f(t_0)}{t_0} \geq a$.

To each $t > 0$, there corresponds an $n \in \mathbb{N}$ such that $t_0 = nt + \delta$ with $0 \leq \delta < t$. Since f is subadditive, $f(t_0) \leq nf(t) + f(\delta)$, and therefore

$$a \leq \frac{nt}{t_0} \frac{f(t)}{t} + \frac{f(\delta)}{t_0},$$

which implies

$$a \leq \liminf_{t \downarrow 0} \frac{f(t)}{t}.$$

($\delta \to 0$, and therefore $f(\delta) \to 0$, as $t \to 0$; also $\frac{nt}{t_0} \to 1$ as $t \to 0$.) Therefore,

$$a \leq \liminf_{t \downarrow 0} \frac{f(t)}{t} \leq \limsup_{t \downarrow 0} \frac{f(t)}{t} \leq q,$$

from which the result follows, since a can be chosen arbitrarily close to q (arbitrarily large, when $q = \infty$). $\qquad\square$

A.2   Linear Algebra

A.2.1   Eigenvalues and Eigenvectors

The basic results of the theory of matrices relative to eigenvalues and eigenvectors will now be reviewed.[1]

Let A be a square matrix of dimension $r \times r$, with complex coefficients. If there exists a scalar $\lambda \in \mathbb{C}$ and a column vector $v \in \mathbb{C}^r$, $v \neq 0$, such that

$$Av = \lambda v \ (\text{resp.}, \ v^T A = \lambda v^T),$$

---

[1]See for instance [Gantmacher, 1959] or [Horn and Johnson, 1985].

then v is called a right-eigenvector (resp., a left-eigenvector) associated with the eigenvalue λ. There is no need to distinguish between right- and left-eigenvalues because if there exists a left-eigenvector associated with the eigenvalue λ, then there exists a right-eigenvector associated with the same eigenvalue λ. This follows from the facts that the set of eigenvalues of A is exactly the set of roots of the *characteristic equation*

$$\det(\lambda I - A) = 0\,,$$

where I is the $r \times r$ identity matrix, and that

$$\det(\lambda I - A) = \det(\lambda I - A^T)\,.$$

The *algebraic multiplicity* of λ is its multiplicity as a root of the *characteristic polynomial* $\det(\lambda I - A)$.

If $\lambda_1, \cdots, \lambda_k$ are *distinct* eigenvalues corresponding to the right-eigenvectors $v_1, \cdots, v_k$ and the left-eigenvectors $u_1, \cdots, u_k$, then $v_1, \cdots, v_k$ are independent, and so are $u_1, \cdots, u_k$.

Call R_λ (resp. L_λ) the set of right-eigenvectors (resp., left-eigenvectors) associated with the eigenvalue λ, plus the null vector. Both L_λ and R_λ are vector subspaces of $\mathbb{C}^r$, and they have the same dimension, called the *geometric multiplicity* of λ. In particular, the largest number of independent right-eigenvectors (resp., left-eigenvectors) cannot exceed the sum of the geometric multiplicities of the distinct eigenvalues.

The matrix A is called *diagonalizable* if there exists a nonsingular matrix Γ of the same dimensions such that

$$\Gamma A \Gamma^{-1} = \Lambda, \tag{A.2}$$

where

$$\Lambda = \ \text{diag} \ (\lambda_1, \cdots, \lambda_r)$$

for some $\lambda_1, \cdots, \lambda_r \in \mathbb{C}$, not necessarily distinct. It follows from (A.2) that with $U = \Gamma^T$, $U^T A = U^T \Lambda$, and with $V = \Gamma^{-1}$, $AV = V\Lambda = \Lambda V$, and therefore $\lambda_1, \cdots, \lambda_r$ are eigenvalues of A, and the ith row of $U^T = \Gamma$ (resp., the ith column of $V = \Gamma^{-1}$) is a left-eigenvector (resp., right-eigenvector) of A associated with the eigenvalue λ_i. Also, $A = V\Lambda U^T$ and therefore

$$A^n = V\Lambda^n U^T\,.$$

Clearly, if A is diagonalizable, the sum of the geometric multiplicities of A is exactly equal to r. It turns out that the latter is a sufficient condition of diagonalizability of A. Therefore, A is diagonalizable if and only if the sum of the geometric multiplicities of the distinct eigenvalues of A is equal to r.

EXAMPLE A.2.1: DISTINCT EIGENVALUES. By the last result, if the eigenvalues of A are distinct, A is diagonalizable. In this case, the diagonalization process can be described as follows. Let $\lambda_1, \cdots, \lambda_r$ be the r distinct eigenvalues and let $u_1, \cdots, u_r$ and $v_1, \cdots, v_r$ be the associated sequences of left and right-eigenvectors, respectively. As mentioned above, $u_1, \cdots, u_r$ form an independent collection of vectors, and so do $v_1, \cdots, v_r$. Define

$$U = [u_1 \cdots u_r], V = [v_1 \cdots v_r].$$

Observe that if $i \neq j$, $u_i^T v_j = 0$. Indeed, $\lambda_i u_i^T v_j = u_i^T A v_j = \lambda_j u_i^T v_j$, which implies $(\lambda_i - \lambda_j) u_i^T v_j = 0$, and in turn $u_i^T v_j = 0$, since $\lambda_i \neq \lambda_j$ by hypothesis. Since eigenvectors are determined up to multiplication by an arbitrary non-null scalar, one can choose them in such a way that $u_i^T v_i = 1$ for all $i \in [1, r]$. Therefore,

$$U^T V = I, \tag{A.3}$$

where I is the $r \times r$ identity matrix. Also, by definition of U and V,

$$U^T A = \Lambda U^T, AV = \Lambda V.$$

In particular, by (A.3), $A = V \Lambda U^T$. From the last identity and (A.3) again, we obtain for all $n \geq 0$,

$$A^n = V \Lambda^n U^T,$$

that is,

$$A^n = \sum_{i=1}^{r} \lambda_i^n v_i u_i^T.$$

A.2.2 Exponential of a Matrix

Theorem A.2.2 *Let A be a finite $r \times r$ matrix with complex elements. For each $t > 0$, the series*

$$\sum_{n=0}^{\infty} t^n \frac{A^n}{n!} \tag{A.4}$$

converges componentwise, and it is denoted by e^{tA}. For any $r \times r$ complex matrix B such that

$$AB = BA, \tag{A.5}$$

we have

$$e^{t(A+B)} = e^{tA}e^{tB} \tag{A.6}$$

for all $t > 0$. Also,

$$\frac{d}{dt}e^{tA} = Ae^{tA} = e^{tA}A. \tag{A.7}$$

Proof. Let $a_{ij} = a_{ij}(1)$, where $a_{ij}(n)$ is the general entry of A^n. Let $\Delta = \max|a_{ij}|$. A straightforward inductive proof shows that

$$a_{ij}(n) \leq r^{n-1}\Delta^n \tag{A.8}$$

for all i, j. Therefore, the general term of (A.4) is bounded in absolute value by the convergent series

$$1 + \sum_{n=1}^{\infty} t^n \frac{r^{n-1}\Delta^n}{n!},$$

and the componentwise convergence of (A.4) is proven.

Using the commutativity hypothesis (A.5),

$$\sum_{k=0}^{n} t^k \frac{(A+B)^k}{k!} = \sum_{k=0}^{n}\sum_{j=0}^{k} t^j \frac{A^j}{j!} t^{k-j} \frac{B^{k-j}}{(k-j)!}$$

$$= \left(\sum_{j=0}^{n} t^j \frac{A^j}{j!}\right)\left(\sum_{l=0}^{n} t^l \frac{B^l}{l!}\right) + C_n,$$

where it is easy to bound C_n and show that $\lim_{n\uparrow\infty} C_n = 0$, so that (A.6) follows by letting $n \to \infty$. To prove (A.7), write

$$\frac{e^{(t+h)A} - e^{tA}}{h} = e^{tA}\frac{e^{tA} - I}{h} - \frac{e^{tA} - I}{h}e^{tA},$$

and

$$\frac{e^{tA} - I}{h} = A + h\left(\sum_{n\geq 2} h^{n-2}\frac{A^n}{n!}\right).$$

In view of (A.8), each element of the latter matrix series is bounded, and the result follows. $\qquad\square$

EXAMPLE A.2.3: THE DIAGONALIZABLE CASE. If A is diagonalizable, that is, if

$$A = V \Lambda U^T,$$

where $\Lambda = \text{diag}\{\lambda_1, \dots, \lambda_r\}$ is the diagonal eigenvalue matrix, $V = [v_1, \dots, v_r]$, $U = [u_1, \dots, u_r]$, v_i (resp., u_i) is a right- (resp., left-) eigenvector corresponding to λ_i, and $U^T V = I$, then

$$A^n = V \Lambda^n U^T,$$

so that

$$e^{tA} = \sum_{n=0}^{\infty} t^n \frac{A^n}{n!} = V \sum_{n=0}^{\infty} t^n \frac{\Lambda^n}{n!} U^T.$$

But

$$\sum_{n=0}^{\infty} t^n \frac{\Lambda^n}{n!} = \text{diag}\left\{e^{\lambda_1 t}, \dots, e^{\lambda_r t}\right\}.$$

Finally,

$$e^{tA} = V \,\text{diag}\left\{e^{\lambda_1 t}, \dots, e^{\lambda_r t}\right\} U^T.$$

A.2.3 Gershgorin's Bound

Theorem A.2.4 *Let A be a finite $r \times r$ matrix with complex elements. Then for any eigenvalue λ, and all $k \in [1, r]$,*

$$|\lambda - a_{kk}| \leq \min(r_k, s_k),$$

where $r_k = \sum_{j=1, j \neq k}^{r} |a_{kj}|$ and $s_k = \sum_{j=1, j \neq k}^{r} |a_{jk}|$.

Proof. Let $v = (v(1), \dots, v(r))^T$ be a right-eigenvector corresponding to the eigenvalue λ, and let k be such that

$$|v(k)| = \|v\|_\infty := \max_i |v(i)|.$$

The kth equality in $Av = \lambda v$ is

$$\sum_{j=1}^{r} a_{kj} v(j) = \lambda v(k),$$

or equivalently,

$$(\lambda - a_{kk})v(k) = \sum_{j=1, j \neq k}^{r} a_{kj}v(j) \,.$$

Therefore,

$$|\lambda - a_{kk}||v(k)| \leq \sum_{j=1, j \neq k}^{r} |a_{kj}||v(j)| \leq r_k \, \|v\|_{\infty} \,.$$

Dividing by $\|v\|_{\infty} = |v(k)|$ yields

$$|\lambda - a_{kk}| \leq r_k \,.$$

By considering the left-eigenvector u associated with λ, we obtain the *Gershgorin bound*:

$$|\lambda - a_{kk}| \leq s_k \,.$$

□

Corollary A.2.5 *Let* $\mathbf{P}$ *be an* $r \times r$ *stochastic matrix. Then, for all eigenvalues* λ,

$$|\lambda - p_{kk}| \leq 1 - p_{kk}. \tag{A.9}$$

In particular, when the smallest eigenvalue λ_r *is real,*

$$\lambda_r \geq -1 + 2\min_{k}(p_{kk}) \,.$$

Proof. For a stochastic matrix,

$$r_k = \sum_{j=1, j \neq k}^{r} p_{kj} = 1 - p_{kk} \,.$$

□

A.3   Probability and Expectation

A.3.1   Expectation Revisited

There are two basic technical tools in probability theory that are indispensable, in this book and elsewhere: the monotone convergence theorem and the dominated convergence theorem. These results give sufficient conditions that allow interchange of limit and expectation.

For most purposes in this book, we need only the versions of the Lebesgue theorems given in Section A.1.4. However, on a few occasions, the more general statements in terms of expectations are necessary. To state them in full generality, we need to briefly summarize the definition of expectation for arbitrary random variables, which are neither discrete nor absolutely continuous. The detailed proofs are outside the scope of this book.

Let X be a non-negative random variable. Its expectation, denoted by $E[X]$ as usual, is defined by

$$E[X] = \lim_{n\uparrow\infty} \left\{ \sum_{k=0}^{n2^n-1} \frac{k}{2^n} P\left(\frac{k}{2^n} \leq X < \frac{k+1}{2^n} \right) + nP(X > n) \right\}. \tag{A.10}$$

The general term of the right-hand side is clearly increasing with n, and therefore the limit is well defined, possibly infinite, however.

From (A.10), it is clear that

$$E[1_A] = P(A). \tag{A.11}$$

For a random variable X that is not non-negative, the procedure already used to define $E[X]$ in the discrete case and in the probability density case is still applicable, that is, $E[X] = E[X^+] - E[X^-]$ if not both $E[X^+]$ and $E[X^-]$ are infinite. If $E[X^+]$ *and* $E[X^-]$ are infinite, the expectation is not defined. If $E[|X|] < \infty$, X is said to be integrable, and then $E[X]$ is a finite number.

The basic properties of the expectation so defined are linearity and monotonicity: If X_1 and X_2 are random variables with expectations, then for all $\lambda_1, \lambda_2 \in \mathbb{R}$,

$$E[\lambda_1 X_1 + \lambda_2 X_2] = \lambda_1 E[X_1] + \lambda_2 E[X_2]$$

whenever the right-hand side has meaning (that is, is not an $\infty - \infty$ form). Also, if $X_1 \leq X_2$, P-a.s., then

$$E[X_1] \leq E[X_2].$$

It follows from this that if $E[X]$ is well-defined, then

$$|E[X]| \leq E[|X|].$$

These properties will be accepted without proof. It is clear from (A.10) and (A.11) that if X is a simple random variable, that is,

$$X(\omega) = \sum_{i=1}^{N} \alpha_i 1_{A_i}(\omega),$$

where $\alpha_i \in \mathbb{R}$, $A_i \in \mathcal{F}$, then

$$E[X] = \sum_{i=1}^{N} \alpha_i P(A_i).$$

The definitions of the mean, the variance, and the characteristic function of a random variable are the same as in the discrete case and the probability density case:

$$\begin{aligned}
m_X &= E[X], \\
\sigma_X^2 &= E[(X - m_X)^2] = E[X^2] - m_X^2, \\
\varphi_X(u) &= E[e^{iuX}],
\end{aligned}$$

where $E[e^{iuX}] = E[\cos(uX)] + iE[\sin(uX)]$.

Now let $g : \mathbb{R} \to \mathbb{R}$ be some function such that $E[|g(X)|] < \infty$ where X is a real random variable with CDF $F(x)$. *By definition* of the symbol on the left-hand side of (A.12),

$$\int_{\mathbb{R}} g(x)\, dF(x) := E[g(X)], \qquad (A.12)$$

and $\int_{\mathbb{R}} g(x)dF(x)$ is called the Stieltjes–Lebesgue integral of $g(x)$ with respect to $F(x)$. It can be shown that if X admits a PDF $f(x)$, that is, $F(x) = \int_{-\infty}^{x} f(y)\, dy$, then

$$\int_{\mathbb{R}} g(x) dF(x) = \int_{\mathbb{R}} g(x) f(x)\, dx$$

in accordance with our previous definition of expectation for absolutely continuous random variables.

Clearly, the Stieltjes–Lebesgue integral inherits from expectation the properties of linearity and monotonicity.

A.3.2   Lebesgue's Theorem for Expectation

The most important technical results relative to expectation are the following ones giving general conditions under which the limit and expectation symbols can be interchanged, that is,

$$E\left[\lim_{n\uparrow\infty} X_n\right] = \lim_{n\uparrow\infty} E[X_n]. \qquad (A.13)$$

Theorem A.3.1 *Let $\{X_n\}_{n\geq 1}$ be a sequence of random variables such that for all $n \geq 1$,*

$$0 \leq X_n \leq X_{n+1}, \quad P\text{-}a.s.$$

Then (A.13) holds true.

Theorem A.3.2 *Let $\{X_n\}_{n\geq1}$ be a sequence of random variables such that for all ω outside a set $\mathcal{N}$ of null probability there exists $\lim_{n\uparrow\infty} X_n(\omega)$ and such that for all $n \geq 1$*

$$|X_n| \leq Y, \ P\text{-}a.s.,$$

where Y is some integrable random variable. Then (A.13) holds.

EXAMPLE A.3.3: A COUNTEREXAMPLE. We shall see that (A.13) is not always true when $\lim_{n\uparrow\infty} X_n$ exists. Indeed, take the following probabilistic model: $\Omega = [0, 1]$, and P is the Lebesgue measure on $[0, 1]$ (the probability of $[a, b] \subset [0, 1]$ is the length $b - a$ of this interval). Thus, ω is a real number in $[0, 1]$, and a random variable is a real function defined on $[0, 1]$. Take for X_n the function whose graph is a triangle with base $[0, \frac{2}{n}]$ and height n. Clearly, $\lim_{n\uparrow\infty} X_n(\omega) = 0$ and $E[X_n] = \int_0^1 X_n(x)\mathrm{d}x = 1$, so that $E[\lim_{n\uparrow\infty} X_n] = 0 \neq \lim_{n\uparrow\infty} E[X_n] = 1$.

EXAMPLE A.3.4: SUM OF A SERIES, TAKE 1. Let $\{S_n\}_{n\geq1}$ be a sequence of non-negative random variables. Then

$$E\left[\sum_{n=1}^{\infty} S_n\right] = \sum_{n=1}^{\infty} E[S_n]. \tag{A.14}$$

It suffices to apply the monotone convergence theorem, with $X_n = \sum_{k=1}^{n} S_k$.

EXAMPLE A.3.5: SUM OF A SERIES, TAKE 2. Let $\{S_n\}_{n\geq1}$ be a sequence of real random variables such that $\sum_{n\geq1} E[|S_n|] < \infty$. Then (A.14) holds. It suffices to apply the dominated convergence theorem with $X_n = \sum_{k=1}^{n} S_k$ and $Y = \sum_{k=1}^{n} |S_k|$. By the result of the previous example, $E[Y] = \sum_{k=1}^{n} E[|S_k|] < \infty$.

EXAMPLE A.3.6: THE PRODUCT FORMULA. As another illustration of the mono-tone convergence theorem we shall now prove the *product formula*

$$E[XY] = E[X]E[Y],$$

where X and Y are independent integrable random variables or independent non-negative random variables. We do the proof in the non-negative case, from which the integrable case follows easily, using the decomposition of a random variable into its non-negative and non-positive parts.

First notice that for any non-negative random variable Z, if we define

$$Z_n = \sum_{k=1}^{n2^n-1} \frac{k}{2^n} 1_{\{\frac{k}{2^n} \leq Z < \frac{k+1}{2^n}\}} + n1_{\{Z \geq n\}},$$

then $\lim_{n\uparrow\infty} Z_n = Z$, P-a.s. Thus, by the Lebesgue monotone convergence theorem, $\lim_{n\uparrow\infty} E[Z_n] = E[Z]$. With obvious notations, $\lim_{n\uparrow\infty} X_n Y_n = XY$, and therefore $\lim_{n\uparrow\infty} E[X_n Y_n] = E[XY]$. Also, $\lim_{n\uparrow\infty} E[X_n]E[Y_n] = E[X]E[Y]$. The result will then be proved if

$$E[X_n Y_n] = E[X_n]E[Y_n]$$

is shown to be true for all $n \geq 1$. But then, this verification amounts to observing that for all i, j,

$$E[1_{\{\frac{j}{2^n} < X \leq \frac{j+1}{2^n}\}} 1_{\{\frac{i}{2^n} < Y \leq \frac{i+1}{2^n}\}}] = E[1_{\{\frac{j}{2^n} < X \leq \frac{j+1}{2^n}\}}]E[1_{\{\frac{i}{2^n} < Y \leq \frac{i+1}{2^n}\}}],$$

that is,

$$P\left(X \in \left(\frac{j}{2^n}, \frac{j+1}{2^n}\right], Y \in \left(\frac{i}{2^n}, \frac{i+1}{2^n}\right]\right)$$
$$= P\left(X \in \left(\frac{j}{2^n}, \frac{j+1}{2^n}\right]\right) P\left(Y \in \left(\frac{i}{2^n}, \frac{i+1}{2^n}\right]\right).$$

The latter equality is just the independence of X and Y, and the proof is therefore complete.

———

Bibliography

Aldous, D. and P. Diaconis, "Shuffling cards and stopping times", American Mathematical Monthly, 93, 333–348, 1981.

Aldous, D. and P. Diaconis, "Strong uniform times and finite random walks", Adv. Appl. Math., 8, 69–97, 1987.

Aldous, D. and J.A. Fill, *Reversible Markov Chains and Random Walks on Graphs*, draft available at http://www.stat.Berkeley.EDU/users/aldous/book.html.

Anantharam, V. and P. Tsoucas, "A proof of the Markov chain tree theorem", Statist. Probab. Lett., **8** (2), 189–192, 1989.

Anderson, W.J., *Continuous Time Markov Chains: An Applications Oriented Approach*, Springer, 1991.

Anily, S. and A. Federgruen, "Simulated annealing methods with general acceptance probabilities", J. Appl. Probab., 24, 657–667, 1987.

Asmussen, S., *Applied Probability and Queues*, Wiley, 1987.

Athreya, K. and P. Ney, *Branching Processes*, Springer, 1972.

Bartlett, M.S., "On theoretical models for competitive and predatory biological systems", Biometrika, 44, 1957.

Barker, A.A., "Monte Carlo calculations of the radial distribution functions for a proton-electron plasma", Austral. J. Phys., 18, 119–133, 1965.

Baxter, R.J., *Exactly Solved Models in Statistical Mechanics*, Academic Press, 1982.

Besag, J., "Spatial interaction and the statistical analysis of lattice systems", Journal of the Royal Statistical Society, B-31, 192–236, 1974.

Bhattacharya, R.N. and E.C. Waymire, *Stochastic Processes with Applications*, Wiley, 1990.

© Springer Nature Switzerland AG 2020
P. Brémaud, *Markov Chains*, Texts in Applied Mathematics 31,
https://doi.org/10.1007/978-3-030-45982-6

Bratley, P., B.L. Fox and L. Schrage, *A Guide to Simulation*, Springer, 1987.

Capetanakis, J.I., "Tree algorithm for packet broadcast channels", IEEE Transactions on Information Theory, IT-25, 505–515, 1979.

Chandra, A.K.P., P. Raghavan, W.L. Ruzzo, R. Smolenski and P. Tiwari, "The electrical resistance of a graph captures its commute and cover times", Comp. Complexity, 6, 4, 312–340, 1996.

Chung, K.L., *Markov Chains with Stationary Transition Probabilities*, Springer, 2-nd ed., 1967.

Cohn, H., "Finite non-homogeneous Markov chains: asymptotic behaviour", Adv. Appl. Probab., 8, 502–516, 1976.

Cohn, H., "Countable non-homogeneous Markov chains: asymptotic behaviour", Adv. Appl. Probab., 9, 542–552, 1977.

Diaconis, P., *Group Representations in Probability and Statistics*, Institute of Mathematical Statistics, 1988.

Diaconis, P., "The Markov chain Monte Carlo revolution", Bull. Amer. Math. Soc. 46 , 179–205, 2009.

Diaconis, P. and J.A. Fill, "Strong stationary times via a new form of duality", The Annals of Probability, 18, 4, 1483–1522, 1991.

Diaconis, P. and L. Saloff-Coste, "Comparison theorems for reversible Markov chains", The Annals of Probability, 3,3 696–730, 1993.

Diaconis, P. and L. Saloff-Coste, "What do we know about the Metropolis algorithm?", J. Comp. System Sci., 57, 1, 20–36, 1998.

Diaconis, P. and M. Shahshahani, "Generating a random permutation with random permutations", Z. für W., 57, 2, 159–179, 1981.

Diaconis, P. and D. Stroock, "Geometric bounds for eigenvalues of Markov chains", The Annals of Applied Probability, 1, 1, 36–61, 1991.

Doeblin, W., "Sur les propriétés asymptotiques de mouvements régis par certains types de chaînes simples", Bulletin Mathématique de la Société Roumaine des Sciences, 39, No.1, 57–115, No.2, 3–61, 1937.

Dobrushin, R.L., "Central limit theorems for non-stationary Markov chains II", Theory of Probability and its Applications, 1, 329–383 (English translation), 1956.

Dobrushin, R.L., "Existence of a phase transition in two and three-dimensional Ising models", Theory of Probability and its Applications, 10, 193–213, 1965.

Douc, R., E. Moulines, P. Priouret and P. Soulier, *Markov Chains*, Springer, 2019.

Doyle, P.G. and J.L. Snell, *Random Walks and Electrical Networks*, 2000. arXiv: math/0001057 v 1 [math. PR], 11 Jan 2000.

Durrett, R., *Essentials of Stochastic Processes*, Springer, 1999, 2016.

Fayolle, G., "Étude du comportement d'un canal radio partagé entre plusieurs utilisateurs", Thèse de Docteur–Ingénieur, Université Paris 6, 1975.

Fayolle, G., V.A. Malishev and M.V. Menshikov, *Topics in the Constructive Theory of Countable Markov Chains*, Cambridge University Press, 1995.

Feller, W. and S. Orey, "A renewal theorem", J. Math. Mech. 10, 619–624, 1961.

Fill, J.A., "Eigenvalue bounds on convergence to stationarity for non-reversible Markov chains, with an application to the exclusion process", The Annals of Applied Probability, 1, 1, 62 87, 1991.

Fill, J.A., "An interruptible algorithm for perfect sampling via Markov chains", Annals of Applied Probability, 8, 131–162, 1998.

Fishman, G.S., *Monte Carlo*, Springer, 1996.

Foss, S. and R.L. Tweedie, "Perfect simulation and backwards coupling, Stochastic Models", 14, 187–203, 1998.

Foster, F.G., "On the Stochastic Matrices Associated with Certain Queuing Processes", The Annals of Mathematical Statistics 24 (3), 1953.

Fréchet, M., *Recherches Théoriques Modernes sur le Calcul des Probabilités. Second livre.* Hermann, 1938.

Frigesi, A., C.R. Hwang and L. Younès, "Optimal spectral properties of reversible stochastic matrices, Monte Carlo methods and the simulation of Markov random fields", The Annals of Applied Probability, 2, 3, 610–628, 1992.

Frobenius, G., "Über Matrizen aus nicht negativen Elementen", Sitzungsber. Knigl. Preuss. Akad. Wiss., 456–477, 1912.

Frobenius, G., "Über Matrizen aus positiven Elementen, 1", Sitzungsber. Knigl. Preuss. Akad. Wiss.:, 471–476, 1908.

Frobenius, G., "Über Matrizen aus positiven Elementen, 2", Sitzungsber. Knigl. Preuss. Akad. Wiss., 514–518, 1909.

Gantmacher, F.R., *Applications of the Theory of Matrices*, Inter Science, 1959.

Geman, D., *Random Fields and Inverse Problems in Imaging.* In: Lecture Notes in Mathematics, 1427, 113–193, Springer, 1990.

Geman, S. and D. Geman, "Stochastic relaxation, Gibbs distributions, and the Bayesian restoration of images", IEEE Transactions of Pattern Analysis and Machine Intelligence, 6, 721–741, 1984.

Gibbs, W., *Elementary Principles of Statistical Mechanics*, Yale University Press, 1902.

Glauber, R., "Time-dependent statistics of the Ising model", J. Math. Physics, 4, 294–307.

Gordon, W.J. and G.F. Newell, "Closed queueing systems with exponential servers", Operations Research, 15, 224–25, 1967.

Griffeath, D, "A Maximal coupling for Markov chains", Z. für W., 31, 95–106, 1975.

Grimmett, G.R., "A Theorem on random fields", Bulletin of the London Mathematical Society, 81–84, 1973.

Grimmett, G.R. and D.R. Stirzaker, *Probability and Random Processes*, Clarendon Press, Oxford, 1992.

Grinstead, C.M. and J.L. Snell, *Introduction to Probability*, American Mathematical Society, 1997.

Gross, D. and C.M. Harris, *Fundamentals of Queueing Theory*, Wiley, 1975, 1984.

Gut, A., *An Intermediate Course in Probability*, Springer, 1995.

Häggström, O. and K. Nelander, "On exact simulation of Markov random fields using coupling from the past", Scand. J. Statist. 26, 395–411, 1999

Häggström, O., *Finite Markov Chains and Algorithmic Applications*, London Mathematical Society Student texts, 52, Cambridge University Press, 2002.

Hajek, B., "Cooling schedule for optimal annealing", Mathematics of Operations Research 13, 311–329, 1988.

Hajnal, J., "The ergodic properties of nonhomogeneous finite Markov chains", Proc. Cambridge Philos. Society, 52, 67–77, 1956.

Hajnal, J., "Weak ergodicity in non-homogeneous Markov chains", Proc. Cambridge Philos. Society, 54, 236–246, 1958.

Hammersley, J.M. and P. Clifford, "Markov fields on finite graphs and lattices", unpublished manuscript, 1968.

Harris, T.E., *The Theory of Branching Processes*, Dover, 1989.

Hastings, W.K., "Monte Carlo sampling methods using Markov chains and their applications", Biometrika, 57, 97–109, 1970.

Horn, R.Aand C.R. Johnson, *Matrix Analysis*, Cambridge University Press, 1985.

Huber, M., "Perfect sampling using bounding chains", The Annals of Applied Probability, 14, 2, 734–753, 2004.

Iosifescu, I., *Finite Markov Chains and their Applications*, Wiley, 1980.

Isaacson, D.L. and R.W. Madsen, "Strongly ergodic behavior for nonstationary Markov processes", Annals of Probability, 1, 329–335, 1973.

Isaacson, D.L. and R.W. Madsen, *Markov Chains*, Wiley, 1976.

Ising, E., "Beitrag zur Theorie des Ferromagnetismus", Zeitschrift für Physiks, 31, 253–258, 1925.

Jackson, J.R., "Networks of waiting lines", Operations Research, 5, 518–521, 1957.

Jerrum, M. and A. Sinclair, "Approximating the permanent", SIAM Journal of Computing, 18, 1149–1178, 1989.

Kac, M., *Random Walk and the Theory of Brownian Motion*, American Mathematical Monthly, 54, 369–391, 1947.

Kakutani, S., "Markov processes and the Dirichlet problem", Proc. Jap. Acad., **21**, 227–233, 1945.

Karlin, S. and M.H. Taylor, *A First Course in Stochastic Processes*, Academic Press, 1975.

Karlin, S. and M.H. Taylor, *A Second Course in Stochastic Processes*, Academic Press, 1981.

Keilson, J., *Markov Chain Models, Rarity and Exponentiality*, Springer, 1979.

Kelly, F.P., *Reversibility and Stochastic Networks*, Wiley, 1979.

Kelly, F.P. and B.D. Ripley, "Markov Point Processes", Journal of the London Mathematical Society, 15, 188–192, 1977.

Kemeny, J.G. and J.L. Snell, *Finite Markov chains*, Van Nostrand, 1960.

Kemeny, J.G., J.L. Snelland A.W. Knapp, *Denumerable Markov chains*, Van Nostrand, 1960.

Kemeny, J.G., "Generalization of fundamental matrix", Linear Algebra and its

Applications, 38, 193–206, 1991.

Kinderman, R. and J.L. Snell, *Markov Random Fields and their Applications*, Contemporary Math., Vol. 1, Providence, RI: Amer. Math. Soc., 1980.

Kingman, J.F.C., *Regenerative phenomena*, Wiley, 1972.

Kirkpatrick, S., C.D. Gelatt and M.P. Vecchi, "Optimization by simulated Annealing", Science, 220, 671–680, 1982.

Klenke, A., *Probability Theory: A Comprehensive Course*, Springer, 2008 (2nd ed. 2014).

Kolmogorov, A., "Anfangsgründe der Theorie der Markoffschen ketten mit unedlich vielen möglichen Zuständen", Rec. Math. Moskov (Mat. Sbornik), 1, 43, 607–610, 1936.

Lanford, O.E. and D. Ruelle, "Observables at infinity and states with short range correlations in statistical mechanics", Communications in Mathematical Physics, 13, 194–215, 1969.

Levin, D.A., Y. Peresand E.L. Wilmer, *Markov Chains and Mixing Times*, American Mathematical Society, 2009.

Lindvall, T., "Probabilistic proof of Blackwell's renewal theorem", Annals of Probability, 5, 482–485, 1977.

Lindvall, T., *Lectures on the Coupling Method*, Dover, 2002.

Liu, J., "Eigenanalysis for a Metropolis sampling scheme with comparisons to rejection sampling and importance sampling", Statistics and Computing, 1995.

Liu, J., *Monte Carlo Strategies in Scientific Computing*, Springer Series in Statistics. Springer, 2001.

Lovasz, L., "Random walks on graphs: a survey", Combinatorics, Paul Erdös is eighty, 1–46, 1993.

Lyons, R. and Y. Peres, *Probability on Trees and Networks*, Cambridge University Press, 2016.

Markov, A.A., "Extension of the law of large numbers to dependent events"(in Russian), Bull. Soc. Phys. Math. Kazan, 2, 15, 155–156, 1906.

Metropolis, N., M.N. Rosenbluth, A.W. Rosenbluth, A.H. Teller and E. Teller, "Equations of state Calculations by fast computing machines", J. Chem. Phys., 21, 1087–92, 1953.

Meyn, S.P. and R.L. Tweedie, *Markov Chains and Stochastic Stability*, Springer, 1993.

Mihaïl, M., "Combinatorial aspects of expanders", Ph.D. dissertation, Department of Computer Science, Harvard University, 1989.

Mitzenmacher, M. and E. Upfal, *Probability and Computing*, Cambridge University Press, 2005.

Montenegro, R. and P. Tetali, "Mathematical Aspects of Mixing Times in Markov Chains", Foundations and Trends in Theoretical Computer Science: Vol. 1: No. 3, pp 237–354, 2006.

Nash-Williams, C.S.J.A., "Random walk and electric currents in networks", Proceedings of the Cambridge Philosophical Society, 55, 181–194, 1959.

Norris, J.R., *Markov chains*, Cambridge University Press, 1997.

Orey, S., *Limit Theorems for Markov Chain Transition Probabilities*, Van Nostrand, 1971.

Pakes, A.G., "Some conditions for ergodicity and recurrence of Markov chains", Oper. Res. 17, 1058–1061, 1969.

Perron, O., "Zur Theorie der Matrices", Mathematische Annalen 64 (2): 248–263, 1907.

Peskun, P.H., "Optimum Monte Carlo sampling using Markov chains", Biometrika, 60, 3, 607–612, 1973.

Peierls, R., "On Ising's model of ferromagnetism", Proceedings of the Cambridge Philosophical Society, 32, 477–481, 1936.

Pitman, J., "Uniform rates of convergence for Markov chains transition probabilities", Z. für W., 29, 193–227, 1974.

Pitman, J., "On coupling of Markov chains", Z. für W., 35, 4, 315–322, 1976.

Propp, J.G. and D.B. Wilson, "Exact sampling with coupled Markov chains and applications to statistical mechanics", Rand. Struct. Algs, 9, 223–252, 1996.

Resnick, S., *Adventures in Stochastic Processes*, Birkhäuser, 1992.

Revuz, D., *Markov Chains*, North-Holland, 1984.

Ripley, B.D., *Stochastic Simulation*, Wiley, 1987.

Ripley, B.D. and F.P. Kelly, "Markov point processes", *Journal of the London Mathematical Society*, 15, 188–192, 1977.

Robertazzi, T.G., *Computer Networks and Systems: Queueing Theory and Performance Evaluation*, Springer, 1990.

Rom, R. and M. Sidi, *Multiple Access Protocols, Performance and Analysis*, Springer, 1990.

Ross, S., *A Course in Simulation*, Macmillan, 1987.

Rudin, W., *Real and Complex Analysis*, McGraw-Hill, 1966.

Seneta, E., *Nonnegative Matrices and Markov Chains*, 2nd edition, Springer, 1981.

Shiryaev, A.N., *Probability*, Springer, 1987.

Sinclair, A. and M. Jerrum, "Approximate counting, uniform generation and rapidly mixing Markov chains", Inform. and Comput., **82**, 93–133, 1989.

Sinclair, A., "Improved bounds for mixing rates of Markov chains on combinatorial structures", Tech. Report, Dept. Computer Science, U. of Edinburgh, 1990.

Sinclair, A., *Randomness and Computation*, Lecture notes CS271 Fall 2011, http://www.cs.berkeley.edu/ sinclair/cs271/f11.html

Smith, W.L., "Regenerative stochastic processes", Proc. R. Soc. A 232, 6–31, 1955.

van Laarhoven, P.J. and E.H. Aarts, *Simulated Annealing: Theory and Applications*, Springer, 1987.

Walrand, J., *An Introduction to Queueing Networks*, Prentice-Hall, 1988.

Williams, D., *Probability and Martingales*, Cambridge University Press, 1991.

Winkler, G., *Image Analysis, Random Fields and Dynamical Monte Carlo Methods*, Springer, 1995.

Wolff, R.W., *Stochastic Modeling and the Theory of Queues*, Prentice-Hall, Englewood Cliffs, 1989.

Yan, D. and H. Mukai, "Stochastic discrete optimization", SIAM J. on Control and Optimization, 30, 3, 549–612, 1992.

Index

Printed in the United States
by Baker & Taylor Publisher Services